Fundamental Aspects of Inert Gases in Solids

NATO ASI Series

Advanced Science Institutes Series

A series presenting the results of activities sponsored by the NATO Science Committee, which aims at the dissemination of advanced scientific and technological knowledge, with a view to strengthening links between scientific communities.

The series is published by an international board of publishers in conjunction with the NATO Scientific Affairs Division

A	Life Sciences	Plenum Publishing Corporation
B	Physics	New York and London
C	Mathematical and Physical Sciences	Kluwer Academic Publishers
D	Behavioral and Social Sciences	Dordrecht, Boston, and London
E	Applied Sciences	
F	Computer and Systems Sciences	Springer-Verlag
G	Ecological Sciences	Berlin, Heidelberg, New York, London,
H	Cell Biology	Paris, Tokyo, Hong Kong, and Barcelona
I	Global Environmental Change	

Recent Volumes in this Series

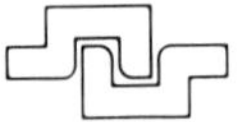

Series B: Physics

Fundamental Aspects of Inert Gases in Solids

Edited by

S. E. Donnelly

The University of Salford
Salford, United Kingdom

and

J. H. Evans

AEA Technology
Harwell, United Kingdom

Plenum Press
New York and London
Published in cooperation with NATO Scientific Affairs Division

Proceedings of a NATO Advanced Research Workshop
on Fundamental Aspects of Inert Gases in Solids,
held September 16-22, 1990,
at Bonas, France

Library of Congress Cataloging-in-Publication Data

NATO Advanced Research Workshop on Fundamental Aspects of Inert Gases
   in Solids (1990 : Bonas, France)
    Fundamental aspects of inert gases in solids / edited by S.E.
  Donnelly and J.H. Evans.
        p.   cm. -- (NATO ASI series. Series B, Physics ; v. 279)
    "Published in cooperation with NATO Scientific Affairs Division."
    Includes bibliographical references and index.
    ISBN 0-306-44051-2
    1. Solids--Effect of radiation on--Congresses. 2. Materials-
  -Fffect of radiation on--Congresses. 3. Gases. Rare--Congresses.
  4. Surface chemistry--Congresses.   1. Donnelly. S. E. (Steve E.)
  II. Evans, J. H. (John H.) III. North Atlantic Treaty Organization.
  Scientific Affairs Division. IV. Title. V. Series.
  QC176.8.R3N36 1990
  530.4'1--dc20   91-31217
                                                             CIP

ISBN 0-306-44051-2

© 1991 Plenum Press, New York
A Division of Plenum Publishing Corporation
233 Spring Street, New York, N.Y. 10013

Printed in the United States of America

PREFACE

The NATO Advanced Research Workshop on *Fundamental Aspects of Inert Gases in Solids*, held at Bonas, France from 16–22 September 1990, was the fifth in a series of meetings that have been held in this topic area since 1979. The Consultants' Meeting in that year at Harwell on *Rare Gas Behaviour in Metals and Ionic Solids* was followed in 1982 by the Jülich International Symposium on *Fundamental Aspects of Helium in Metals*. Two smaller meetings have followed—a CECAM organised workshop on *Helium Bubbles in Metals* was held at Orsay, France in 1986 while in February 1989, a Topical Symposium on *Noble Gases in Metals* was held in Las Vegas as part of the large TMS/AIME Spring Meeting.

As is well known, the dominating feature of inert gas atoms in most solids is their high heat of solution, leading in most situations to an essentially zero solubility and gas-atom precipitation. In organising the workshop, one particular aim was to target the researchers in the field of inert-gas/solid interactions from three different areas—namely metals, tritides and nuclear fuels—in order to encourage and foster the cross-fertilisation of approaches and ideas. In these three material classes, the behaviour of inert gases in metals has probably been most studied, partly from technological considerations—the effects of helium production via (n, α) reactions during neutron irradiation are of importance, particularly in a fusion reactor environment—and partly from a more fundamental viewpoint. The tritides (and metals in which tritium has been dissolved) are of interest because of the decay product ^{3}He and its precipitation in the absence of irradiation-produced displacement damage. The third materials class, that of nuclear fuels such as UO_2, has been studied from the earliest days of nuclear power to understand the behaviour and release of the fission gases, mainly xenon, produced during the burn up of the uranium atoms.

The Workshop contributions adequately covered the topic areas and reflected the overall fields of interest, both theoretical and experimental. In addition, there was a good balance between the quarter or so papers presented with a review-type emphasis and those which dealt with more specific aspects. One interesting feature was the number of new techniques which have been recently applied to the area. Among these, nuclear magnetic resonance (NMR), extended X-ray absorption spectroscopy (EXAFS), glancing angle X-ray diffraction, Mössbauer spectroscopy, small angle neutron scattering (SANS), and high resolution dilatometry are all covered by articles in this book—in addition to reviews involving the better established techniques such as transmission electron microscopy and diffraction, positron annihilation spectroscopy, ion channelling and thermal desorption spectroscopy.

There seems no doubt that new studies and several of these new techniques have been inspired by the recent discovery of solid phase precipitation of the heavier inert gases in metals at ambient temperatures. (For the general reader it might be mentioned that hitherto, solid inert gas formation at ambient temperatures could only be knowingly achieved using high pressure diamond anvils). On the theoretical side, both the physics of precipitate growth, and the importance of correct equations of state to translate from measured gas atom packing densities to the ultra-high pressures involved, are discussed.

In the area of high-temperature annealing, where the coarsening mechanisms of bubbles has long been a fertile field for investigation in both metals and UO_2, several articles document the various aspects of the progress made in establishing the rôle of thermal resolution—and hence Ostwald ripening—in the bubble coarsening processes in metals. There is now no doubt that at high temperatures, a zero inert gas solubility can no longer be tacitly assumed.

Although many of the techniques and topics above apply to all three material classes targeted in the workshop, several articles cover the aspects specific respectively to tritides and nuclear fuels.

Of the meetings mentioned in the opening paragraph, only those at Harwell, 1979, and Jülich, 1982, were published. The present book, bringing together the papers presented at Bonas, is therefore timely and should provide an invaluable up-to-date snapshot of both recent advances and current topics of major interest.

We would like to thank the Workshop Committee—Helmut Trinkaus, Claude Templier, George Thomas and Tom van Veen—for their help and advice. We are also grateful to all the authors for their efforts in preparing the manuscripts and to Richard Lyst for assistance in typesetting the book. Finally, we should like to express our gratitude to the NATO Scientific Affairs Division for their generous support which enabled us to hold this workshop, and to the staff at the Château de Bonas for helping the participants to gain the most from the congenial surroundings.

Steve Donnelly   **John Evans**
Salford   Harwell

CONTENTS

INERT GAS BUBBLES IN METALS: HIGH-TEMPERATURE BUBBLE EVOLUTION

BUBBLE GROWTH MECHANISMS

INERT GASES IN NUCLEAR FUELS

INDEXES

ATOMISTIC THEORY

THEORETICAL STUDIES OF HELIUM IN METALS

J. B. Adams,[1] W. G. Wolfer,[2] S. M. Foiles,[2] C. M. Rohlfing[2] and C. D. Van Siclen[3]

[1] *Dept. of Materials Science and Engineering*
University of Illinois, Urbana
IL 61801, USA

[2] *Theoretical Division*
Sandia National Laboratories Livermore, P.O. Box 969
Livermore, CA 94551-0969, USA

[3] *Idaho National Engineering Laboratory*
Idaho Falls
ID 83415, USA

ABSTRACT
Several complementary theoretical techniques are used to study the effect of He in metals. The Embedded Atom Method (EAM) has been applied to the study of the solubility of He in liquid and solid Ni. The EAM and continuum elasticity theory were used to study dislocation loop punching by the growth of He bubbles in Ni. Several diffusion mechanisms of He in Ni were investigated with the EAM. The energetics of He atoms in small vacancy clusters were investigated with the EAM, and used as input data in a kinetic rate equation model of void growth in rapidly-quenched steel. Ab initio methods were used to study the electron energy levels of dense He in Al.

1.   Introduction

During the last several years, the theoretical group at Sandia National Laboratories has investigated several aspects of He in metals. This paper represents a summary of these efforts, some of which have been described in more detail elsewhere [1-6]. Much of this work has involved the Embedded Atom Method (EAM) developed by Daw and Baskes [7]. Semi-empirical EAM functions for He in Ni were developed as described in Section II; Ni was chosen as a model system because many of its properties are similar to those of FCC steels, and it is easier to treat from a theoretical point of view. Although the EAM functions are not expected to be as accurate as *ab initio* techniques, they have been critical in providing much qualitative and semi-quantitative information, particularly in the modelling of large systems of thousands of atoms.

The EAM was used for the study of the solubility of He in liquid and solid Ni [1], the diffusion of He in Ni [2], the energetics of He atoms in vacancy clusters [3], and dislocation loop punching by the growth of He bubbles. These studies were complemented by continuum elasticity models of loop punching [4]. Also, the results of the EAM studies were used as input data to a detailed model of the effects of entrapped He on the formation of voids in rapidly-crystallized steels [5]. Finally, *ab initio* models were used to study the shift of electronic energy levels in high-pressure He in Al [6].

Fundamental Aspects of Inert Gases in Solids
Edited by S.E. Donnelly and J.H. Evans, Plenum Press, New York, 1991

2. Applications of the Embedded Atom Method to Helium in Nickel

The interatomic interactions are modeled using the embedded atom method (EAM) due to Daw and Baskes[7]. This approach has been used successfully to study a wide variety of structural properties in metals[8]. In the EAM, the energy is written as the sum of two terms. First, the energy associated with placing each atom into the local electron density. (The electron density is modeled by the superposition of atomic electron densities.) The second term is a sum of pair-wise interactions which account for electrostatic interactions. In particular, the total energy of an arbitrary arrangement of atoms is written:

$$E = \sum_i F_i \left(\sum_{j \neq i} \rho_j^a (R_{ij}) \right) + \frac{1}{2} \sum_{ij} \phi_{ij}(R_{ij})$$

where $\rho_j^a (R_{ij})$ is the atomic electron density due to atom j, $F_i(r)$ is the energy to place atom i into the electron density r, and $F_{ij}(R_{ij})$ is the pair interaction between atoms i and j. In practice these functions are determined empirically by fitting to known properties of the metals and impurity in question. The Ni embedding function, F_{Ni}, and the Ni-Ni pair interaction are determined from the elastic constants, lattice constant, sublimation energy and vacancy formation energy of pure Ni. In particular, the pair interaction between the Ni atoms, $F_{Ni\text{-}Ni}$, is modeled by a Morse potential:

$$\Phi_{Ni\text{-}Ni} (R) = A \{\exp[-2b(R-R_e)] - 2\exp[-b(R-R_e)]\} \qquad (1)$$

with the values A=0.40 eV, b = 2.146 Å^{-1}, and R_e = 2.420 Å. The electron density, $\rho_j^a (R_{ij})$ is obtained from the Hartree-Fock results of Clementi and Roetti [9]. (In practice, only the two longest ranged s functions and the longest ranged d function were used for practical convenience.) The embedding function is determined by requiring the equation of state to be reproduced as described in Foiles, Baskes, and Daw [10].

The He embedding function is taken from a linear fit to the first-principles results of Puska, et al [11]. ($F_{He}(r) = 42.56$ ev-Å^3 r) The Ni-He pair interaction is determined by requiring that the effective potential between the Ni and He atoms fits the interaction deduced by Melius, Bisson and Wilson [12] from fits to the results of quantum cluster calculations. Thus, the Ni-He interaction is given by:

$$\phi_{Ni\text{-}He} (R) = B \frac{e^{-\alpha R}}{R} - F'_{Ni} \rho_{He}^a (R) - F'_{He} \rho_{Ni}^\alpha (R) \qquad (2)$$

with B = 45.9 eV-Å and a = 2.03 Å^{-1}. In this expression F′ is the derivative of the embedding functions evaluated at densities found in the ideal bulk crystal. Finally, the He-He pair interaction is determined by requiring that the energy of an fcc lattice of He fits the equation of state of solid He due to Wolfer.

3. Helium Solubility in Solid and Liquid Nickel

Although it is generally assumed that inert gases are insoluble in metals with high melting points, equilibrium solubility measurements have been attempted for He in solid Ni [13] and solid Au [14]. The atomic concentrations were found to be 10^{-10} and 10^{-7}, respectively. Other experiments on the rapid quenching of liquid steels [15] have found He concentrations on the order of 10^{-6}, presumably due to solubility in the liquid phase.

To determine if the above experiments were actually measuring equilibrium solubilities,

Temp (K)	V($\mathring{A}^3$)/atom	γ (J/m^2)	E_{hole}	Heat of Solution
1726	15.6	1.78	3.36	2.91
2000	17.5	1.70	3.46	2.98
3000	24.5	1.40	3.56	3.25

we have used the EAM to determine the heats of solution of solid and liquid Ni [1]. For the liquid state, a series of Molecular Dynamics simulations were carried out at different pressures, He concentrations (0, 1, or 2 He per 500 atoms), and temperatures (1726 K = T_{mp}, 2000 K, and 3000 K). These simulations yielded the heat of solution of the system as a function of pressure and temperature, and it was possible to extrapolate to zero pressure for the temperature range 1726 K to 3000 K.

The He atoms created a small hole around themselves in the liquid. The heat of solution was found to be well-approximated by the energy required to form the hole, namely:

$$E_{hole} = 4.836\, \gamma(T)\, [V(T)]^{2/3}$$

where γ is the surface energy and V is the volume of the hole, or the excess volume of He in liquid Ni. Table 1 compares the heat of solution (determined from the EAM) with E_{hole}.

Using a model for the solubility of gases in liquids developed by Neff and McQuarrie [16], it is then possible to determine the excess entropy of solution from knowledge of the effective hard sphere diameters of Ni and He. Determination of the enthalpy and entropy of solution then yields the equilibrium solubility of He in Ni, as shown in Fig.1 for a pressure of 1 kbar.

Similarly, the heat of solution of He in solid Ni was determined using the EAM. The computer simulations consisted of placing a single He atom in an interstitial site and a substitutional site, in a crystal of 256 or 255 Ni atoms, respectively. The heats of solution at

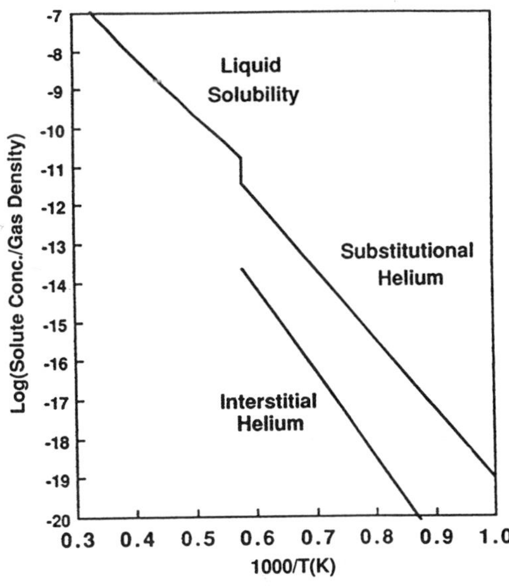

Fig. 1   Predicted solubility of helium in solid and liquid nickel for a helium gas density of 2.4 10^{22} cm^{-3} which corresponds to a gas pressure of 1 kbar (100 MPa) at room temperature.

0 K were found to be:

$$E^I_{He} = 4.1 \text{ eV}$$

$$E^S_{He} = 3.30 \text{ eV}$$

The vibrational frequencies of He atoms in the interstitial and substitutional sites were determined (from a harmonic oscillator approximation) to be 1.6×10^{13} and 1.7×10^{13} sec^{-1}, respectively. Knowing these vibrational frequencies, it is then possible to determine the entropy and enthalpy of solution as a function of temperature [1], which then yields the solubility, as shown in Fig. 1.

The solubility in both the liquid and solid state is seen to be highly temperature dependent. The liquid state solubility is higher than that in the solid state. The substitional sites are seen to be significantly more favorable than the interstitial sites. It is interesting to note that the equilibrium solubility of He in solid Ni is significantly below that measured by Driesch and Jung [13], who found a solubility of 10^{-10}, independent of temperature over the range from $T_{mp} = 1726$ K to approximately 1100 K.

4. Dislocation Loop Punching by Helium Bubble Growth

4.1. Computer simulations of bubble growth

Computer simulations were performed to determine the equilibrium configuration of small He bubbles in Ni. The simulations are not designed to determine the kinetics of the bubble growth process, but rather to determine the equilibrium structure of the bubbles. The simulations are performed using Monte Carlo simulation techniques with the energetics coming from the EAM as described above. The simulations considered a periodic cubic cell with sides of 13Å where a is the lattice constant. (This contains 8788 Ni lattice sites.) So the simulations really consider a simple cubic array of identical bubbles separated by 13Å. This is a bubble density of 10^{19} cm^{-3} which is close to the range of bubble densities found experimentally. A single vacancy was created in this lattice and a few He atoms were placed in the vacancy. The simulations are then allowed to run until equilibrium is established. Then a few (~10–20) more

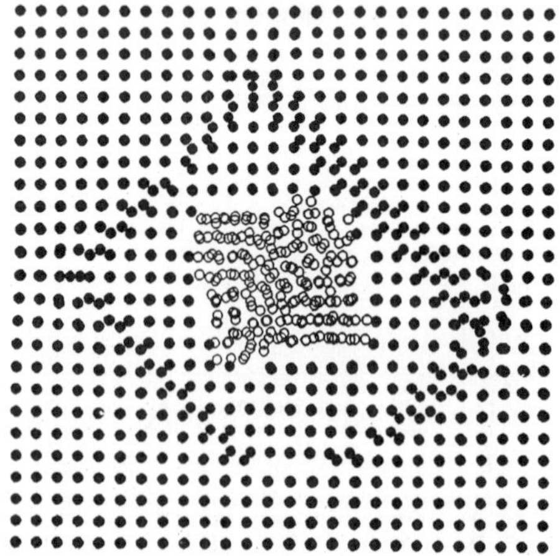

Fig. 2   Cross-section (2Å wide) through the center of a helium bubble of 369 He atoms (open circles) in a nickel crystal of 8787 Ni atoms (filled circles). The dislocation loop surrounding the bubble was spontaneously 'punched out' by the bubble during the Monte Carlo computer simulation.

He atoms are added to the cluster and the system is again equilibrated by the Monte Carlo simulations. This process was repeated several times until bubble sizes of about 370 He atoms were obtained. Larger bubbles were not considered so as to avoid large interactions between the bubble and its periodic images.

Three different results can be obtained from these simulations as a function of bubble size. First, the qualitative structure of the small bubble can be examined. Also, the density of He in the bubble can be determined and also the overall swelling of the metal lattice. All of these predictions can then be compared with the results of elasticity treatments.

The qualitative structure of a bubble containing 369 He atoms is shown in Figure 2. This figure shows a projection of the atoms contained in a slice 2Å thick which passes through the center of the bubble. Note that the metal lattice is normal both far from the bubble and adjacent to the bubble, but there is a rectangle of defected material around the bubble. These defects are dislocation loops that have been pushed out from the bubble. This qualitative picture of the structure is consistent with the elasticity treatment which will be discussed below. The average density of the He in the bubble can also be estimated. (It is hard to get a unique value due to the ambiguity of defining the boundary between bubble interior and metal.) The He densities that are obtained in the simulations 6-7 Å^3 per He atom which gives of He/vacancy ratio of 1.6 to 1.8. The overall swelling of the metal is determined from the volume of the system containing the bubbles. (The Monte Carlo simulations are performed at constant pressure so that the system is allowed to expand so that zero total pressure is achieved.) The swelling found in the simulations is $dV/dN_{He} = 9.6$ Å^3 or $S = 0.87$ C_{He}. How this compares with the elasticity results and experiments will be discussed later.

4.2. *Continuum elasticity theory of loop punching*

Both in the original model of Greenwood, Foreman and Rimmer (GFR Model) as employed by Evans [17] and in the computer simulation results discussed above, the process of dislocation loop punching is not analyzed in a continuous manner. Rather, one compares the energies before and after the loop is punched out and placing the loop at a sufficiently large distance from the bubble so it no longer interacts with it.

As a result, the condition derived for loop punching is only a necessary one. It would also be a sufficient one if no "activation barrier" exists for this loop punching process.

In order to determine both necessary and sufficient conditions for bubble growth by loop punching, a detailed energetic analysis has been carried out by Wolfer [4]. In this analysis, for a given bubble radius and initial helium density, a dislocation loop is placed at various distances from the bubble, and the change in total energy is evaluated. Now, loop punching occurs only when this change in energy is negative for all separation distances between the bubble and the dislocation loop.

The gradual completion of the dislocation loop during the process of loop punching is schematically illustrated in Fig. 3. The initial stage (not modelled in detail) may be viewed as a small number of surface atoms being squeezed into interstitial positions between the two atomic planes closest to the equator of the bubble. This interstitial ledge is then completed by more surface atoms being pushed between these two planes. The resulting interstitial ring is then the first stage of the dislocation loop punching process. Because the energy of this attached loop is proportional to its line length, it will be completed first before it or any segments move away from the bubble surface.

The energy of the system as a function of the bubble-loop separation distance is shown in Fig. 4 for various initial helium densities in the bubble. In order for this energy to become all negative, the initial helium density must be equal to or greater than about 2 helium atoms per vacant site.

At this critical density or helium pressure, the total energy change as shown in Fig. 4 is composed of four major contributions:

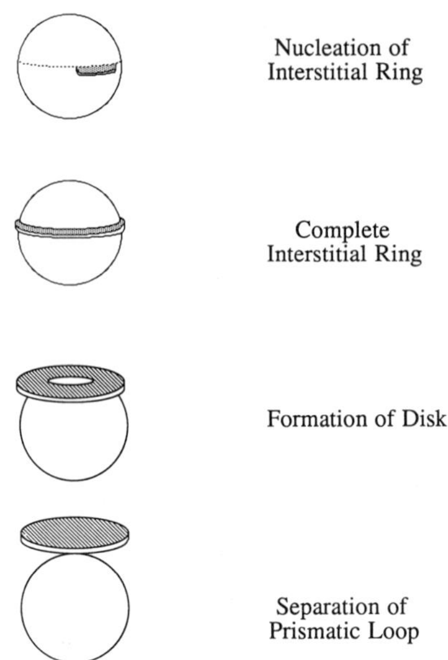

Fig. 3   Stages in the nucleation and growth of a dislocation loop at a helium bubble.

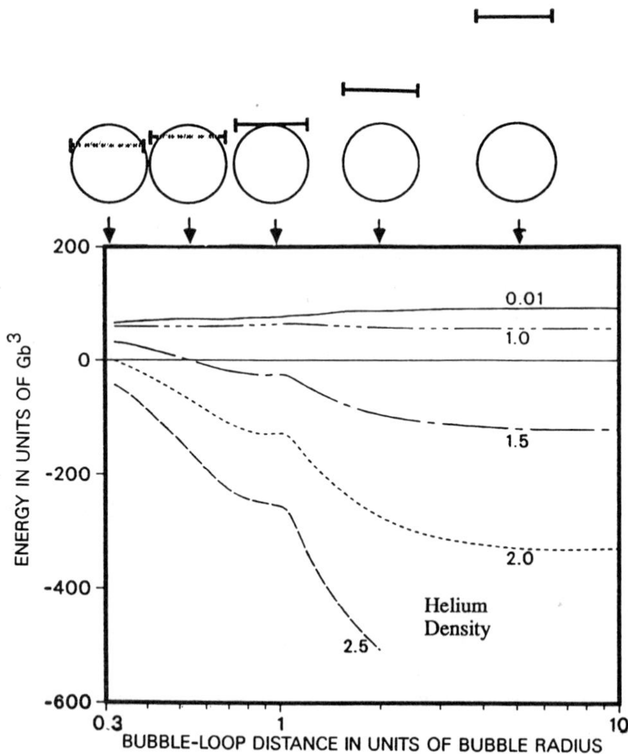

Fig. 4   Interaction energy of a helium bubble and a dislocation loop (G is the shear modulus and b is the Burgers vector).

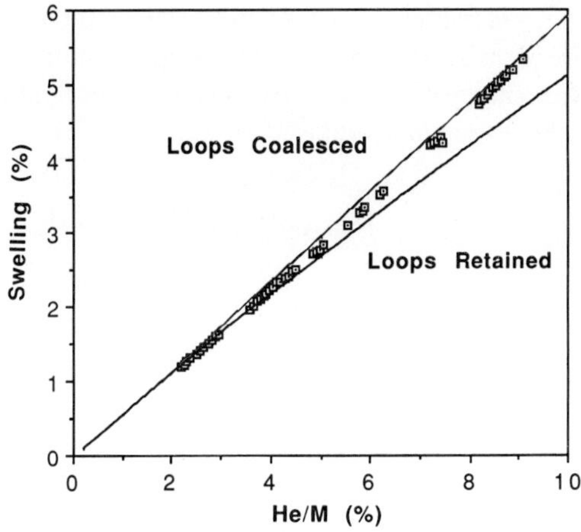

Fig. 5   Comparison of theoretical and experimental results for swelling of $PdT_{0.6}$ containing He bubbles.

1) the loop strain energy which is nearly a constant
2) the loop-cavity interaction energy
3) the bubble strain energy which decreases somewhat as the loop recedes from the bubble and its volume increases, and
4) the Helmholtz free energy of the gas which also decreases significantly as the loop recedes.

The change in contributions 3) and 4) is due to the gradual bubble volume expansion. It is found that the initial bubble volume V_o has fully increased by the loop volume V_{loop} only when the loop is at a distance of 5 bubble radii from the bubble center. At smaller distances, the distortion field of the dislocation loop squeezes the bubble to a smaller volume.

The critical helium density for loop punching or the critical helium pressure change little with the initial bubble radius, in contrast to the GFR condition for loop punching.

The remaining variation of the critical density with bubble radius can probably not be observed. The reason is that the helium density in the bubble after the expulsion of the loop drops. Since in an actual sample, one probably has bubbles which have densities between these two limits, the average helium density is indeed independent of the bubble radius. However, there is one exception. If one has a high bubble density ($> 10^{18}$ per cc) and if the dislocation loops expelled accumulate between the bubbles, it becomes more difficult for the bubbles to punch out more loops as they grow to diameters greater than about half the interbubble (center-to-center) distance. The helium density rises then dramatically. Experimental evidence [18] on the swelling of $PdT_{0.60}$ has however recently revealed that loops are not retained between the bubbles, but seem to coalesce. In this case, the percentage of swelling for the high bubble density (10^{18} per cc) coincide with those for the low bubble densities (10^{15} per cc).

We can therefore reach the conclusion that helium bubbles in nickel which can only grow by loop punching have a helium density of about 1.8 per vacant site. It follows then immediately that the sample exhibits a volume expansion or swelling of:

$$S = 1/1.8 \ C_{He} = 0.55 \ C_{He}$$

where C_{He} is the helium content of the sample in terms of atom fractions.

The swelling predicted from elasticity theory thus appears to occur at a different rate than that predicted by the atomistic model. However, the elasticity model assumes that the bubbles are near the critical He density (i.e., just before loop punching). The atomistic model determines the lowest free energy state (i.e., after loop punching). The elasticity model relates the swelling before and after loop punching:

$$S_{before} = S_{after} \{1-3b/(4R)\}$$

where b is the Burger's vector of the loop and R is the bubble radius. *After this correction, the simulation results and elasticity calculations are in good agreement.*

The agreement of the swelling predictions for $PdT_{0.6}$ with the experimental data is demonstrated in Fig. 5. Clearly, the experimental results for helium concentrations greater than 5% indicate that previously punched-out loops do not impede further coalescence. A preliminary analysis indicates that the mutual attraction of loops on the same habit plane forces them to climb by pipe diffusion, and thereby induce the coalescence.

The rate by which the fractional loop volume coalesces depends critically on the activation energy for pipe diffusion. In order for this rate to keep up with the swelling rate, an activation energy for pipe diffusion in $PdT_{0.6}$ of about 1.2 eV is required. This represents about 45% of the value for the self-diffusion energy of Pd, and is therefore in line with traditional estimates of pipe diffusion energies.

5. Diffusion Mechanisms for Helium in Nickel

Several experiments have been carried out to determine the long-range diffusion rate of He in metals (Ni, Au, Ag, Al) [19-24]. These experiments have revealed activation energies which correspond to vacancy mechanisms for He diffusion in Au, Ag, and Al [19], but for He in Ni an unusually low activation energy has been found [20], suggesting a dissociative mechanism.

He release experiments by Thomas *et al.* [23] found two activation energies, 0.2 eV and 0.35 eV, which they interpreted in terms of pipe diffusion and interstitial migration, respectively. However, it is possible that the 0.2 eV measurement was instead due to surface effects

Table 2 Comparison of experimental and theoretical activation energies, although it is not well-known which physical process may be associated with the experimental energies.

Measured Activation Energies (eV)	Activation Energies Calculated in This Work (eV)	Physical Process
0.20 [23]		Unknown; possibly due to pipe diffusion or surface effects
0.35 [23]	0.53	Interstitial migration
0.81 [20]		Unknown; possibly dissociative mechanism
1.3 [24]	1.29	Dissociative Mechanism
2.3 [24]	2.49	Vacancy Mechanism
2.7 [24]		Unknown; possibly due to self-diffusion
3.1 [24]	3.1	Exchange Mechanism

10

(release of He from the oxide layer). High temperature He release experiments by Reed et al. [24] provide estimates of activation energies for several other diffusion and detrapping processes, which we discuss later.

The purpose of our work was to determine the activation energies for several possible diffusion mechanisms of He in Ni. To accomplish this, we used the EAM to determine the optimal migration path and migration energy [2]. We began with a study of interstitial migration, and found that the migration energy for this process was:

$$E^{m(I)}_{He} = 0.53 \text{ eV},$$

somewhat higher than Thomas *et al.*'s value of 0.35 eV [23].

However, He atoms could only diffuse for a short distance until they encountered vacancies, to which they would be strongly bound. Thus, long-range diffusion would involve a dissociative mechanism, in which the He first jumps out of the vacancy, and then diffuses a long distance until it encounters another vacancy. Under conditions of thermal equilibrium, and ignoring the activation entropy of migration, the diffusion rate for dissociative helium migration is:

$$D^d_{He} = r^2/6 \, \nu \exp(-S_v^f/k) \exp(-E^d/kT),$$

where r is the distance between interstitial sites, ν is the vibrational frequency, S_v^f is the entropy of vacancy formation, and:

$$E^d = E^{m(I)}_{He} + E^I_{He} - E^S_{He} - E_v^f = 1.29 \text{ eV},$$

assuming that $E_v^f = 1.8$ eV [25]. If we further assume that $S_v^f = 2k$, then $D^d_{He} = .01 \exp(-1.29/kT)$ cm^2/sec. Comparing this result with that of Phillipps and Sonnenberg [20], $D_{He} = 0.0063 \exp(-0.81/kT)$ cm^2/sec, we find agreement on the pre-exponential factor, but significant disagreement on the activation energy. The disagreement is presumably due to the approximate nature of the EAM model.

Another possible mechanism for diffusion is a vacancy mechanism, whereby a substitutional He atom encounters a vacancy, and migrates much as a Ni atom would. Using the five-frequency model and the results of several EAM calculations, the diffusion rate for this mechanism was found to be:

$$D^s_{He} = 1.83 \exp(-2.49 \text{ eV}/kT) \text{ cm}^2/\text{sec}$$

which has a much higher activation energy than that for the dissociative mechanism.

Finally, He diffusion by an exchange mechanism was investigated, whereby a substitutional He atom exchanges positions with an adjacent Ni atom. The activation energy for this process was found to be 3.1 eV, the highest of any of the mechanisms considered. Table 2 summarizes the calculated activation energies for the different diffusion mechanisms, and compares them with the experimental data. The comparison is difficult, since the diffusion mechanism associated with the experimental results is not known. However, it appears that the EAM significantly overestimates the activation energies for the interstitial mechanism and especially the dissociative mechanism. The calculated activation energy for the vacancy mechanism is probably fairly reasonable, since it is similar to self-diffusion which is well-described by the EAM [26]. The most important result of the calculations is the clear ordering of the activation energies for the different diffusion mechanisms, which indicates that the dissociative mechanism is much favored over the vacancy mechanism.

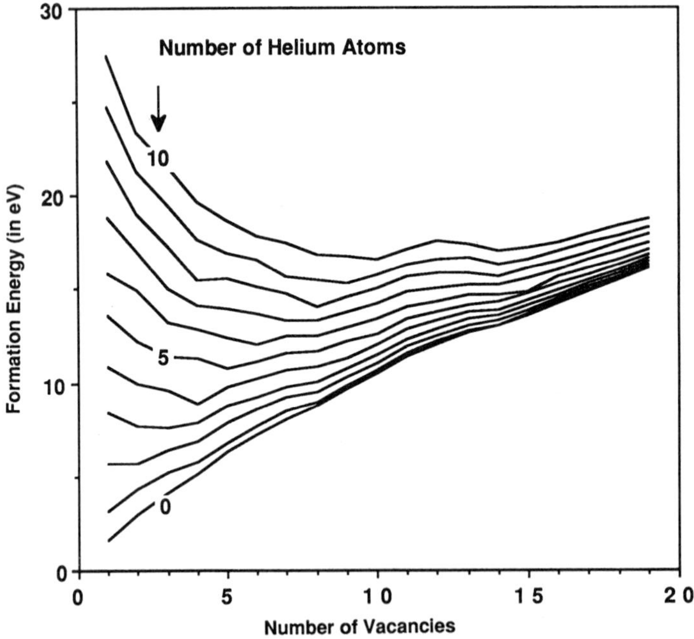

Fig. 6 Formation energy of clusters of He atoms and vacancies in Ni, relative to a perfect Ni crystal with the He far outside the crystal.

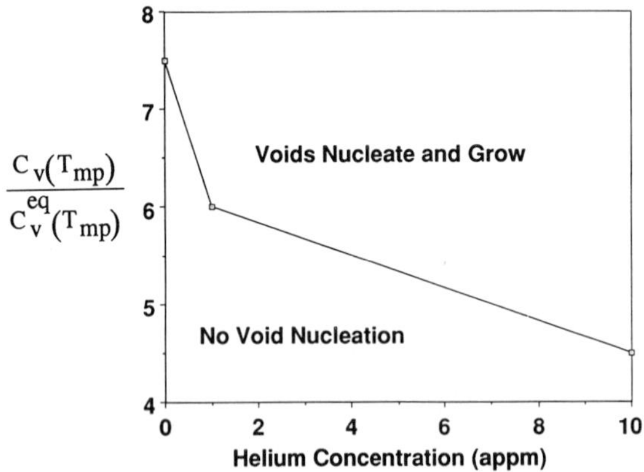

Fig. 7 Conditions for void nucleation. Void nucleation occurs only if a critical supersaturation of vacancies is quenched into the material. The critical supersaturation depends on the helium content.

12

6.   Energetics of Helium Atoms in Small Vacancy Clusters

The structure and formation energies of small clusters of He atoms and vacancies were determined using the EAM [3]. The simulations were carried out for clusters of 0-10 He atoms and 1-19 vacancies, a total of 200 structures ranging from high-density He bubbles to pure voids. These defects were placed in a Ni crystal of 250-1000 atoms, with periodic boundary conditions. Larger crystals were used for the larger defects, to minimize size effects.

The results of the calculations are shown in Fig. 6. It has been previously shown that the formation energies of small vacancy clusters can be well-approximated by the surface energy of the cluster, with small corrections for the curvature [27]. The effect of He is to raise the formation energy of the system, since the formation energy is relative to He atoms existing far outside the crystal. The increase in energy is quite dramatic for large numbers of He atoms in a few vacancies (dense He), and almost insignificant for a single He atom in 19 vacancies.

However, the formation energy of He on an interstitial site in Ni is 4.1 eV, much greater than the increase in energy if the He were to join a low-density He-vacancy cluster. Thus, although it is very unlikely for He to enter the crystal (low solubility), any He in the crystal will tend to segregate to He-vacancy clusters, particularly the low-density ones.

7.   Effect of Entrapped He on the Formation of Voids in Rapidly-Crystallized Steels

It is well-known that the rapid-quenching of solids from high-temperature to low temperature can result in a vacancy supersaturation at the lower temperature. However, it is also possible to develop an even higher vacancy supersaturation by rapid crystallization of a liquid into a solid phase. Extremely rapid cooling of liquids can result in the formation of amorphous solids. Somewhat slower cooling rates allow liquids to crystallize, but may not allow the crystal's vacancies sufficient time to reach their equilibrium concentration [28]. For those rates, the following inequality may hold:

$$C_v(T) > C_v^{eq}(T_{mp})$$

Our purpose was to develop a detailed model of the dynamic nucleation and growth of voids in rapidly-quenched steel powders containing entrapped He. He contents on the order of 10^{-5} have been reported in such steels [29], and it was suspected that the He was critical in stabilizing the voids which eventually formed. These voids act as nucleation sites for the precipitation of carbides, which are stable up to 1300 °C and significantly increase the yield stress of the material.

The model involves a large set of coupled kinetic rate equations, which describe the change in the number of He-vacancy clusters of all sizes. The clusters are allowed to grow or decay by absorbing or emitting He atoms and/or vacancies, and new clusters may nucleate. The effect of dislocations as a source/sink for vacancies is included. The model depends on several atomistic parameters, such as the diffusion rate of vacancies and helium and the binding energies of He-vacancy clusters. These parameters were determined previously [2,3].

The kinetic rate equations were solved iteratively using a finite-difference scheme. The He content, initial vacancy supersaturation, and quench rate were varied to determine their effect on the nucleation and growth of voids. It was found that high vacancy supersaturations of $\sim 8 \times C_v^{eq}(T_{mp})$ were required for void nucleation without He (see Fig. 7). However, He concentrations as low as 10^{-5} were found to significantly lower the barrier for void nucleation, so that vacancy supersaturations of only $\sim 4 \times C_v^{eq}(T_{mp})$ were required. For vacancy concentrations below that size, voids did not nucleate, and excess vacancies were eventually absorbed by dislocations. The quench rate did not significantly affect the results directly, but it would of course control the initial vacancy supersaturation due to imperfect rapid crystallization.

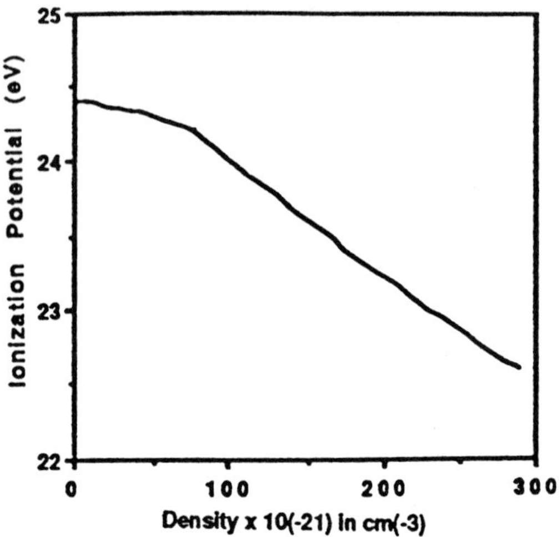

Fig. 8   Calculated dependence of He ionization potential (eV) on density within a He$_{13}$ cluster.

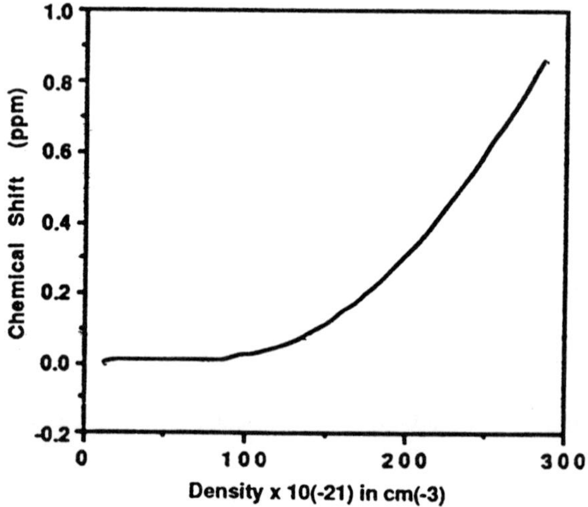

Fig. 9   Calculated dependence of isotropic ^{3}He chemical shift (ppm) on density within a He$_{13}$ cluster, relative to the gas-phase atomic value.

8.   Shift of Electronic Energy Levels of Dense Helium in Aluminum

VUV absorption and EELS have been performed on rare gas bubbles implanted in thin Al films, and the spectra interpreted in terms of blue-shifted atomic transitions [30-32]. The magnitude of the shift was thought to increase with increasing density within the bubbles. *Ab initio* quantum chemical calculations [33] have predicted maximum blue shifts of 1 - 1.5 eV in the ^{1}S - ^{1}P transition as rare gas atom clusters are compressed, in agreement with experiment for all but He. In the case of He-implanted Al films, broad spectral features in the region of 21-25 eV have been interpreted as blue shifts up to 3.3 eV in the He atomic transition of

Table 3 Experimental data for rare gas atoms.

	1S —>1P Transition (eV)	First Ionization Potential (eV)
He	21.2	24.6
Ne	16.9	21.6
Ar	11.8	15.8

21.2 eV. *Ab initio* quantum chemical results on He clusters [33] indicate, however, that the maximum blue shift is 1.5 eV at a density of 6 x 10^{22} cm^{-3} and a red shift occurs at densities above 2 x 10^{23} cm^{-3}.

The influence of the surrounding metal on the He cluster, neglected in this earlier study, is addressed here. The model cluster consists of a cage of twelve Al atoms, fixed at their bulk fcc sites, with a He atom at the center. Moderately sized basis sets at the self-consistent-field (SCF) level of approximation can address the nature of spectral line shifts semiquantitatively, i.e., whether to the blue or red, and to what extent. Exciting a He 1s electron into a He 2p orbital results instead into the promotion of the electron to a previously unoccupied valence Al orbital. That is, the He atom is *ionized*, with its electron entering the Al conduction band. This result suggests that He ionization is occuring experimentally, and also offers an explanation of why He behaves differently from other rare gases. For the first three rare gases, Table 3 lists the ionization potential (IP) and ^{1}S -> ^{1}P excitation energy. The difference between these is smallest for He (3.4 eV), and substantial blue-shifting of the ^{1}S $\rightarrow$ ^{1}P transition clearly approaches the ionization threshold.

Density-induced IP changes were calculated for a model system of a central He atom in a cubo-octahedron of twelve He atoms. The results are shown in Fig. 8, in which the IP drops sharply with increasing density. The effect of an Al host will further lower this transition. This suggests that the experimentally observed bands in the 22.5-24.5 eV range result from density-induced red-shifting of the He atomic ionization rather than blue-shifting of the He atomic ^{1}S - ^{1}P transition. The width of these same peaks follows from the fact that the ionized electrons can occupy a range of low-lying Al orbitals. Thus, the present calculations provide an alternative interpretation of the VUV and EELS spectra. The relatively sharp peak of 21.2 eV, (the gas-phase atomic transition), observed at low bubble densities merely disappears at higher densities, rather than being broadened and blue-shifted.

Spectroscopic techniques, such as solid-state nuclear magnetic resonance (NMR), may be more suited to the determination of average densities within He bubbles. Chemical shifts obtained by magic angle spinning or multiple pulse NMR methods on powder or single crystal samples can be assigned unequivocally to the ^{3}He isotope. NMR is a promising indirect experimental approach for the extraction of density information since accurate *ab initio* quantum chemical predictions of chemical shifts are possible [34]. Such calculations are presented in Fig. 9, for a central He atom surrounded by a cubo-octahedron of twelve He atoms. The chemical shift of the central He (relative to gas-phase) steeply increases with increasing density within the He cluster. A chemical shift of approximately 1 ppm is found at a density of 3 x 10^{23} cm^{-3}, the highest density considered here. Thus it appears that average densities within He bubbles can be extracted from solid-state NMR spectra, provided the latter are of sufficiently high resolution.

REFERENCES

1. W.G. Wolfer, C.D. Van Siclen, S.M. Foiles and J.B. Adams, Acta metall. **37**, 579 (1989).

2. J.B. Adams and W.G. Wolfer, J. Nuc. Mater. **158**, 25 (1988).

3. J.B. Adams and W.G. Wolfer, J Nuc. Mater. **166**, 235 (1989).

4. W.G. Wolfer and W.G. Drugan, Phil. Mag **A57**, 923 (1988).
   W. G. Wolfer, Phil Mag **A58**, 285 (1988); W. G. Wolfer, Phil. Mag **A59**, 87 (1989).

5. J.B. Adams and W.G. Wolfer, submitted to Acta metall.

6. C.M. Rohlfing, J. Nuc. Mater. **165**, 84 (1989).

7. M. S. Daw and M. I. Baskes, Phys. Rev. **B29**, 6443 (1984).

8. M. S. Daw in *Reconstruction of Solid Surfaces*, edited by K.Christman and K. Heinz (Springer Verlag, Berlin, in press); S. M. Foiles in *Surface Segregation and Related Phenomena*, edited by P. A. Dowben and A. Miler (CRC Press, 1990).

9. E. Clementi and C. Roetti, At. Data Nucl. Data Tables **14**,17(1974).

10. S. M. Foiles, M. I. Baskes, and M. S. Daw, Phys. Rev. **B33** 7983 (1986).

11. M. J. Puska, R. M. Nieminen and M. Manninen, Phys. Rev. **B24**, 3037 (1980).

12. C. F. Melius, C. L. Bisson, and W. D. Wilson, Phys. Rev. **B18**, 1647 (1978).

13. H.J. von den Driesch and P. Jung, High Temp.-High Press. **12**, 635 (1980).

14. J. Laakmann, P. Jung and W. Uelhoff, Acta metall. **35**, 2063(1987).

15. J.E. Flinn, private communication.

16. R.O. Neff and D.A. McQuarrie, J. Phys. Chem **77**, 413 (1973).

17. J.H. Evans, J. Nuc. Mater. 68, 129 (1978); J.H. Evans, J. Nuc. Mater. **76-77**, 228 (1978).

18. S.E. Guthrie, private communication.

19. V. Sciani and P. Jung, Radiat. Eff. **78**, 87 (1983).

20. V. Phillipps and K. Sonnenberg, J. Nucl. Mater. **114**, 95 (1983).

21. V. Phillipps and K. Sonnenberg, J. Nucl. Mater. **107**, 271 (1982).

22. D.B. Poker, Radiat. Eff. **78**, 101 (1983).

23. G.J. Thomas, W.A. Swansiger and M.I. Baskes, J. Appl. Phys. **50**, 6942 (1979).

24. D.J. Reed, F.T. Harris, D.G. Armour and G. Carter, Vacuum **24**, 179 (1974).

25. L.C. Smedskjaer, M.J. Fluss, D.G. Legnini, M.K. Chason and R.W. Siegel, J. Phys. **F11**, 2221 (1981).

26. J.B. Adams, S.M. Foiles and W.G. Wolfer, J. Mater. Res. **4**, 102 (1989).

27. A. Si-Ahmed and W.G. Wolfer, in: Proc. 11th Conf. on Effects of Radiation on Materials, ASTM STP 782, Eds. H.R. Brager and J.S.Perrin Eds. (American Society for Testing and Materials, 1982) p. 1008.

28. W.W. Webb, J. Appl. Phys. **33**, 1961 (1962).

29. J.C. Bae, T.F. Kelly, J.E. Flinn and R.N. Wright, Script Met. **22**, 691(1988).

30. a) S. E. Donnelly, Rad. Effects **90**, 1 (1985); b) A. A. Lucas, J. P. Vigneron, S. E. Donnelly, and J. C. Rife, Phys. Rev. **B28**, 2485 (1983); c) J. C. Rife, S. E. Donnelly, A. A. Lucas, J. M. Gilles, and J. J. Ritsko, Phys. Rev. Lett. **46**, 1220 (1981); d) S. E. Donnelly, J. C. Rife, J M. Gilles, and A. A. Lucas, J. Nucl. Mat. **93-94**, 767 (1980).

31. a) W. Jäger, R. Manzke, H. Trinkaus, R. Zeller, J. Fink, and G. Crecelius, Rad. Effects **78**, 315 (1983); b) R. Manzke, G. Crecelius, W. Jäger, H. Trinkaus, and R. Zeller, Rad. Effects **78**, 327 (1983); c)W. Jäger, R. Manzke, H. Trinkaus, G. Crecelius, R. Zeller, J. Fink, and H. L. Bay, J . Nucl. Mat. 111-112, 674 (1982).

32. A. vom Felde, J. Fink, T. Müller-Heinzerling, J. Pflüger, B. Scheerer, G. Linker, and D. Kaletta, Phys. Rev. Lett. **53**, 922 (1984).

33. P. R. Taylor, Chem. Phys. Lett. **121**, 205 (1985).

34. C. M. Rohlfing, L. C. Allen, and R. Ditchfield, Chem. Phys. **87**, 9 (1984).

MOLECULAR DYNAMICS SIMULATIONS OF RARE GASES IN METALS: INTERACTIONS

R.M. Nieminen

Center for Scientific Computing
P.O. Box 40
02101 Espoo, Finland
  and

Laboratory of Physics
Helsinki University of Technology
02150 Espoo, Finland

ABSTRACT

The construction of interatomic forces for inert gas-metal systems is discussed, with particular emphasis on force laws applicable in molecular dynamics simulations. It is argued that the effective-medium theory, based on the simple concept of atoms embedded in electron gas, provides a useful framework for the construction over a wide range of energies. The constant-pressure simulations of pressure-volume isotherms of dense Ne and Ar are presented.

1. Introduction

Molecular dynamics (MD) is the code word for atomistic simulation of materials[1]. With the increase in computational power, this technique is gaining wide popularity in computational physics of materials. The principle is very simple: the classical Newton´s equations of motion for the individual atoms are integrated numerically under the influence of interatomic and external forces. The simplest MD technique is one where the equations of motion are solved for a fixed number N of particles in constant volume and with constant total energy, and the interactions are given by a pairwise force law. Such microcanonical simulations can be extended to various other statistical ensembles. An important extension is (i) the constant-pressure simulation [2], which allows both the volume and the shape (symmetry) of the system to fluctuate and seek its equilibrium values. Another extension is the constant-temperature simulation[3], where a "thermostat" is applied to system so as to keep its temperature (mean kinetic energy) at a prescribed value. One can also impose constraints for some degrees of freedom (for example, fix bond distances or angles if the system is composed of inert molecules).

For simulations of extended systems one uses periodic boundary conditions in one or more directions. The fundamental unit cell can contain up to several thousand atoms in present day supercomputer simulations. Near equilibrium, the timestep for the numerical integration is typically a small fraction of lattice vibration times, i.e. of the order of 10^{-14} s. If the simulation involves energetic atoms far from equilibrium, a much shorter timestep maybe necessary. The total feasible simulation time depends naturally on the system size and computer resources available, but seldom exceeds 10^{-8} s. Even with supercomputers, one has a long way to go to reach timescales relevant to, say, activated diffusion or annealing at low temperatures! Fortunately, however, one can learn a great deal about phenomena with short relaxation times.

Fundamental Aspects of Inert Gases in Solids
Edited by S.E. Donnelly and J.H. Evans, Plenum Press, New York, 1991

Apart from thermodynamic equilibrium situations, where MD will provide information of the ground-state structure and phase diagram of the system, the technique can also treat nonequilibrium processes with complicated temporal behavior. Such processes include, for example, kinetics of phase transitions (quenching, nucleation) or collision cascades of energetic atoms in solids. In the case of nonequilibrium simulations, it is often necessary to mimic dissipation by a suitable coupling of the computational system to its environment. This can be accomplished by introducing Langevin-type viscous forces acting on the atoms near the cell boundary or by some other heat-bath algorithm. Moreover, there can be important inelasticity in the interactions (collisions) between atoms, as well as "friction" due to coupling to conduction electrons. The means to handle such problems are still rather primitive.

Roughly speaking, MD simulations provide two kinds of information. At or near equilibrium, time averages over the atoms' motion yield the expectation values of any desired thermodynamic observables, including the structural order parameters. For example, the diffusion constant for labelled particles can be obtained from the average mean-square displacement as:

$$D = \lim_{t \to \infty} \langle \Delta \mathbf{R}(t)^2 \rangle \frac{1}{6t} \tag{1}$$

Secondly, one can obtain dynamic or temporal characteristics by calculating averages over event histories in nonequilibrium cases. For example, by averaging over physically relevant initial conditions one can obtain statistical distributions for equilibrating scenarios in many-atom scattering sequences. The amount of information from such studies is often overwhelming and requires sophisticated post-processing techniques, such as visualisation and animation using computer graphics methods.

Finally, there is the all-important question of interatomic interactions [4]. There are cases where the information obtained from an MD simulation seems to be fairly insensitive to the choice of the force law, i.e. the simulated properties are generic to a large class of systems. For example, the structure factor of a quenched, amorphous solid seems to depend little on the interatomic potential. However, there are obviously cases where a simple classical, pairwise potential fails to give a correct description of the condensed system. The equation of a metal or a tetrahedrally coordinated semiconductor is a case in point. Then one has to use more sophisticated recipes for the interactions. Recently, the technique of first-principles MD, where the electron-mediated interactions are calculated "on the fly" has become feasible for some systems, due to the ingenious algorithms invented [5]. Such simulations are still, however, restricted to a few small systems. I discuss the interactions relevant to the rare gas-metal systems in the next section.

2.   Interactions Between Atoms

2.1. Interactions among inert gas atoms

Inert gas systems can rather well be described by pairwise, rigid interactions, i.e. the density-dependent, many-atom interactions are small. The conventional way to describe the interaction is to divide it into a repulsive part, due to Pauli orthogonalisation, and an attractive tail due to the van der Waals interaction. For practical purposes this situation is often modelled by the Lennard-Jones potential:

$$V(R) = 4\varepsilon \left(\left(\frac{\sigma}{R} \right)^{12} - \left(\frac{\sigma}{R} \right)^6 \right) \tag{2}$$

where R is the internuclear distance. The $-1/R^6$ tail arises from the fluctuating dipoles, but the analytic form of the repulsive part has no physical meaning. The Lennard-Jones potential can be fitted rather well to the equilibrium cohesive properties of inert gas solids, but the description fails at high pressures or short interatomic distances (high energies). In section 3, a summary of a comparison of different empirical potentials in constant-pressure MD simulations is presented. A general feature of the Lennard-Jones potential is that it is too repulsive at short distances.

First-principles calculations for the inert gas potentials require the introduction of electron correlation effects, as the Hartree-Fock level results give purely repulsive interactions. The perturbative configuration interaction (CI) calculations are very costly, especially so for the heavier systems. The work by Gordon and Kim [6] showed that the incorporation of correlation effects enables an accurate description already within a very simple approximation, superimposing free atom electron densities and using a strictly local formula for all parts of the energy, including the kinetic energy. Recently, Harris and others [7] have shown that the Gordon-Kim model is an approximative total energy functional in density-functional theory, and can be systematically improved.

We have recently approached [8] the inert gas potentials from a completely new viewpoint. The approach is based on the concept of embedding a rare gas atom in homogeneous electron gas. The associated energy and its density dependence, as well as the structure of the induced electron screening cloud can be used to construct a parameter-free interatomic potential over a wide range of distances. The homogeneous electron gas in the model can be thought to provide the electrons, which in a real, condensed inert gas system are the bound electrons of the neighboring atoms. When the gas density is high, the embedding energy is repulsive as the major contribution is the Pauli repulsion (increase in kinetic energy) due to orthogonalisation. For low gas densities the interaction becomes attractive due to the Coulomb and exchange-correlation energy contributions. The link between the model problem of a single atom embedded in electron gas and the many-atom interactions in condensed matter is the effective medium theory (EMT) [9].

The central concept in EMT is the so-called cohesive function $E_c(n)$. It is a function of the background electron density n and is defined as:

$$E_c(n) = \Delta E^{hom}(n) - \alpha(n) \, n \tag{3}$$

where $\Delta E_{hom}(n)$ is the embedding energy of a free atom to the electron gas. It is calculated by solving the Kohn-Sham density-functional equations numerically [10]. The function α is the volume integral of the induced Coulomb potential $\Phi(r)$ and reads:

$$\alpha(n) = \int_0^{s(n)} \Phi(r) \, 4\pi r^2 dr \tag{4}$$

where the upper limit s(n) is the so-called neutral sphere radius, defined by requiring that the total electron density inside this sphere neutralizes the nuclear charge Z. Thus:

$$\int_0^{s(n)} n(r) \, 4\pi r^2 dr = Z. \tag{5}$$

These definitions lead to a unique dependence between the radius (distance) s and the background density n.

The experience with EMT has shown that the cohesive function $E_c(n)$ and the neutral sphere radius s(n) accurately predict the cohesive properties of sp-bonded *metals* of fcc symmetry. If one eliminates the background density, one obtains the usual cohesive energy *vs.* volume description. The minimum of $E_c(n)$ and the corresponding s(n) give the equilibrium cohesive energy and the Wigner-Seitz radius, respectively. The bulk modulus is proportional to d^2E_c/ds^2 *etc*. What is somewhat surprising is that this simplest form of EMT also gives a good description of the equation of state of inert gas solids, good *insulators* with very narrow bands.

As the EMT most naturally describes solids with fcc symmetry (it can be shown that in this closed-packed structure the construction of neutral, space-filling atomic spheres needs no Coulomb corrections), we can convert the total energy data into a pair potential between inert gas atoms by simple bond counting. In a nearest neighbor model, the pair potential is simply:

$$V(R) = \frac{E_c(n)}{6} \tag{6}$$

where the distance R is calculated from the neutral sphere radius corresponding to the same background density (in fcc structure):

$$R = \frac{1}{\sqrt{2}} \left(\frac{16\pi}{3}\right)^{\frac{1}{3}} s(n) \tag{7}$$

The omission of other than nearest-neighbor contributions is not a severe simplification because the interactions are short-ranged. For example, in the case of Ar the contribution of the second-nearest neighbors is to decrease the near-neighbor attraction by around 3 %.

The cohesive energy functions $E_c(n)$ are shown in Fig. 1 for the inert gases He to Kr in Fig. 1. Apart from He, there is a minimum which signals that EMT predicts these systems to be bound. The cohesive properties are compared with experiment in Fig. 2. The noticeable feature is the overall good agreement, given that the approach is based on the concept of atoms embedded in electron gas! The calculated bulk moduli are too large which reflects the fact that the curvatures of $E_c(n)$ are too large near the minimum: the curves rise too quickly as the density decreases. This is due to the fact the local-density approximation (LDA) used in calculating the Kohn-Sham embedding energies does not give the right van der Waals tail to

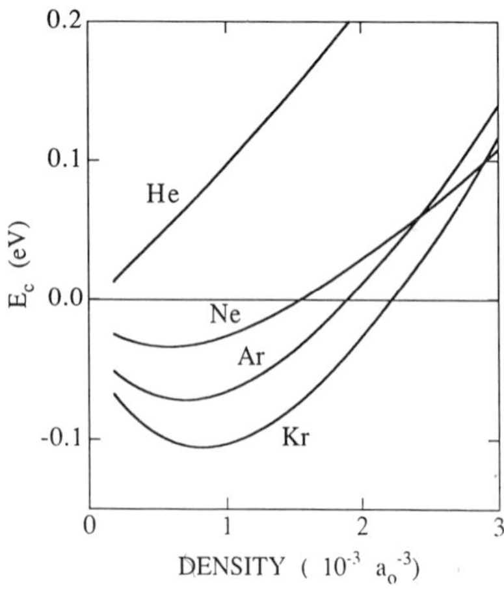

Fig. 1   The cohesive function $E_c(n)$ vs. electron density n for He, Ne, Ar and Kr.

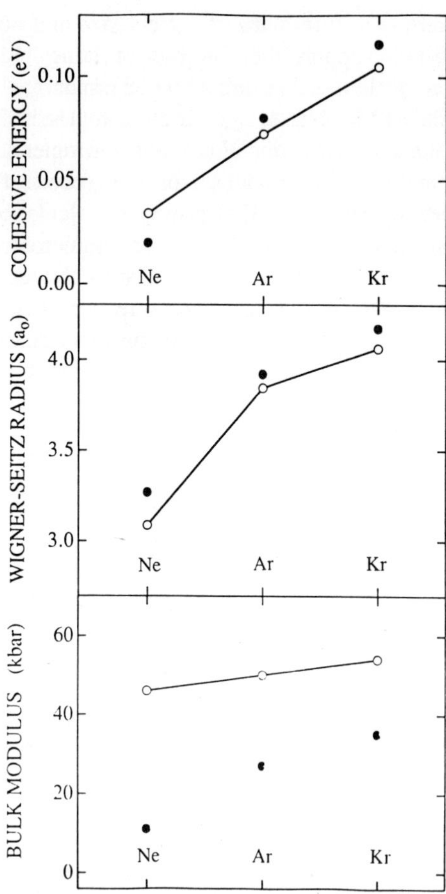

Fig. 2   Cohesive properties of of solid Ne, Ar and Kr calculated using EMT (open circles) in comparison with the experimental values (full circles).

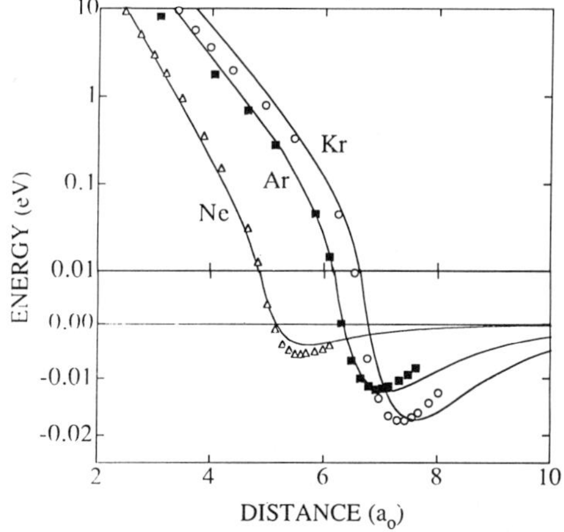

Fig. 3   Pair potentials for inert gases. The EMT results for Ne (triangles), Ar (black squares) and Kr (circles) are compared with the semiempirical pair potentials [18] (solid lines). Note the change of scale from linear to logarithmic at 0.01 eV.

the interaction, even if there is a minimum of correct size and position in the embedding function. Overall, the results support the findings of Lang [11] who has studied the physisorption of rare gas atoms in metals using a related model.

The pair potentials calculated for Ne, Ar and Kr are compared with recent semi-empirical fits in Fig. 3. Apart from the lack of van der Waals tails, the general agreement is very good . In particular, the present parameter-free model seems to be very useful at high energies (short distances). In fact, a close comparison with first-principles calculations for Ar show excellent agreement. The agreement of the present results with the Gordon-Kim values persists to energies near 1000 eV. This is in fact fairly surprising, since the repulsion in the present case is dominated by the kinetic energy term (there is no screened nucleus-nucleus repulsion). In the Gordon-Kim model about a third of the repulsion at these energies results from the nucleus-nucleus interaction. Thus our results strongly suggest that the role of the screened Coulomb repulsion, certainly dominant at very high energies, may have been systematically overestimated in interatomic potentials in the energy region below 1000 eV.

2.2. Interactions between metal atoms

As already mentioned above, the EMT philosophy has developed furthest in describing metallic interactions. A recent review of the ideas and applications has been given by Jacobsen [9], and no comprehensive survey is done here. The basic idea is to incorporate density-dependent, many-atom interactions in a simple transferrable way, and yet avoid any parametrisation based on empirical results. The generic form for the total energy of a system comprising N atoms reads:

$$E_{tot}(N) = \sum_{i=1}^{N} F_i(n_i(\mathbf{R}_i)) + \frac{1}{2}\sum_{i \neq j}^{N} \phi(\mathbf{R}_i - \mathbf{R}_j) + E^{bs} \tag{8}$$

where $\mathbf{R}_i$ denote the atomic positions and $F(n)$ is a density-dependent embedding function, closely related to $E_c(n)$ above. The second term is a pair-potential type correction, which accounts for the short-range electrostatic interactions between the atoms. The third term is a "band-structure" term, which can within EMT be related to the change in the one-electron eigenvalues in going from the reference system (homogeneous electron gas, jellium) to the real one.

There are several other schemes which can be expressed in the generic form of eqn. (8), but where the functions F and ϕ are obtained from other arguments, typically from empirical fits to cohesive and structural properties of the system under study. These schemes include the embedded atom method (EAM) [12] and the glue [13] models. The first one, in particular, has been widely applied in molecular dynamics simulations. The important point is that the form of eqn. (8), while physically motivated and by far more realistic in many situations, is basically as easy to adopt in a molecular dynamics simulation as are the traditional pair-potential schemes.

2.3. Inert gas-metal interactions

The embedding or effective-medium arguments can also be extended to deal with the case of an inert gas atom interacting with metallic media. The repulsive region of the interaction, which arises from the overlap of the bound electron states at the atom with the metal valence electrons, corresponds to the high-density region in the embedding problem. There the embedding energy is linearly proportional to the density (function F in eqn. (8) is linear). If (as is often the case) the electron density in the system can be expressed as a superposition of localized, atomic-like distributions, the total energy of interaction reduces to a pairwise summation over inert-gas/(pseudo)atom pairs. This means that the (repulsive) gas-metal interaction can be expressed as:

$$V_{\text{met-gas}}(\mathbf{R}) = \alpha_{\text{eff}} n_{\text{at}}(\mathbf{R}) \tag{9}$$

where α_{eff} is a constant [14] specific to the inert gas in question and n_{at} is the electron density distribution of the new metal atom. This form of the potential has proved useful in e.g. analysing helium and neon diffraction from metal surfaces. However, its range of validity in the high-energy side has not been fully explored.

In the attractive region, a van der Waals contribution has to be added to eqn. (9). One way of accomplishing this [15] is to add a term proportional to $-1/R^6$ to each pair potential and sum over half-space to recover (i) the Lifshitz form $-C/z^3$ asymptotically valid for a rare gas atom at a distance z from a metal surface, and (ii) the depth and position of the physisorption minimum known experimentally in many cases. The constant C can be obtained from the known polarizability of the inert gas and the plasma frequency of the metal in question.

3. Constant-Pressure Simulations of PV Isotherms

We have recently [16] carried out constant-pressure molecular-dynamics simulations of the pressure-volume isotherms of Ne and Ar, in order to test the validity of the various proposed potentials especially in the high-pressure region. The tested potentials included the Lennard-Jones potential and those proposed by Ross *et al.* [17] and by Siska *et al.* [18]. The computational unit cell comprised 864 atoms, and the investigated pressure range was from 1 to 600 GPa at room temperature. The equations of motion were solved using the Gear predictor-corrector algorithm with a typical timestep of 10 fs. The total equilibration times included several thousand timesteps, whereby the statistical uncertainty in the calculated values was

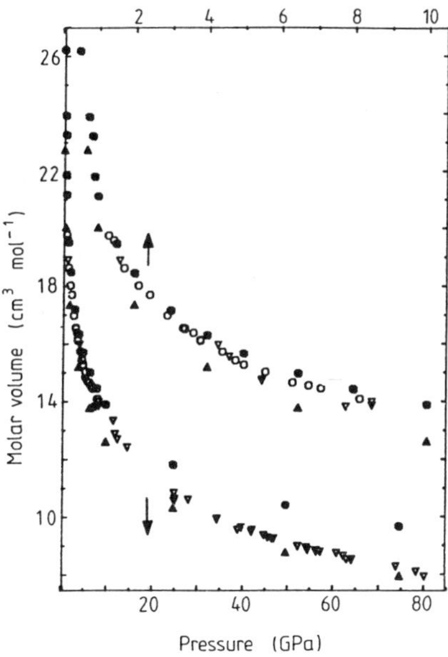

Fig. 4 Pressure-volume isotherms for Ar at T = 293 K. The upper curve corresponds to the pressure region 0 to 10 GPa (upper scale) and the lower one to 0 to 80 GPa (lower scale). The constant-pressure MD simulations are denoted by full symbols (circles: Lennard-Jones potential, triangles: potential from Ref.[17]). The open symbols are experimental values [17].

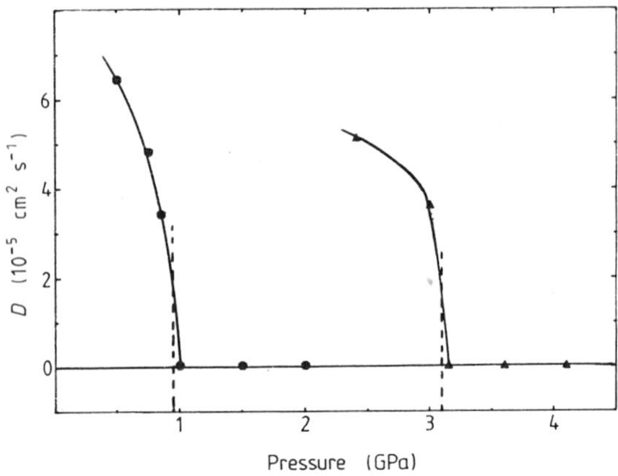

Fig. 5 The self-diffusion coefficient from MD simulations as a function of pressure for Ne (triangles) and Ar (circles). The dotted lines denote the melting pressures.

clearly less than 0.1%. The extracted quantities include, in addition to the isotherms, radial pair distribution functions (RDF), structure factors, self-diffusion coefficients and crystallisation (melting) pressures.

Fig. 4 shows the simulated pressure-volume isotherms for Ne using different potentials, compared with experimental data, and Fig. 5 displays the behavior of the self-diffusion coefficient as a function of pressure for Ne and Ar. Below 10 GPa the isotherms are rather well reproduced by the Lennard-Jones potential, but at higher pressures this form leads to serious overestimation of the molar volumes. Thus the other forms, which typically have exponential repulsion instead of a power law, are preferrable. The EMT potentials discussed above fall into this category. They have not yet been quantitatively tested for the high-pressure equations of state.

The melting pressure can be monitored from the rapid change of the self-diffusion constant. The room-temperature melting pressures were simulated to be 3.10 GPa and 0.95 GPa for Ne and Ar, respectively. These pressures correspond to the loss of mechanical stability of the crystalline phase. The values are somewhat lower than the respective experimental values of 4.74 and 1.15 GPa. However, there is hysteresis in the constant-pressure simulation of melting and crystallisation, and the true melting point should be monitored by imposing genuine two-phase coexistence.

4. Conclusions

The powerful method of molecular dynamics simulation is now being applied to rare gas precipitates in metals. It seems that all the ingredients in the simulation are relatively well understood. Indeed the first applications to Kr precipitates in Cu [15] are most interesting and encouraging.

In this paper we have discussed a general approach to describing the interactions of rare gas atoms with *any* electronic environment, both with metallic hosts and with themselves. The approach is based on the concept of embedding an inert gas atom into a simple reference system, the homogeneous electron gas. It is shown that this provides a widely useful concept and recipe for describing the interactions.

ACKNOWLEDGMENTS

It is a pleasure to acknowledge the long and most fruitful collaboration with Martti Puska on many issues discussed in this paper.

REFERENCES

1. See e.g. G. Ciccotti, D. Frenkel and I.R. McDonald, eds., *Simulation of Liquids and Solids*, North-Holland, Amsterdam (1987).

2. M. Parrinello and A. Rahman, Phys. Rev. Lett. **45**, 1196 (1980).

3. S. Nose, J. Chem.Phys. **81**, 511 (1984); W. Hoover, Phys. Rev. A **31**, 1695 (1985).

4. For a summary of recent developments, see *Many-Atom Interactions in Solids*, R.M. Nieminen, M.Puska and M.Manninen, eds., Springer, Heidelberg (1990).

5. R. Car and M. Parrinello, Phys. Rev. Lett. **55**, 2471 (1985).

6. R.G. Gordon and Y.S. Kim, J. Chem. Phys. **56**, 3122 (1972).

7. J. Harris, Phys. Rev. B **31**, 1770 (1985); W.M. Foulkes and R.Haydock, Phys. Rev. B **39**, 12520 (1989).

8. M.J. Puska and R.M. Nieminen, J. Phys. CM: Condensed Matter (to be published).

9. See e.g. K.W. Jacobsen, in Ref. [4], pp. 34-47, and references therein.

10. M.J. Puska, R.M. Nieminen and M. Manninen, Phys. Rev. B **24**, 3037 (1981).

11. N.D. Lang, Phys. Rev. Lett. **46**, 842 (1981).

12. M.S. Daw, in Ref. [4], pp. 48-63.

13. F. Ercolessi, E. Tosatti and M. Parrinello, Phys. Rev. Lett. **57**, 719 (1986).

14. M. Manninen, J.K. Nørskov, M.J. Puska and C. Umrigar, Phys. Rev. B **29**, 2314 (1984); M.J. Puska, unpublished.

15. K.O. Jensen and R.M. Nieminen, Phys. Rev. B **35**, 2087 (1987).

16. V. Karttunen, J. Ignatius, J. Keinonen and R.M. Nieminen, J. Phys. CM: Condensed Matter **1**, 4885 (1989).

17. M. Ross, H.K. Mao, P.M. Bell and J.A. Xu, J. Chem. Phys. **85**, 1028 (1986).

18. P.E. Siska, J.M. Parson, T.P. Schafer and Y.T. Lee, J. Chem. Phys. **55**, 5762 (1971) ; R.A. Aziz and M.J. Slaman, J. Chem. Phys. **92**, 1030 (1989); U. Buck, M.G. Dondi, U. Valbusa, M.L. Klein and G. Scoles, Phys. Rev. A **8**, 2409 (1973).

THEORETICAL DESCRIPTION OF THE GROWTH AND STABILITY OF HELIUM PLATELETS IN NICKEL

M. D'Olieslaeger, G. Knuyt, L. De Schepper and L. M. Stals

Materials Research Institute
Limburg University Centre
University Campus
B-3590 Diepenbeek, Belgium

ABSTRACT

A static computer simulation is performed in order to study the nucleation and the two-dimensional growth mode of helium-filled aggregates in nickel. The results indicate that a helium-filled aggregate expels neighbouring atoms at its outer edges. These atoms are displaced in the direction of a nearest octahedral position and they are bound to the cluster. The helium-filled aggregate grows in a planar way and an interstitial loop is formed by the emitted atoms. The two dimensional growth occurs parallel with the close-packed planes. To explain the collapse of the platelets into clusters of smaller aggregates, a theoretical model is developed to describe the total energy of a platelet. It is based partly on the results of the simulations. The variation of the total energy gives information regarding the platelet stability and is calculated in both a numerical and an analytical way. It is shown that small platelets are stable and that, once a critical radius is attained, the platelet becomes unstable and a transformation into a group of smaller aggregates takes place. The theoretical model can also explain the fact that two-dimensional helium aggregates are observed to be more stable in bcc than in fcc structures.

1. Introduction

Helium in metals precipitates into bubbles due to its insolubility. These bubbles do not always have the expected three-dimensional morphology. In molybdenum samples, irradiated with sub-damage helium ions, helium aggregates with a planar morphology, parallel with the close-packed planes, were observed [1-4]. Also planar groups of small clusters were observed. The reason for this became apparent after annealing studies on the specimens containing platelets: the platelets transformed into similar groups of bubbles between 600°C and 800°C. The assumption was made that the planar groups of small clusters had developed from platelets during the helium filling at room temperature. For the fcc metal nickel, platelet-like aggregates [5] and planar groups of small bubbles [6] were reported. It was established that, for low implantation doses, the aggregates have a real platelet structure, while for higher doses, the aggregates consist of planar groups of small helium bubbles [7].

Up till now, no theoretical explanation could be given to explain in a quantitative way the two-dimensional growth mode of helium aggregates. Inspired by the TEM results, Caspers *et al* in Delft, Holland [8] performed atomistic computer simulations on molybdenum using the RELAX code. They studied a modified version of the trap-mutation process, which involves the stepwise creation of multi-vacancy traps by emission of interstitials. The modification of the original trap-mutation process is based on the assumption that the emitted interstitial (I) remains bound to the mutation product formed. The underlying thought for postulating this reaction is that self-interstitials are known to cluster two-dimensionally, which could lead to a platelet structure containing helium. The authors preliminary conclusions were that the modified trap mutation process could trigger the formation of platelets. The Delft group performed further calculations on helium precipitation in alpha-iron [9,10] and molybdenum [10]. In the first paper they tentatively conclude that the bound-interstitial model, if applied over many mutations, could lead to (110)-platelet formation, but this is based on calculated mutations up to only four vacancies V_4. A firm conclusion could not be drawn since only a few possible mutation reactions were studied. In the second paper they investigate again the alternative mutation sequence. For alpha-iron, it appears that it is energetically most favourable that the emitted interstitials cluster in a two-dimensional way in a (110)-plane with increasing binding energy. This could cause the helium to precipitate in the same plane. In the calculations on molybdenum, the dynamical relaxation code MOLDY was used. It was found that, for the Mo-Mo potential of Van Heugten [11], it is energetically favourable that for n >15, $He_n V$ complexes bind subsequent helium atoms interstitially outside the vacancy in a (110) plane. The authors state that this would seem to give a favourable indication for the stability of [110] directionally oriented cracks extending from vacancies. These could eventually, possibly in conjunction with self-interstitial emission, lead to the early formation of (110) oriented platelets as observed in TEM experiments. More recently, D'Olieslaeger *et al.* [12] performed atomistic computer simulations,the results of which show that the two-dimensional growth of the aggregates can be explained under the assumption that the emitted interstitials are not bound to the cluster.

Finnis *et al.* [13,14] developed theoretical models in order to understand the instability of the two-dimensional aggregates. Simplified models of platelets and bubbles filled with helium were discussed. By comparing energies over a range of sizes they found that the bubble is always the preferred configuration, as the calculations indicated that platelets should have higher energy. The question remains as to why platelets are observed at all, since they seem to be unstable against collapse into smaller aggregates. Finnis *et al.* [13,14] state that the answer may be found partially in the combination of the favourable kinetics of helium absorption at the platelet edge with a trap-mutation process, which generates [110] interstitials at the edge. Competing diffusional processes would lead to three dimensional growth. However, no quantitative theory is yet available. Finally, Finnis *et al.* [13,14] show that, under conditions of high internal pressure, a cluster of several bubbles has a lower total energy than one single large bubble. This can be understood on the basis of the elastic interactions between bubbles placed on a lattice, if one assumes that this is approximately the same in magnitude as for a void lattice. In the present paper the results of an atomic computer simulation are presented briefly. A more detailed description of the calculations was published earlier [15,16]. Also a theoretical model to describe the stability of the platelets is explained. An extended description of the model was published elsewhere [17].

2.   Atomistic Computer Simulation

2.1. Procedure

A static computer model is developed to study the nucleation and growth of two-dimensional helium-filled aggregates in the fcc metal nickel.The calculations are performed using the embedded atom theory of Daw and Baskes [18] (Ni-Ni interaction), the He-Ni potential of Wilson and Johnson [19], and the He-He potential of Gaydaenko and Nikulin [20]. The nickel matrix is represented by a spherical crystallite with radius $4a_0$ ($a_0 = 3.52$ Å), containing 1058 atoms.

In a first part of the simulation, we calculate the binding energy of the expelled interstitial to a hexagonal helium-filled vacancy cluster. The outer boundary is left completely free and the system is relaxed using the method of conjugate gradients [21]. The vacancy cluster is filled with helium according to the same procedure used in reference [12].

In the second part, we study the nucleation and growth of a helium-filled platelet. For this part of the simulation, a fixed boundary is used. The reason for this change in boundary condition is as follows. When the boundary of the crystallite is free, an interstitial, formed at the boundary of the helium cluster, can push subsequent atoms further away in the direction of the outer boundary along close-packed directions. Finally, the interstitial will be situated at the surface of the spherical matrix. Besides another boundary condition, we also use another relaxation procedure for the second part of the calculations since it is established that the conjugate gradient method is not suitable for high helium densities. The relaxation method of Knuyt *et al.* [22] is used. In this method the displacement of each atom is proportional to the local force acting on it (proportionality constant = 0.0182 $Å^2eV^{-1}$). Also the helium filling procedure is slightly different from the one used in the first part of the calculations. Instead of an adjustable cell, a cubic cell with fixed dimensions is used. The cell has edges of $a_0 = 3.52$ Å and is centred around the lattice position (0,0,0) (centre of the crystallite).

2.2. Nucleation and growth simulations of helium platelets in nickel

Recently, D'Olieslaeger *et al.* [12] reported on the results of calculations in which the emitted interstitial was not bound to the cluster. However, later calculations [16] showed that the emitted interstitials are bound to the cluster. In order to study the growth mechanism of the cluster, we fill a vacancy with helium, and contrary to previously published results [12], the expelled interstitials are no longer taken away.

A vacancy, placed at the central lattice position with coordinates (0,0,0), is filled with helium. In Figs. 1 and 2, (next page) the atomic positions are plotted under various projections. For the case of 226 helium atoms inside the cluster, it appears that the helium cluster grows parallel to a (111) plane. Upon calculating the largest displacements of the atoms compared to their original position, it is established that the six largest displacements are all of atoms which lie originally on the (111) plane. The positions shown in Fig. 1 are the projected positions of those matrix atoms, the undisturbed initial positions of which are lying in a slice with a thickness of 7Å, centred around the original vacancy and perpendicular to the (111) plane. The coordinates of the atoms are projected in a plane perpendicular to the (111) plane. The empty circles (o) represent the original positions of the matrix atoms, while the final relaxed positions of these atoms are indicated by solid circles (●). The positions of the helium atoms are represented by small squares (□). Fig. 2 shows the projection of the matrix atoms with an initial position in the (111) plane containing the vacancy. It can be seen that some of the atoms bordering to the platelet are expelled to octahedral interstitial positions, as was the case in the previous calculations. When investigating the projected atom positions, it can be seen that there are dislocation-like regions, which seem to remain in the near vicinity of the helium cluster, instead of gliding through the crystallite.

This can be understood by the interaction of the strain fields of the two defects, i.e. the overpressurized cluster and the loop, which lowers the total elastic energy. Another reason can be found in the properties of the mobility of dislocation loops in fcc structures. It is known that in an fcc structure Frank loops, i.e. dislocation loops with Burgers vector [111]a/3, are sessile. Further calculations were performed in order to investigate the behaviour of the dislocation loop when the pressure in the helium aggregate decreases. Therefore, the relaxed positions of the matrix atoms in the case of 226 helium atoms present in the cluster, are taken as the input data of an iteration cycle in which the helium atoms are not included. It appears that when removing the helium atoms from the cluster, the nickel atoms return approximately to their undisturbed positions. The perfect lattice positions, however, are not reached exactly due to the presence of the empty space left by the helium atoms and to the use of a static computer model.

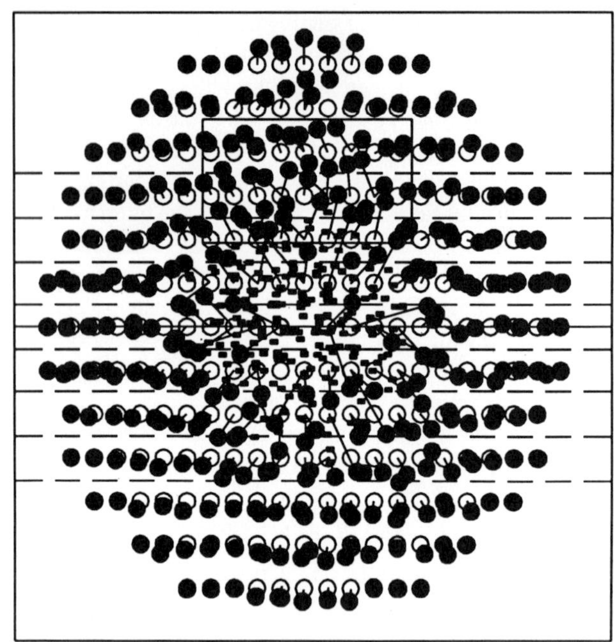

Fig 1 Projection in a plane perpendicular to the (111) plane of the original and final position of all matrix atoms for the case of 220 helium atoms in the cluster. The empty circles (o) represent the original positions of the matrix atoms, while the final relaxed positions of these atoms are indicated by solid circles (●). The position of the helium atoms are represented by small bars. A dislocation-like region, as discussed in the text, is indicated.

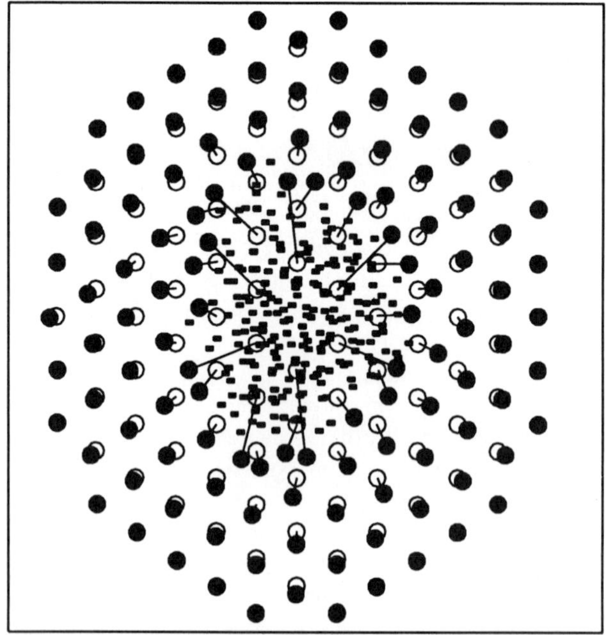

Fig. 2 Projection in the (111) plane of the platelet of the matrix atoms lying in the (111) plane. The empty circles (o) represent the original positions of the matrix atoms, while the final relaxed positions of these atoms are indicated by solid circles (●).

The computer simulations presented here are made for a static lattice. In a molecular dynamics calculation, which would require much more computer time, the effect of temperature might be to lower the number of helium atoms in the cluster, needed to push out the interstitials. Another consequence might be a more homogeneous helium gas. However, in our opinion, the effect of temperature would not affect the final results.

3. Stability of Helium-Filled Platelets

3.1. Internal energy of a gas-filled platelet

If the pressure in a gas-filled cavity is not too large, the total energy consists of the following terms :
   (1) the elastic energy stored within the bulk of the metal : E_{el}
   (2) the surface term in the energy of the metal where the interaction between the gas and the metal is included : E_{sur}
   (3) the interaction energy between the gas atoms: E_{gas}
   At very high pressures there will be further contributions to the energy (see section 3.3). We now describe contributions (1) to (3) in more detail :

(1) for the elastic energy stored within the bulk of the metal, we use the results of isotropic linear elasticity theory [23] for a circular platelet with radius r and internal pressure p, and with a negligible thickness d (i.e. d << r):

$$E_{el}^{pl} = \frac{8(1 - v^2)}{3Y} r^3 p^2 \tag{1}$$

with Y Young's modulus and v Poisson's ratio. The value of Y for nickel is taken to be 1.24 eV Å^{-3} = 1.99 10^{11} Pa [24]. Poisson's ratio is taken as 0.3 [24]. The elastic anisotropy of nickel is not likely to affect eqn. (1) in such a way that the conclusions drawn from the calculations are no longer valid. The volume change of the platelet corresponding to a pressure p, is given by:

$$\Delta V_{el} = \frac{2E_{el}}{p} = \frac{16(1 - v^2)}{3Y} r^3 p^2 \tag{2}$$

(2) The surface term E_{sur} is taken proportional to the total internal surface of the defect. It is known that the surface energy density of a metal depends on the crystallographical structure of the surface. The value of the surface energy of (111) planes in nickel was calculated by Daw and Baskes [17]; they found $\gamma\{111\}$ = 0.08 eV Å^{-2} = 1.28 J m^{-2}. Due to an extra interfacial energy term which takes into account the interactions between the nickel atoms and the helium atoms, the value of the surface energy of the gas-metal interface is probably somewhat higher as compared with the surface energy of a free surface. In the present calculations, we take γ_1 equal to 0.1 eV Å^{-2} = 1.6 J m^{-2}.

(3) In order to calculate the interaction energy between the gas atoms, we assume that the latter are located on a fcc lattice and interact with the He-He potential of Gaydaenko and Nikulin [20]. If we truncate the helium interaction potential after first neighbours, the total energy of the gas can be written as:

$$E_{gas} = 6N_{He} A \exp\left(2b\left(\frac{3V}{4\pi N_{He}} \right)^{1/3} \right) \tag{3}$$

with A = 197.65344 eV, b = 4.24603175 Å^{-1}, V the total volume available for gas atoms and N_{He} the total number of helium atoms.

By neglecting interactions between second neighbours, we introduce an error which is small compared with the error due to the assumption that the gas atoms are located on an fcc lattice and that due to to the fact that edge defects are neglected. The pressure of the inert gas is given by:

$$p = -\frac{\partial E_{gas}}{\partial V} = \frac{3Ab}{\pi}\left(\frac{3V}{4\pi N_{He}}\right)^{-2/3} \exp\left(-2b\left(\frac{3V}{4\pi N_{He}}\right)^{1/3}\right) \tag{4}$$

3.2. Theoretical model

Consider N_s satellite platelets with radius r_s growing at the border of a two-dimensional helium filled vacancy cluster with radius r_p, lying on a close-packed plane in a fcc structure (see Fig. 3). The use of satellite platelets instead of satellite bubbles is inspired by the micrographs of planar bubble arrays [6], where it seems that the small aggregates do not have a spherical shape. The radius r_s of these satellites is initially considerably smaller than the radius r_p of the "host-platelet". In the further discussion, the host platelet, together with its satellites, will be called "the system". The problem of calculating the elastic energy of such a system is complicated. Even for a regular bubble lattice it can only be calculated approximately [27]. We therefore introduce the following simplification illustrated in Fig. 3.

We assume that the elastic energy of the system can be written as (see eqn. (1)) :

$$E_{el} = C\,r_{eff}^3\,p^2 \tag{5}$$

with $C = 8(1-v^2)/3Y$ and r_{eff} an effective radius. The latter is determined as follows. In a good approximation, the projection of a satellite platelet on the circumference of the host platelet is $2r_s$ (see Fig. 3). For a total of N_s satellites, the total length of the projection is thus $2N_s r_s$. For a part of the system, the distance from the centre of the host platelet to the edge of the system is approximately given by $r_p + 2r_s$, while for the remaining part this distance is simply r_p. This way of reasoning results in the following expression for the effective radius r_{eff}:

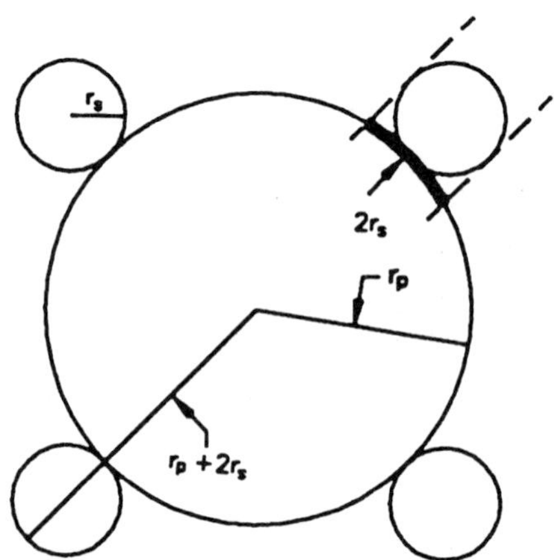

Fig. 3 Determination of the effective radius of a host-platelet with radius r_p and four satellite platelets with radius r_s.

$$r_{eff}^3 = (r_p + 2r_s)^3 \frac{N_s r_s}{\pi r_p} + r_p^3 \left(1 - \frac{N_s r_s}{\pi r_p}\right) \tag{6}$$

Before writing down an expression for the total energy of a helium filled system, we have to remark that eqn. (1), which is the elastic energy stored in the bulk of the metal, is only valid when the thickness of the platelet is small compared to its radius, i.e. for large platelet radii. For small radii, the platelet thickness becomes comparable to its radius and eqn. (1) needs to be adapted.

The results of atomistic computer simulation show that for small helium-filled aggregates under high pressure large lattice deformations occur. The atoms which are originally located in the close-packed plane of the platelet are pushed into octahedral positions in the neighbourhood of the cavity and form an interstitial loop.

An acceptable approximation for the elastic energy of the system with non-negligible thickness is:

$$E_{el}^{pl} = Br^2 + Cr^3 p^2 \tag{7}$$

For large platelet radii, the first term in the right hand side of eqn. (7) can be neglected compared to the second term and the expression for the elastic energy is reduced to eqn. (1). However, for small platelet radii, the term in r^2 cannot be neglected at all. In order to determine the order of magnitude of the parameter B, we consider the platelet with a small radius as the analogue of a small dislocation loop. The energy $W(r)$ of a dislocation loop [27] can, for small values of r, be approximated by a parabola. This yields a quadratic term Br^2 which can be identified with a surface energy term $2\pi\gamma^* r^2 p$. This yields $\gamma^* = 0.13$ eV Å^{-2}. This is an acceptable value compared to normal values of surface energies which are in the order of 0.1 eV Å^{-2}. Besides the correction for the elastic energy, an additional gas volume needs to be introduced for small platelet radii, since it is obvious that a platelet has a minimal thickness. The thickness d_o of a planar helium aggregate must be clearly at least the interplanar distance between two close-packed planes, since the helium atoms between these planes will push them apart. The value of d_o is taken as 2.5 Å, which is somewhat larger than the interplanar distance of (111) planes in nickel (2.03 Å). Due to the thickness d_o an extra surface appears and consequently an additional surface energy term must be taken into account, with surface energy γ_3'. The atomistic computer simulation shows that for small helium-filled aggregates an interstitial loop is formed. Once a certain loop-radius is attained, the energy of a dislocation loop is proportional to the radius. A fit of a straight line through the curve $W(r)$ at large r-values, determines the value of this energy. One finds:

$$E_{loop} = Ar = 8.9 \ 10^{-3} r \tag{8}$$

Finally, we can write the total energy of a helium-filled platelet with small radius (SR) as:

$$E_{SR} = Cr_{eff}^3 p^2 + 2\pi\gamma_1(r_p^2 + N_s r_s^2) + 2\pi\gamma^*(r_p^2 + N_s r_s^2) + E_{loop} + 2\pi\gamma_3' d_o(r_p + N_s r_s) + E_{gas}(V) \tag{9a}$$

and for the volume:

$$V = 2Cr_{eff}^3 p + \pi d_0(r_p^2 + N_s r_s^2). \tag{9b}$$

Incorporating the energy of the dislocation loop in the surface energy term due to the minimal thickness of the platelet and combining similar terms, we have:

$$E_{SR} = Cr_{eff}^3 p^2 + 2\pi\gamma_2 (r_p^2 + N_s r_s^2) + 2\pi\gamma_3 d_0 (r_p + N_s r_s) + E_{gas}(V) \qquad (10a)$$

with $\gamma_2 = \gamma_1 + \gamma^*$ and $\gamma_3 = \gamma_3' + \dfrac{A}{2\pi d_0}$

and for the volume :

$$V = 2Cr_{eff}^3 p + \pi d_0 (r_p^2 + N_s r_s^2) \qquad (10b)$$

Both γ_2 and γ_3 can be considered as a surface energy density. The interaction energy between platelet and dislocation loop is unknown. However, it is acceptable that this interaction lowers the value of the surface-energy density to be used. Therefore, we take γ_3 as 0.1 eV Å^{-2}; γ_2 is taken as 0.23 eV Å^{-2}.

The total energy of a helium-filled platelet with a large radius is much simpler since we do not need to adapt the expression for the elastic energy (eqn. 2). Also it was shown in the simulation that for larger platelets, the dislocation loop formed in the initial stage of the growth process, glides back into the helium cluster. This means that the dislocation loop needs not to be incorporated in the expression for the total energy. The energy of a platelet with large radius (LR) can then be written as:

$$E_{LR} = Cr_{eff}^3 p^2 + 2\pi\gamma_1 (r_p^2 + N_s r_s^2) + E_{gas}(V) \qquad (11a)$$

and:

$$V = 2Cr_{eff}^3 p^2 \qquad (11b)$$

3.3. Analytical stability calculations

To study the stability of helium-filled platelets, we investigate the variation of the total energy if satellite platelets nucleate at the border of the host platelet. For large platelets, the thickness of the platelet is small compared to its radius. The total energy of the system is then given by eqn. (11a). The variation of this total energy, if N_s satellite platelets grow at the outer boundary of the host platelet can be written as:

$$\delta E_{LR} = 3Cr_{eff}^2 p^2 \delta r_{eff} + 2Cr_{eff}^3 p \delta p + 2\pi\gamma_1 \delta(r_p^2 + N_s r_s^2) + \frac{\partial E_{gas}}{\partial V} (6Cr_{eff}^2 p \delta r_{eff} + 2Cr_{eff}^2 \delta p)$$

$$(12)$$

Expanding r_{eff}^3 in a Taylor series in r_s and neglecting terms in r_s^3 and r_s^4, differentiation of r_{eff}^3 yields:

$$3r_{eff}^2 \delta r_{eff} = \left(3r_p^2 + \frac{6}{\pi} r_s^2 N_s \right) \delta r_p + \frac{6}{\pi} r_s^2 N_s r_p \qquad (13)$$

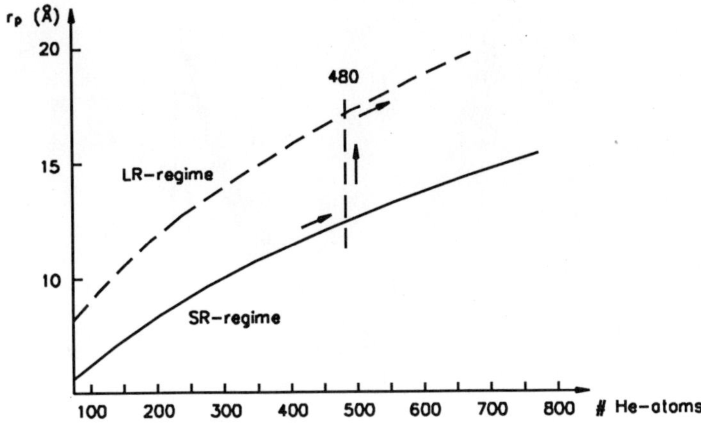

Fig. 4   Host-platelet radius versus number of helium atoms for both regimes. At the transition point of 480 atoms, the radius increases from 12.55 Å to 17.27 Å .

We assume that the total surface of the system remains constant during satellite growth, since this does not limit the generality of the calculations. Combining (12) and (13), we then have:

$$\delta E_{LR} = -3Cr_{eff}^2 p^2 \, \delta r_{eff} \tag{14}$$

Since under the assumption of a constant surface, δr_p is given by:

$$\delta r_p = \frac{-N_s r_s^2}{2r_p} \tag{15}$$

we finally obtain :

$$\delta E_{LR} = -Cp^2 N_s r_s^2 r_p \left(0.41 - \frac{3N_s r_s^2}{\pi r_p^2} \right) \tag{16}$$

This means that, for large platelet radii, the total energy of the system decreases if satellite platelets nucleate at the border of the host platelet. For larger rs values, eqn. (16) can become positive, indicating an increase of the total energy of the system. However, for this region in the (r_p, r_s)-plane, the equations describing the total energy for large platelets are no longer valid. Conclusively, large platelets can lower their energy by nucleating satellite platelets. This means that the transformation into a group of smaller aggregates can take place. For small platelet radii, it can be shown that, under the same assumption of constant surface, the variation of the total energy is positive if satellite platelets nucleate at the border of the platelet.This shows that under certain conditions satellite growth is energetically unfavourable, but it does not exclude the existence of a path in the (r_p, r_s)-plane for which the energy would decrease. More elaborate analytical calculations might reveal a true stability of small platelet radii. It is however more straightforward to calculate numerically the energy surface $E(r_p, r_s)$ for both small and large platelet radii.

35

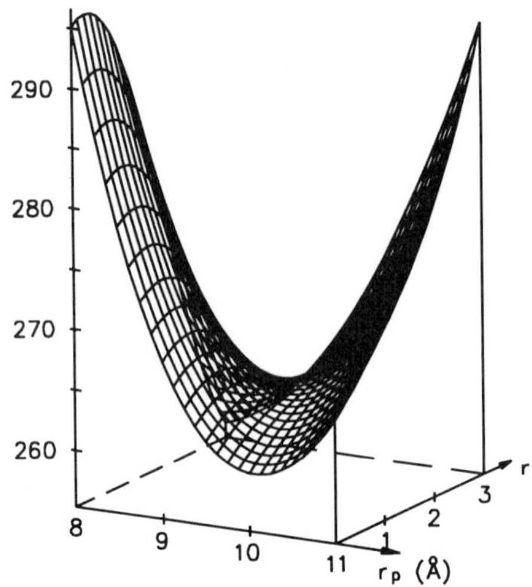

Fig. 5 Energy surface ESR(r_p,r_s) for 300 He-atoms. The host platelet radius r_p varies from 8 to 11 Å , while the radius of the satellite platelets varies from 0 to 3 Å .

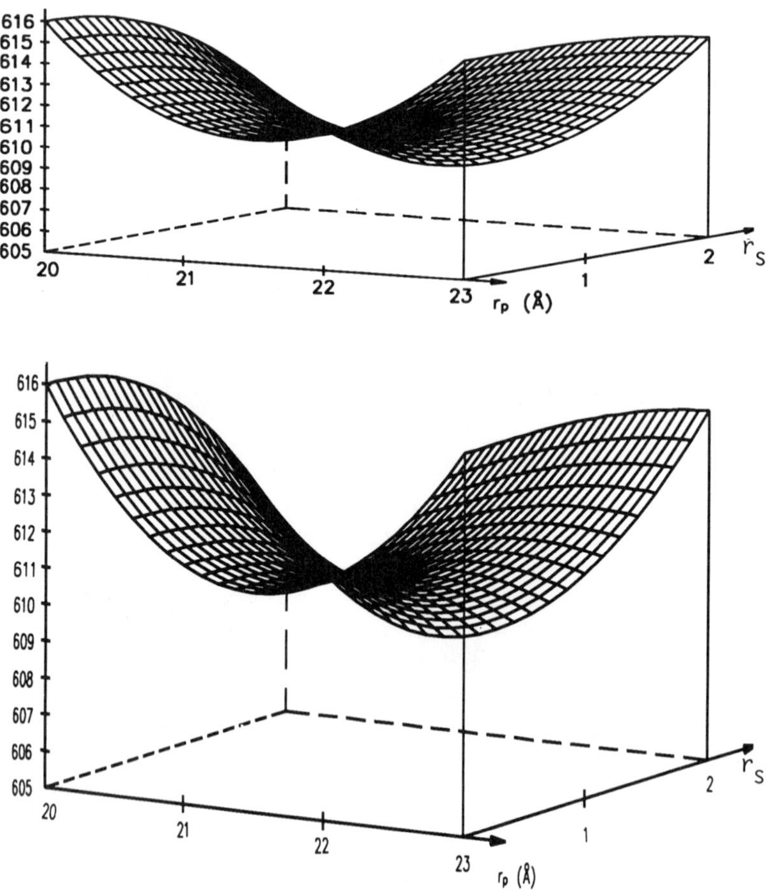

Fig. 6 Energy surface ELR(r_p,r_s) for 800 He-atoms. The host platelet radius ranges from 20 to 23 Å , while the radius of the satellite platelets varies from 0 to 2 Å .

3.4. Numerical stability calculations

<u>Total energy</u>

The total energy can only be calculated as a function of the number of helium atoms present in the cluster and not as a function of the platelet radius. This is due to the fact that the gas term depends on the number of helium atoms. At the point of 480 helium atoms, both regimes are energetically equal, while for a higher number of gas atoms (i.e. larger platelets), the LR-regime becomes favourable. The radius of the host platelet as a function of the number of helium atoms is plotted in Fig. 4. The SR-regime remains energetically favourable until a platelet radius of 12.55 Å is attained at 480 helium atoms. At that moment, the LR-regime becomes favourable and the platelet radius increases abruptly to 17.27 Å. We have to remark here that neither the results of the atomistic computer simulation, nor the energy calculation allows us to establish whether the transition between the two regimes is discontinuous. Probably the transition takes place smoothly. However, this would not have any effect on the final conclusions regarding platelet stability.

<u>Energy surfaces</u>

To investigate the stability of helium platelets, the energy surfaces for both SR- and LR-regimes are calculated. In Fig. 5 the energy surface ESR (r_p, r_s) is shown. This surface is calculated using eqn. (10). The number of helium atoms is 300 and the number of satellite platelets is 3. From a certain value of r_s on, the energy starts to decrease. From the numerical calculations of the energy surface ESR we can conclude that a platelet with small radius is stable against nucleation of satellite platelets and thus against transformation into a group of smaller aggregates. The energy surface $ELR(r_p, r_s)$ is shown in Fig. 6 for the case of 800 helium atoms. It is calculated using equation (11). It is clear that this energy surface does not show a saddle point, as was already known from analytical calculations. This means that even the smallest nucleus of a satellite platelet makes the platelet transform into a group of smaller aggregates.

4.   Conclusions

It has been established by computer simulation that a helium-filled aggregate grows in a two-dimensional way. The growth plane is the close-packed plane in an fcc structure. The planar growth of a helium filled platelet proceeds by the emission of self-interstitial atoms at the boundaries of the platelet. These interstitials cluster together in the vicinity of the platelet, forming a dislocation loop. This loop is unable to glide through the crystal and remains close to the cluster. When the pressure in the platelet decreases, the dislocation loop glides back into the helium cluster. In order to study the stability of helium filled platelets in fcc structures, a theoretical model of the total energy of a platelet was developed. From the atomistic computer simulation it is known that for small platelet radii a dislocation loop is formed in the vicinity of the helium cluster. As long as a certain critical radius is not attained, the energy of the dislocation loop and of the lateral surface energy stabilizes the platelet. Consequently the transformation into a group of smaller platelets cannot take place. At the transition radius, the dislocation loop disappears. Due to both the decrease of the pressure and the decrease of the lateral surface energy, the platelet becomes unstable.

ACKNOWLEDGEMENTS

Thanks are due to the Belgian Science Supporting Institute (Interuniversitair Instituut voor Kernwetenschappen, Brussels) for financial support under contract number 4.0002.75 N.

REFERENCES

1. J. H. Evans, A. van Veen and L. M. Caspers, Nature **291**, 310 (1981).
2. J. H. Evans, A. van Veen and L. M. Caspers Scripta Met.**17**, 549 (1983).

3. J. H. Evans, A. van Veen and L. M. Caspers, Rad.Eff. **78**, 105 (1983).

4. A. van Veen, L. M. Caspers and J. H. Evans, J.Nucl. Mater **103 &104**, 1186 (1981).

5. A. van Veen, J. H. Evans, L. M. Caspers and J. Th. De Hosson, J.Nucl. Mater. **122 &123**, 560 (1984).

6. M. D'Olieslaeger, L. De Schepper, G. Knuyt and L. M. Stals, J.Nucl.Mater. **138**, 27 (1986).

7. M. D'Olieslaeger, PhD. Thesis, Diepenbeek, Belgium (1990).

8. L. M. Caspers, M. R. Ypma, A. van Veen and G. J. van der Kolk, Phys. Stat. Sol. **A63**, K183 (1981).

9. L. M. Caspers, A. van Veen, M. R. Ypma and G. J. van der Kolk, Phys. Stat. Sol. **A70**, 109 (1981).

10. L. M. Caspers, A. van Veen and T. J. Bullough, Rad.Eff. **78**, 67 (1983).

11. W. F. W. M. van Heugten, Phys. Stat. Sol. **B82**, 501 (1983).

12. M. D'Olieslaeger, G. Knuyt, L. De Schepper and L. M. Stals, J. Nucl. Mater. **144**, 200 (1987).

13. M. W. Finnis, A. van Veen and L. M. Caspers, AERE Harwell TP Report 927 (1982).

14. M. W. Finnis, A. van Veen and L. M. Caspers, Rad. Effects **78**, 121 (1983).

15. M. D'Olieslaeger, G. Knuyt, L. De Schepper and L. M. Stals, Nucl. Instr. Meth. Phys. Res. **B39**, 445 (1989).

16. M. D'Olieslaeger, G. Knuyt, L. De Schepper and L. M. Stals, submitted to Phil. Mag.

17. M. D'Olieslaeger, G. Knuyt, L. De Schepper and L. M. Stals, submitted to Phil. Mag.

18. M. S. Daw and M. I. Baskes1984, Phys. Rev. **B29**, 6443.

19. W. D. Wilson and R. A. Johnson, *Interatomic Potentials and the Simulation of Lattice Defects* p.375, Eds. P. G. Gehlen, J. R. Beeler Jr. and R. I. Jaffee, Plenum Press (1972).

20. V. I. Gaydaenko and V. K. Nikulin, Chem. Phys. Lett. **7**, 360 (1970).

21. R. Fletcher and C. M. Reeves, Computer Journal **7**, 149 (1964).

22. G. Knuyt, L. De Schepper and L. M. Stals, J.Phys. **F16**, 1989 (1986).

23. R. A. Sack, Proc. Roy. Soc. **58**, 729 (1948).

24. W. Simmons and H. Wang, *Single Crystal Elastic Constants and Calculated Aggregate Properties: A Handbook.* MIT-Press, Cambridge, Massachusetts, USA, (1971).

25. Metals Handbook, Vol. 1 *Properties and Selection of Metals*, ed. T. Lyman, American Society for Metals, Ohio, USA (1961).

26. A. M. Stoneham, J. Phys. **F1**, 778 (1971).

27. J.P. Hirth and J. Lothe, *Theory of Dislocations*, p. 144 McGraw-Hill (1968).

HELIUM DEFECT INTERACTIONS
AND DIFFUSION

HELIUM DEFECT INTERACTIONS IN METALS AND SILICON

A. van Veen

*Interfaculty Reactor Institute, Delft University of Technology
Mekelweg 15, Nl 2629JB Delft, The Netherlands*

ABSTRACT

An overview is given of the results obtained with Thermal Helium Desorption Spectrometry (THDS) and other techniques on the interaction of helium with defects in metals and silicon. The majority of results concerning point defects and small defect complexes in this overview have been obtained by THDS. TEM observations have been used as a complementary technique to determine the morphology of the microscopically visible helium defect complexes. Atomistic defect simulations e.g. by the Embedded Atom method or by the Many Body Potential approach have been used to provide theoretical support. THDS studies were originally devoted to defects in metals only, but are now applied also to silicon and ceramics. Helium interactions with defects in silicon differ considerably from those obtained for metals. Entropy effects seem to play an important role.

1. Introduction

Interest in the interaction of helium with defects in materials originates in nuclear materials research. Helium production by (n, α) neutron capture reactions is a considerable source of helium in the metal containers of reactor fuels in fission-type reactors. A strong impetus for helium-defect studies has also come from the development of plasma fusion devices. In addition to helium generation in the structural components, considerable amounts of helium are expected to be introduced into the first wall materials either by surface injection of helium particles from the plasma or by radioactive decay of dissolved tritium [1].

A second field of interest involves the use of helium as a probe particle to recognise defects in different materials. Thermal helium desorption spectrometry has been developed as a useful technique for quantitative analysis of small concentrations of defects.

A third area where helium is employed deals with the controlled modification of defects by helium decoration. Decoration may serve the purpose of defect analysis by other techniques e.g. the enhancement of the effects observed in hyperfine studies [2-5] or the creation of porous layers as has been shown for helium implanted silicon [6].

In this overview, results obtained in these studies are summarized and discussed. Activities in helium desorption research have always been limited to a few specialized groups so that the results do not cover all metals. However, for the bcc metals Mo and W, data are fairly complete. Earlier reviews have been given by van Veen [7], Buters *et al* [8], and van der Kolk and van Veen [9]. Atomistic calculations for helium related defects have been pursued by many groups. Initially calculations were based on pair potentials [10], but since about 1985 improved calculations were based on many body interaction potentials; see the overview by

Nieminen [11]. Recently calculation schemes have also been developed for helium in silicon and ceramics e.g. UO_2 [12]. In this contribution the emphasis will be mainly on the binding, clustering, and desorption processes at various defects in metals and silicon. A few results on selected ceramics will also be described. A short introduction will be given on the experimental methods employed.

2. Experimental Methods

2.1. Defect production and helium introduction

In experiments on helium defect interactions the production of defects and the introduction of helium is generally carried out in two successive steps. The ideal experiment requires that the introduction of the helium causes no extra damage. It can be calculated that the kinetic energy of the helium should be below the threshold for damage production as follows: $E_{th} = E_d / G(m,M)$, where E_d is the displacement energy and $G=4m\,M/(m+M)^2$ is the factor for maximum energy tranfer in a single collision of helium (mass m) with the atoms (mass M) of the material. Thus introduction by helium ion irradiation is limited to a certain maximum energy. In Table 1 these threshold energies have been given for a number of metals. When light impurities are present in the metal, threshold energies may be appreciably lower because collision sequences may occur in which the impurity is involved. In the Table data are given for double collisions involving oxygen: $He \rightarrow O \rightarrow M$. For helium energies higher than the threshold energy, Frenkel pairs are formed which give rise to the build up of vacancies. Measurements of the vacancy production yield values of up to 1 vacancy per helium ion for metals irradiated with 3 keV helium at room temperature (see Table 1). A consequence of low energy helium irradiation is that the helium is introduced at rather shallow depth below the surface (typically 1 nm deep). However at low defect concentrations the helium can reach deeper-lying defects by diffusion. The diffusion length in material with 1 appm vacancies amounts to 100 nm. This region matches rather well with the damage profile of keV heavy ions and with the analysis depth of techniques such as TEM, Nuclear Reaction Analysis, and Thermal Desorption Spectrometry. For defect analysis using techniques which require a uniform defect population in the bulk this manner of introduction is not suitable. The only way to introduce non-damaging helium in these cases is by dissolving tritium which decays to ^{3}He while transferring a recoil energy much lower than the displacement energy [14]. In some cases the damage created by helium introduction is at a much lower level than the primary damage e.g. caused by irradiation with fast neutrons or energetic heavy ions. Then the extra damage can probably be neglected. Defect generation techniques used in association with

Table 1 Threshold energies E_{th} for damage production in metals by helium and room temperature vacancy production n_{He} at different energies. Values of E_{th} between brackets are calculated for metals containing oxygen impurities (see text).

Metal	E_d (eV)	E_{th} (eV)	ν_{He} (vacancy/ion)			
			Ion Energy (keV)			
			0.5	1	2	3
Mo(100)	35	227 (112)	0.035	0.1	0.29	0.50[a]
Mo(210)			0.015	0.1	0.43	0.60[a]
W (100)	40	470 (212)				
Au(100)	30	384 (168)				
Ni(110)	25	105 (58)	0.23	0.3	0.75	1.0
$FeCr_{17}Ni_{13}$	25	100 (57)				
316 L	25	100 (57)				

[a]) ref 13.

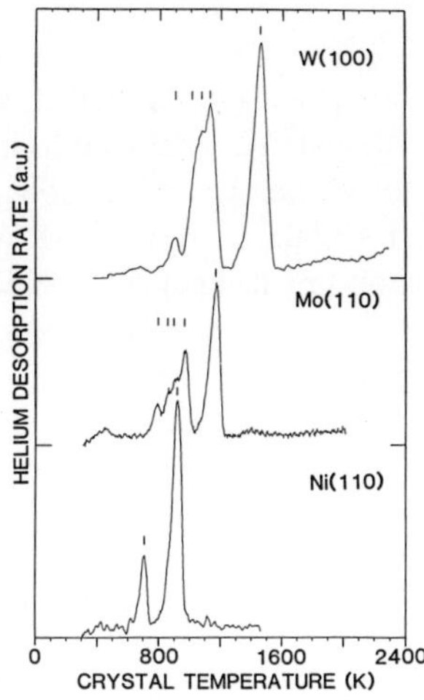

Fig. 1 Desorption spectra for different metals obtained after irradiation of 10^{12} He/cm^2, ion energy 1 keV. Note the different peak structure for the bcc metals compared with that for the fcc metal.

helium are ion-implantation in the energy range from keV to MeV, neutron irradiation, mechanical deformation and laser-irradiation.

2.2. *Thermal helium desorption spectrometry (THDS)*

The principles of this technique have been described by Kornelsen and van Gorkum [15,16] and by van Veen *et al* [17,18]. A typical experiment proceeds as follows:

> 1. defect production by keV ions
>   —damage production step.
>
> 2. annealing to temperature T_A
>   —defect anneal step.
>
> 3. sub-threshold implantation of low energy helium ions
>   —helium decoration step.
>
> 4. monitoring of the helium release rate while heating the sample with
>   a constant heating rate dT/dt (typically 40 K/s)
>   —helium desorption step yielding the desorption spectrum.

In Fig. 1 examples are given of desorption spectra obtained following 1 keV He irradiation of different metals. Note that in this case, only steps 1 and 4 were needed to obtain the spectra. From the desorption peaks contributing to the desorption spectrum the population of defects can be derived in a quantitative way. Analysis of the desorption peaks yields values of defect detrapping enthalpies, jumping frequencies, diffusion enthalpies etc [7,15,19]. The method gives unambiguous results when defect densities are low and therefore the helium diffuson length is larger than the average distance between the defects. At higher densities retrapping of detrapped helium plays a role. A description of the trapping and de-trapping processes has been given in reference [20]. For defect assignment, use is made of the opportunity offered by low energy sub-threshold helium irradiation to decorate defects in a controlled way with an increasing amount of helium. Also a method is employed where self-interstitials are injected into the sample by irradiation with low energy heavy ions. Interaction of the self-interstitial with vacancy complexes leads to size reduction or removal of the complexes [21].

2.3 *Other techniques*

Helium decoration methods similar to those used in THDS have been employed to decorate defect sites populated by radioactive atoms employed in hyperfine interaction studies. The presence of helium attached to these sites has been shown to cause drastic changes in the angular frequencies in a Perturbed Angular Correlation (PAC) measurement [2,3]. Studies of helium mobility have been based on this technique [4]. Nuclear Reaction Analysis methods [22] have been used to perform depth profiling on helium and to determine the lattice location of helium in defects [23]. Positron lifetime measurements on ion-irradiated samples have been employed to perform annealing studies on helium-vacancy complexes [24]. Slow positron beams are applied to study helium associated defects created by keV helium ion beams in metals and silicon [25]. It should be noted that positrons have a higher sensitivity for empty vacancies and vacancy complexes than for helium saturated vacancy complexes. It has been shown that the positron lifetime is reduced considerably when the helium pressure inside a cavity increases [26].

3.   Helium Release from Defects

The detrapping of helium from defects can be described in terms of the detrapping rate:

$$\frac{dN_{He}}{dt} = -4\pi r D c_{He} N_o$$
$$= -4\pi r D N_o \exp\left(\frac{\mu_{He}}{kT} - \frac{G_{He}}{kT}\right) \tag{1}$$

where c_{He} is the concentration (atomic fraction) of He, μ_{He} is the chemical potential of the helium, G_{He} is the Gibbs free energy of the bulk dissolved helium, D is the diffusivity of the helium in the bulk, and r is the radius of the defect. This equation has been derived with the assumption of (quasi-) thermal equilibrium between the helium in the defects and the helium in the bulk of the solid.

For *small defects*—the size of a monovacancy or smaller—the trapped state of the helium can be considered as a particle vibrating with a frequency ω_t in a three-dimensional potential well. With the assumption that the bulk dissolved helium—in a substitutional or interstitial site— vibrates with a frequency ω_s, eqn. 1 can be rewritten as follows:

$$\frac{dN_{He}}{dt} = -4\pi r D c_{He} N_o N_{He} \left(\frac{\omega_t}{\omega_s}\right)^3 \exp\left(-\frac{E^{B,He}}{kT}\right) \tag{2}$$

with $E^{B,He}$ the binding energy of helium at the defect. For example the detrapping rate for He from a monovacancy in fcc metals can be written as follows

$$\frac{dN_{He}}{dt} = -\frac{\pi}{3} r_v \nu_o \exp\left(\frac{\Delta S}{k} - \frac{E_{He}^{D,He} V}{kT}\right) \tag{3}$$

when the diffusivity

$$D = \frac{1}{6} \lambda^2 \nu_o \exp\left(-\frac{\Delta S}{k} - \frac{E_{He}^m}{kT}\right) \tag{4}$$

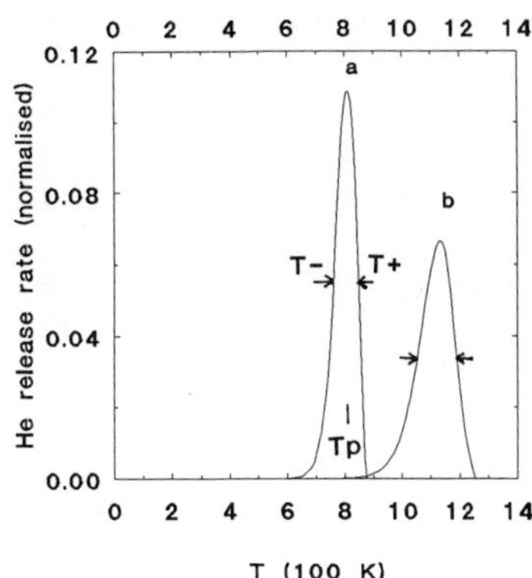

Fig. 2 Calculated desorption spectra for cavities in silicon with a) r=0.6 nm and P=15 kbar, and b) r = 10 nm and P = 1 kbar. The helium dissociation energy is 1.7 eV and the heating rate dT/dt = 10 K/s.

is substituted in eqn. 2. The radius r_v is the effective radius of the vacancy in units of the lattice constant. This expression for the detrapping rate yields values which correspond reasonably well with the experimental data observed for detrapping from small defects.

The detrapping rate for detrapping from *micro-cavities* or pressurised *bubbles* is derived by substituting in eqn. 2 the term containing the chemical potential of the helium which can be written as follows when the ideal gas law P = nkT applies:

$$\exp\left(\frac{\mu_{He}}{kT}\right) = \frac{P}{kT}\left(\frac{2\pi\hbar^2}{mkT}\right) \tag{5}$$

For pressures higher than 100 MPa the chemical potential must be derived from the equation of state of high pressure helium. Trinkaus [27] has presented μ_{He} as a function of temperature and pressure. In the present paper μ_{He} has been derived by the EOS given by Mills [28].

Fig. 2 shows some results of detrapping calculations. The helium release rates from bubbles of different size and pressure have been plotted vs the temperature. Differences observed for the temperature T_p at which the maximum release occurs and the FWHM-width, $T_+ - T_-$, is due to size effects and differences in the chemical potential of the gas in the bubbles. Fig. 3 shows the values T_p, T_+, and T_- for varying pressure in a 2 nm diameter bubble (the helium dissociation energy is fixed at $E^{D,He} = 1.7$ eV).

It is observed that beyond a pressure of 1 kbar (100 MPa) the temperatures calculated with the Mills EOS deviate strongly from the temperatures obtained for ideal gas, which are independent of the pressure. In Fig. 3 release temperatures have also been indicated for detrapping from a mono-vacancy calculated according to eqn. 3 with $\Delta S = 0$, $E^{D,He} = 1.7$, 1.5 and 1.3 eV respectively, and with $n_0 = 10^{13}$ s^{-1} and $r_v = 1$. Apparently the increase of the pressure in a bubble considerably enhances the detrapping probability. At very high pressures in Fig. 3 it is seen that the release temperature has been reduced to the temperature at which helium detraps from a monovacancy. The effect can mainly be attributed to a decrease in the entropy of the gas in the bubble at high pressures. The entropy difference $\Delta S = S$(dissolved He)-S(He in bubbles) is increased causing a higher value of the detrapping rate. Further size and pressure effects are shown in Fig. 4 for bubbles obeying the pressure-radius relation P= 1/r Pa (r in metres). Release temperatures increase with size and can be significantly higher than the release temperature calculated for a vacancy with identical helium dissociation energy. The FWHM value of the desorption peak increases faster than T_p causing extra broadening of the peak.

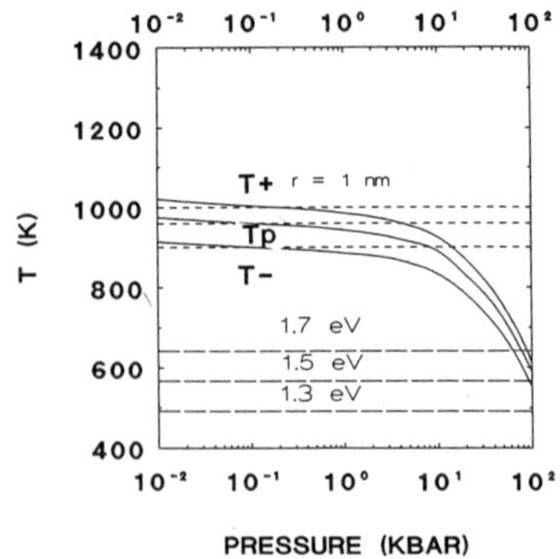

Fig. 3 The temperature at peak maximum and the FWHM temperature interval (T_+, T_-) of calculated desorption peaks for bubbles with $r = 1$ nm and varying pressure. The solid lines indicate results of calculations using Mills EOS. The dotted lines represent results for ideal gas. At the bottom the detrapping temperature for a mono-vacancy is indicated (see text).

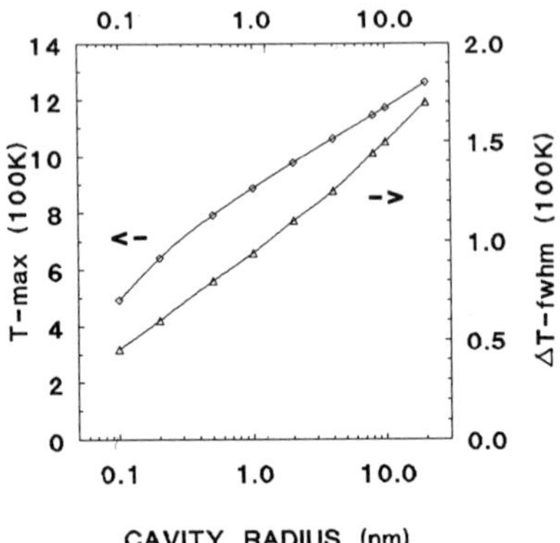

Fig. 4 T_{max} (T_p) and the FWHM vs cavity radius (see text).

In the above calculations it has been assumed that the cavity remains unchanged during the detrapping of the helium. Though this has been observed for bubbles in silicon (see section 5) it is generally not true for bubbles in metals. At the temperatures where helium detraps from bubbles in metals the bubble volume is in dynamic equilibrium with the thermal vacancy concentration [27] so that $P=2\gamma/r$ where γ is the surface energy. Detrapping rates and bubble shrinkage rates for this case have been reported by Evans *et al* [29].

4. Defects in Metals

4.1. *General*

In Fig. 5 a selection is given of THDS results obtained for defects in various metals. Temperatures or temperature intervals are indicated in which desorption is observed for simple defects and defect-agglomerates. Defect agglomerates are indicated by $He_n XV_m$ with n the number of helium atoms, X an impurity or noble gas atom, and m the number of vacancies in the agglomerate. A few interesting trends follow from the scheme in Fig. 5:

1. On a relative temperature scale the desorption from monovacancies (E, F, G, H desorption peaks) for two different bcc metals (Mo and W) is found to occur at nearly identical values of T/T_m $(T_m=T_{melting})$. Furthermore the same peak structure is found; for increasing helium occupancy of the monovacancy the

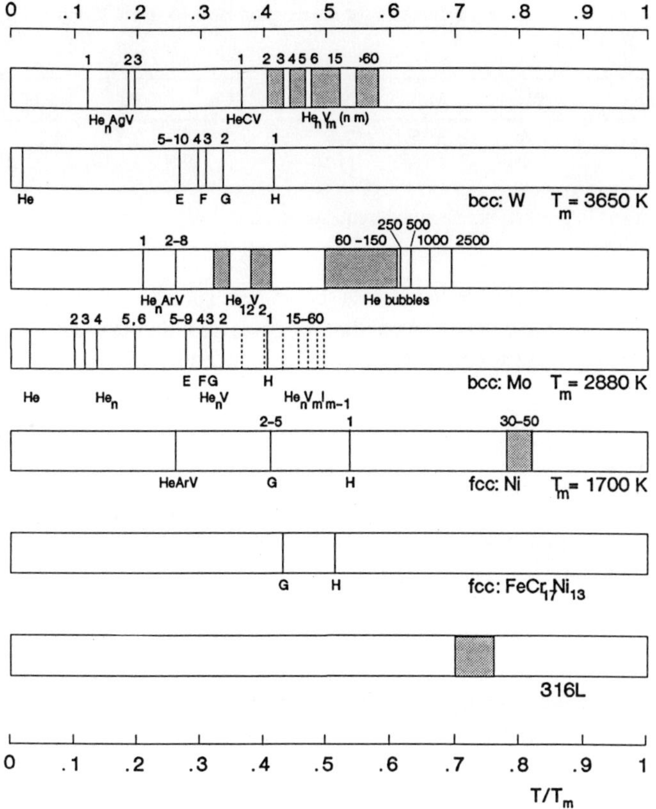

Fig. 5 Helium release temperatures for helium release from different defects indicated on a relative temperature scale (T/T_{melting}).

desorption temperature decreases until the filling level reaches 5–10 helium atoms. Further filling then leads to an increase in the desorption temperature.

2. For the fcc metals Ni and FeCrNi, but also for Cu and Pd (not shown here) the relative temperatures of helium degasing from monovacancies is considerably higher and the pattern of desorption peaks for multiply occupied mono-vacancies is different; only two peaks H and G are present.

3. For all metals the sequence of degasing from defects starting at low temperature is:

    - mobility of interstitial helium
    - interstitial helium precipitates
    - He at substitutional metallic impurities and at substitutional noble gas atoms
    - helium filled vacancies
    - helium from light impurities associated with vacancies
    - small helium vacancy complexes
    - helium bubbles in thermal equilibrium.

It should be noted that for all metals, vacancy-assisted self-diffusion becomes an important transport process at relative temperatures in the range from $T/T_m = 0.5$ to 0.6. Therefore defects surviving to these temperatures might be altered by vacancy capture or by diffusion

Table 2 Helium dissociation energies for the reaction $HeV \rightarrow He + V$ derived from TDS desorption results.

Metal	W	Mo	Ni	Cu	Pd	$FeCr_{17}Ni_{13}$
$E^{D,He}$ (eV)	4.6	3.75	2.4	2.1	2.3	2.3
$\nu(s^{-1})^a$	$2\ 10^{15}$	$3\ 10^{15}$	$(1\ 10^{13})$	$(1\ 10^{13})$	$5\ 10^{12}$	$(1\ 10^{13})$

[a]) numbers in brackets are assumed frequencies.

through the metal. Apparently for the metals shown here such an effect is not observed for the HeV defects. However, the work showed that, for gold, HeV defects start to diffuse before the helium dissociates from the vacancy as was also concluded by Jung for the fcc metals Au, Al and Ag [30].

4.2. Mono-vacancies

In Table 2 data on helium release from mono-vacancies are collected. The so called H- peak release in THDS spectra has been assigned to the dissociation reaction:

$$HeV \rightarrow He + V \tag{6}$$

i.e. release of a helium atom from a helium occupied mono-vacancy [7]. For Mo, W, and Ni it has been proved that the defect giving rise to the H-peak could be removed by a so called reduction or "kick-out" reaction [21] as follows:

$$HeV + I \rightarrow He \tag{7}$$

The dissociation energies, which are determined by the formation energy of interstitial helium E^f_{He} and the formation and migration energies of helium in a vacancy, are well predicted by atomistic calculations—see, for definitions, the potential energy diagram in Fig. 6. For bcc metals (period 6B) the migration energy is very small (<0.2 eV) and therefore the dissociation energy is mainly governed by the two other quantities. From atomistic calculations it follows that $E^{f,He}_{HeV}$, the formation of helium in an existing vacancy, is virtually zero for vacancies in fcc metals [31,32] and about 20% of $E^{f,He}_{HeV}$ in bcc metals. Further the calculated $E^{f,He}_{HeV}$ values scale rather well with the cohesive energy. Thus it follows that helium is relatively stronger bound to vacancies in fcc metals than in bcc metals. In addition it is remarkable that helium migration energies predicted are much larger for fcc metals, 0.5–1.0 eV [31,32], than for bcc metals: 0–0.2 eV. Annealing studies by Viswanathan *et al* [33] on helium implanted nickel show a drastic reduction of the population of mono-vacancies at a temperature corresponding with the helium detrapping temperature in the above table, thus confirming THDS results on the reaction in eqn. (5).

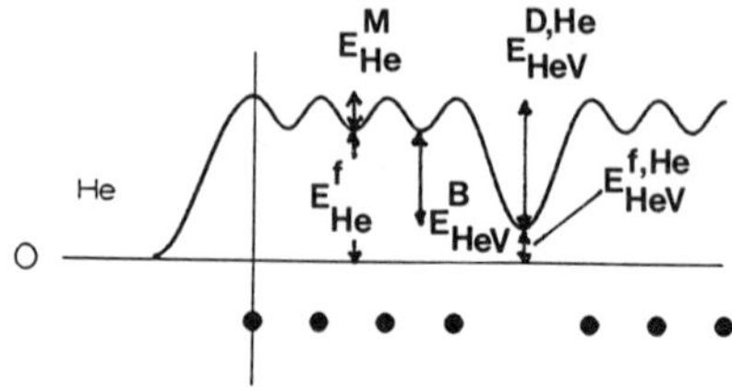

Fig. 6 Potential energy diagram for the helium interaction with a metal including a vacancy. Definitions of interaction energies are indicated.

Table 3 Helium dissociation energies for the reaction HeXV → XV + He, with X = noble
gas atoms or metallic impurity atoms, in tungsten.

$X^{a)}$	Ne	Ar	Kr	Xe	
$E^{D,He}$ (eV)	2.6	2.1	1.8	1.17	

$X^{b)}$	Ag	Cu	Al	Mn	Cr	In
$E^{D,He}$ (eV)	1.03	1.48	0.99	1.13	1.02	<0.90

[a)] ref. 34 ; [b)] ref. 35.

4.3. Substitutional atoms

Binding of helium has been found to substitutional impiurities in the form of both metallic
atoms and noble gas atoms. In general binding increases when the difference in size between
the matrix atoms and the impurity atom increases. In Table 3 desorption temperatures and
desorption energies demonstrate this effect for Xe, Kr,Ar and Ne in tungsten with Xe the
lowest binding energy [34]. It can be easily understood that undersized atoms leave some
space which the helium atom may occupy, and that oversized atoms e.g. Xe create a strain
field in the surrounding lattice where helium can be bound more strongly than in the matrix.
For metallic impurities [35] given in Table 3 the interactions are more complicated and must
be discussed in terms of embedded atom theory (See contributions by Adams and Nieminen
this volume). Roughly the binding energy should decrease as the solubility of the impurity in
the matrix increases—as can be seen by, for instance, comparing the binding energies at Cu
and Cr respectively.

4.4. Light impurities

Only a few studies have been made of the interaction of helium with C, N and O impurities. It
is known however from desorption spectra taken during different stages of cleaning tungsten
that the presence of carbon causes the H-peak to shift to a lower temperature. Tentatively we
assign the shifted peak to the reaction HeCV → He + CV i.e. helium desorbs from the carbon
decorated vacancy. The binding energy is estimated to be reduced by 0.3 eV. For molybdenum
a rather detailed study was performed [36] which indicated the following dissociation sequence HeNV → N+HeV→ He + V, thus first dissociation of the nitrogen followed by
dissociation of helium. The dissociation of nitrogen at about 900 K from vacancies was also
found in positron annihilation studies [37]. Recent helium desorption experiments in our
group on vanadium containing oxygen impurities showed a rather different behavior than
found before in the other bcc metals W and Mo, which we ascribe to interaction of oxygen
with helium and vacancies. Vacancies are immobilizied by the oxygen to a temperature of
about 400 K; stage III in pure V is at 250 K [38]. From the measurements we infer that helium
vacancy complexes nucleate preferentially at interstitial oxygen.

4.5. Interstitial clusters and dislocations

There are no quantitative experimental data known for the binding of helium at interstitial
clusters, loops and dislocations. However TEM work in molybdenum revealed that bubbles
develop at small interstitial loops created during loop punching processes and at dislocations
[45]. For a dislocation array forming a low angle tilt boundary it was found that ribbonlike
cavities develop along the dislocation line. Results of computer simulations support the
observed behaviour [39]. From THDS results on deformed molybdenum it was concluded by
Buters *et al* that helium in that case is drained at room temperature to the surface by pipe

diffusion along the dislocations [40]. Calculations show that binding of helium to single interstitials and small interstitial clusters is of the order of 0.4 eV in molybdenum. Therefore these helium trapping sites will play a role at low temperatures.

4.6. Defects in alloys

In Fig. 7 desorption spectra are shown from nickel, monocrystalline FeNiCr alloy and austenitic steel samples irradiated with 1 keV He ions at doses varying from 10^{12} to 10^{15} He/cm^2. From desorption experiments taken for a sub-threshold helium ion-irradiated $FeCr_{17}Ni_{13}$ mono-crystal it was found that structural defects were present in this sample, at least in the subsurface zone where helium was implanted. The defects give rise to helium desorption observed from 400 to 800 K, between the G (He_2V) and H (HeV) peak temperature. This indicated that defects with a slightly smaller size than that of a monovacancy had been present. When decorated with helium, the defects could not be reduced by self-interstitials as was shown to happen for the vacancies in this sample [7], nor could they be removed by prolonged heating at 1400 K. The presence of these defects delayed the formation of large defect cluster as was observed for nickel—see for example in Fig. 7a the fast development of desorption peaks at temperatures beyond the H-peak temperature in nickel.

In the 316L sample, containing even larger numbers of impurities and structural defects than the FeCrNi crystal, distinct desorption peaks indicating the presence of mono-vacancies were not observed (see Fig. 7c). Although some desorption was observed near the temperature of the G and H peaks in the FeCrNi crystal, most of the implanted helium was released in the temperature interval between 1000 and 1400 K. The results might be explained by supposing that the implanted heliun and the formed self-interstitials are both trapped by the impurities and the structural defects; thus a larger fraction of vacancies than in nickel remain empty or survive recombination. During heating, vacancies become mobile and are trapped at the same defects resulting in a high concentration of relatively small clusters. For stainless steel type 304, helium release was found in the same temperature interval as found here [41].

4.7. Helium precipitates

Results of atomistic calculations by Wilson et al.[42] and Caspers et al [43] have indicated that clusters of interstitial helium He_n can be formed with increasing binding energy when the cluster size increases. Recently we repeated the calculations with the many body potential, proposed by Finnis and Sinclair [44], trather than the pair potentials used earlier. The results for molybdenum are shown in Table 4. Experimental evidence for homogeneous precipitation has been found from TEM work on tritium loaded samples. THDS work on molybdenum irradiated with low energy helium at low temperatures gave evidence for the early stage of precipitation. Fig. 8 (spectrum a) shows the desorption of subthreshold (150 eV) implanted helium at implantation temperature 60 K. Desorption temperatures of He_n (n>1) clusters are found from 100 to 350 K. Heterogenous precipitation has been found to occur at virtually all defects. It has been studied for monovacancies decorated with helium, metallic impurities and noble gas impurtiites [7,18,34,35]. In Table 4 helium dissociation energies for He_nV_m precipitates are given. An interesting aspect is that all precipitates when they reach a certain size convert to complexes which offer more room for the helium by pushing away a matrix atom; often this is called trap mutation as follows:

$$He_n\, V \rightarrow He_n\, V_2 + I$$

The self interstitial atom that has been created remains bound to the complex. However at elevated temperatures the self-interstitial might dissociate from the complex before helium is released. Calculations (see Table 4) show that this is a feasible process. THDS experiments on

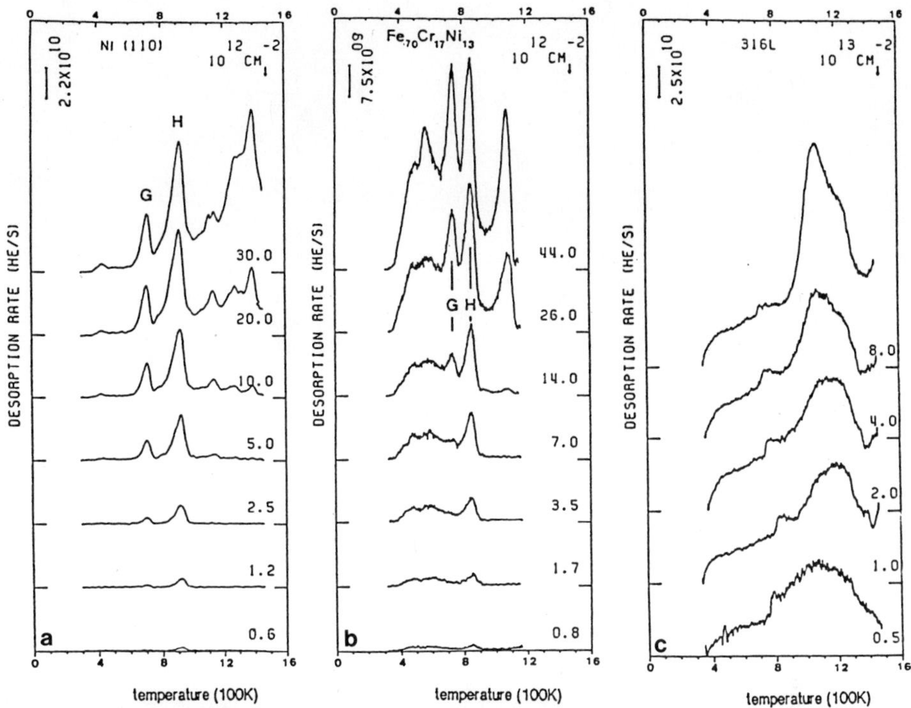

Fig. 7 Desorption of helium from (a) Ni(110), (b) FeCrNi, and (c) stainless steel 316L after irradiation with 1 keV He-ions at the indicated doses. The peaks marked G and H are ascribed to release from mono-vacancies.

Table 4 Calculated dissociation sequence of $He_{10}V$ and $He_{12}V_2I$ complexes.

Reaction	E^D (eV)	Reaction	E^D (eV)
$He_{10}V \to He$	1.99	$He_{12}V_2I \to I^{b)}$	1.33
$He_9 V \to 3\,He$	<1.99[a)]	$He_{12}V_2 \to He$	2.34
$He_6 V \to 4\,He$	2.24	$He_{11}V_2 \to 6He$	2.41
$He_2 V \to He$	2.54	$He_5 V_2 \to 2He$	2.63
$He\,V \to V,He$	3.77	$He_3 V_2 \to He$	3.20
$He_2 V_2 \to V,He$	<3.20[a)]		
$He\,V \to V,He$	3.77		
Average $E^{D,He}$	2.32		2.68

a) reactions at a lower dissociation energy than the preceding reaction; thus occurring simultaneously with the preceding reaction.

helium-filled monovacancies have shown that at n=10 SIA dissociation takes place giving rise to creation of helium filled divacancies. Similarly it has been observed that He_n clusters convert to helium filled monovacancies. The spectra c and d give evidence of the creation of the first and second vacancy respectively by trap mutation processes. When the precipitates grow at low temperature, self-interstitials remain bound until they form a small dislocation loop which can can be emitted in a loop punching process. For molybdenum we estimate this to occur when more than 20 SIAs are formed. TEM observations for Ni and Mo irradiated with subthreshold helium to very high doses have revealed platelet like precipitates which grow by loop punching [45].

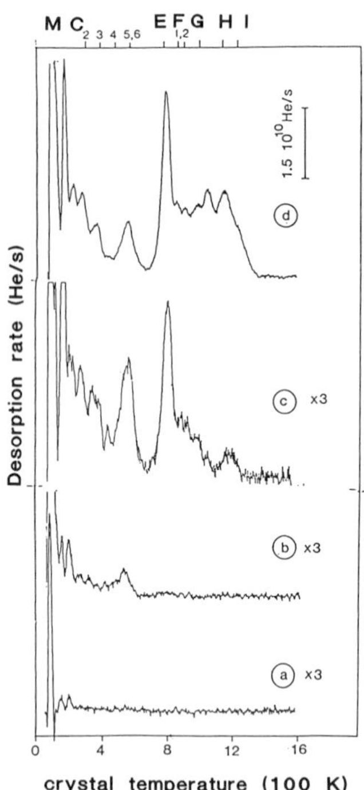

Fig. 8 Evolution of desorption spectra obtained for molybdenum irradiated with different doses subthreshold helium (150 eV), increasing from a through d, at an irradiation temperature of 60 K. M indicates helium migration, C indicates a group of peaks due to degassing of interstitial helium clusters, E, F, G, H indicate helium release from monovacancies, and I from divacancies.

4.8. Helium bubbles

Helium bubbles are usually studied by implantation of helium ions to high doses so that a high density of small bubbles is created. This condition is unfavourable for investigation of the binding energy of helium to bubbles because coalescence may occur and detrapping helium is then redistributed over other bubbles. A method for creating a low density of small bubbles is low energy implantation, followed by annealing so that the helium precipitates convert to small bubbles. Evans *et al.* [29] showed that, under certain conditions in gold, faceted bubbles survive the annealing to relatively high temperatures. It was found that the bubbles disappear by detrapping of the helium and not by bubble motion. A similar study of helium bubbles in molybdenum by THDS also indicates helium detrapping from bubbles [46]. The detrapping at 2000 K can be described with a desorption energy of 5 eV, which is close to the calculated heat of solution of helium in molybdenum.

4.9. Surface effects

There is experimental and theoretical evidence for specific trapping sites near surfaces. Edwards *et al.* [47] describe desorption from vacancies in the neigbourhood of a Ni(100) surface. Fig. 9 shows the evolution of near surface helium trapping with increase of the dose of subthreshold helium. At the highest dose of $1.65 \ 10^{15}$ /cm^2, helium defect clusters have developed which bind helium as strongly as monovacancies in the bulk. It can be envisaged that similar effects will occur when helium arrives at interfaces i.e. lattice sites neighbouring the interfacial plane will act as nucleation sites for bigger helium agglomerates. Experimental

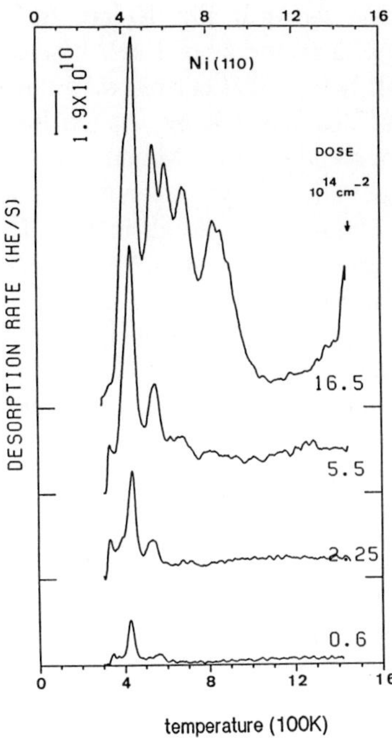

Fig. 9 Desorption of helium from Ni(110) after irradiation with 50 eV He ions at the indicated doses.

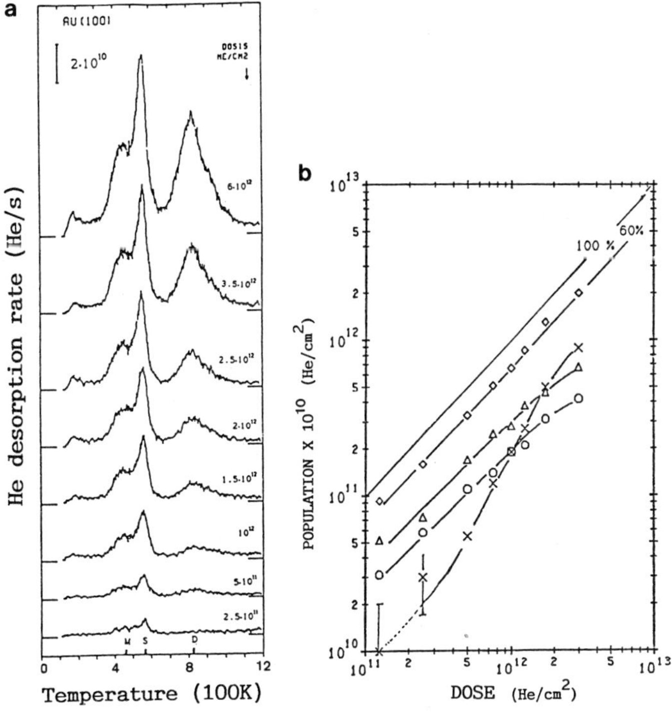

Fig. 10 a) Desorption spectra obtained for Au(110) irradiated with the indicated doses of 1 keV He. M indicates migration, S surface related release (self-trapping) and D vacancy assisted diffusion of substitutional helium (HeV). b) Peak populations from (a) vs the irradiation dose.

results on helium implanted gold shown in Fig. 10 indicate rather strong trapping near an Au(100) surface [48]. Helium with an energy of 1 keV has been injected in the gold sample kept at 150 K. It seems that all helium that entered the sample i.e. 60%, see Fig. 10b, has remained inside the sample at that temperature. At the lowest dose the majority of the implanted helium desorbs in desorption peaks M and S at 420 and 550 K respectively. At higher doses the vacancy concentration grows and an increasing number of helium atoms get trapped by vacancies; these helium atoms desorb in peak D which has been assigned to diffusion of substitutional helium (see section 4.1). Tentatively peak M has been assigned to migration of interstitial helium followed by free release from the surface. Peak S, we assign to helium in traps close to the surface which may have been caused by a self-trapping process near the gold surface. Self-trapping i.e He $\rightarrow$ HeV + I in the bulk is energetically unfavourable but near a surface this reaction requires much less energy because the formed self-interstitial converts immediately to a surface adsorbed gold atom which has a strong binding to the surface. Calculations on these reactions yield an activation energy of 3.7 eV for helium self-trapping in the bulk and 1.7 eV [48] for self-trapping near a surface.

5.   Defects in Silicon and Ceramics

5.1.   Helium from cavities in silicon

In contrast to metals the solution enthalpy of interstitial helium is low, E = 0.5 eV, and the helium migration energy is large, E = 1.2 eV. These values have been obtained by van Wieringen and Warmoltz [49] from permeation measurements. A direct consequence is that the dissociation energy of helium from any defect in silicon should not exceed 1.7 eV.(assuming that the van der Waals binding of helium with silicon can be neglected). Applying simple first order detrapping mechanisms with attempt frequencies of the order of the Debye frequency (10^{13} s^{-1}) and neglecting entropy effects, desorption temperatures no higher than 450 K would be expected. However desorption spectra shown in Fig. 11 for a number of defects reveal helium desorption to temperatures as high as 1200 K. Thus entropy

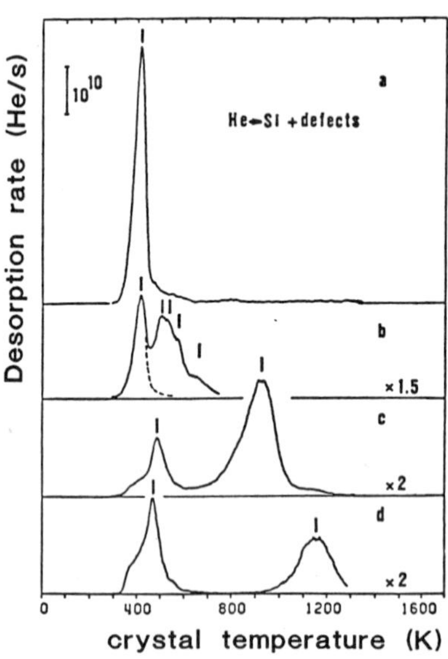

Fig. 11 Helium desorption spectra obtained for Si(111) containing defects: a) defect-free, b) 5x10^{12} cm^{-2} 6 keV Ar, c) 10^{15} cm^{-2} 3 keV Ar + annealing to 1350 K (Ar-bubbles), and d) helium produced voids. For a) and b) 100 eV He (2x10^{13} cm^{-2}) and for b) and c) 2.5 keV He (5x10^{13} cm^{-2}) was used for defect probing.

effects seem to be important. Analysis of the desorption peak found for helium desorption from cavities gave an entropy of solution S = − 8 K, in line with the results of van Wieringen and Warmoltz on the solubility of helium in silicon. The negative entropy change has been explained in section 3; the helium goes from the gas phase with high entropy to the interstitial state where it can be considered as an harmonical vibrating particle in a deep potential well with a low entropy. Only for monovacancies and small vacancy clusters will the entropy change be small, so that for these defects desorption is found in the expected temperature region as, for instance, in Fig. 11a. Note the desorption peaks at slightly higher temperatures than the temperature (350 K) where interstitial helium is released via diffusion.

For cavities in metals similar entropy changes will occur but then the effect on the desorption rate is smaller e.g. a dissociaton energy of 5 eV in Mo vs dissociation energy of 1.7 eV in Si. Furthermore, interstitial sites in metals are shallower potential wells than in silicon and thus the entropy changes are smaller.

5.2. *Voids and bubbles*

An interesting difference in silicon, relative to metals, is the stability of voids to relatively high temperatures so that helium desorption can be measured for non-equilibrium helium decorated cavities (low pressure). Because the helium dissociation energy is small, desorption measurements can also be performed on bubbles of the heavier noble gases. In Fig. 11c an example is shown of helium desorption from a sample with argon bubbles. The argon bubbles do not release any argon until temperatures of 1400 K but helium is released at about 900 K. The desorption peak is similar to the desorption from voids but is shifted to lower temperature. In line with the discussion in section 5.1, it must be assumed that the negative entropy change is smaller for bubbles than for voids. Thus the entropy of helium in the bubble is larger than in a void, which can be explained by assuming that argon is at a high pressure. From Fig. 4 in section 3 it follows that a kbar pressure in a 1nm radius bubble corresponds with the observed helium release temperature. The FWHM value of the measured desorption peak is larger than expected. However this can be accounted for by a 10% spread in bubble radius.

5.3. *Hydrogen related defects and amorphous silicon*

Desorption experiments have been performed on 1.5 keV hydrogen irradiated silicon. Defects were probed by 100 eV helium and the desorption revealed desorption peaks at 450 and 550 K indicating that vacancies or small vacancy clusters had been formed. Annealing of the hydrogen irradiated silicon led to the disappearance of the two peaks after annealing to 800 K. Thus the following reactions might be proposed:

$$\text{He H V} \xrightarrow{450K} \text{H V} + \text{He}$$

$$\text{H V} \xrightarrow{800K} \text{V} + \text{H}$$

Similar reactions might occur for He and H in divacancies with release at respectively 550 K and 800 K. Thus hydrogen stabilizes vacancies and divacancies in silicon. The temperature of 800 K corresponds with release of hydrogen from the sample. This was confirmed by direct desorption measurements on implanted deuterium. The major release peak was found at 800 K. The Si-H or Si-D binding amounts to 3.09 eV and the binding of atomic interstitial hydrogen to silicon amounts to 1 eV (see the discussion by Johnson *et al.* [50]), so that for dissociation 2.09 eV is required. This corresponds rather well with the observed release temperature.

Helium diffusion through amorphous silicon has been studied by experiments in which helium was implanted in an amorphised layer on a silicon substrate. The layer had been amorphised by 3 keV Ar ion bombardment at a dose higher than the critical dose for amorphisation (10^{14} /cm^2). The release occurred at T= 500 K, which corresponds with an effective migration energy of about 1.5 eV (c.f. the value for c-Si of 1.2 eV).

5.4. Helium in ceramic materials

Recently results have been published on helium desorption from helium irradiated TiN and TiC crystals [51,52]. It appears that low energy helium is trapped in structural defects caused by a slightly off-stoichiometric composition of the crystals. In TiC structural vacancies are present on the carbon sub-lattice and in TiN on the titanium sub-lattice. Release temperatures are rather high, e.g. 1000 K for TiC, but are comparable with release temperature measured for helium from monovacancies in Ni, Cu and Pd (see Table 2).

At high implantation energies, displacement damage on the Ti sub-lattice was created by the ions to give a population of Ti vacancies. Higher desorption temperatures resulted from the increase in binding energy at these defects.

6.   Final Remarks

In this contribution an overview has been given of recent developments in helium desorption spectrometry. Though helium defect interactions in certain groups of bcc and fcc metals are fairly well understood, there is still a large number of metals and alloys, which have not yet been sufficiently investigated. It appears that THDS can be applied fruitfully to explore defects in other materials than metals. Vacancies, voids and bubbles in silicon can be studied by helium decoration and desorption methods. A new aspect is that voids and bubbles release the trapped helium at temperatures lower than the dissociation temperature of these cavities. In ceramics strong helium trapping centers have been found, some of which have a structural nature.

ACKNOWLEDGEMENTS

Dr. J.H. Evans is acknowledged for his stimulating activities in many of the topics described in the above contribution.

REFERENCES

1. P. Schiller and J. Nihoul, J. Nucl. Mat. **155-157**, 41 (1988).
2. Hyperfine interaction Investigations of Helium Trapping in Metals, H. de Waard , in: Nuclear and Electron Resonance Spectroscopies Applied to Materials Science, ed Kaufmann and Shenoy, (North-Holland, Amsterdam (1981)).
3. A.R. Arends and F. Pleiter, Hyperfine Interactions **10**, (1981).
4. T.H. Wichert, Radiation Effects **78**, 177 (1983.)
5. T.H. Wichert, Hyperfine Interactions **15-16**, (1983).
6. A. van Veen, C.C. Griffioen, and J.H. Evans, Mat. Res. Soc. Symp. Proc. **107**, 449 (1988).
7. A. van Veen, Materials Science Forum **15-18**, 3 (1987).
8. G.J. van der Kolk and A. van Veen, Physica Scripta **T13**, 53 (1986).
9. W.Th.M. Buters, J.H. Evans, A. van Veen, A. van den Beukel, Defect and Diffusion Forum **57/58**, 75 (1988).
10. W.D. Wilson, Radiation Effects **78**, 11 (1983).
11. R.M. Nieminen, this volume.
12. R. Grimes, this volume.
13. M. Hou, A. van Veen, L.M. Caspers and M.R. IJpma, Nucl. Instr. Meth. **209/210**, 19 (1983).
14. W. Jäger, R. Lässer, T. Schober and G. J. Thomas, Radiation Effects **78**, 165 (1983). See also T. Schober and G.J. Thomas, this volume.
15. A.A. van Gorkum and E.V. Kornelsen, Vacuum, **31**, 89 (1981).

16. E.V. Kornelsen and A.A. van Gorkum, Vacuum **31**, 99 (1981).

17. A. van Veen, A. Warnaar and L.M. Caspers, Vacuum **30**, 109 (1980).

18. A. van Veen and L.M. Caspers, Harwell Symposium on Inert gases in Metals and Ionic solids, ed. S.F. Pugh, AERE Report 9733, 1980 p 517.

19. J.E. Hoogenboom, W. de Vries, J.B. Dielhof and A.J.J. Bos, J. Appl. Phys. **64**, 3193 (1988).

20. W.Th.M. Buters, J. H. Evans, A. van Veen and A. van den Beukel, J. Nucl. Mat. **148**, 17 (1987).

21. A. van Veen, W.Th.M. Buters, T.R. Armstrong, B. Nielsen, K.T. Westerduin, L.M. Caspers and J.Th.M. de Hosson, Nucl. Instr. Meth.**209/210**, 1055 (1983).

22. B.M.U. Scherzer, J. Ehrenberg and R. Behrisch, Radiation Effects **78**, 17 (1983).

23. S.T. Picraux, Nucl. Instr. Meth. **182/183**, 413 (1981).

24. D. Segers, I. Lemahieu, L. de Schepper, M. Dorikens, D. Geshef, L.Dorikens van Praet, L.M. Stals, G. Severne and A. Hermanne, in *Positron Annihilation*, ed L. Dorikens van Praet, M. Dorikens and D.Segers, World Scientific Singapore 1989, p 416. See also articles by K. Jensen and B. Viswanathan, this volume.

25. A. Uedono, S. Tanigawa and H. Sakairi, in: *Positron Annihilation*, ed L. Dorikens van Praet, M. Dorikens and D.Segers, World Scientific Singapore 1989, p 413.

26. H.E. Hansen, R.M. Nieminen, and M.J. Puska, J. Phys. F: Met. Phys. **14**,1299 (1984).

27. H. Trinkaus, Radiation Effects **78**, 189 (1983).

28. R.L. Mills, D.H. Liebenberg and J.C. Bronson, Phys. Rev. **B21**, 5137 (1980).

29. J.H. Evans, A. van Veen and M.W. Finnis, J. Nucl. Mater. **168**, 19 (1989).

30. P. Jung and K. Schroeder, J. Nucl. Mat. **155-157**, 1137 (1988). See also P. Jung, this volume.

31. M.I. Baskes and C.F. Melius, Phys. Rev. **B20**, 3197 (1979).

32. J.B. Adams and W.G. Wolfer, J. Nucl. Mater. **158**, 25 (1988).

33. B. Viswanathan, this volume.

34. E.V. Kornelsen and A.A. van Gorkum, J. Nucl. Mater. **92**, 79 (1980).

35. G.J. van der Kolk, A. van Veen, L.M. Caspers and J.Th.M. de Hosson, J. Nucl. Mater. **127**, 56 (1985).

36. H.A. Filius and A. van Veen, J. Nucl. Mater. **114**, 1 (1987).

37. B. Nielsen, A. van Veen, L.M. Caspers, H.A. Filius, H.E. Hansen and K. Petersen: in *Positron Annihilation*, eds. P.G. Coleman, S.C. Sharma and L.M. Diana, North-Holland, Amsterdam 1982, p 438.

38. H. Schultz in: *Point Defects and Defect Interactions in Metals*,eds. Jin-Ichi Takamura, Masao Doyama, Michio Kiritani, University of Tokyo Press and North Holland Publishing Co. Amsterdam 1982, p 183.

39. J.Th.M. de Hosson, J.R. Heringa, F.W. Schapink, J.H. Evans and A. van Veen, Surface Science **144**, 1 (1984.)

40. W.Th.M. Buters and A. van den Beukel, J. Nucl. Mater. **137**, 57 (1985).

41. D.S. Whitmell and R.S. Nelson, Radiation Effects **14**, 249 (1972).

42. W.D. Wilson, C.L. Bisson and M.I. Baskes, Phys. Rev. **B24**, 5616 (1981).

43. L.M. Caspers, A. van Veen and T.J. Bullough, Radiation Effects **78**, 67 (1983).

44. M.W. Finnis and J.E. Sinclair, Phil. Mag. **A50** 45 (1984).

45. J.H. Evans, A. van Veen and L.M. Caspers, Radiation Effects **78**, 105 (1983). See also M. d'Olieslager, this volume.

46. A. van Veen and J.H. Evans, to be published. See also J.H. Evans, this volume.

47. D. Edwards, and E.V. Kornelsen, Radiation Effects **52**, 25 (1980).

48. W.B. Zeper, unpublished results.

49. A. van Wieringen and N. Warmoltz, Physica **22**, 849 (1956).

50. N.M. Johnson, C. Herring, and D.J. Chadi, Phys. Rev. Letters **56**, 769 (1986).

51. L.C. Seijbel, W.H.B. Hoondert, T.P. Huijgen, B.J. Thijsse, A. van Veen and A. van den Beukel, Nucl. Inst. and Methods B, in press.

52. W.H.B. Hoondert, W.Th.M. Buters, B.J. Thijsse and A. van den Beukel, Nucl. Inst. and Methods B, in press.

DIFFUSION AND CLUSTERING OF HELIUM IN NOBLE METALS

P. Jung

*Institut für Festkörperforschung, Forschungszentrum Jülich
Postfach 1913, D-5170 Jülich, Germany Asscoiation EURATOM–KFA*

ABSTRACT

Migration of helium in gold was investigated by electrical resistivity measurements after implantation at 5 K with energies below the threshold for defect production. Annealing experiments indicated mobility of interstitial helium in Au already at 5 K. After room temperature implantation, diffusion coefficients of substitutional helium in Cu, Ag and Au were determined by thermal helium desorption spectroscopy during and after implantation. By comparison to self diffusion data and to experimental dissociation energies the operating diffusion mechanisms were determined. These are the vacancy mechanism in gold and probably also in silver, and the dissociative mechanism in copper. At temperatures below 673 K and 775 K, helium diffusion in gold and silver is promoted by implantation induced vacancies. Retention of helium by clustering was quantitatively analyzed in terms of stability of small helium clusters.

1.   Introduction

For most metals the operative mechanism of helium diffusion has not yet been identified. Due to its low solubility, helium must be introduced into metals by implantation. Only at low implantation temperatures and energies, will helium reside in interstitial positions, while at higher temperatures and energies when vacancies are available, substitutional helium will prevail.

2.   Low Temperature Implantation

Because of the strong interaction of helium with vacancies, its high mobility, and its strong tendency to form clusters, interstitial diffusion can only be studied when the implantation procedure meets the following requirements:

   1) low temperatures, to avoid migration during implantation,
   2) energies below the threshold for atomic displacement to avoid defect production,
   3) low concentrations to avoid clustering when the helium becomes mobile.

 On the other hand, the measuring technique should meet the following demands:

   1) specific, to discriminate between helium and other impurities or defects,
   2) sensitive, to work at low concentrations, and
   3) it should detect the helium inside the specimen in order to avoid surface effects.

In the present experiment, electrical resistivity measurement (ERM) of thin films was used to monitor changes both during implantation and during a subsequent isochronal annealing experiment. Table 1 compares ERM to other techniques which have been used in previous studies on interstitial helium in metals.

Table 1 Comparison of different experimental techniques.

Method	Specific	Sensitive	Inside	Metal [Ref.]
THDS	+	+	-	Au [1], Ni [2,3,4,5]
FIM	+	-	±[a]	W [6,7]
PAC	±[b]	-	±[c]	Au [8], Cu [9]
TEM	-	-	+	Au [1]
ERM	-	+[d]	+	Pt [10], Lu[11]

[a] detection at, or after leaving the surface
[b] subject to calibration
[c] detection at impurity atoms (e.g. In)
[d] depending on specimen geometry

The only shortcoming of ERM is to be non-specific. Therefore annealing experiments were made before implantation to discriminate effects of impurities, and also at energies above displacement threshold to identify possible contributions of displacement defects.

ERM was applied to study interstitial helium diffusion in fcc (Au and Pt), bcc (Ta and W), and in hcp (Re and Lu) metals. Results on Pt [10] and Lu [11] have been published previously, while results on gold and a summary on the other metals are given in the following.

2.1. Experimental details

Gold films of 80 to 320 nm thickness were evaporated on sapphire substrates and implanted in an evaporation cryostat at 5 K with He^+ ions of energies from 0.2 to 3 keV. More details on specimen preparation and implantation apparatus are given elsewhere [12,13].

The displacement energy in gold is 34 eV, corresponding to a helium implantation energy of about 440 eV. The range of such ions is less than about 5 nm, according to Monte Carlo calculations [13,16]. No coherent specimens of this thickness can be produced, and implantation in the present films will be inhomogeneous. But it can be shown [10] that for small relative resistivity changes, the measured resistivity change is proportional to the resistivity change in the implanted region. After implantation to different doses, specimens were annealed in isochronal annnealing experiments, with logarithmic temperature steps of typically $\Delta T/T=0.5$.

2.2. Results and discussion

The results for He in gold are shown in Fig.1. Included in the figure are recovery experiments after 1.2 [14] and 3 MeV [15] electron irradiation. The solid lines connect recovery data after low dose implantation below (∇) and above ($\blacktriangledown$) the displacement threshold. The given concentrations are average concentrations in the implanted region, derived from the measured doses $\Delta\emptyset$ by:

$$c = \frac{\Delta\emptyset}{n_0\,w} \tag{1}$$

n_0 is the atomic density of the target and w is the width of the implanted region. w and the correction of $\Delta\emptyset$ for backscattering were taken from Monte Carlo calculations [16]. For more details see refs. [13,17].

Fig. 1 indicates mobility of helium in gold at rather low temperatures. Even mobility during implantation at 5 K cannot be excluded. The onset temperature of helium mobility in the other metals are given in Table 2. On the other hand the rather gradual recovery at low doses and the increasing retention at higher doses indicate strong trapping and clustering. TEM [1] and

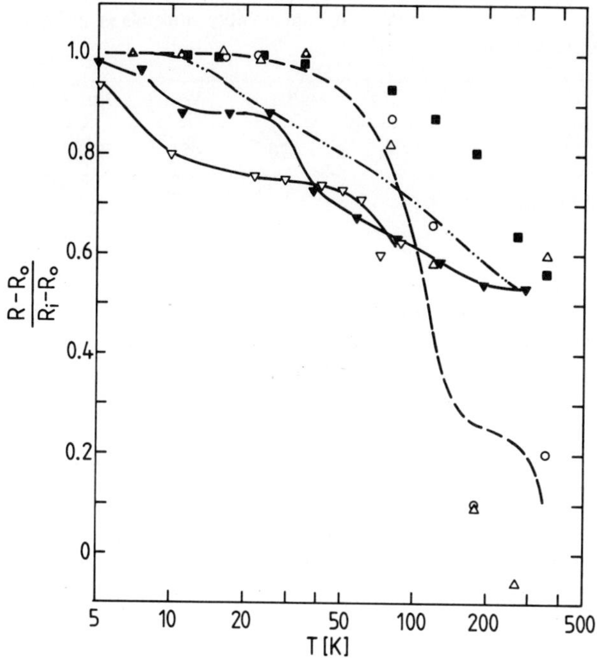

Fig. 1 Resistivity recovery of gold after helium implantation at 5K at energies [keV] and average concentrations [atppm] in the implanted region as follows: 0.25, 1.1 (∇); 0.25, 28 ($\triangle$); 0.44, 45 ($\bigcirc$); 3.0, 0.9 ($\blacktriangledown$); 3.0, 250 ($\blacksquare$). The dashed and dash–dotted lines give annealing after electron irradiation below [14] and above threshold [15] energy, respectively. More data are given in Refs.[13] and [17].

PAC [8] studies on gold indicated no helium mobility below 100 and 250 K, respectively. According to equation (1) the mean concentration in those experiments amounted to about 22 at% and 1 at%, respectively. The data in Fig. 1 show suppression of helium mobility up to 50 K already at concentrations of 28 atppm. It cannot be excluded that the retarded mobility found in Refs.[1] and [8] is due to the much higher concentrations .

3. Room Temperature Implantation

After implantation at not too low energies and temperatures, helium atoms are assumed to be trapped by thermal or by irradiation-induced vacancies, i.e. they are in substitutional positions. For substitutional helium under non-irradiation condition, two mechanisms are most frequently taken into consideration. The first one is the "vacancy mechanism", as observed for most impurity diffusion in metals, in which the helium jumps from its lattice site into a neighbouring vacancy. The activation energy needed for this jump is most probably lower than that to reach an interstitial position. If the vacancy concentration is in thermal equilibrium, it can be estimated [18] that the diffusion energy for this process, E_{VAC}^{diff}, is bracketed by the following values:

$$Q_{2V} - E_v^f < E_{VAC}^{diff} < Q_{1V} \tag{2}$$

Q_{1V}, Q_{2V}, and E_v^f are the self diffusion energies by one- and two vacancy mechanism, and the vacancy formation energy, respectively.

Table 2   Onset temperatures of helium mobility in metals as derived from resistivity annealing experiments.

	Au	Pt[a]	Ta	W	Re	Lu[b]
T[K]	≤5	17	≤20	≤5	>300	26

[a] Ref.[10]
[b] Ref.[11]

On the other hand if the thermal activation of the helium atoms is sufficiently high to reach the interstitial position, the helium can migrate quickly over large distances. The energy necessary to dissociate from the substitutional site is the dissociation energy E^{diss} which is the sum of binding energy E^b and interstitial migration energy E^m. E^{diss} is observed experimentally when the helium atom is detected, e.g. by thermal desorption, before it is retrapped in another vacancy. In this case, which may occur for low implantation depths, desorption follows first order reaction kinetics. If the helium is retrapped frequently before detection, diffusion kinetics are observed ("dissociative mechanism"). In a desorption experiment the dissociative diffusion mechanism slows down the dissociated helium atoms above a vacancy concentration c_V approximately given by:

$$c_V > 2 \, / \, \pi \, R \, d^2 \tag{3}$$

This estimate is obtained by comparing the trapping rate (R=trap radius) to the diffusion time needed to reach the surface (d=specimen thickness). The effective diffusion energy E_{DIS}^{diff} of the dissociative mechanism depends on the retrapping probability, i.e. on the vacancy concentration. When the vacancies are in thermal equilibrium, their concentration is determined by E_v^f and one obtains [18,19]:

$$E_{DIS}^{diff} = E^{diss} - E_v^f \tag{4}$$

On the other hand, after room temperature implantation, the vacancy concentration may exceed the equilibrium concentration by orders of magnitude, at least for a transient period of time. It can be estimated that about 100 vacancies are produced per implanted helium atom in the 10 MeV range. Only a few of these will survive recombination with interstitials or annihilation at dislocations. If the number of surviving vacancies is proportional to the implanted helium concentration c_0, for the vacancy mechanism a diffusion coefficient will be obtainedwhich is proportional to c_0, with an activation energy equal to the vacancy migration energy. In the case of the dissociative mechanism, D will be inversely proportional to c_0 with an activation energy E^{diss}.

Two methods were applied to study diffusion of substitutional helium in metals. Both were thermal desorption experiments (THDS), i.e. the helium was detected after leaving the surface by mass spectrometry.

3.1. Post-implantation desorption

The first method comprised homogeneous implantation of foils of 2 to 50 μm thickness at room temperature to concentrations c_0 from 10^{-3} to 100 atppm. Some of these specimens were ramped up to the melting point at rates of typically 1 K/s, but most were analyzed in isothermal thermal desorption experiments. Under the latter condition the diffusion coefficient D is obtained from the released fraction $(c_0-c)/c_0$ at time t by:

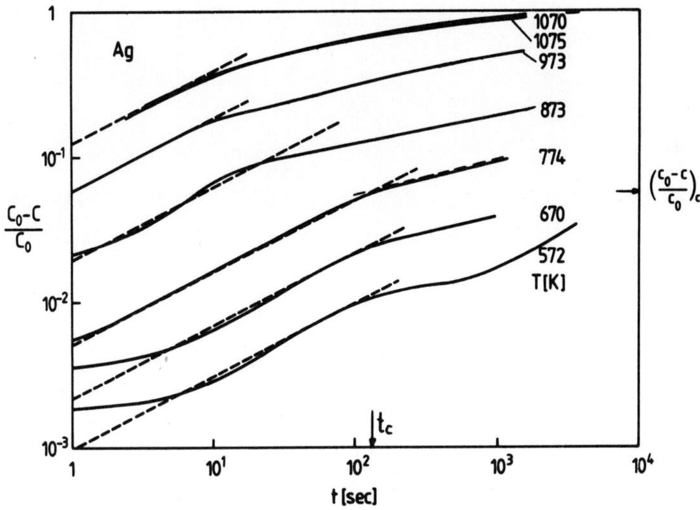

Fig. 2 Fractional release of helium from 13 µm silver foils, containing 10^{-2} atppm helium, during isothermal desorption experiments.

$$\frac{c_0-c}{c_0} = \left(\frac{16Dt}{\pi d^2}\right)^{\frac{1}{2}} \quad \text{for} \quad \frac{c_0-c}{c_0} \leq 0.5 \tag{5}$$

The $\sqrt{t}$ dependence is indicative of diffusion kinetics, while a first order dissociation process would give a linear time dependence. Fig. 2 shows isothermal desorption curves of silver.

Fig. 2 shows that free diffusion is only observed during a period t_c. After that the helium is immobilized by trapping and/or clustering. The fact that the remaining fraction after clustering depends on concentration indicates that clustering is more important than trapping. Numerical calculations [20] give a dependence of the released fraction on d and c_0 which is only slightly influenced by the detailed assumptions on the stability of the small helium clusters. Only the prefactor depends on the assumption at what size helium clusters become stable. Equation (6) describes the most simple case, when di-helium complexes are stable:

$$\left(\frac{c_0-c}{c_0}\right)_c = 0.6\left(\frac{\Omega_0}{2Rd^2c_0}\right)^{\frac{1}{2}} \tag{6}$$

Experimental data at various temperatures are compared to equation (6) for $R=R_{nn}$ (nearest neighbour distance) in Fig. 3 for copper. It can be seen that only at the highest temperature (1170 K) does the released fraction significantly exceed equation (6), indicating the beginning of di-helium instability at these temperatures. Results for the other metals are summarized in Table 3.

3.2. In-situ desorption

In the second type of experiment, 40 to 100 µm thick foils were implanted at high temperatures at beam current densities j_0 from 2 to 200 nA/cm^2, i.e. fluxes of 6 to 600 10^{13} He/m^{2}s. Fixed implantation energies between 10 and 24 MeV gave a narrow profile in a depth equal to

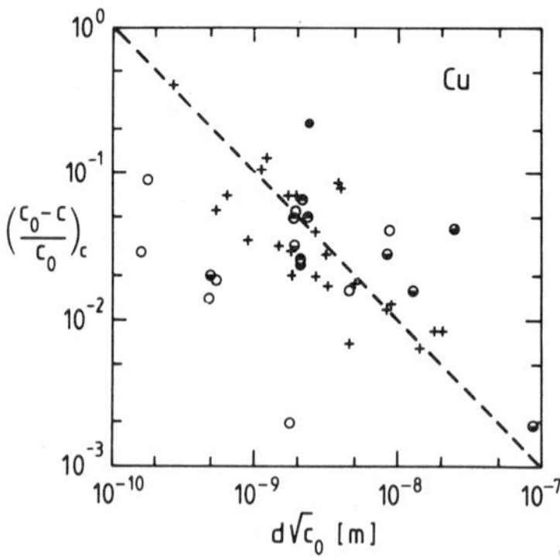

Fig. 3   Fractional release of helium from copper in isothermal experiments at 673 K (O), 773 K (◑), 860 – 1070 K (◓), 1170 K (●) and in linear heating experiments, as a function of $d\sqrt{c_0}$. The dashed line gives equation (6) for $R=R_{nn}$.

the range r of the α-particles. Helium desorption was recorded during and after implantation at the rear surface of the hot foil, giving a diffusion path x = d-r.

Fig. 4 shows the evolution of the normalized helium current at the rear surface, starting from a leakage current j_l, until a maximum current j_m is reached. The time $\tau_{1/2}$ between starting the implantation and reaching $(j_l+j_m)/2$ is used to determine the diffusion coefficient according to diffusion theory:

$$D = \frac{x^2}{0.97\, f\, \tau_{1/2}} \tag{7}$$

The factor f is unity if no trapping or agglomeration of helium is taking place in the specimen, i.e. if $j_m - j_l$ is equal to $j_0 r / d$. Numerical calculations for homogeneously distributed traps of constant concentration and trapping radius give:

$$f = 1 - \frac{4}{\sqrt{\pi}}\, \ln\!\left(\frac{j_m - j_l}{j_0}\right) \tag{8}$$

If it is assumed that the trapping radius grows during trapping, a prefactor 6 instead of 4 is obtained. As $(j_m - j_l)/j_0$ is typically in the range of a few percent, the correction factor f is of the order of 10. The $\tau_{1/2}$ values after shutting off the beam give similar diffusion coefficients.

3.3. Results and discussion

In Fig. 5 helium diffusion coefficients in Cu, Ag and Au are given. In gold the activation energy as well as the absolute values are in close agreement with self diffusion (dashed lines). In silver, helium diffusion is significantly faster than self diffusion and has a somewhat smaller activation energy (1.5 versus 1.76 eV). In copper the discrepancy is huge, and the activation energy is even below the lower limit $Q_{2V}-E_v^f$ conceivable for diffusion by a vacancy mechanism, see equation (2) and Table 3. We may therefore assign the vacancy mechanism to helium diffusion in gold and probably also in silver, while in copper the dissociative mechanism is operating. This statement is backed up by the prefactors of the diffusion coefficients, which are about 10^{-5} m²/s in Au and Ag, i.e. again in close agreement with self diffusion values. On the other hand, for copper a value of only about 10^{-6} m²/s is

64

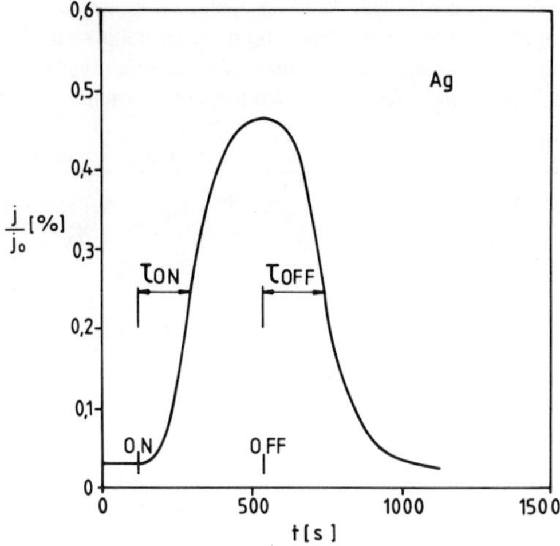

Fig. 4 Helium current desorbing during and after implantation of helium in silver at P 948 K. Diffusion length was 18 μm in a 72 μm thick foil.

derived. This smaller value can be understood as in the dissociative process the diffusion steps are shorter and the vacancy formation entropy occurs in the denominator (equation (4)).

At low temperatures the data of gold and silver deviate from Arrhenius behaviour and show large scatter. The scatter of the silver data can be substantially reduced (triangles) when D is divided by the helium concentration c_0—see right hand side ordinate. This indicates a vacancy mechanism driven by implantation induced vacancies. In this case the activation energy should equal the vacancy migration energy. The slope of the dash-dotted line gives 0.73 eV in reasonable agreement with $E_V^m = 0.66$ eV [21]. The pre-exponential factor is by about three orders of magnitude higher than typical pre-exponentials of vacancy diffusion coefficients. On the other hand, the number of implantation-induced vacancies, which are eventually available for the diffusion process, is estimated to be roughly equal to the number of implanted helium atoms. This discrepancy may be reconciled by assuming that vacancies are strongly correlated to substitutional helium at these lower temperatures.

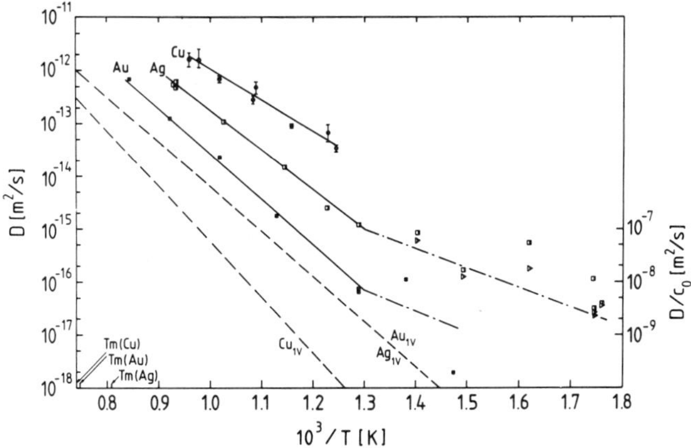

Fig. 5 Diffusion coefficients of Cu, Ag and Au as a function of reciprocal temperature. Part of the Ag and Au data have been published previously [18]. The dashed lines give self diffusion by the single-vacancy process [25]. The scatter of the low temperature (10^3/T > 1.3) silver data is reduced by dividing by c_0, see right hand side ordinate.

Table 3 Helium diffusion coefficients E^{diff}, temperature ranges of operating diffusion mechanisms and stability regimes of di-helium complexes in noble metals. $Q_{2V}-E^f_V$, Q_{1V} and $E^{diff}+E^f_V$ give expected activation energies for diffusion by a divacancy mechanism, for single vacancy mechanism and a lower limit for the helium-vacancy dissociation energy, respectively.

Metal	E^{diff} [eV]	$Q_{2V}-E_V^f$ [eV]	Q_{1V} [eV]	$E^{diff}+E_V^f$ [eV]	Mechanism T[K]	Di-He stable T[K]
Au	1.7	1.42	1.76	2.6_5	vac. (673-1183)	not (>750)
Ag	1.5	1.06	1.76	2.6_3	vac. (775-1075)	< 900
Cu	1.1_5	1.31	2.07	2.4_3	diss. (813-1022)	<1070

Assuming that for Au and Ag, $E^{diff}_{DIS} > E^{diff}_{VAC}$ allows us to derive from equation (4) a lower limit of the dissociation energies, giving for both metals a value of ≥ 2.6 eV, see Table 3. Theoretical calculations [22] gave values of 1.97 and 2.30 eV, respectively. The E^{diss} value for Cu (2.4 eV) is to be compared to desorption experiments which gave 1.9 [20] and 2.1 [23], respectively. The fact that these values are slightly lower may be ascribed to contributions from helium dissociating from divacancies with a somewhat lower dissociation energy. Table 3 summarizes diffusion energies and the underlying mechanisms. Results for other metals will be published elsewhere [24].

REFERENCES

1. G.J. Thomas and R. Bastasz, J. Appl. Phys. **52**, 6426 (1981).
2. G.J. Thomas, W.A. Swansinger and M.I. Baskes, J. Appl. Phys. **40**, 851 (1979).
3. D.B. Poker and J.M. Williams, Appl. Phys. Lett. **40**, 851 (1982).
4. V. Philips and K. Sonnenberg, J. Nucl. Mater. **114**, 95 (1983).
5. D.B. Poker, Radiat. Effects **78**, 101 (1983) .
6. J. Amano and D.N. Seidman, J. Appl. Phys. **56**, 983 (1984).
7. A. Wagner and D.N. Seidman, Phys. Rev. Lett. **42**, 515 (1979).
8. M. Deicher, G. Grübel, E. Recknagel, W. Reiner and T. Wichert, Mat. Sci. Eng. **69**, 57 (1985).
9. T. Wichert, M. Deicher, G. Grübel, E. Recknagel and W. Reiner, Phys. Rev. Lett. **55**, 726 (1985).
10. R. Vaßen and P. Jung, Phys. Rev. **B37**, 2911 (1988).
11. P. Jung and R. Lässer, Phys. Rev. **B37**, 2844 (1988).
12. H.J. Odenthal, thesis, RWTH Aachen, Germany, unpublished.
13. R. Vaßen, thesis, RWTH Aachen, Germany, unpublished and Ref. 10.
14. W. Bauer and A. Sosin, J. Appl. Phys. **35**, 703 (1964).
15. F. Dworschak, G. Holfelder and H. Wollenberger, Radiat. Effects **59**, 35 (1981).
16. J.P. Biersack and L.G. Haggmark, Nucl. Instr. Meth. **174**, 257(1980).
17. A.S. Soltan, thesis Assiut University, Egypt, unpublished.
18. V. Sciani and P. Jung, Radiat. Effects **78**, 87 (1983).
19. V. Philipps, K. Sonnenberg and J.M. Williams, J. Nucl. Mater. **107**, 271 (1982).
20. P. Jung and K. Schroeder, J. Nucl. Mater. **155–157**, 1137 (1988).
21. R.W. Balluffi, J. Nucl. Mater. **69 & 70**, 240 (1978).
22. M.I. Baskes and C.F. Melius, Phys. Rev. **B20**, 3197 (1979).
23. A. van Veen, Materials Sci. Forum **15–18**, 3 (1987) .
24. R. Vaßen and P. Jung, to be published.
25. N.L. Peterson, J. Nucl. Mater. **69 & 70**, 3 (1978).

MOBILITY OF HELIUM AND NITROGEN IMPLANTED AT HIGH FLUENCES INTO SOLIDS, AS DERIVED FROM THEIR CONCENTRATION PROFILES

D.Fink,[1] L.Wang[1†] and J.Martan[2]

[1] Hahn-Meitner Institut
Glienickerstr. 100
D- 1000 Berlin 39, Germany

[2] Instytut Technologii Elektronowej
Politechniki Wroclawskiej
Pl- 50-372 Wroclaw, Poland

ABSTRACT

It has recently been shown that the depth profiles of gases such as N and He broaden with increasing high fluence, after implantation into metals at high energies. Those depth profiles can be well simulated by an analytic approach, which assumes simultaneous implantation and radiation induced mobility of the implants. A universal relation is derived for the depth profile broadening as a function of the average implanted concentration. The comparison of simulated and measured distributions yields numerical values for the radiation induced diffusion coefficient of the implanted gas as a function of the fluence. Remarkable differences are found for both He and N mobilities. Measured changes of these depth profiles by thermal annealing are compared to results of a diffusion simulation program which includes trapping and detrapping at radiation induced defects. For all high fluence He and N implanted samples, the thermal gas mobility shows rapid onset at some specific temperatures, in contrast to samples implanted at low fluences. For He implanted systems, it was possible to identify several stages of gas release from He/defect clusters. The He and N depth profile shapes change with temperature in different manners, which may be understood by different atomistic migration mechanisms.

1.   Introduction

For the prediction of depth distributions of low fluence implanted ions in solids, several theoretical approaches (see for instance [1-3]) are available, which usually have proven to be quite realistic. Little is known however at present about the shapes of the depth profiles of implanted ions in solids at elevated fluences. In this paper, we concentrate on gas ion implantation, comparing the behaviour of He and N implanted into solids at room temperature, on the basis of the present- still somewhat poor- knowledge about the shapes of their depth distributions, see refs. [4] for He and [5] for N.

Those depth profiles show a considerable increase in profile height even far beyond the stoichiometric limit, with simultaneously rapidly increasing broadening. On the other hand, some authors [6-8] report a dramatic decrease in concentration for the gas implanted samples, when a certain threshold has been reached. Finally, for the case of O implanted into Si, a

---

†On leave from the Institute of Semiconductors, Academia Sinica, Beijing, China

plateau formation is reported [9] at the exact stoichiometric concentration, and excess ions are found at the profile sides, leading to a gradual broadening of this rectangular-like depth profile.

A reasonable explanation for the depth profile broadening in the case of high fluence He and N implantation in metals may be given by the assumption of radiation induced gas bubble migration. In fact, the existence of He and N gas bubbles was verified long ago by, for instance, TEM [10,11]. Instead of this mechanism, the driving force for the profile broadening could also be a chemical potential which is built up by the high dose implantation. We tend however to drop this explanation, as the broadening shows up for noble gases in solids (He) as well as for reactive ones (N). Apart from this, volume swelling (i.e. density changes by the implanted gas) has always to be taken into account in the case of high fluence gas implantation, if proper depth profiles are required.

The aim of the present work is to study the depth profile broadening of high fluence He and N in metals in some more detail, and to derive some information about the underlying gas mobility mechanism. Finally, we treat the case of thermal mobility of He and N in solids for the sake of comparison to their radiation induced mobility.

2. Experimental Database for High Fluence Gas Implantation and Theoretical Description

In order to gain a reliable understanding of the processes going on during high fluence ion implantation, and hence to avoid too great complexity of the registered depth profiles, we tried to select "simple" systems where, for instance, surface recession due to sputtering and depth profile broadening due to mixing effects could be neglected. This can easily be verified by the choice of sufficiently light projectiles at sufficiently high implantation energies.

Another fundamental problem in the evaluation of high fluence gas implantation profiles from energy loss spectra is the change of the target density with fluence ("volume swelling"), which affects the depth scales via the particles' stopping powers during measurement. As those density changes are usually not precisely known, we prefer to present areal density scales (units in [mg/cm]) instead of depth scales (units in μm), thus largely avoiding eventual inaccuracies. (The additional depth scales in μm refer to the (here somewhat unrealistic) constant target density at low implantation fluences, only to give the reader an idea of the order of magnitude of depth with which we are dealing here.)

A great number of our high fluence depth distributions were measured by the neutron depth profiling (NDP) technique. This is a technique which exploits nuclear reactions with thermal neutrons, such as ^{3}He(n,p)^{3}H or ^{14}N(n,p)^{14}C. It is described in detail elsewhere [12]. It is one of the most reliable and sensitive techniques presently available, and has especially the advantage of zero impulse transfer from the probing neutrons to the target atoms during the measurement. This means that the measuring results are neither affected by introduction of radiation damage in the sample, nor by radiation enhanced diffusion of the nuclides to be measured by the analyzing particle beam.

A number of depth profiles of He in metals such as Ni, Cu, and Nb [4] and of N in Fe [5] have been obtained by this technique for various implantation fluences and energies. Some other results, gained by different techniques [6], [13] are additionally taken from the literature in order to enlarge the available data base. Finally, some new measurements were undertaken by ourselves in order to clarify some special questions. The procedure for those experiments is the same as described in refs. [4] and [5], respectively.

The shapes of all measured profiles are compared to the analytical theory of Ryssel [14] and Zorin *et al.* [15] which takes into account the broadening of the implantation profiles by simultaneous diffusion according to:

Table 1 Fluence dependence of parameters of 100 keV He implanted into Cu. Ballistic parameters according to TRIM [1]: R_p = 0.32 μm, dR_p = 0.09 μm, implanted ion flux ≈5 μA/cm².

No. in Fig. 1	Fluence [ions/cm²]	Max. Conc. [atomic %]	Profile Width (σ) [mgcm⁻²]	Profile Width (FWHM) [mgcm⁻²]	Derived D value [cm⁻²s⁻¹]
not shown	8.0 10¹³	0.0020	0.31	0.226	not evaluated
not shown	8.0 10¹⁴	0.018	0.31	0.226	not evaluated
not shown	3.6 10¹⁵	0.081	0.31	0.226	not evaluated
----0----	6.3 10¹⁵	0.14	0.31	0.226	not evaluated
----1----	6.6 10¹⁶	1.5	0.33	0.226	7 10⁻¹⁴
----2----	4.2 10¹⁷	13.0	0.33	0.234	5 10⁻¹⁴
----3----	1.6 10¹⁸	35.0	0.45	0.344	8 10⁻¹⁴

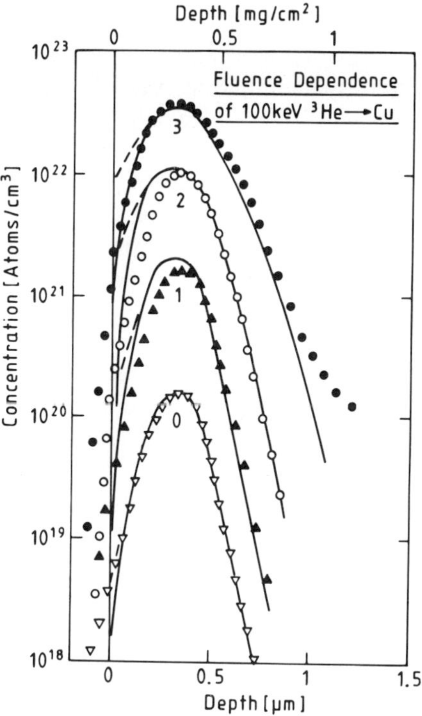

Fig. 1 Depth distributions of 100 keV He in Cu as a function of implanted fluence, comparison of experimental results (symbols) to simulation (curves). Dashed curve—calculation for an infinite medium, solid curve—correction for surface influence, see text. The data points for negative depths arise from the finite detector resolution of the measurement (≈300 Å). The upper depth scale in [mg/cm] is an universal one insofar as it is independent of the volume swelling during implantation. The lower depth scale in μm is calculated for an unchanged target density, to give a rough approximation.

$$C(x,t) = \frac{\Phi}{D \cdot t} \left(\sqrt{\frac{2D \cdot t + dR_p^2}{2 \cdot \pi}} \cdot e^{-\left(\frac{(x-R_p)^2}{4D \cdot t + 2dR_p^2}\right)} - \sqrt{\frac{dR_p^2}{2\pi}} \cdot e^{-\left(\frac{(x-R_p)^2}{2dR_P^2}\right)} - \right.$$

$$\left. \frac{(x-R_p)}{2} \cdot \left(\mathrm{erfc}\left(\frac{x-R_p}{\sqrt{4Dt+2dR_p^2}}\right) - \mathrm{erfc}\left(\frac{x-R_p}{\sqrt{2dR_p^2}}\right) \right) \right) \tag{1}$$

where $C(x,t)$ is the depth and time dependent concentration of the nuclides under consideration, Φ is the fluence of implanted ions, and D is the diffusion coefficient. The implantation profile $g(x)$ itself is assumed to be Gaussian in this calculation, i.e:

$$g(x) = \frac{\Phi}{\sqrt{2\pi} \cdot dR_p} \cdot e^{-\left(\frac{(x-R_p)^2}{2dR_p^2}\right)} \tag{2}$$

The latter assumption implies that we deal with symmetric curves here, thus neglecting the non-zero higher moments of the real cases.

This solution holds only for sufficiently large R_p. In the low R_p case, degassing through the surface cannot be neglected, and $C(x,t)$ has to be replaced by the expression: $[C(x,t)- C(-x,t)]$, so that $C(x=0,t)=0$. An example (He in Cu) for the latter case is depicted in Fig. 1 and listed in Table 1. The above derived formulae are only valid if sputtering can be neglected, i.e. preferentially for the case of light ion implantation with energies exceeding some 10 keV. (The additional incorporation of sputtering would lead to a much more complicated situation which can be described by analytical approaches only for less realistic assumptions -e.g. the replacement of Gaussian functions (2) by δ-functions or exponential functions [16].) It is seen that the measured profile shapes are described well by the present theory. Nevertheless, there remains some deviation in the near-surface region where we find slightly less He than expected. The reason for this deficiency is still unknown. The model according to equation (1) holds also for systems such as He in In, Ge, Nb, Ag, Ta, Tb, Dy and Mn above fluences of 6 10^{16} ions/cm^2.

Fig. 2 compares the dependence of the maximum concentration of He and N implantation profiles on the mean normalized implanted concentration $C_{gen} = \Phi/\sigma N_{[]})$. (Here, σ is the second moment of the distribution, and $N_{[]}$ is the atomic density of the implanted target matrix.) We see that in both cases, the maximum concentrations are proportional to the fluences until, at elevated mean concentrations above $\approx$20 at.%, the curves level off towards saturation concentrations in the order of some 35 at.% for He in Cu and some 60 %atomic for N in Fe.

Surface deformations due to blistering do not influence our depth profile shapes noticably, as we measured the depth profiles perpendicularly to the sample's surface (in contrast to conventional RBS or ERDA, where the analysing beam usually impinges at a large- or even glancing-angle, thus probing the surface structure quite sensitively!). Exfoliation or flaking however lead to deviations from depth profiles according to equation (1), by superposition of different depth profile segments from different flaking generations, which might give rise to step-like depth profiles (observed for e.g. 5 10^{17} to 2 10^{18} cm^{-2} 50 keV He in Zn, Cd and Sb, see for instance Fig. 3c in ref. [17]). Also in (exceptional) cases when blister covers of different blister generations are not lost at increasing fluences but remain partly connected to the sample (see, e.g. He in Co, Figs. 3b and e in ref. [17], or 50 keV He in Bi and Ti at fluences between 5 10^{16} and 4 10^{17} ions cm^{-2}) , equation (1) is not applicable. In those cases, the "cabbage-like" surface geometry leads to broad depth profiles which begin at the surface and extend far into the bulk, with the implantation profile being superposed.

The dramatic decrease in He content after reaching some critical maximum concentration, due to outdiffusion through open channels, which was reported by Terreault [6], Johnson [7]

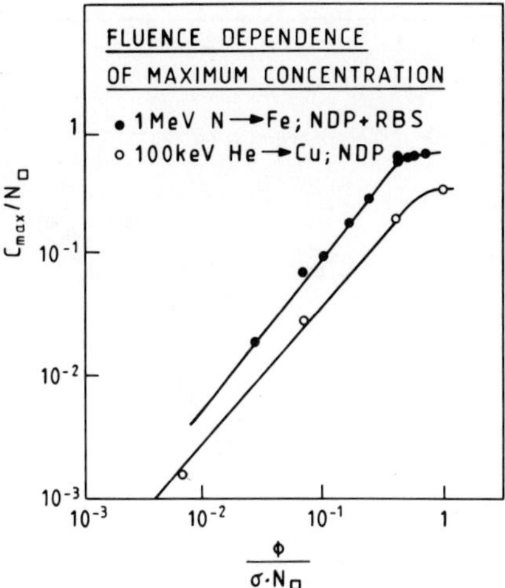

Fig. 2 Dependence of the depth profile maximum concentration on the averaged implanted ion concentration $\Phi/\sigma\,N_{\square}$, for 100 keV He in Cu and 1 MeV N in Fe.

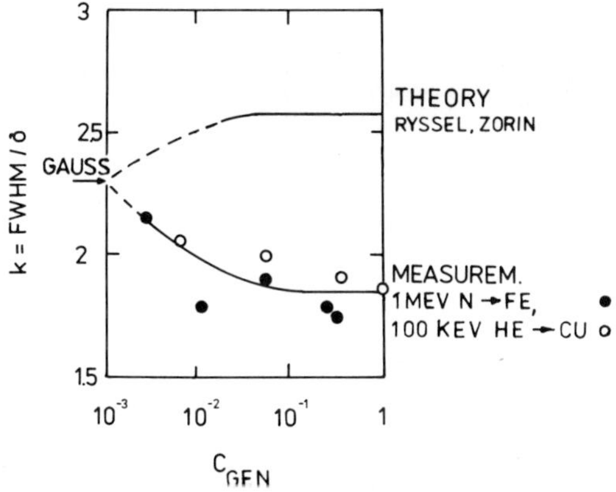

Fig. 3 The ratio k= FWHM/σ in terms of generalized coordinates for theory [14],[15] and measurements.The systematic deviation for higher fluences indicates that, in spite of the general usefulness of the theory for the description of the depth profiles and derivation of approximate diffusion coefficients, this analytic theory is good only as a first-order approximation of our findings.

and Paszty [8] could not yet be reconfirmed by us for any system. The explanation for this disagreement may be found in heating up to different target temperatures during the high fluence ion implantations, due to different ion fluxes and different target holder geometries (i.e. different heat dissipation during the implantations), which might modify the samples in uncontrolled ways. Future measurements should clarify this point better.

Calculating the second moment σ—which is by definition:

$$\sigma^2 = \frac{\displaystyle\int_{-\infty}^{+\infty}(x-R_p)^2\cdot C(x,t)dx}{\displaystyle\int_{-\infty}^{+\infty}C(x,t)dx} \tag{3}$$

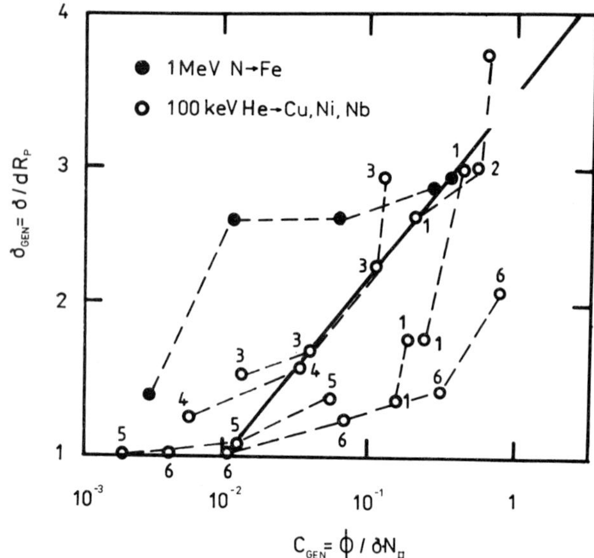

Fig. 4 Dependence of the normalized profile width σ/dR_p on the averaged implanted ion concentration $\Phi/\sigma N_\square$. All values of the same experimental system are connected with dashed lines. Full circles: 1 MeV N in Fe [5], open circles: 50-300 keV He in Ni, Cu and Nb. The numbers mark the individual systems. 1: He in Cu, (Terreault *et al* [6]), 2: He in Nb according to Ehrenberg [13], 3: He in Nb (Fink *et al.* [18]), 4: He in Ni (Fink *et al.* [18]), 5: He in Nb (Fink, unpublished), and 6: He in Cu (this work).

for the distribution C(x,t) according to equation (1), we obtain:

$$\sigma^2 = Dt + dR_p^2 \tag{4}$$

Equation (4) is a universal relation insofar as it holds for any system which can be described by eqn. (1). Table 1 shows the experimental verification for the 100 keV He in Cu.

Further, we can deduce from eq. (1) that the relation: σ = FWHM/2.55 (FWHM is the full profile width at half maximum concentration value) holds for the given theoretical distribution C(x,t) for sufficiently large Dt values. This is a very peculiar case, and should be compared to Gaussian profiles (eq. (2)) with σ = FWHM/2.35. In contrast, the experimental values yield σ = FWHM/1.8 for very large Dt (Fig. 3). This again indicates that the above applied analytic theory may only be regarded as a good first-order approximation to the real case, but should not be overestimated in its applicability.

In order to derive further universal relations, the hitherto examined systems, N in Fe and He in Cu, were plotted, together with other systems, in universal scales, i.e. in the plot of generalized depth profile broadening $\sigma_{gen} = \sigma/dR_p$ versus the normalized mean implanted concentration: $C_{gen} = \Phi/\sigma N_\square$), Fig. 4. We see that all systems (within about half an order of magnitude, independently of the examined target system) roughly scale according to a linear relation in the lin/log plot of Fig.4:

$$\sigma_{gen} = a + b \, \log C_{gen} , \tag{5}$$

with the parameters a = (3.6 ± 0.8) and b = (1.3 ± 0.5). This indicates that the individualities of the different systems are ruled out by the dominating basic mechanism of bubble migration in solids.

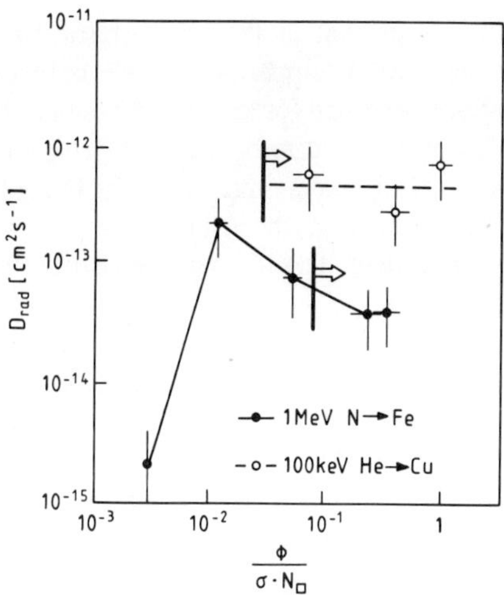

Fig. 5 Dependence of radiation induced diffusion coefficient D_{rad} on the generalized concentration for N in Fe and He in Cu. Thick line with arrow: blister threshold. This value is, in the case of N in Fe, additionally the stoichiometric limit. For He in Cu this value additionally marks the onset of radiation induced bubble mobility.

Inserting equation (5) into (4), we obtain a general relation for the impurity mobility:

$$Dt = dR_p^2 \, (a^2(1 + b/a \log(\Phi/(\sigma N_\square))) - 1) \qquad (6).$$

Division by the corresponding implantation times yields the relation for the diffusion coefficients shown in Fig. 5. We see that He and N behave differently—below the blistering threshold, no depth profile broadening is observed in the case of He in Cu, above this threshold the broadening proceeds regularly, due to a constant He mobility. In contrast, mobility sets in for N in Cu already far below the blistering threshold. The N mobility, as observed via the depth profile broadening, increases dramatically at increasing fluences, until it reaches a maximum value, still below the blistering threshold. Further implantation leads to a slight decrease of the mobility, until above the blistering threshold (which coincides in this case with the stoichiometry limit for the depth profile maximum) the N mobility is roughly constant, as in the case of He implantation. The N and He mobilities at very high fluences coincide within one order of magnitude.

3.   Discussion of Gas Atom Mobility after High Fluence Implantation

When interpreting the experimental findings for the systems 1 MeV N in Fe and 100 keV He in Cu, we should first state that regular thermal mobility at room temperature can be excluded. High fluence implanted N in Fe remains immobile up to $\approx 370°C$ anneal, and the mobility of low-fluence implanted N in Fe is estimated to be around 10^{-16} cms^{-1}—for details see next section. Also in the case of high fluence implantation of He in Cu, it was shown in a previous work [4] that up to 300°C annealing, nearly no He is lost from the sample.

Hitherto the underlying mechanism for impurity depth profile broadening is not at all clear. Strain fields around He bubbles are not capable of preventing the capture of nearby He interstitials [19], which otherwise would be forced to migrate for longer distances, thus leading to an overall depth profile broadening. As the majority of the implanted gases at high fluences most probably exist in the form of gas bubbles, we have to attribute the profile broadening essentially to bubble mobility.

73

However, if the impurity is present in the form of solid overpressurized bubbles (for instance, small He precipitates at room temperature), these bubbles are immobile. On the other hand, bubble mobility has been observed for liquid and gas bubbles (e.g. for large He bubbles at low pressures or bubbles at elevated temperatures). The underlying mechanism of gas bubble mobility may be, for instance, surface diffusion of matrix atoms along the bubble-matrix interface. Recently, it has been proposed by Trinkaus [19] that the bubble dynamics in a highly radiation-damaged medium might be described by the model of a bubble in a viscosous liquid medium.

The complicated fluence dependence of the observed nitrogen mobility (Fig. 5) might qualitatively be understood as follows: at low fluences, all implanted N atoms combine with Fe to form nitride. Thus, the Fe matrix transforms to a mixture of Fe and Fe_2N. When the internal strain fields exceed a certain threshold value, the matrix reorganizes spontaneously under formation of Fe_2N precipitates. This is visible as a dramatic onset of N mobility. Subsequently implanted N bonds with the residual Fe matrix atoms to Fe_2N molecules, their rearrangement to precipitates macroscopically showing up again as enhanced N mobility. As the remaining volume fraction of the precipitates decreases with increasing fluence, we observe a gradual decrease in N mobility. This mechanism is superposed with the onset of bubble formation after N implantation into the Fe_2N precipitates, where no more free bondings are available for the implanted N. The nitrogen bubbles then undergo the same type of diffusion (see discussion above) as observed for the He bubbles.

4.   Thermal Mobility of He and N in Solids, Implanted up to High Concentrations

In this section, we present some examples of the thermal behavior of gases implanted into solids up to high concentrations. They include isothermal annealing experiments with He-doped metals and with N doped Fe. Some information about the first systems has been given already previously [4]. Here, we show measured He depth profiles in Pt and Bi targets for different stages of isochronal (1 hour annealing time each) thermal treatment, see Figs. 6a and 7. The corresponding total amount of retained He after implantation and annealing as a function of temperature is shown in Figs. 6b and 8 for the examples He in Pt and He in Si. The latter graph illustrates that the high-fluence He-implanted samples are characterized by increasingly rapid He degassing with increasing implanted He fluence even at low temperatures (see also ref. [4]).

In order to extract further information from the experimental findings, we compared them to results of theoretical diffusion simulations, which have been done by means of a computer program working after the method of finite differences [20]. The basic concept of this program extends far back to our early studies of diffusion of implants in solids [21], [22]. The program has been refined meanwhile insofar as it now takes into account trapping at and detrapping from any given defect distribution [23]. This is done by simultaneously solving two differential equations, one for the fraction of mobile particles in the target, and one for the immobile (i.e. trapped) fraction. Both equations are coupled by the trapping (A), respectively detrapping (B) probabilities.

The basic problem of such computer simulations is that their result depends sensitively on the underlying model. In the example chosen here, 200 keV He in Pt, two possible models are discussed. In the first model, we assume that at a given temperature, practically all implanted He is released and becomes mobile, so that it either leaves the sample or is trapped at defects. The majority of those defects are assumed to result from the previous implantation, i.e. distributed according to the depth distribution of nuclear energy transfer. This implies a gradual transition from the shape of the He range profile (immediately after the implantation) to the shape of the nuclear energy transfer profile (damage distribution), which in fact is observed (see. ref. [4]).

74

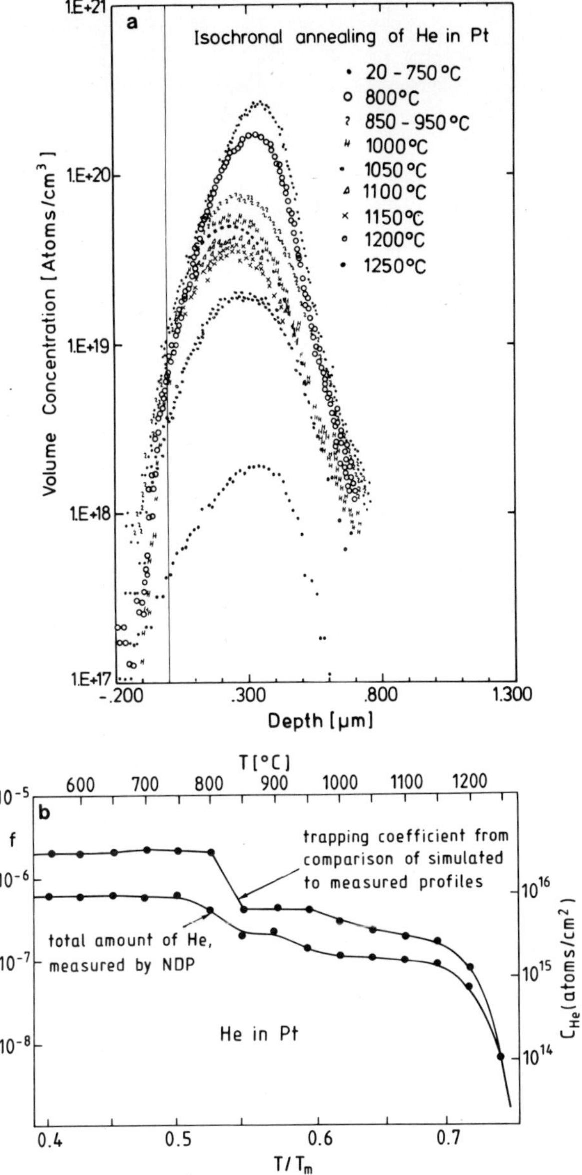

Fig. 6 a) Measured depth profiles of 200 keV He in Pt (implanted ion fluence $\approx 9\ 10^{15}$ ions cm^{-2}), after isochronal annealing in 50°C steps for one hour each. Concentrations given in [atoms cm^{-3}]. b) Total amount $C_{He}(T)$ of implanted He in Pt, as a function of the isochronal annealing temperature T (annealing time 1 hour each). The fitting factor f(T) which describes the ratio of trapping centers (for the mobile He) to the total number of created vacancies (according to TRIM [1]) is additionally included.

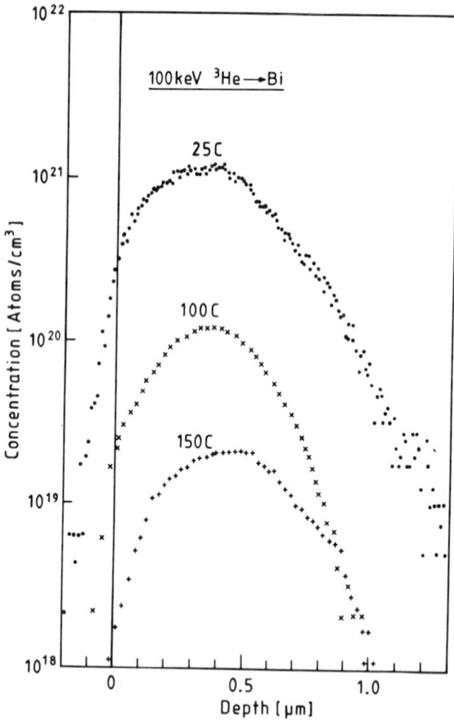

Fig. 7 Measured depth profiles of 100 keV He implanted into Bi at high fluence (ion flux ≈10 μAcm⁻²), after isochronal annealing in 50°C steps for one hour each.

Hence, in our calculations, we describe the depth distributions of the trapping probability A(x) by the depth distributions of vacancies created by nuclear collisions V(x) (as given by the TRIM Monte Carlo code [1]), times a proportionality factor f. This factor hence determines the fraction of all defects, which act as traps for the moving gas atoms at a given temperature, i.e. f contains the combined effects of a) spontaneous and b) thermal defect recombination, c) the neglect of shallow traps (if present) which are not capable of trapping the mobile impurities permanently, and d) of the trapping efficiency of each defect (telling us how many He atoms can be trapped by one defect). The detrapping probability B(x) was set to zero, to describe the stationary state of permanent trapping at the given temperature.

By performing a best fit between the measured concentration distributions and the above described computer simulations for various annealing stages, one obtains the temperature dependence of the proportionality factor f(t), see Fig. 6b. We see that the shape of f(T) closely resembles the measured curve N(T), which describes the total amout of implanted He remaining immobile at the given temperatures. This indicates that the underlying model is consistent with the experimental findings. It is seen that the He release occurs in several distinct steps at different temperatures, each being representative for the decay of a special type of He/defect complex. The factor f may serve us as a characteristic parameter to describe the zoo of available traps at a given temperature T. The difference $f(T_1)-f(T_2)$ for two temperatures T_1 and T_2 which describe two neighbouring annealing stages yields the fractional abundance of the special type of trap which decays between T_1 and T_2, thus leading the way to more detailed information on the behavior of He in solids (this has long been known as Thermal Desorption Spectrometry, see e.g. [24] or article by van Veen in this volume).

Another possible way of interpretating the experimental findings might be to assume that above the Pt recrystallization temperature (≈800°C), the fraction of He in bubbles becomes mobile and hence leads to depth profile broadening. A careful analysis of the depth profiles shows however that their broadening is essentially restricted to the temperature range around 800-950°C, above which the depth profile width no longer changes. This implies that we can exclude bubble mobility here, as this mechanism would lead to a continuously increasing depth profile broadening up to the melting temperature of the sample. The observed depth profile broadening near the recrystallization temperature indicates a He mobility of the order of $(2.0 \pm 0.5)\ 10^{-9}\ \mathrm{cm^2 s^{-1}}$, and—according to the above discussion—should better be ascribed to hindered He atom mobility than to bubble mobility.

The higher the implanted He fluence, the more rapid is the He release (see for instance Fig. 7). In contrast to this, the depth profiles of He, implanted into metals at lower concentrations, exhibit less dramatic changes upon thermal annealing, see for instance Fig. 8. At even lower implantation fluences (typically some 10^{15} ions $\mathrm{cm^{-2}}$ for energies around 50 to 300 keV), the profile shapes tend to remain unchanged [4] nearly up to the sample's melting temperature, so that we cannot draw any more conclusions about the atomic mobility from these findings only.

The He depth profile shapes usually do not broaden during thermal annealing but remain constant in width or even narrow somewhat, while decreasing in height. This indicates the rapid loss of a fraction of the implanted He, the residual He remaining rather immobile. The mobile He is thought to originate from small He/defect complexes which are not stable at elevated temperatures. The released He atoms leave the sample readily as highly mobile interstitials. Additionally, for very high He fluences, new ways open up for He to escape by the formation of a porous network of interlinked He bubbles, due to the onset of crack formation during annealing [4].

In the case of N implanted into Fe the low fluence implanted gas atoms behave completely differently from high fluence implanted N during thermal annealing [5]. In the former case,

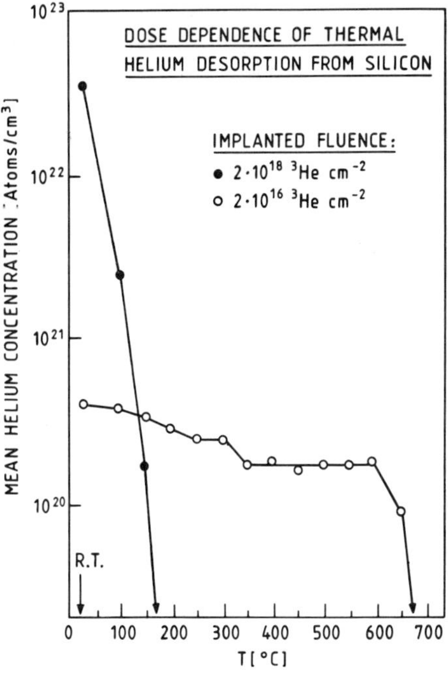

Fig. 8   Mean concentration of 100 keV He implanted into Si (averaged over the implanted region, not corrected for swelling) as a function of the isochronal annealing temperature T (annealing time 1 hour each). Samples, implanted at higher fluences exhibit much faster degassing than samples implanted at lower fluences.

the depth profiles exhibit changes during annealing which can be explained well by regular thermal diffusion; in the latter case, they remain constant up to at least 370°C, before they vanish rapidly below 450°C. For the regular diffusion of implanted N at low fluences in the low temperature range of 300 to 400°C, an Arrhenius plot yields an activation energy of about 0.4 eV and a pre-exponential of about 5.3 10^{-12}cms^{-1} (For comparison, literature values for *unimplanted* N in Fe are E = 1.87 eV in the temperature interval 400 to 500°C [25] and E = 3.27 eV in the temperature interval 800 to 1200°C[26]). Both the low activation energy and the order of magnitude of the pre-exponential indicate interstitial thermal diffusion without any retarding influence by transient trapping at irradiation induced defects.

This interstitial N mobility is however lower by orders of magnitude than that of purely interstitially migrating He (which, by the way, could never be determined properly as hitherto-known mobility values (see, e.g. [27]) refer only to hindered diffusion in a defect rich matrix, and hence yield only lower limits of the true interstitial mobility). This explains why we can observe a depth profile broadening in the case of N but not for He in solids: Whereas N interstitials move slowly enough so that we can make 'snapshots' of the changing profile by our measuring technique, He interstitials are so fast that they have vanished completely before we could ever start our measurement. The reason for the rapid He interstitial mobility (in contrast to N interstitials) is found in the lack of chemical activity of this noble gas atom.

5.   Conclusions

The experimentally observed broadening of depth profiles of gases such as He and N, implanted into solids at elevated fluences and energies (to avoid noticeable sputtering) can well be simulated by a simple analytic approach, which assumes simultaneous implantation and radiation induced mobility of the implants, and additionally the gas loss through the surface in the case of shallow implantation depths. Comparison of simulated and measured distributions yields the radiation induced diffusion coefficients of the implanted gas as a function of the fluence. For He and N at high fluences, this mobility is best described by gas bubble diffusion in the highly distorted medium. The observed mobility of N at lower fluences might be ascribed to the formation of nitride precipitates.

The changes of these depth profiles during thermal annealing are compared with theoretical simulations. For both He and N implanted samples, the thermal gas mobility shows a rapid onset at some specific temperature in the case of high implanted gas concentrations, in contrast to low gas concentrations. In addition, it was possible to identify—for He implanted systems—several stages of gas release from He/defect clusters. For N implanted in Fe, only one annealing stage was found. Differences in the thermal annealing behaviour of implanted high fluence He and N are essentially understood in terms of highly differing interstitial mobilities.

ACKNOWLEDGEMENTS

This work has been enabled by financial support from NATO and the ILL Grenoble (for D.F.), from the HMI Berlin (for J.M.), and from the MPG Bonn (for L.W.). We are indebted to Dr. P. Jung and many other participants of the NATO Advanced Research Workshop on *Fundamental Aspects of Inert Gases in Solids* for discussions and valuable contributions to this paper. Finally, we thank Miss Lang for drawing the figures.

REFERENCES

1. J.P.Biersack and L.G.Haggmark, Nucl. Instr. Meth. **178**, 257 (1980).
2. J.P.Biersack, Z.Phys. **A305**, 95 (1982).
3. J.P.Biersack, Nucl. Instr. Meth. **B35** 205 (1988), and L. Wang, Thesis, Free University Berlin, 1990.
4. D.Fink, Radiation Eff. **106**, 231 (1988), and references therein.

5. D.Fink, M.Mueller, L.Wang, J.Siegel, A.Vredenberg, J.Martan, and W.Fahrner, Radiation Effects and Defects in Solids, **115**, 121 (1990), and references therein.

6. B.Terreault, J.G.Martel, R.G.St.Jacques, G.Vieilleux, J.L.Ecuyer, C.Brussard, C.Cardinal, L.Dechenes, and P.L.Labrie, J.Nucl. Mat. **63**, 106 (1976) and B.Terreault, G.Abel, J.G.Martel, R.G.St.Jacques, J.P.Labrie and J.L.Ecuyer, J.Nucl. Mater. **76/77**, 249 (1978).

7. F. Paszti, personal communication and article in this volume.

8. P. Johnson, personal communication and article in this volume.

9. See, e.g. H.U. Jaeger, Solid State Phenomena, **1&2**, 11 (1988) or H.U.Jaeger, Nucl. Instr. Meth. **B15**, 748 (1986).

10. D.Kaletta and J.Stubbins, J.Nucl. Mat. **74**, 93 (1978).

11. P.Catterjee and A.K.Batabyal, Thin Solid Films **169**, 79 (1989).

12. J.P.Biersack, D.Fink, R.Henkelmann and K.Mueller, Nucl. Instr. Meth. **149**, 93 (1978).

13. J.Ehrenberg, IPP 9140 (1982).

14. H. Ryssel, thesis, Technical University Muenchen (1973).

15. E.I.Zorin, P.V.Pavlov and D.I.Tetelbaum, *Ionnoe Legirovanie Polprovodnikov*, Ed. Energia, Moskva (1975) (in Russian).

16. J.P.Biersack, Radiation Eff. **19**, 249 (1973).

17. D.Fink, J.P.Biersack, K.Tjan and V.K.Cheng, Nucl. Instr. Meth. **194**, 105 (1982).

18. D.Fink, J.P.Biersack, M.Staedele and V.K.Cheng, Radiation Eff. **104**, 1 (1987).

19. H.Trinkaus, personal communication and article in this volume.

20. For a recent description, see: G.D.Smith, *Numerical Solutions of Partial Differential Equations: Finite Difference Methods*, 2nd. ed., Clarendon Press, Oxford (1978).

21. J.P.Biersack and D.Fink, in: *Radiation Effects and Tritium Technology for Fusion Reactors*, CONF-750989, Vol. II, p. 363 (1978), and Proc. Symp. Fus. Technol., 8th, 1974 (EUR 5182), p.907.

22. J.P.Biersack and D.Fink, in: *Ion Implantation in Semiconductors*, S.Namba, ed., Plenum Publ. Corp.,New York (1975) p. 211.

23. The basic outline of this program was derived by J.P. Biersack, further details have been developed by K.Tjan and D.Fink. For a general description, see: K.Tjan, Thesis, Free University Berlin, Germany (1985) or L. Wang, Thesis, Free University Berlin, Germany (1990).

24. A. van Veen, personal communication and article in this volume.

25. J.Hirvonen and A. Antilla, Appl. Phys. Lett. **46**, 835 (1985).

26. R. Hales and A.C. Hill, Metall. Soc. **7**, 241 (1977).

27. V. Philips, Technische Hochschule Aachen, Germany, thesis (1980).

HELIUM IN METAL TRITIDES

^{3}He EFFECTS IN TRITIDES

T. Schober and H. Trinkaus

Institut für Festkörperforschung
Forschungszentrum Jülich, D-5170 Jülich, Germany

ABSTRACT

Experiments performed on metal tritides in the tritium laboratory of KFA Jülich are reviewed. A theoretical interpretation of the main results is provided. After reviewing TEM results on V, Zr and Ti, dilatometry of tritides is discussed which gives rather precise values for the ^{3}He-density in bubbles. Data on X-ray and γ-ray diffraction provide information on the evolution of the dislocation network. Experiments on the acoustic emission and the time dependent hardness of tritides are also discussed. Finally, room temperature creep, resistivities and time-dependent diffusivities are covered.

1. Introduction

A quite frequently used technique for the introduction of He into metals is the "tritium trick". Metals which occlude tritium endothermically are soaked at elevated temperatures in tritium gas for a certain time span. As a consequence of radioactive transmutation described by:

$$T \rightarrow \ ^3He + e^- + \bar{v} \tag{1}$$

(e^- is a β–particle, $\bar{v}$ an anti-neutrino) an increasing amount of ^{3}He is generated in the metal. After thorough degassing only ^{3}He is left in the matrix. In exothermic occluders, tritium again is introduced at elevated temperatures. After cooling, the dilute or concentrated tritide is aged at a certain reference temperature for a time t. The ^{3}He concentration, c_{He}, in the sample is then given by:

$$c_{He} = c_{T,0} \left\{ 1 - \exp(-\lambda t) \right\} \tag{2}$$

where λ the decay constant and $c_{T,0}$ the initial tritium concentration.

In the tritide case ^{3}He is generated in the material itself. The key difference between implantation of He ions and ^{3}He generated by decay is that the latter case is not associated with lattice damage (the average He energy of about 1 eV and the mean electron energy 5.7 keV are too low for displacements). He injected into metals, in contrast, produces displacement damage in the form of Frenkel defects or, at higher energies, displacement cascades. Obviously, the vacancies or vacancy clusters in the implantation case will influence He diffusion and may serve as suitable nucleation sites for He bubble nucleation. In tritium trick doping with ^{3}He no such instantaneous nucleation sites are formed. Here, agglomeration of several ^{3}He interstitial atoms results in "self-trapping" by the spontaneous formation of a Frenkel defect once more than 5–7 ^{3}He atoms are clustered together [1]. This Frenkel pair/^{3}He atom cluster is the nucleation site for a bubble. When trapping further ^{3}He atoms it continously ejects metal atoms either in form of single atoms, or else, interstitial loops (prismatic punching).

2. Recent TEM Results

2.1. Aging at room temperature

The first direct observation of ^{3}He bubbles in metal tritides was reported by Thomas and Mintz [2]. In TEM work on vanadium sufficiently charged to form the β–tritide phase embedded in the dilute α–phase, a rather dense distribution of small interstitial loops was observed in β–phase areas after a few weeks of aging. Again in β–phase patches large numbers of small 1.2 nm diameter ^{3}He bubbles were observed after about 80 days. The bubble density was about 3.5 10^{23} m^{-3} [3]. α–phase areas only displayed occasional ^{3}He bubbles with punched prismatic dislocation loops in their vicinity. The loops were emitted along projections of the bcc Burgers vector a /2[111].

In quite extensive TEM work on Zr-tritides [4] the microstructure consisting mainly of precipitates of $ZrT_{1.6}$ was followed over several years. In this case, the bubble density reached a value of approximately 5 10^{23} m^{-3} after about 2-3 weeks and then remains constant. The mean bubble diameter of the ^{3}He bubbles increases with time as $t^{1/3}$. The bubbles display very strong strain contrast under dynamical diffraction conditions indicating high internal gas pressures above a few GPa. A dislocation network is formed and reached a density between 10^{15} and 10^{16} m^{-2} after aging for about 18 months. The dislocations also become decorated with He bubbles. After about 2 years an interconnected network of ^{3}He filled channels evolves from the dislocation network, dislocation loops and decorated internal boundaries such as grain or phase boundaries.

Fig. 1 shows a typical micrograph of the microstructure after aging for approximately 4 years. The observation of an interconnected ^{3}He network provides a plausible explanation for the phenomenon of the "state of accelerated release". In the latter state, aging samples lose substantial amounts of ^{3}He to the surrounding atmosphere. (Prior to that state the ^{3}He born in the sample essentially becomes trapped). It was argued [4] that the high pressure may open a part of the interconnected ^{3}He network to the outside leading to huge outbursts of ^{3}He.

An analysis of the observed bubble evolution in the light of bubble nucleation and growth models [5] suggests the following conclusions: (1) the time independence of the bubble density after about 50 d indicates within experimental accuracy that, during the very early

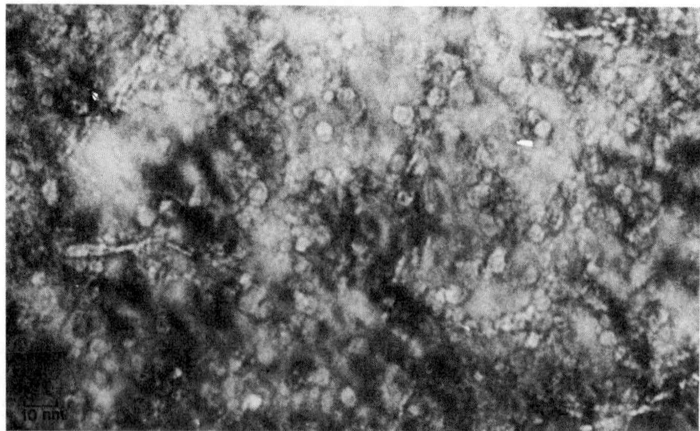

Fig. 1 TEM micrograph of a Zr tritide area aged for about 4 years. Visible are highly pressurized ^{3}He-bubbles and ^{3}He-decorated dislocations. The latter are part of the interconnected network of ^{3}He-filled channels permeating the tritide area.

nucleation stage, the interstitial clusters forming the bubble nuclei consist of several but more than two ^{3}He atoms; (2) from the observed bubble density, the energy for ^{3}He dissociation from such a nucleus is estimated to be about 1 eV; (3) the observed $t^{1/3}$ -dependence of the mean diameter reflects an approximate constancy of the ^{3}He density in the bubbles.

A schematic picture of the evolution of the microstructureof two-phase Zr tritides is depicted in Fig. 2. After aging times substantially longer than the half-life time, a carcass remains at the site of the original tritide plate. It is filled to saturation with He bubbles and its state corresponds to a highly cold-worked condition.

A similar, but less extensive TEM study on Ti [6] supported and corroborated the conclusions of the above work.

2.2. *Aging at elevated temperatures*

Aging of $TiT_{0.1}$ -samples at 300°C for a few weeks resulted in the formation of platelike ^{3}He precipitates on {0001} planes of the matrix (Fig. 3) [7] similar to those observed previously in Mo after He ion implantation [8]. Under dynamical diffraction conditions these platelets displayed spectacular strain contrast again indicative of a high internal pressure. Obviously, thermal vacancies are not yet sufficiently available to alleviate the stresses incurred during growth of the platelets. The platelike shape is probably due to a corresponding shape of the critical nuclei.

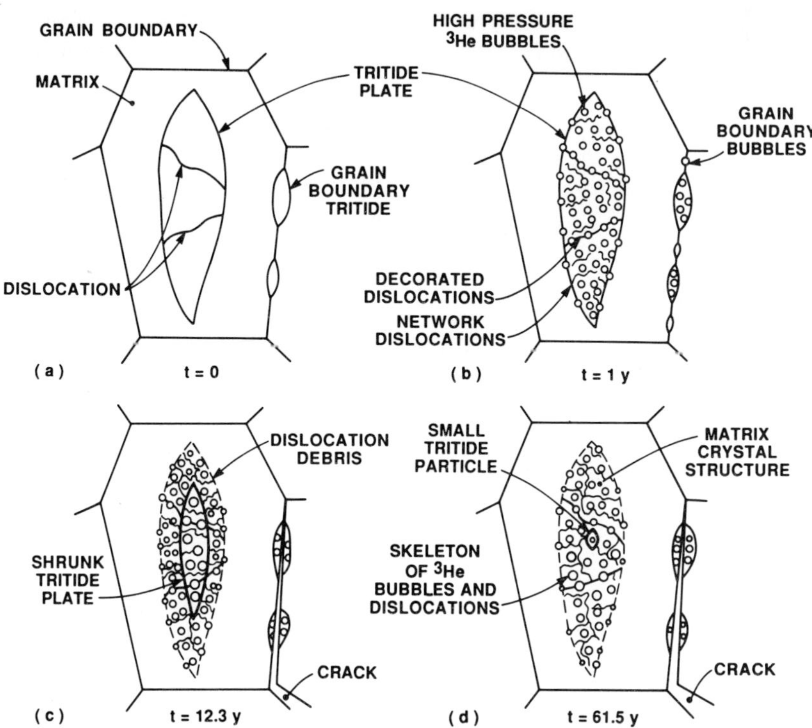

Fig. 2   Model of the likely microstructure of an aged two-phase alloy where a tritide plate is embedded into a matrix which dissolves no tritium. (a) t=0 (b) t=1 year. A dense dislocation network and a high density of ^{3}He-bubbles have formed. Dislocations become decorated with ^{3}He. (c) t=12.3 y. The tritide plate is only half as large as initially. The location of the initial tritide plate is marked by ^{3}He-bubbles and dislocations and is in a highly cold-worked state. (d) t=61.5 years. Only a small tritide particle is remaining. A skeleton of ^{3}He bubbles and dislocations remains.

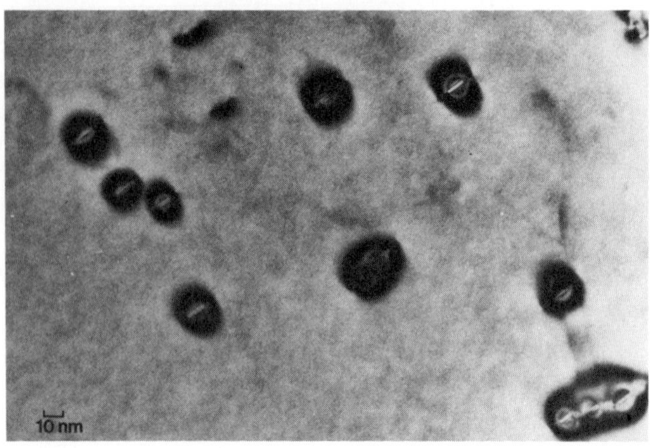

Fig. 3   Platelike ^{3}He-precipitates on {0001}-planes in Ti after aging at 300 °C. Dynamical contrast, s≈0. Note the very strong diffraction contrast indicative of high internal ^{3}He pressure.

Aging of similar TiT-samples at 550 °C produced the more usual spherical, or flattened ^{3}He bubbles which were often facetted. Apparently, thermal vacancies are already sufficiently available at this temperature and establish the usual pattern of relatively large bubbles which are not too far from the thermal equilibrium state. From the observed bubble densities, energies for He dissociation from critical bubble nuclei around 2 eV are estimated.

3.   Swelling of Tritides

Early work on tritide films showed that their dimensions increased roughly linearly with time [9]. More recently, first generation studies performed in this laboratory focussed on Ta [10], Ta and Nb [11] and on V, Lu and Nb [12]. Second generation studies will be outlined in more detail below and were carried out on Nb, Ta and Lu [13].

Swelling of tritides is a consequence of the increased space requirement of a ^{3}He atom compared to the one of a tritium atom from which it originates. Swelling data represent the most direct information on ^{3}He densities and associated pressures [11]. Useful information can also be obtained using NMR techniques [14].

There are several dilatometric procedures to measure the swelling of tritides. We have selected strain gauge techniques and density measurements. Strain gauges may directly be bonded to the surface of bulk tritides. The resistivity change in the strain gauge is linearly related to the dimensional change of the tritide, for details see Ref.[10]. Density measurement is the classic technique to investigate the swelling behavior of irradiated materials. In our tritide case, the density of a swelling tritide decreases almost linearly with time.

Figs. 4 and 5 show examples of strain gauge and density measurements, respectively. In both cases, an approximately linear behavior is noted. The data deduced from this second generation swelling study are presented in Table 1 [13].

The primary quantity resulting from measurements of relative length or density changes, $\Delta L/L$ or $\Delta\rho/\rho$, is the volume change per transmutation event and host atom volume, $(\Delta v/\Omega)_{T\to He}$ which is given by

$$3\Delta L/L \approx 3\Delta V/V \approx -\Delta\rho/\rho \approx c_{He}\,(\Delta v/\Omega)_{T\to He} = c_{T,0}\,(\Delta v/\Omega)_{T\to He}\,[1- \exp(-\lambda t)] \qquad (3)$$

Note that the volume per host atom increases with tritium concentration. Assuming in a first approximation that essentially all of the ^{3}He produced is contained in bubbles and neglecting

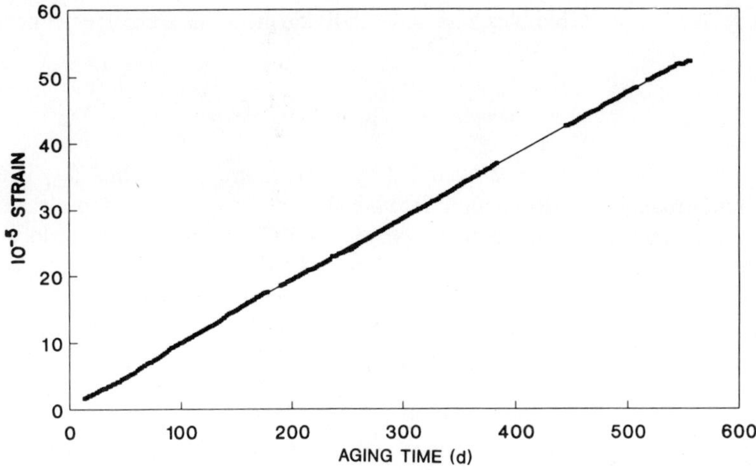

Fig. 4 Relative length change of a $LuT_{0.15}$ sample versus aging time as measured at -195 °C using a strain gauge with a known k factor at that temperature. Thin film strain gauge is solidly bonded to the tritide plate.

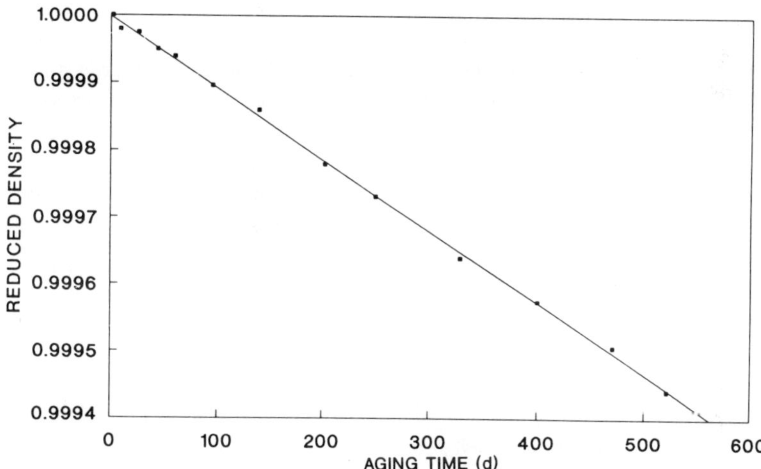

Fig. 5 Reduced density of the tritide $NbT_{0.0253}$, $\rho(t)/\rho(t=0)$, versus aging time. The density measurement was carried out using a high precision buoyancy technique capable of an accuracy in the low 10^{-5} region. The immersion bath (dibromoethane) had to be thermostated to ±0.01°C.

Table 1 Results for the end of the swelling measurements [13].

Specimen (Temperature)	$NbT_{0.0253}$, 299 K	$TaT_{0.0744}$, 299 K	$LuT_{0.15}$, 78 K
$(\Delta v/\Omega)_{T\rightarrow He}$	0.285 ± 0.017	0.255 ± 0.008	0.178 ± 0.007
v_{He} (10^{-30} m^3)	7.89 ± 0.32	7.1 ± 0.21	8.0 ± 0.48
p(GPa)	9.2 ± 1.5	12.8 ± 1.5	7.4 ± 1.5
p/μ	0.196 ± 0.031	0.203 ± 0.024	0.264 ± 0.053
Δp/p	−6%	−10%	−30%

the elastic relaxation of the bubbles, one may subdivide the volume change per decay event as [10,13]:

$$\Delta v_{T \to He} = (\tilde{v}_{He} / \Omega) \, \tilde{\Delta v}_I - \Delta v_T . \tag{4}$$

Here, $\tilde{v}_{He}$ is the effective volume occupied by a ^{3}He atom in a bubble, Δv_T is the volume change upon introducing a tritium atom into the alloy and $\tilde{\Delta v}_I$ is the volume change per SIA transferred from a bubble to an interstitial position or to SIA sinks such as dislocations, grain boundaries and surfaces. For the latter cases, $\tilde{\Delta v}_I \approx \Omega$ to a first approximation. Since Δv_T is available for the metals considered, $\tilde{v}_{He}$ can be determined from length or density measurements using equations (3) and (4). From the atomic volume of ^{3}He, the pressures within the bubbles, p , can be deduced by using an appropriate equation of state [15].

A necessary correction to the first approximation values of $\tilde{v}_{He}$ and $\tilde{p}$ is associated with the elastic relaxation of the bubbles. The values of v_{He} and p given in Table 1 for the end of our measuring periods are corrected with regard to this. Other possible sources of errors in v_{He} and p, in addition to the experimental uncertainties, are the assumptions that essentially all of the ^{3}He is contained in bubbles and that most of the SIAs are transferred to sinks with $\tilde{\Delta v}_I \approx \Omega$. Estimated errors arising from this are included in the uncertainties of the v_{He} and p values given in Table 1. In the p - errors, uncertainties in the equation of state are included.

With the exception of the initial phases, the accuracy is even sufficient to identify changes of Δv_{He} and p with time. In fact, Δv_{He} increases slightly with time. The associated relative decrease of the pressure during the last 90 % of the measuring periods is included in Table 1.

We draw the following conclusions from the data in Table 1: The values for the volume of a ^{3}He atom in a bubble are in a narrow range from 7 to 8 10^{-30} m^{-3}. The corresponding pressure values are around 10 GPa. The ratios of the pressure to the shear modulus of the metal, μ, controlling shear along the glide cylinder of an expected loop punching process is found to be about 0.2. The correlation between p and μ becomes even more spectacular when including the estimated relative decrease of p with time. The magnitude of $\Delta p/p$ is observed to decrease with decreasing p/μ, indicating that the latter quantity converges to a limiting value somewhat below 0.2, possibly close to the value of $1/2\pi$ for the theoretical shear strength. This is in agreement with a recent detailed analysis of dislocation loop punching from bubbles [16]. The implication of this for bubbles forming under inert gas ion implantation is discussed in a further contribution to this volume [17].

4.   X-ray, γ-ray and Neutron Diffraction of Aging Tritides

In X-ray diffraction studies on the tritide TaT$_{0.164-c}$He$_c$ it was observed that the lattice parameter a increased slightly during the first 400 days [18] followed by a rather steep decrease with time (Fig. 6a). Note that $\Delta a/a_o \approx -\Delta G/G_o$ where G_o is the magnitude of the reciprocal lattice vector. Using the Simmons-Balluffi relation and the above dilatometry results it was concluded that for the initial phase where the lattice parameter increases, $\approx 70\%$ of the SIA's produced upon bubble formation are incorporated into the dislocation network. After about 400 days- corresponding to 1 at.% of ^{3}He–100% of the SIA's appear to be incorporated into the network.

There is also a pronounced broadening of the measured rocking curves with time (Fig. 6b) [19]. The broadening accelerates around 1 at.% ^{3}He in correlation with the reversed tendency in the evolution of the lattice parameter. An analysis of the broadening in terms of a dislocation system evolving upon SIA production confirms the conclusions drawn from the evolution of the lattice parameter. The formation of the dislocation network starts surprisingly early, even at about $c_{He} \approx 2 \, 10^{-4}$. At the end of the measuring period, the dislocation density reaches a value of about $2 \, 10^{16}$ m^{-2}.

In γ-ray diffraction [18] it was also observed that there is peak broadening with aging time

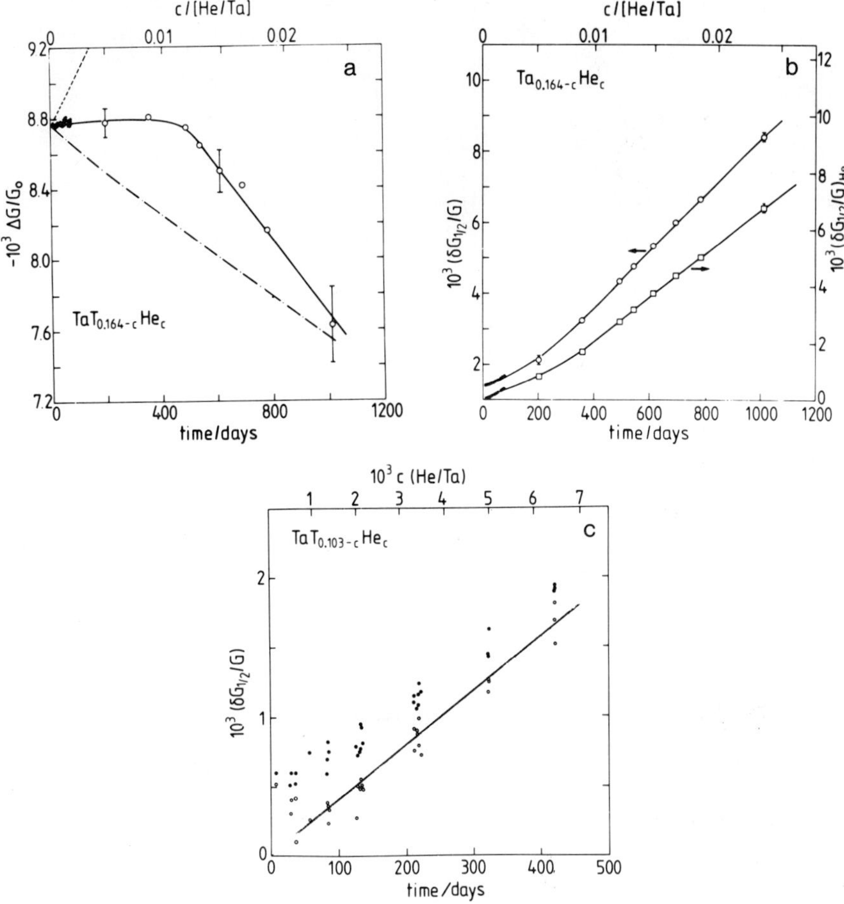

Fig. 6 X-ray and γ-ray diffraction of aging Ta tritides.

a) Evolution of the (222) X-ray rocking curve for $TaT_{0.164-c}{}^3He_c$. Full circles: short term results; open circles: long-term study; decreasing dotted-dashed line: decay produced decrease.

b) Relative broadening of original (circles) and deconvoluted (squares) peaks

c) Tritide $TaT_{0.103}$: Relative broadening of γ-ray rocking curves for a number of different reflections (deconvoluted data). Line: linear fit.

and that there is an almost linear increase of the deconvoluted widths with time (Fig. 6 c). The agreement between the X-ray and γ-ray rocking curve data in the range considered in both studies (up to 1 % of ^3He) confirms that the distribution of dislocations is the same for the crystal regions probed by either form of radiation.

We note that neutron diffraction of aging tritides also yielded a small increase of the lattice parameter and peak broadening with time [20]. The neutron results are qualitatively (but not quantitatively) consistent with the conclusions in the above X–ray and γ–ray studies [18–19].

5.   Acoustic Emission

Two tritide samples, $TaT_{0.103}$ and $NbT_{0.59}$, were examined with a standard acoustic emission (AE) apparatus over a period of three months [21]. Rather strong, irregularly spaced accumulations of acoustic events occurred in both samples 6 to 10 days after preparation of the samples (Fig.7). These AE signals ceased after a time span of 1–2 months for both tritides. Thus, AE only occurs when the ^3He concentration is between 10^{-4} and 10^{-2} which coincides

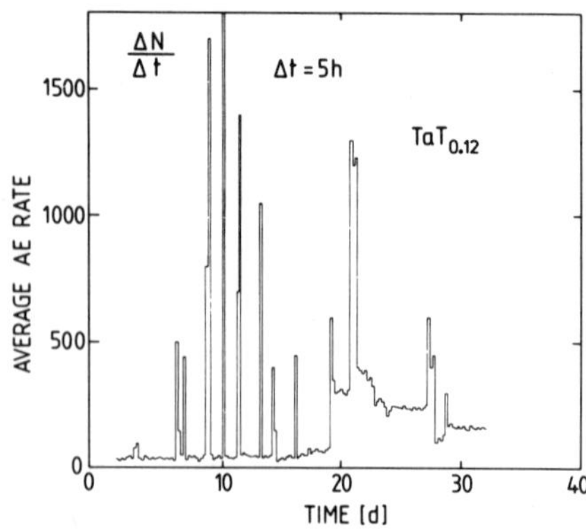

Fig. 7　Average acoustic emission rate versus aging time for the tritide $TaT_{0.12}$. Note strong activity between days 6 and 30.

with the initial stage of loop punching. Since a single loop process does not provide enough acoustic energy we concluded that large collectives of correlated loop punching processes must be considered as the origin. It was speculated that a single loop punching event could trigger a whole bunch of loop punching avalanches. This is consistent with the idea that the binding between bubbles and loops is weakest for small bubble/loop sizes.

6.　The Time-Dependent Hardness of Tritides

Vickers hardness measurements on tritides with the initial composition $NbT_{0.0225}$ and $TaT_{0.097}$ performed over the first 20 months showed that the hardness, HV, increased monotonically, but with a decelerating rate (Fig. 8a, b) [22]. Possible reasons for this increase in hardness include the contributions from the ^{3}He-filled highly pressurized bubbles, dislocation loops and network dislocations. The data in Fig. 8 may be approximated by

$$HV \approx \text{const.} \times c_{He}{}^a \tag{5}$$

with a = 0.7 for Nb and a = 0.76 for Ta. From an analysis of the data it was concluded that for Nb, hardening seems to be dominated by dislocation loops. For Ta, the ^{3}He-bubbles constitute the dominant factor.

For data on the increase of HV in a concentrated tritide phase, in this case the phase β-$NbT_{0.7}$, see Ref. [7]. Again a rather steep monotonic increase of HV with time is observed.

7.　Room Temperature Creep of Tritides

It is conventional wisdom that in standard materials creep is only observed if the temperature is above 0.3 T_M. As shown in Ref.[23] tritides of Ta display creep even at room temperature since decay of tritium produces defects which can interact with the applied stress leading to an effective elongation of the sample. This effect is analogous to irradiation creep where a stress induced preferential adsorption of SIA's by dislocations occurs (SIPA).

Three polycrystalline Ta strips charged with tritium to between 6 and 8% of tritium were stressed at a low, an intermediate, and a high stress level, all within Hooke's linear range. For the intermediate and high stress levels, strain rates were observed experimentally which exceeded substantially the ever present strain rate due to swelling. Fig. 9 illustrates this behavior for the intermediate stress level. It is seen that the strain rate in the stressed state exceeds the one in the unstressed state by the factor 1.3 to 1.5. At the highest stress level somewhat below the yield stress the slopes differ by a factor of 1.8.

90

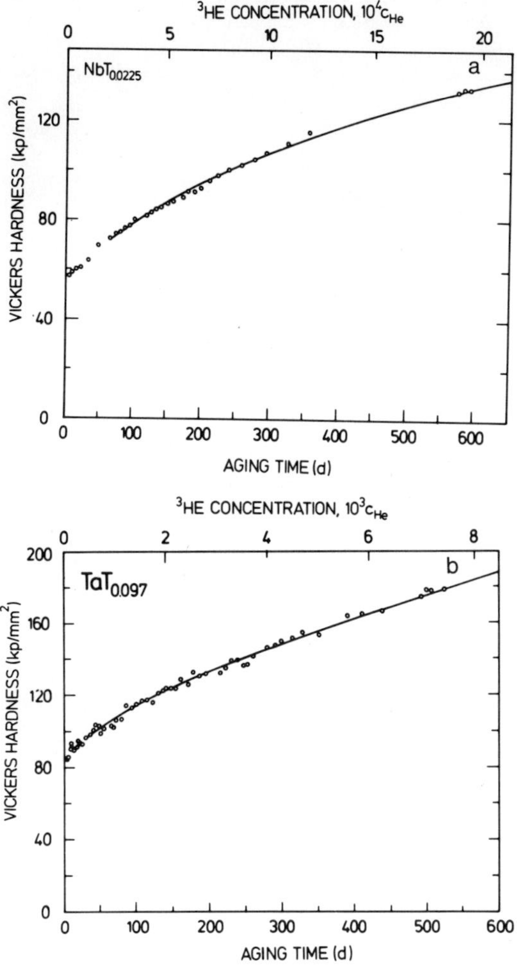

Fig. 8   Hardness versus time (lower abscissa) and ^{3}He concentration (upper abscissa) for the tritides with the following initial concentrations: a) NbT$_{0.0225}$, b) TaT$_{0.097}$. The lines represent least squares fits.

A theoretical treatment of this effect [23] started from the premise that stress introduces an anisotropy in the dislocation loop punching (DLP) process. It is argued that it is energetically most favorable to emit interstitial dislocation loops in a direction as close as possible to the stress axis. This would lead to preferential elongation of the sample in that direction. The above premise is schematically depicted in Fig. 10 where the DLP process is shown for $\sigma = 0$ and a large σ.

8.   Miscellaneous

8.1.   *Time dependent tritium diffusivity in metal tritides [24]*

The long range diffusivity of tritium in Nb and Ta was measured at room temperature via the Gorsky effect. The Nb sample had a concentration of 1.28%, the Ta sample 4.9% of tritium. In both cases the diffusivity dropped by a few percent in the course of 2-3 months. As in the above examples of aging effects the trapping of tritium is attributed to ^{3}He bubbles, or alternatively, to trapping by the evolving dislocation network. Both defect types can, in principle,

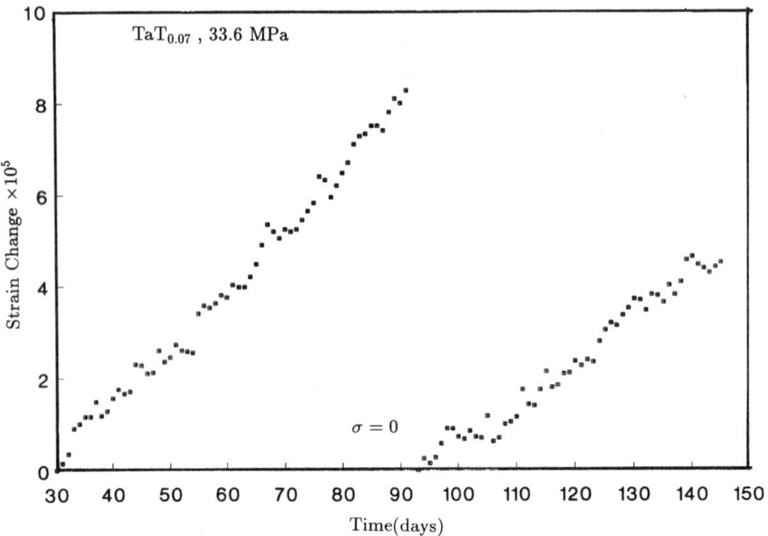

Fig. 9   Strain change under a load of 33.6 MPa versus aging time. Curve to the right labelled σ=0: above load was taken off; pure isotropic swelling was measured. Slope in creep case is 1.3 to 1.5 times higher than in the unloaded case depending on what portions of the curves are analyzed.

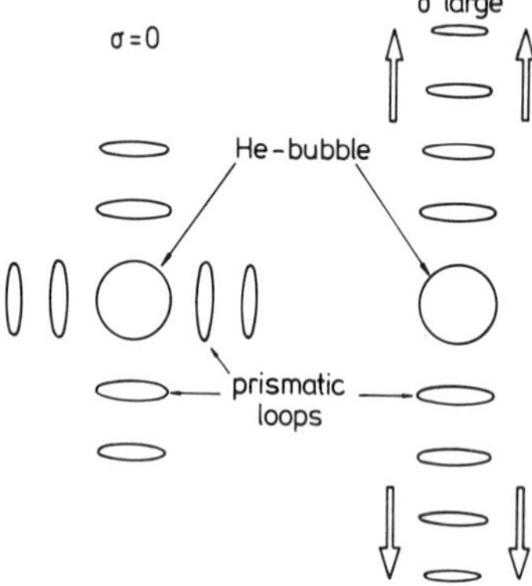

Fig. 10   Schematic of stress induced preferential loop punching (SIPP) by highly pressurized bubbles. Right hand side: in the presence of a large stress, loop punching occurs only in directions which are close to the stress axis. The sample expands more in this than in perpendicular directions.

account for the trapping effect. Theoretical modelling, though, of the time dependence of the trapping processes showed that the ^{3}He-bubbles are most probably the dominant trapping sites in this experiment.

8.2.   Low-temperature resistivity measurements [25]

In the course of this experiment on Lu-tritides, which are in the disordered α-phase even at cryogenic temperatures, it was found that ^{3}He atoms only become mobile around 26 K. Aging below that temperature leads to a linearly rising resistance which is due to the atomically dispersed ^{3}He-atoms. The main finding was that the resistivity per gas atom is almost exactly doubled during the transmutation process: $\rho_{He}/\rho_T = 2.03 \pm 0.16$. For the resistivity per unit concentration of ^{3}He, $\rho_{He} \approx 5.8$ μΩ m / atom, was obtained.

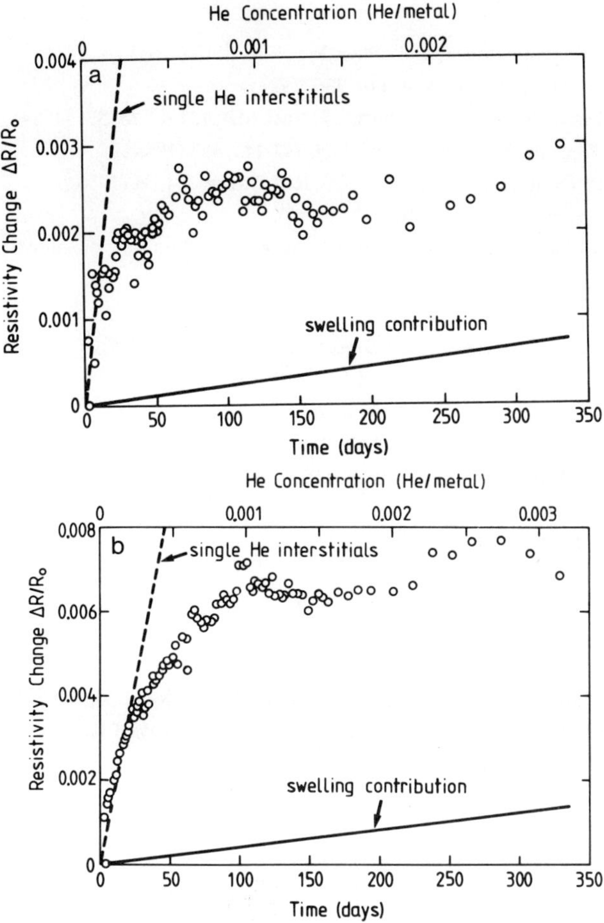

Fig. 11   Relative change in resistivity as a function of time for dilute tritides.(a) Nb (b) Ta. The dashed lines are the expected change for interstitially dissolved and dispersed ^{3}He. The solid lines represent the slope expected due to bubble growth.

8.3. *Room temperature resistivity data [26]*

Precise resistivity measurements on dilute Nb and Ta tritides over roughly one year at room temperature showed that there is a steep increase of the resistivity in the first few days of aging followed by a reduced increase up to a maximum around 100 d and again a slight increase after about 200 d (Fig. 11 a, b). The steep increase at the beginning is attributed to ^{3}He interstitials and interstitial clusters, the maximum to bubble formation and the slight increase in the late stage to bubble growth.

8.4. *Differential thermal analysis (DTA) of aging Ta tritides [27]*

As aging progressed, DTA peaks of selected phase transitions in Ta tritides were recorded. It was found that the height decreased whereas the peak width increased. Also, small shifts of the peaks from their original positions were noted. The above effects were again ascribed to the precipitation of the ^{3}He–bubbles and the formation of the dislocation network.

ACKNOWLEDGEMENT

Fruitful comments by R. Lässer, H. Wenzl and J. B. Condon are gratefully acknowledged.

REFERENCES

1. W.D. Wilson, C.L. Bisson, M.I. Baskes, Phys. Rev. **B24**, 5616 (1981).
2. G.J. Thomas, J.M. Mintz, J. Nucl. Mat. **116**, 336 (1983).
3. T. Schober, R. Lässer, W. Jäger, G.J. Thomas, J. Nucl. Mat, **122 & 123**, 571 (1984).
4. T. Schober, H. Trinkaus, R. Lässer, J. Nucl. Mat, **141-143**, 453 (1986).
5. H.Trinkaus, J.Nucl.Mat. **133 & 134**, 105 (1985); Rad. Effects **101**, 91 (1986).
6. T. Schober, K. Farrell, J. Nucl. Mat, **168**, 171 (1989).
7. T. Schober KFA Jülich Report 2340 (1990). T. Schober and H.Trinkaus, submitted to Phil Mag. A.
8. J.H. Evans, A. van Veen, L.M. Caspers, Nature **291**, 310 (1981).
9. L.C. Beavis, C.J. Miglionico, J. Less-Comm. Met, **27**, 201(1972).
10. T. Schober, R. Lässer, J. Golczewski, C. Dieker, H. Trinkaus, Phys. Rev. **31**, 7109 (1985).
11. T. Schober, J. Golczewski, R. Lässer, C. Dieker, H. Trinkaus, Z. Phys. Chem. NF **147**, 161 (1986).
12. T. Schober, R. Lässer, C. Dieker, H. Trinkaus, J. Less-Comm. Met, **131**, 293 (1987).
13. T. Schober, C. Dieker, R. Lässer, H. Trinkaus, Phys. Rev. **B40**, 1277 (1989).
14. G.C. Abell and A. Attala, Phys. Rev. Lett. **59**, 995 (1987).
15. H. Trinkaus, Rad. Effects **78**, 189 (1983).
16. W.G. Wolfer, Phil. Mag. **A58**, 285 (1988).
17. H. Trinkaus, this volume.
18. R. Lässer, H. Trinkaus, Z. Phys. Chem. NF **163** 19 (1989). R. Lässer, K. Bickmann, H. Trinkaus, Phys. Rev. **B40**, 3306 (1989).
19. R. Lässer, L.Gain, H. Trinkaus, Nucl. Instr. Meth. in Phys. Res. **B43**, 67 (1989).
20. O. Blaschko, G. Ernst, P. Fratzl, G. Krexner, P. Weinzierl, Phys. Rev **B34**, 4985 (1986).
21. T. Schober, J. Golczewski, R. Lässer, C. Dieker, H. Trinkaus, Z. Phys. Chem. NF **147**, 799 (1986).
22. T. Schober, C. Dieker, H. Trinkaus, J. Appl. Phys, **65**, 117 (1989).
23. T. Schober and H. Trinkaus, J. Appl.Phys. **67**, 7583 (1990).
24. T. Schober, C. Dieker, K. Schroeder, J. Nucl. Mat. **165**, 205 (1989).
25. P. Jung, R. Lässer, Phys. Rev. **B37**, 2844 (1988).
26. T. Schober, J.B. Condon, H. Trinkaus, Appl. Phys. **69**, 2961 (1991).
27. R. Lässer, J, Nucl, Mat. **160**, 63 (1988).

FUNDAMENTAL PROPERTIES OF HELIUM IN METAL TRITIDES

G.J. Thomas

*Sandia National Laboratories
Livermore, CA 94551-0969*

ABSTRACT

Tritides can be a useful tool for studying the fundamental properties of helium in metals and offer certain advantages over ion implantation methods. In this paper, recent studies on a number of different tritides are described which have increased our understanding of helium mobility, bubble nucleation, bubble growth and helium densities in bubbles.

1. Introduction

The study of rare gas properties in solids has technological as well as fundamental interest. Since these species are often produced by nuclear transmutation in metals, they can play a major role in influencing material behavior and integrity in radiation environments. Helium, in particular, is a common byproduct in fission reactor materials that is known to affect mechanical properties and cause swelling through void nucleation and growth. In future fusion reactors, copious quantities of helium will be produced as the "ash" of D-T reactions and could find its way into the surrounding structure. Furthermore, the tritium used as a fuel in the fusion process will either be absorbed into the walls of storage containers or be stored on hydride beds, and the decay of this hydrogen isotope would result in the production of ^{3}He within these materials.

Much has been learned about the properties of helium and other rare gases in metals within the last twenty years. Theoretically, the closed electronic shell structure of rare gases can be treated using a number of approximations, allowing predictions to be made of basic properties, such as mobility and gas-defect interactions. Furthermore, properties of materials are affected by the presence of inert gas bubbles, and these can be treated reasonably well with continuum models. Experimentally, however, while some properties have been investigated successfully, others continue to be elusive. One experimental technique to study helium effects in metals has been to utilize the natural decay of tritium to ^{3}He. Tritium, a hydrogen isotope, is soluble to varying degrees in metals and the beta decay process only produces a maximum beta energy of 18.6 keV and about 1 eV recoil energy, insufficient to damage metal lattices. Hence, intrinsic point defects are not concurrently produced. This technique has the further advantages of uniformly doping a macroscopic sample with helium and of allowing one to follow the development, or time evolution, of helium-induced property changes within a given sample.

A critical step in understanding the experimental observations in tritides was the concept of self trapping [1]. In this process, interstitial helium atoms migrate and cluster to form He-He pairs. Then, additional He atoms become bound to the cluster with successively greater binding energies. At some critical number of He atoms (of the order of 5-10), the defect converts to a stable He-vacancy complex by forming a Frenkel pair. The displaced metal atom

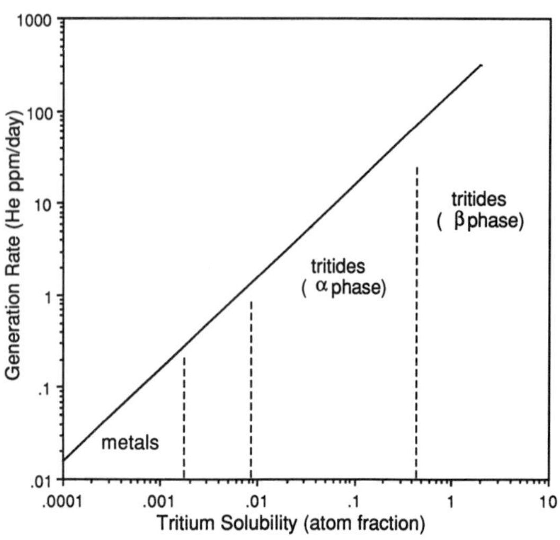

Fig. 1 The rate of helium generation as a function of tritium concentration. Most non-hydride-forming metals have very low tritium solubility and hence slow helium generation. Significant quantities of helium can be formed in tritides relatively rapidly.

can remain bound to the complex. These defects then become the embryo for bubble growth by the further accretion of He atoms and ejection of metal atoms. Thus, a similar microstructure, with bubbles and dislocations, is produced by helium generated in the tritium decay process as in energetic ion implantation. The formation of bubbles in tritides is, in essence, the same as the precipitation of supersaturated solutions of other elements in metals, but can occur at room temperature or below because of the higher rare gas mobility.

The purpose of this paper is to describe the usefulness of tritides as a tool for studying the fundamental properties of helium in metals. A number of recently published studies will be used in this description, as well as some recent unpublished work. Helium effects have been studied in several tritides, including Pd [2, 3], Ti [4], Zr [5], V [6, 7], Nb [8, 9], Ta [10 – 12] and Lu [9, 13]. The same qualitative behavior has been found in all cases; that is, bubbles form by the self-trapping process and grow to cause volumetric swelling and lattice strains. The following sections will discuss two main topics: (a) helium mobility and bubble nucleation, and (b) helium density and bubble growth.

2.   Helium Mobility and Bubble Nucleation

2.1. *Helium production rate*

The half-life of tritium, 12.5 years, results in a generation rate of about $1.8 \ 10^{-9} \ s^{-1}$ per tritium atom. In metals with low tritium solubility, such as Cu, Au, Fe or Al, helium accumulation occurs slowly and long times are needed to attain experimentally significant concentrations. Also, at low helium generation rates in a solid, bubble nucleation is determined by pre-existing defects, such as dislocations and grain boundaries, and not by He-He interactions. The instantaneous helium generation rate in metals is plotted in Fig. 1 as a function of tritium concentration. The generation rate is given in terms of ppm He/day since this is a useful quantity in determining experimental conditions. Typical ranges of tritium solubilities are also shown on the plot for various hydride and non-hydride forming metals. One can see that 2 or 3 orders of magnitude faster helium accumulation occurs in ordered hydride phases compared to non-hydride forming metals. In fact, even the solid solution phases of hydrides can have significantly higher tritium solubilities over other metals.

96

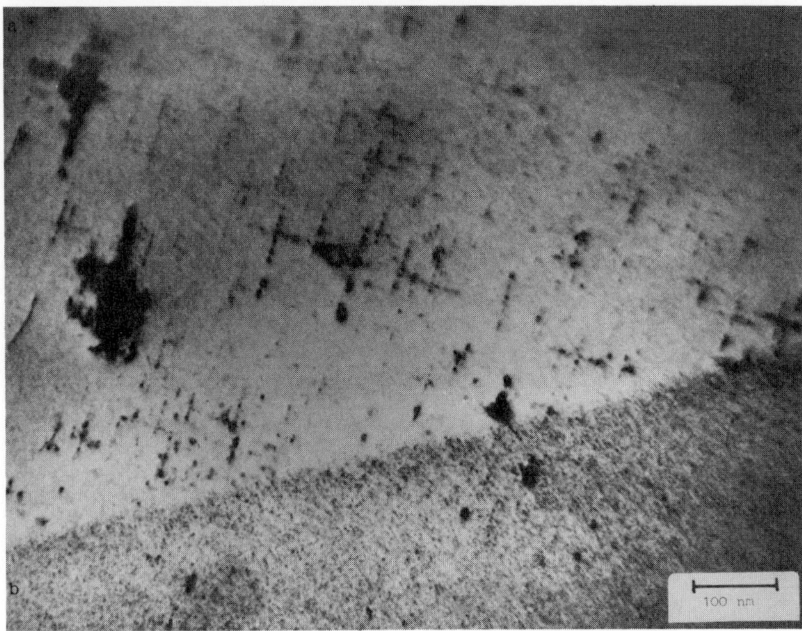

Fig. 2 Transmission electron micrograph of a vanadium tritide sample showing both phases — the solid solution (a) phase in the upper half and the ordered hydride (b) phase in the lower half of the figure. There are few bubbles formed in the a phase and they appear to be heterogeneously nucleated. The linear features are due to prismatic loop punching by the helium bubbles. In collaboration with T. Schober, W. Jäger, and R. Lässer.

Some metals, such as Pd, require an overpressure gas to maintain the tritide phase at room temperature, while others, such as Nb, are usually tritided at elevated temperatures and retain the tritium at room temperature and below. Unless it is continuously re-supplied by, for example, an overpressure gas, the tritium content of the sample will decrease as the helium increases. That is, the change in state of the sample corresponds to one He atom increase and one T atom decrease for each decay event.

2.2. Helium mobility

One fundamental property of helium in metals which remains elusive is the activation energy for migration. There are a number of similarities in the properties of interstitial helium and the self-interstitial atom (SIA) in pure metals: both defects have high energies of formation, large defect volumes (large induced lattice strains), and both have theoretically predicted low activation energies for interstitial diffusion. These physical properties may well be the reason that the unambiguous determination of interstitial mobility for both of these defects has proved to be very difficult. That is, they result in a tendency for the defects to bind with other lattice defects or with themselves, and these interactions will dominate long range transport phenomena. Experimental measurements, then, are typically dominated by trapping-detrapping effects. Even in dilute tritium concentrations, almost all of the tritium generated in solid samples has been found to be trapped, with only a small fraction being evolved [14].

There are some upper bound determinations of helium diffusional activation energies, E_a, in metals. The observations of bubbles in tritides after aging at room temperature clearly indicate that helium atoms are mobile by 300 K, and sub-threshold-energy helium implantations also show clustering and defect formation below room temperature [15]. These results suggest an upper bound of about 1 eV for E_a. Other measurements suggest a much lower

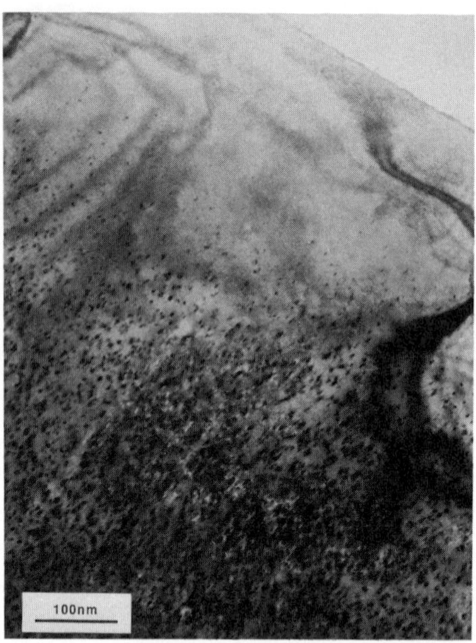

Fig. 3   Transmission electron micrograph of a palladium tritide sample containing 0.01 atom fraction of helium. The sample was pre-thinned and an overpressure of tritium maintained a uniform tritide phase. Measurements of the denuded zone at the foil edge indicates a helium escape depth of approximately 5nm.

energy. In early work on tritium-charged Ni aged at 77 K, thermal desorption data was interpreted to indicate an activation energy of about 0.35 eV [14]. However, as previously mentioned, trapping effects dominated the helium release so that this value may be the sum of diffusional plus trap binding energies. Recent resistivity measurements by Jung and Lässer in Lu-T indicate that helium atoms are mobile at 26 K in this hydride [13].

The helium generation rate in high concentration (0.5 to 2.0 T/M) β phase hydrides is sufficiently great that bubbles should be nucleated homogeneously. Electron microscopy evidence for this is shown in Figs. 2 and 3. The transmission electron micrograph in Fig. 2 shows a region of a vanadium tritide sample in which the α and β phases coexisted during the helium build-up at room temperature. One can see the high density of defects generated in the β phase, whereas in the α phase (T/M $\approx$ 0.03) somewhat larger bubbles occur only in isolated positions and are often associated with dislocations. The linear structures in the α phase are rows of prismatic dislocation loops emanating from isolated bubbles. The observed difference in bubble densities between the two phases is consistent with homogeneous nucleation in the hydride phase. If the bubbles were heterogeneously nucleated, an unreasonably high density of impurities or other defect sites would be needed in the hydride phase to account for the observed bubble density and, furthermore, the microstructure in the α phase would then also exhibit a fine dispersion of small defects, with much smaller bubbles on the dislocations.

Fig. 3 is an electron micrograph of a palladium tritide with 0.01 atom fraction helium. In this case, the sample was pre-thinned for electron microscopy by electrochemical polishing prior to tritium charging. During helium build-up at room temperature, the entire sample was kept in the hydride phase by an overpressure of tritium. The micrograph shows a region including the foil edge and, as is typical, the sample thickness increases with distance from the edge. A denuded zone, where no bubbles have formed, is clearly seen near the edge. Measurements of stacking fault widths at the onset of the bubble growth regions indicate that bubbles were not nucleated within approximately 5 nm of either sample surface. This denuded

98

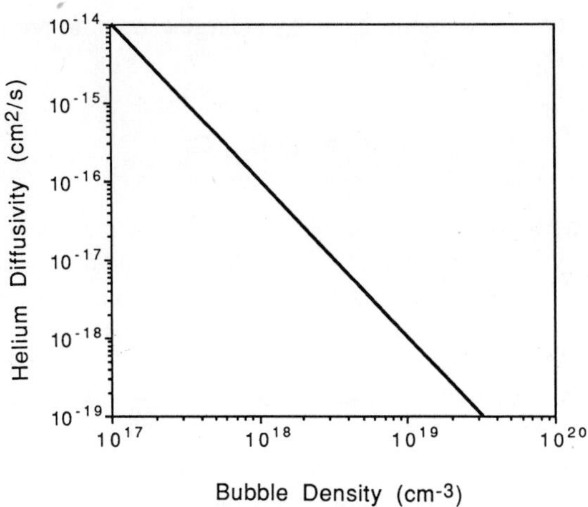

Fig. 4   A plot showing the helium diffusivity required to nucleate an observed density of bubbles, assuming a large He-He binding energy.

zone was observed in all samples examined over a helium concentration range from about 0.001 to 0.02 atom fraction. A distance of 5 nm corresponds to the inter-bubble spacing at a density of $8 \times 10^{18}/cm^3$, somewhat greater but in reasonable agreement with the bubble densities determined by counting in thicker regions of the foils.

Homogeneous nucleation of the bubble precursors can be modelled mathematically, enabling one to determine parameters of the process from measurements of the bubble density. The diffusion equation for He atom clustering into clusters c_j containing j+1 atoms is [16]:

$$\frac{\partial c_{He}}{\partial t} = \frac{\partial}{\partial x}\left(D_{He}\frac{\partial C_{He}}{\partial x}\right) + \lambda C_t + \sum_{j=1}^{N}[v_j C_j - \mu_j C_{He} C_j] \tag{1}$$

where $\quad \mu_j = 4\pi r_j D_{He} \quad$ (accretion rate factor)

$\qquad v_j = v_o j \exp[-Ej/kT] \quad$ (detrapping rate)

and $\quad \lambda C_t$ is the He generation rate.

An additional set of equations is needed to account for the time rate of change of the clusters:

$$\sum_{j=1}^{N}[u_{j+i}C_{j+i} - u_j C_j + \mu_{j-i}C_{j-i} - \mu_j C_j] = 0 \tag{2}$$

in the steady state and with no gradients,

$$\lambda C_t = \sum_{j=1}^{N}[\mu_j C_{He} C_j - u_j C_j] \; .$$

99

In the case where the He-He binding energy is large enough to produce a stable diatomic nucleation center, then:

$$\lambda C_t \cong 4\pi r\, D_{He}\, C_b^2 \tag{3}$$

where C_b = bubble density.

Thus, the diffusivity is proportional to C_b^{-2} in this case and measurements of bubble densities nucleated in tritides would then lead to a determination of the helium diffusivity. This function is plotted in Fig. 4 for $PdT_{0.6}$ and a trapping radius corresponding to the nearest neighbor distance. TEM observations in Zr [5], Pd [2] and V [6] tritides yield bubble number densities from $5\ 10^{17}$ to $5\ 10^{18}/cm^3$. These densities correspond to diffusivities of the order of 10^{-16} to $10^{-17}cm^2/s$. For a typical pre-exponential factor in the range of $10^{-2} - 10^{-3}$, this would mean that the activation energy for helium migration is of the order of 1 eV in these materials. As stated earlier, such a high value would be unexpected from theoretical estimates and the limited experimental data available indicate a lower value as well.

The lack of homogeneously nucleated bubbles in the solid solution phase shown in Fig. 2 gives another estimate for the helium diffusivity. Using the equation above with the vanadium α phase generation rate and assuming a foil thickness of about 50 nm, one obtains an upper bound of about 0.7 eV (for the same pre-exponential factor used earlier), somewhat lower than the hydride phase. It appears, therefore, that de-trapping of He atoms from small clusters (di- or tri-atomic clusters) plays a dominant role in determining homogeneously nucleated bubble density and that density measurements in tritides aged at room temperature cannot be used to deduce helium mobility. Kinetic calculations based on the full set of equations (1) and (2) would depend heavily on the cluster binding energies, which are not well known.

3.   Helium Density and Bubble Growth

Bubble growth in tritides results solely from the accretion of migrating He atoms, without external sources of vacancies. Thus, the pressure that the helium exerts on the metal tells us much about the growth mechanism. This pressure is in turn a function of the helium density within the bubbles. An excellent review article by Donnelly [17] summarized most of the experimental observations of helium densities in bubbles up to 1985. Considerable variation in the density was found between workers using different experimental techniques or even different interpretations of similar data. The range in reported helium densities was from about 1 He per atomic volume to as much as 3 He per atomic volume. The pressures corresponding to these different densities span a large range and, therefore, a growth process cannot be deduced from these data.

As mentioned earlier, helium is generated uniformly within a tritide phase and the total amount of helium increases with time at a rate proportional to the stoichiometry. This behavior makes tritides a good choice for observing macroscopic effects of helium accumulation. In particular, the rate of volumetric swelling can be determined as a function of helium concentration by monitoring its length over a period of time. The swelling is then simply given by $S = 3(\Delta L/L)$, where $\Delta L/L$ is the fractional length change. The change in volume is due to the difference between the partial molar volumes of tritium and helium. Since these volumes are well known for hydrogen isotopes in metal hydrides, measurements of swelling versus helium content give a direct indication of the partial molar volume of helium. Furthermore, since essentially all of the helium resides in bubbles, the volume gives the density of helium within the bubbles. This is, perhaps, one of the most direct methods for determining helium densities in bubbles.

Length change measurements have been made by Schober *et al.* in Ta tritide [11] and by Guthrie in Pd tritide [3] at room temperature. Together, these measurements cover a range of

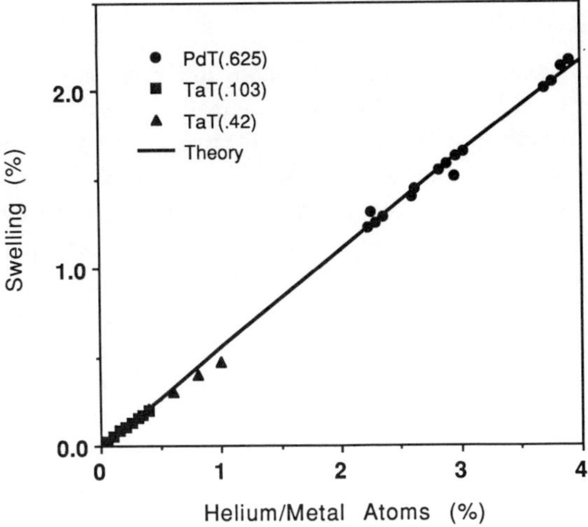

Fig. 5   A plot of swelling versus helium concentration. The points are experimental values and the solid curve is calculated by a loop punching model. References for the data are: Pd [3], Ta [11], Theory [18].

helium concentrations from 150 ppm up to 4% helium to metal ratio. Within this range, the swelling results indicate a constant helium volume of approximately 0.5 atomic volumes (2 He per atomic volume) for both tritides. This is shown in Fig. 5, where the swelling is plotted as a function of helium content. Also plotted is a theoretical prediction by Wolfer [18] based on a loop punching mechanism. It is seen that the agreement is excellent. Wolfer's calculations differ from earlier estimates by including a more detailed treatment of the energetics in the loop punching process. He finds that the pressure required to generate a prismatic loop and expand the bubble is simply equal to one-fifth of the shear modulus of the metal. Note that, in contrast to earlier loop punching models, this pressure is independent of the bubble radius. Thus, a constant swelling rate is predicted, in agreement with the tritide measurements.

Other experimental techniques have been employed to measure helium densities recently which verify the length change measurements. Schober and co–workers measured the density of a Nb tritide using a high precision buoyancy technique [9] and found, as in the length change experiments, a linear relationship between helium concentration and density. Their data indicates a helium volume of 0.46 atomic volumes, essentially the same as the previous experimental values and loop punching model estimate. An entirely different approach, using nuclear magnetic resonance, has recently been employed by Abell and Attalla [19] to determine ^{3}He densities in bubbles in palladium tritide. Here, the NMR relaxation time, T_1, for He-He interactions was measured as a function of temperature from room temperature down to 80 K. A large change in the relaxation time was observed at around 225 K which was interpreted to indicate a liquid-solid phase transition. Using this temperature and an equation of state, a helium density of 2 He per atomic volume was determined. Furthermore, a minimum in the T_1 value at about 125 K could be interpreted to yield a solid density in agreement with the phase transition estimate.

The previous discussion indicates that there is good experimental and theoretical agreement on bubble helium densities in tritides and that growth occurs by prismatic loop punching at room temperature for helium concentrations up to a few percent. Additional NMR measurements in palladium tritide at higher helium concentrations [20] show multiple component relaxation times below room temperature, indicating a distribution of helium densities rather than a unique value. The distribution ranges from the loop punching density to lower values.

4.   Concluding Remarks

It is hoped that the above discussion has shown the value of utilizing metal tritides for studying the fundamental behavior of helium in metals. In particular, from measurements in tritides the density of helium within bubbles now appears to be well established (within the limited helium concentration range studied) and is bounded at the upper end by the pressure required for loop punching. Since this pressure is essentially determined by the theoretical strength of the solid, it is unlikely that a helium density greater than that found in tritides would occur in other metals. Other helium injection techniques, such as ion implantation, produce vacancies along with the helium and these would be expected to reduce the pressure, and hence the density, required for bubble growth. Similarly, dislocation-bubble interactions could occur at higher helium concentrations, again resulting in an easier growth channel and lower helium density.

Information on helium mobility remains limited. The low temperature resistivity recovery in Lu tritide indicates a very low activation energy. However, verification by another technique and in different materials would be a valuable addition to our present knowledge. This would require experiments where samples are maintained at low temperatures, with the inherent difficulties of measuring specific physical properties at that temperature. The complexity of He atom interactions with He clusters and intrinsic defects make observations at room temperature difficult to interpret.

ACKNOWLEDGEMENTS

The author wishes to thank W. G. Wolfer and S. E. Guthrie for many useful discussions and for allowing the use of unpublished data. Fig. 2 was in collaboration with T. Schober, W. Jäger, and R. Lässer. This work was supported by the U. S. Department of Energy under contract number DE-AC04-76DP00789.

REFERENCES

1. W.D. Wilson, C.L.Bisson and M.I. Baskes, Phys. Rev. **B 24**, 5616 (1981).
2. G.L. Thomas and J.M. Mintz, J. Nucl. Matl. **116**, 336 (1983).
3. S.E. Guthrie, (1991), to be published.
4. T. Schober and K. Farrell, J. Nucl. Mater. **168**, 171 (1989).
5. T. Schober, H. Trinkaus and R. Lässer, J. Nucl. Mater. **141-143**, 453 (1986).
6. W. Jäger, R. Lässer, T. Schober and G.J. Thomas, Rad. Effects **78**, 165 (1983).
7. T. Schober, R. Lässer, W. Jäger and G.J. Thomas, J. Nucl. Mater. **122–123**, 571 (1984).
8. T. Schober, C. Dieker and H. Trinkaus, J. Appl. Physics (1988).
9. T. Schober, C. Dieker, R. Lässer and H. Trinkaus, Phys. Rev. **B 40**, 1277–1281 (1989).
10. R. Lässer, K. Bickmann, H. Trinkaus and H. Wenzel, Phys. Rev. **B 34**, 4364 (1986).
11. T. Schober, R. Lässer, J. Golczewski, C. Dieker and H. Trinkaus, Phys. Rev. **B 31**, 7109 (1985).
12. R. Lässer, K. Bickmann and H. Trinkaus, Phys. Rev. **B 40**, 3306 (1989).
13. P. Jung and R. Lässer, Phys. Rev. **B 37**, 2844 (1988).
14. G.J. Thomas, W.A. Swansiger and M.I. Baskes, J. Appl. Phys. **50**, 6942 (1979).
15. G.J. Thomas and R.J. Bastasz, Appl. Phys. **52**, 6426 (1981).
16. M.I. Baskes and W.D. Wilson, Phys. Rev. **B 27**, 2210 (1983).
17. S.E. Donnelly, Rad. Effects **90**, 1–47 (1985).
18. W.G. Wolfer, Phil. Mag. **A 58**, 285 (1988).
19. G.C. Abell and A. Attalla, Phys. Rev. Lett. **59**, 995 (1987).
20. G.C. Abell, this volume (1990).

ELUCIDATION OF FUNDAMENTAL PROPERTIES OF HELIUM IN METALS BY NUCLEAR MAGNETIC RESONANCE TECHNIQUES

G.C. Abell

EG&G Mound Applied Technologies
Miamisburg, Ohio 45342, USA

ABSTRACT

The nuclear magnetic resonance (NMR) properties of very high density ^{3}He in metals are discussed in the context of the corresponding properties in relatively high density bulk ^{3}He. In particular, the effects of ^{3}He diffusion on the contribution of the ^{3}He-^{3}He dipolar interaction to the lineshape and to the spin-lattice relaxation parameter (T_1) are described. It is shown that the temperature dependence of the lineshape and of T_1 are independent sources of information about helium density and also about helium diffusivity. Moreover, T_1 is shown to be a sensitive indicator of melting transitions in bulk ^{3}He. Palladium tritide is presented as a model system for NMR studies of ^{3}He in metals. Experimental NMR studies of this system reveal behaviour analogous to what has been observed for bulk helium. Evidence for a ^{3}He phase transition near 250 K is provided by the temperature dependence of T_1. Assuming this to be a melting transition, a density is obtained from the bulk helium EOS that is in good agreement with theory and with swelling measurements on related metal tritides. ^{3}He NMR measurements have also provided information about the density distribution, helium diffusivity, and mean bubble size in palladium tritide.

1. Introduction

Nuclear magnetic resonance (NMR) is, in principle, an ideal technique for studying ^{3}He in metals. If the helium concentration is sufficient to overcome the inherent sensitivity limitation of NMR ($\approx 10^{18}$-10^{19} spins) and if the host metal is non-magnetic, then NMR experiments can be expected to provide significant information about the properties of ^{3}He in that particular host. Because of the sensitivity limitation, materials containing ^{3}He via ion-implantation are difficult to study. This is illustrated by the work of Weaver et al [1] on ion-implanted Pd, in which a specially designed NMR probe cooled by liquid helium provided a limited amount of information. For this reason, most NMR studies of ^{3}He in metals have been performed on metal tritides, in which the helium is implanted with very little kinetic energy by triton decay ($\approx$12 year half-life). A homogeneous tritide will result in homogeneous deposition of ^{3}He. A typical metal tritide requires about five months of decay before a useful ^{3}He signal is obtained. Because of the radiolytic hazards associated with tritium, either special containment or sample preparation procedures will generally be required.

The ^{3}He NMR studies of Bowman [2] on the tritides of lithium, titanium and uranium illustrate the "tritium trick" approach. However, these materials are not necessarily representative of helium in metals; certainly not LiT, which is an ionic salt. The only one of these for which a temperature-dependent NMR study was performed was UT_3 [3] which, unfortunately, is ferromagnetic below about 200 K. Consequently, the magnetic properties of ^{3}He are obscured in this crucial temperature range. Nonetheless, this work did provide microscopic evidence that the ^{3}He was in bubbles and not on host lattice interstitial sites. The transmission

Fundamental Aspects of Inert Gases in Solids
Edited by S.E. Donnelly and J.H. Evans, Plenum Press, New York, 1991

electron microscopy (TEM) work of Thomas and Mintz [4] on palladium tritide ($PdT_{0.6}$) provided the first direct images of bubbles in a metal tritide. Since then, bubbles have been imaged in several other tritides as well [5].

Subsequent to the TEM work on $PdT_{0.6}$, NMR studies of ^{3}He in palladium tritide by Abell and Atalla [6] provided evidence for a melting transition near 250 K and also provided information about helium diffusivity in solid high-density helium. The density obtained from the observed melting temperature using the bulk helium equation-of-state (EOS) was consistent with theoretical predictions of Wolfer [7] and also with densities inferred from dilatometry measurements on the tritides of Nb and Ta [8]. After additional aging of the palladium tritide material, detailed NMR analysis revealed a range of melting temperatures, from which a distribution of densities could be inferred [9]. This work demonstrated the potency of the NMR technique for providing fundamental information about ^{3}He in metals.

The present article describes the observed NMR behaviour of ^{3}He in palladium tritide, which is used as a model system to illustrate the kind of information about ^{3}He in metals that can be obtained via NMR. The article begins with a description of the NMR behaviour of *bulk* ^{3}He as it relates to the corresponding behaviour of ^{3}He in metals. The underlying theory is briefly reviewed in order the explain the effects of motion on NMR relaxation and lineshape parameters, and to illustrate the different motional regimes. Against this backdrop, the NMR behaviour of ^{3}He in palladium tritide is then described in relation to such fundamental properties as density and self-diffusion. Hydrogen isotope effects on ^{3}He lineshape and T_1 parameters in palladium tritide, due to interaction of ^{3}He with hydrogen at the bubble surface, are also discussed.

2.   NMR Background

If surface effects can be ignored, the NMR properties of ^{3}He in highly pressurized nm-bubbles should be essentially those of bulk ^{3}He of comparable density. For bulk densities greater than about 0.056 moles/cm^3, the helium-helium exchange interaction [10] is unimportant and the bulk NMR behaviour is dominated by the dipolar interaction between the nuclear spins and by the modulation of this interaction due to diffusive motion. The NMR properties of bulk ^{3}He in this diffusive regime have been determined for a range of densities up to a maximum density of only about 0.060 moles/cm^3. (The density of ^{3}He in bubbles is $\approx$0.2 moles/cm^3 [6,8]). Nonetheless, the theoretical understanding of NMR characteristics for this relatively simple system provides the basis for prediction of certain properties at much higher densities than have been observed for bulk ^{3}He. The point of departure for understanding the NMR properties of bulk ^{3}He is the Hamiltonian for a system of interacting spin-1/2 nuclei in a uniform magnetic field [11]:

$$H = H_Z + H_D \tag{1}$$

The term

$$H_Z = -\gamma \hbar H I_Z \tag{2}$$

is the Zeeman interaction of the spin system with an external magnetic field of magnitude H directed along the z-axis. The quantity γ is the gyromagnetic ratio of a ^{3}He nucleus, while I_z represents the component of the total spin angular momentum along the external field direction and $\hbar$ is Planck's constant divided by 2π. The last term in eqn. (1) is the dipolar interaction, i.e., coupling of a spin to the magnetic dipole fields of neighbouring spins, for which the classical expression is:

$$H_D = \Sigma \, [\mu_i \cdot \mu_j/(r_{ij})^3 - 3(\mu_i \cdot r_{ij})(\mu_i \cdot r_{ij})/(r_{ij})^5]. \tag{3}$$

In this equation, the quantity μ_i is the magnetic moment of nucleus i, r_{ij} is the internuclear vector between nuclei i and j, and the sum is over all distinct pairs of nuclei. In most NMR studies, H_D is much smaller than H_Z and is accurately treated by perturbation theory.

It is clear from the form of eqn. (3) that motion of the nuclei will modulate the dipolar interaction. At low enough temperatures, the motion will be sufficiently slow that only the instantaneous value of H_D will be relevant. The effect of H_D in this case is to broaden the spectrum relative to that for free spins. This regime is called the rigid or static-lattice limit; linewidth analysis in this limit provides structural information. For fast isotropic motion, $\langle H_D \rangle = 0$ (the angular brackets here denote an appropriate time-average) in first order and to this extent, there is no broadening. Fluctuations of H_D represent second-order effects which provide a mechanism for spin-lattice (T_1) relaxation. This latter regime is the motionally narrowed limit; a study of T_1 in this limit provides dynamical information.

The way in which these different regimes fall out of the Hamiltonian (1) is best appreciated by expressing the dipolar interaction in terms of spin operators. H_D then consists of six quantum-mechanical operators for each pair of nuclei, having the form:

$$0_{ij} = R_{ij}I_{ij}. \tag{4}$$

These operators have matrix elements connecting nuclear Zeeman states (the eigenstates of H_Z) which differ in total magnetic spin quantum number m by either 0, 1, or 2. Each operator 0_{ij} is a product of a spin function I_{ij} (which is itself a product of a spin operator for nucleus i with one for nucleus j) and a spatial function R_{ij}, which depends on the length and orientation (relative to the external field H) of r_{ij}. For sufficiently slow motion, it is the instantaneous value of H_D that matters; in this static-lattice limit, only the $\Delta m=0$ matrix elements (i.e. the secular or first-order terms) of the 0_{ij} operators are important in the overall Hamiltonian [1]. If there were no dipolar interaction (i.e., $H_D = 0$), the—spectrum which includes all possible transitions with $\Delta m=1$ —would consist of a single line at the Larmor frequency, defined as:

$$\omega_0 = \gamma H \tag{5}$$

When $H_D \neq 0$ (i.e. the case of static dipolar coupling) the spectrum is broadened out about ω_0. The broadening is due to the multiplicity of values that the secular perturbation can assume as a consequence of its dependence on spin operators for each nucleus in the coupled array (the spin quantum number for each ^{3}He nucleus is $\pm 1/2$). Moreover, in a polycrystalline sample, the r_{ij} are distributed randomly in all directions, resulting in a smeared out spectrum even for a two-spin system.

Apart from special cases, there is no exact solution for the Hamiltonian of eqn. (1) in the static-lattice limit. Thus an exact solution of the resonance shape is not possible in this limit. Nonetheless, the method of moments introduced by Van Vleck [12] allows a determination of properties of the resonance line without explicit determination of the eigensolutions. Thus in the case of equivalent ^{3}He spins, the second moment of the resonance lineshape, which is defined very generally as:

$$M_2 = \int_{-\infty}^{\infty} (\omega_0 - \omega_0)^2 \, g(\omega) \, d\omega \tag{6}$$

($\omega g(\omega)$ is the lineshape function) is determined precisely by the method of moments to be:

$$M_2 = (9/16)\, \gamma^4\, \hbar^2 \Sigma_k (1 - 3\cos^2\theta_{jk})^2 / (r_{jk})^6. \tag{7}$$

In this expression, M_2 is in frequency units and θ_{jk} is the angle between $\underline{r}_{jk}$ and the external field.

The lattice sum in eqn. (7) has been determined for various infinite lattice structures [13], and can be expressed either in units of $1/a^6$ (a being the lattice parameter) or in units of ρ^2, where ρ is the ^{3}He density. When expressed in terms of ρ, the lattice sum is nearly independent of structure. The remaining quantities in eqn. (7) are precisely known, and for ^{3}He in a close-packed structure

$$M_2 = 544.5\rho^2 \tag{8}$$

with M_2 in units of Gauss^2 (the conversion between Gauss and frequency units is given by eqn. (5)) and ρ is in mole/cm^3. The experimental determination of M_2 is relatively straightforward, once it is demonstrated that the static-lattice limit has been achieved. A good example of the use of M_2 to obtain structural information is the proton NMR work of Bowman, et al [14] on zirconium hydride.

The effects of motion due to self-diffusion of the ^{3}He nuclei on the Hamiltonian givenby eqn. (1) will now be described, following the work of Schlicter [11]. The motion is characterized by a correlation time τ which is essentially the time between diffusive hops. It is appropriate here to define more precisely the criterion for distinguishing the static-lattice (large τ) and motionally-narrowed (small τ) regimes. Clearly there is an intermediate range for τ corresponding to a crossover between these two motional regimes. This crossover regime, which is difficult to characterize quantitatively, is defined by the condition

$$\omega_d\tau = 1 \tag{9}$$

where $\omega_d = \sqrt{M_2}$ is a frequency characterizing the strength of the dipolar interaction. The changeover from the broad static-lattice line to a relatively narrow line generally occurs over a small range of values for τ. For thermally activated diffusion

$$\tau = \tau_0\, \exp(W/T), \tag{10}$$

where $W = E_a/k$ is the activation energy in units of degrees Kelvin and τ_0 is the correlation time in the high temperature limit. Substitution of eqn. (10) into eqn. (9) gives the temperature, T_N, at which the narrowing condition expressed by eqn. (9) is satisfied:

$$T_N = -W/\ln(\omega_d\,\tau_0). \tag{11}$$

In the case of fast *isotropic* motion satisfying the condition $\tau \ll (w_d)^{-1}$, the secular terms (terms in H_D which produce first order effects) get averaged to zero. Thus, the relevant zero-th order description is one of non-interacting spins, *viz.* the eigensolutions of the Zeeman Hamiltonian (eqn. (2)). Components of the fluctuating dipolar field (the dipolar field is defined by eqn. (3)) which are *transverse* to the static field H, can cause changes in the ensemble-averaged nuclear spin magnetization parallel to the static field, i.e., changes in $<M_z>$. A good analogy for this effect is the manipulation of $<M_z>$ by an applied rf field transverse to the static field (this is the "experimental handle" in pulsed NMR experiments). But this is a resonance phenomenon—the applied rf must provide a magnetization vector which rotates in the transverse plane *at the Larmor frequency* ω_0, defined by eqn. (5), in order to give a finite probability for transitions between Zeeman states. These latter transitions are ultimately behind any changes in $<M_z>$. In the same way, transverse components of the fluctuating dipolar field which are precessing at ω_0 provide a mechanism for changing $<M_z>$. This relaxation process involves a transfer of energy between the spin system and the phonon modes associated with self diffusion; it is thus a spin lattice or T_1 process. The relaxation will be most efficient when the amplitude at ω_0 of the Fourier spectrum of the fluctuating dipolar field is optimum. This occurs when $\omega_0 \tau \approx 1$ and is manifested by a minimum value for T_1 as a function of τ (i.e. as a function of temperature—see eqn. (10)). Quantitative analysis of this relaxation mechanism gives the result [10]:

$$1/T_1 = (2M_2/3)[\tau/(1 + \tau^2 \omega_0^2) + 4\tau/(1 + 4\tau^2 \omega_0^2)]. \tag{12}$$

The magnitude of T_1 at the minimum is determined as an extremum of this equation to be:

$$(T_1)_{min} = 1.05\, \omega_0/M_2. \tag{13}$$

If there are no additional relaxation mechanisms in play, it follows from equations (8) and (13) that determination of $(T_1)_{min}$ provides a measure of the density. Determination of T_1 as a function of temperature provides dynamical information via the parameter τ in eqn. (12). It is clear that a discontinuous change in τ will be manifested as a discontinuity in T_1. A ^{3}He melting transition involves abrupt changes in the concentration and mobility of vacancies, which determine ^{3}He self-diffusion. For bulk ^{3}He, T_1 increases by a factor of $\approx 10^3$ in going from the solid to the liquid phase [15]. It follows that a study of T_1 as a function of the appropriate intensive variable (P or T in the case of ^{3}He) can be used to monitor the melting transition. The work of Holcomb and Norberg [16] illustrates the study of the temperature-dependence of T_1 as a means of obtaining information about self-diffusion in alkali metals, which just as for solid ^{3}He is via a vacancy mechanism. A similar study of self-diffusion in bulk solid ^{3}He [17] shows that the diffusion activation energy W is a monotonically increasing function of density.

3.   NMR Properties of ^{3}He in Palladium Tritide

3.1.   *Effects due to ^{3}He-^{3}He dipolar interaction*

The NMR behaviour of bulk solid ^{3}He as described above provides a background for the following discussion of the observed NMR behaviour of ^{3}He in aged palladium tritide. Figure 1 shows the ^{3}He spin lattice relaxation parameter (T_1), represented by the filled circles,

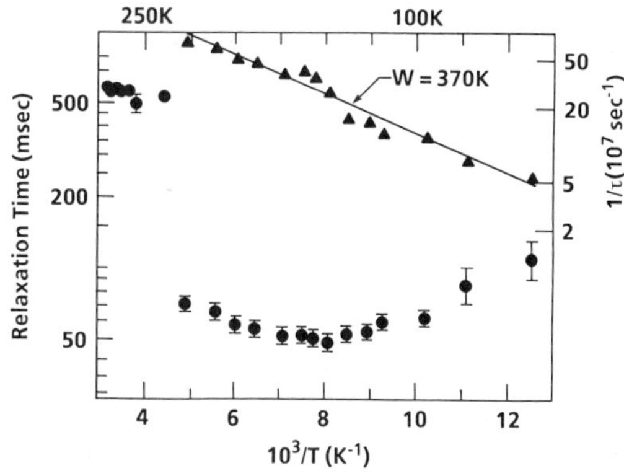

Fig. 1   Temperature dependence in 1 year old PdT$_{0.6}$ of ^{3}He T$_1$ relaxation time (filled circles, left-hand scale) at 25 MHz; and of ^{3}He jump frequency τ^{-1} (filled triangles, right-hand scale) obtained from the T$_1$ data via eqn. (12).

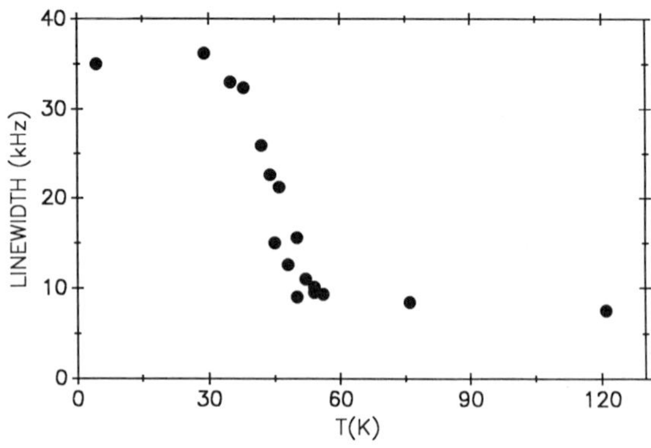

Fig. 2   Spectral line width of ^{3}He in 8 years old PdT$_x$ as a function of temperature ($\omega_0 2\pi = 142.8$ MHz).

as a function of temperature in 1 year old PdT$_{0.6}$ (He/Pd atomic ratio ≈ 0.030). The major feature in Figure 1 is the abrupt change in T$_1$ near 250 K. Assuming that this is due to a solid/fluid melting transition and assuming that the bulk helium equation-of-state (EOS) applies, the helium density (ρ) is found to be 0.20 mole/cm^3. Another important feature in Figure 1 is the existence of a minimum for T$_1$, which could be a consequence of activated diffusive motion modulating the ^{3}He dipolar interaction. If so, then equation (12) allows τ (essentially, the time between diffusive hops) to be determined from the T$_1$ data. The results of this latter analysis are represented in Figure 1 by the solid triangles, which correspond to τ^{-1}, shown on the right-hand scale. From eqn. (10), the slope of the semilog plot of τ^{-1} vs. 1/T gives an activation energy of 0.032 eV (i.e., W = 370 K) for the ^{3}He self-diffusion. This result is consistent with an extrapolation from much lower densities of the density dependence of W determined for bulk solid ^{3}He [6], [17].

A significant test of the hypothesis that the relaxation mechanism is diffusive modulation of the dipolar interaction is provided by eqn. (11), which allows a prediction of the characteristic temperature for motional narrowing, T$_N$. Using the Arhennius parameters obtained from the fit of the solid triangles in Figure 1 to the inverse of eqn. (10), and a value for M$_2$ obtained from eqn. (8) with $\rho = 0.20$ moles/cm^3 (remembering that $\omega_d \equiv \sqrt{M_2}$), eqn. (11) gives the result T$_N = 34 \pm 5$ K. Figure 2 shows the measured ^{3}He line width at $\omega_0/2\pi = 143$ MHz [18] as a function of temperature in palladium containing He/Pd≈ 0.3. (This material was aged ≈ 8 years as a tritide (PdT$_{0.6}$), and then detritided by ambient temperature evacuation [19]. Abrupt narrowing is observed to occur in the range 30-45 K, which agrees with the predicted result. Analysis of the static lattice line shape at 4 K provides a value for M$_2$ which, when substituted in eqn. (8), gives the result $\rho = 0.18$ moles/cm^3. (The correction for finite bubble size is presumed to be small). The accuracy of this result is probably no better than about ± 0.03, due to a fairly large correction for magnetic susceptibility effects in palladium metal. (The broadening due to susceptibility effects scales with ω_0, whereas that due to the dipolar interaction is independent of ω_0). Melting temperature information obtained from a study of T$_1$ vs. temperature for the same sample [18] indicates that the mean ^{3}He density is about 0.15 moles/cm^3, which agrees reasonably well with the determination from the line shape analysis. (Note that while the prediction for T$_N$ given earlier was based on $\rho = 0.20$ moles/cm^3, W is not a strong function of ρ so that the predicted T$_N$ is reduced only to about 30 K using $\rho = 0.15$ moles/cm^3). These results constitute very strong evidence that for T < T$_m$, the NMR behaviour of ^{3}He in palladium tritide is essentially that expected for bulk ^{3}He at a comparable density.

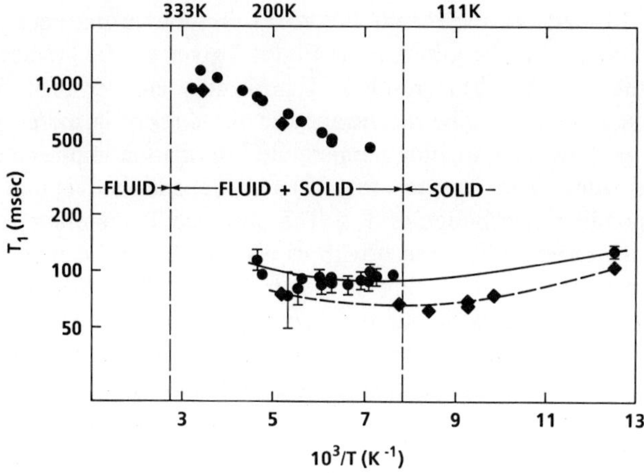

Fig. 3   Temperature dependence of ^{3}He T_1 in 2 years old PdT$_{0.6}$ at 45.7 MHz (filled circles) and 25 MHz (filled diamonds). The curves are from an eyeball fit to the solid phase T_1 data.

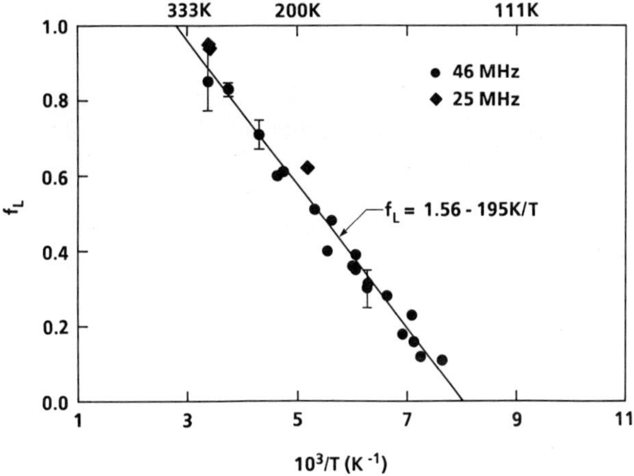

Fig. 4   Fluid phase fraction, f_L, of ^{3}He in 2 years old PdT$_{0.6}$ as a function of temperature.

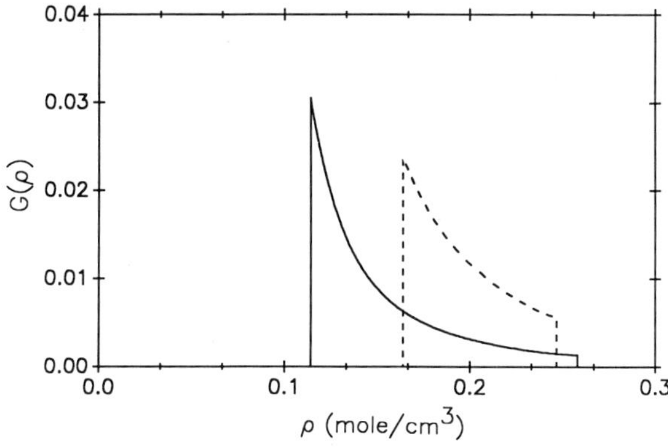

Fig. 5   The dashed curve shows the density distribution for ^{3}He in 2 years old PdT$_x$ derived from the fit to the data shown in Figure 4, using the bulk helium EOS. The solid curve shows the density distribution obtained for ^{3}He in 8 years old PdT$_x$ [18].

There is one difficulty in the results described thus far: using eqns. (8) and (13), the observed magnitude of T_1 at the minimum in Figure 1 gives $\rho \approx 0.13$ moles/cm^3 for the one-year-old palladium sample. This result is significantly smaller than the value of 0.20 moles/cm^3 obtained from T_m. The discrepancy in the density is more apparent than real because, as will be shown, the melting temperature information implies a moderately broad distribution of densities. Thus by eqns. (8), (10) and (12) and the fact that W depends on ρ, there is a corresponding distribution of T_1s. The observed T_1 minimum, obtained from the superposition of the different T_1s—each with its own minimum—can be appreciably larger than the result given by eqn. (13) using the mean density. The difficulty is that eqns. (8) through (13) are based on the assumption of *homogeneity*; thus they require modification to take the density distribution into account. For eqn. (8), this amounts to using $\langle\rho^2\rangle$ in place of ρ^2, but for the other equations the modifications are not so straightforward. Nonetheless, so long as the distribution is not too broad, the behaviour is qualitatively the same as for a homogeneous system.

The evidence for a distribution of densities in the one year old palladium sample is that between 200 and 280 K, T_1 is strongly non-exponential, whereas outside this range it is ostensibly exponential [9]. Moreover, the non-exponential T_1s are well described by a sum of two exponentials having the characteristics of the two phases outside this range. Figure 3 shows T_1 data taken for the same sample shown in Figure 1 after aging for two years (He/Pd $\approx$0.060), and includes the results from a two exponential fit in the mixed phase region. The observed dependence of T_1 on ω_0 in the solid phase is consistent with eqn. (13). From the T_1 analysis, one also gets the relative fraction of ^{3}He in each phase as a function of temperature. Figure 4 shows the fluid phase fraction vs. temperature for the 2-year-old material. Figure 5 compares a distribution function derived from the information in Figure 4 [9] using the bulk EOS with one obtained in a similar manner for the Pd material with He/Pd $\approx$0.3 (the 8-year-old Pd). Because of the approximations involved in the analysis, the distributions shown in Figure 5 are fairly crude representations of the actual distributions. In fact, a bimodal distribution would not be inconsistent with the observed ^{3}He melting behaviour in the two different samples. The existence of a distribution of densities is confirmed for the 8-year-old sample by the observation [18] of coexisting broad and narrow ^{3}He line shape components in the temperature range (30–45 K) where motional narrowing occurs (see Figure 2). This observation is understood by recognizing that eqn. (11) implies a distribution for T_N corresponding to that for ρ.

The evolution of the density distribution with age (i.e. He/Pd, atomic ratio) in palladium tritide is represented in Fig. 6, which shows the range of melting temperatures as a function of

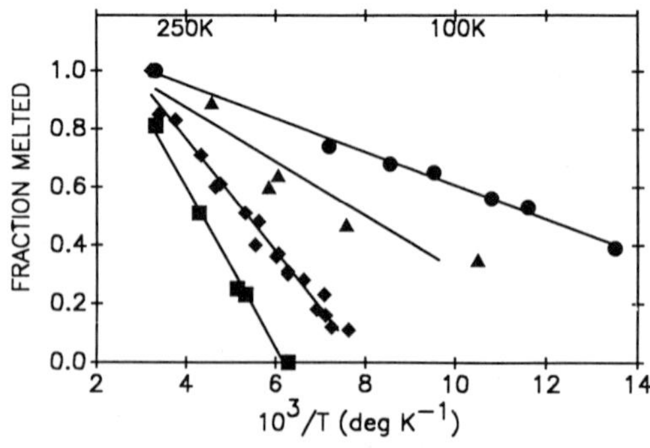

Fig. 6   Age dependence of ^{3}He fluid phase fraction in palladium tritide; squares, 0.55 years old; diamonds, 2 years old; triangles, 4 years old; circles, 8 years old. The solid curves are linear fits to the corresponding data.

age. It is clear from Figs. 5 and 6 that the distribution broadens with age and shifts to lower densities. The mean density obtained from the data in Fig. 6 systematically decreases with age, from a value of 0.21 moles/cm^3 when He/Pd = 0.017 (0.55-year-old sample), to a value of 0.15 moles/cm^3 when He/Pd = 0.3 (8-year-old sample).

3.2. *Effects due to ^{3}He interaction with unlike spins*

Calorimetry measurements reveal [19] that when aged palladium tritide is evacuated at ambient temperature, an amount of tritium remains absorbed in the material considerably in excess of the expected amount based on the known solubility of tritium in unaged palladium [20]. Figure 7 is a plot of the excess absorbed tritium as a function of He/Pd atomic ratio; it reveals that the concentration of excess tritium ($[T]_{xs}$ is defined as the T/Pd atomic ratio in excess of the equilibrium solubility ratio in unaged Pd) scales approximately with the 2/3 power of the ^{3}He concentration. An exact 2/3 power law scaling is of course, not expected given the existence of a distribution of densities (and presumably of bubble sizes) and given that the mean density decreases with the ^{3}He concentration. This result suggests that $[T]_{xs}$ is associated with the bubble surface. Assuming trapping at the bubble/metal interface, $[T]_{xs}$ for the sample with He/Pd = 0.0050 corresponds to a "surface coverage" of the order of one monolayer. This estimate uses the TEM determination of bubble size in a comparably aged palladium tritide [4].

If a large fraction of the excess tritium is in close proximity to the bubbles, then the tritium nuclei can influence the ^{3}He NMR behaviour through the ^{3}H-^{3}He dipolar interaction (given also by eqn. (3)). An important factor in this regard is that ^{3}H has the largest magnetic moment of any nucleus. But how can one determine whether or not the ^{3}He NMR is influenced by the ^{3}H spins? Exchange of hydrogen isotopes in palladium at near ambient temperatures is straightforward [21] and, because $\gamma(^2H)/\gamma(^3H) \approx 0.2$, replacement of tritium by deuterium will greatly diminish the hydrogen-helium dipolar coupling.

Figure 8 compares the ^{3}He line shape at 300 K (8a) and the temperature dependence of the ^{3}He T_1 (8b) before and after replacement of tritium by deuterium in PdT$_{0.07}$ ^{3}He$_{0.3}$ (8-year-old palladium tritide sample). These results provide strong microscopic evidence that a significant fraction of the tritium is in close proximity to the helium bubbles. The observation at 300 K of significant line shape effects due to ^{3}H-^{3}He dipolar coupling is not inconsistent with rapid ^{3}He motion, given the highly anisotropic nature of the motion.

An analysis, using a continuum representation, of the problem of ^{3}He spins interacting with ^{3}H spins trapped at the bubble surface [22] predicts that if both spin species are highly mobile, the secular terms from the dipolar interaction average to zero. Thus, in this limit, a hydrogen isotope effect on the ^{3}He line shape is not expected. However, in the limit of highly mobile ^{3}He spins and static ^{3}H spins (relative to an appropriate time-scale analogous to that given in eqn. (9)), the analysis predicts a residual broadening with a second moment given by:

$$<(M_2)_{IS}> = (8\pi^2/15)\,\hbar\,\gamma_I\,\gamma_S)^2\,(N_S/N_I)\,\rho_I/r^3, \tag{14}$$

where I, S refer to ^{3}He, ^{3}H spins respectively and r is the radius of the shell of trapped tritium spins (presumed to be trapped at the bubble surface). The angular brackets on the left-hand side of eqn. (14) denote a time averaging over the motion of the I spins. Using, in eqn. (14), the values of $N_S/N_I = 0.25$ (from Figure 7), $\rho_I = 10^{23}$ spins/cm^3 (^{3}He atomic density), and $<(M_2)_{IS}> = 6 \pm 4\,10^7$ sec^{-2} obtained from an analysis of the line shape information in Figure 8a, gives the result r = 1.0 $\pm$ 0.2 nm for the 8-year-old material. This value of r seems to be on the low side; it may be that the ^{3}He mobility is not large enough to give complete averaging. This question can presumably be resolved by a temperature dependent study of the line shape effect. Given that $\rho_I/N_I \propto r^{-3}$ and assuming that $N_S \propto r^2$ (see Fig. 7), eqn. (14) predicts that $<(M_2)_{IS}>$ scales with $1/r^4$. The residual dipolar broadening of the ^{3}He lineshape by surface

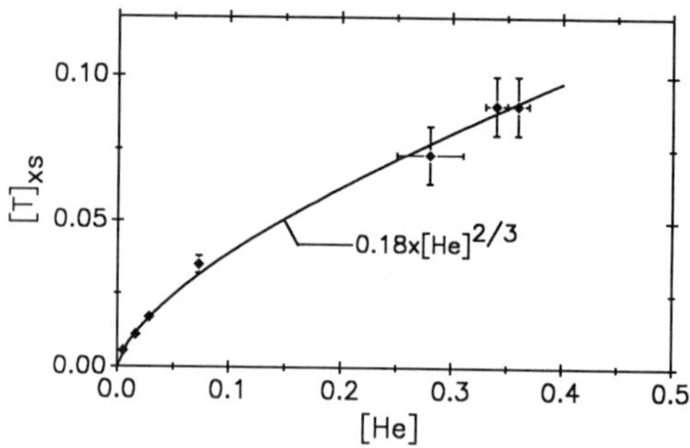

Fig. 7   Excess absorbed tritium (see text for definition) as a function of ^{3}He concentration. The curve is from a least-squares fit, assuming a 2/3 power law relationship.

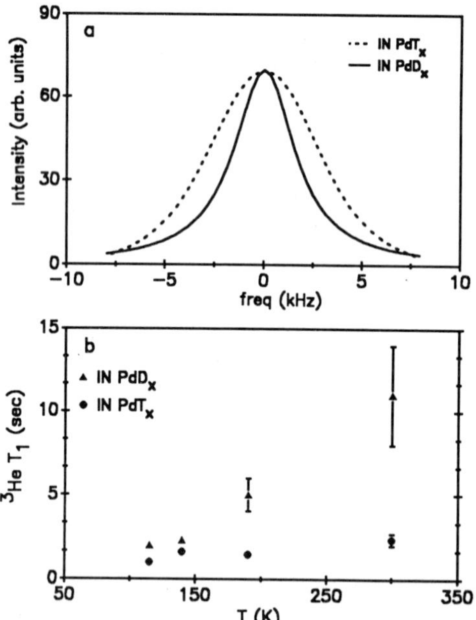

Fig. 8 ^{3}He line shape at 300 K (a) and spin lattice relaxation time (b) before and after replacement of ^{3}H by ^{2}H in PdT$_{0.07}$ ^{3}He$_{0.3}$.

trapped hydrogen should therefore provide a fairly sensitive probe of mean bubble size.

The continuum analysis also provides an estimate, similar to eqn. (13), for the optimum spin lattice relaxation available from the IS-dipolar interaction:

$$(T_1)_{IS} > \omega_0 / (M_2)_{IS}. \tag{15}$$

In this expression, $(M_2)_{IS}$ is the contribution of the helium-hydrogen dipolar interaction to the second moment of the helium lineshape in the limit when both spin species are static. It is given by [22]:

$$(M_2)_{IS} \approx (\pi / 30) \, (\hbar \, \gamma_I \, \gamma_S)^2 (N_S/N_I)\rho_I / d^3, \tag{16}$$

where d is the separation between the outermost ^{3}He layer and the innermost shell of trapped

112

tritium spins. The observed behaviour of $(T_1)_{IS}$, deduced from Figure 8b using

$$1 / T_1(^3H) \approx 1 / T_1(^2H) + 1 / (T_1)_{IS}, \qquad (17)$$

is consistent with d < 0.25 nm, which supports the hypothesis that the tritium is trapped at the bubble/metal interface.

The above results indicate that experimental determination of the effects of either ^3H or ^1H spins (the magnetic properties of ^1H are very similar to those of ^3H), on ^3He line shape and relaxation parameters can provide information about bubble geometry and about dynamics of hydrogen species trapped at the bubble surface. However additional experimental work is needed to establish these results.

4.   Conclusions

This review has attempted to demonstrate the utility of ^3He NMR for providing fundamental information about helium in metals. Information about helium density and mobility has been obtained from measurements of ^3He line shape and spin lattice relaxation time parameters. Bubble size information and information related to hydrogen trapped at the bubble/metal interface have been obtained by studying hydrogen isotope effects on the ^3He NMR behaviour. While fundamental characterization of helium in metals via ^3He NMR has for the most part been limited to the palladium tritide system, the technique should be useful for other systems as well. The tritides of group IV and V metals in particular should be viable candidates. Because of the inherent sensitivity limitation of NMR, study of ion implanted materials will be more difficult. Application to ferromagnetic materials will be limited to temperatures above the Curie transition. There is probably much that can be gained by extending the technique. In particular, implementation of magic angle sample spinning and double resonance techniques may allow more detailed characterization of helium in favourable systems such as palladium tritide.

ACKNOWLEDGEMENTS

The author acknowledges the contributions of D. West, D. Kirk, L. Matson, W. Tadlock, W. Rodenburg, R. Yauger, R. Baker and A. Attalla of EG&G Mound to various aspects of the experimental effort. EG & G Mound Applied Technologies is operated by EG&G for the US Dept. of Energy under Contract No. DE-AC04-88DP43495.

REFERENCES

1. H.T. Weaver and W.Beezhold, Appl.Phys.Lett.24, 522 (1974).

2. R.C. Bowman Jr., Nature **271**, 531 (1978).

3. R.C. Bowman Jr., and A. Attalla, Phys. Rev. **B16**, 1828 (1977).

4. G.J. Thomas and J. M. Mintz, J. Nucl. Mater. **116**, 336 (1983).

5. T. Schober and R. Lasser, J. Nucl. Mat. **120**, 137 (1984); see also R. Lasser in *Tritium and Helium-3 in Metals*, Springer-Verlag, Berlin, (1989) ch. 6.

6. G.C. Abell and A. Attalla, Phys. Rev. Lett. **59**, 995 (1987).

7. W.G. Wolfer, Phil. Mag., **A58**, 285 (1989).

8. T. Schober, J. Golczewski, R. Lasser, C. Dieker and H. Trinkaus, Z Phys. Chem., **147**, 161 (1986).

9. G.C. Abell and A. Attalla, Fusion Technol. **14**, 643 (1988).

10. A. Abragam and M. Goldman *Nuclear Magnetism: Order and Disorder*, Clarendon, Oxford (1982), ch. 3.

11. C.P. Slichter, *Principles of Magnetic Resonance*, Springer-Verlag, Berlin (1978), ch. 3, ch. 5.

12. J.H. Van Vleck, Phys Rev **74**, 1168 (1948).

13. H.S. Gutowsky and B.R. McGarvey, J Chem Phys **20**, 1472 (1952).

14. R C. Bowman Jr, E.L. Venturini and W-K. Rhim, Phys Rev **B26**, 2652 (1982).

15. H A. Reich, Phys Rev **129**, 630 (1963); R.L. Garwin and A. Landesman, Phys. Rev. **133**, A1503 (1964).

16. D.F. Holcomb and R.E. Norberg, Phys. Rev. **98**, 1074 (1955).

17. N. Sullivan, G. Deville and A. Landesman, Phys. Rev. **B11**, 1858 (1975).

18. G.C. Abell and D.F. Cowgill, submitted for publication.

19. R.C. Bowman Jr, G. Bambakidis, G.C. Abell, A. Attalla and B.D. Craft, Phys Rev **B37**, 9447 (1988).

20. R. Lasser, J. Less-Common Met. **131**, 263 (1987); R. Lasser, Phys. Rev. **B29**, 4765 (1984).

21. W.M. Rutherford, in *Hydrogen Storage Materials*, edited by R.G. Barnes, Mater Sci Forum Series, Vol. 31 (Trans Tech Publications, Aedermannsdorf, Switzerland, 1988) p. 19.

22. G.C. Abell and G. Bambakidis, submitted for publication.

INERT GAS BUBBLES IN METALS FOLLOWING ROOM-TEMPERATURE IMPLANTATION: ELECTRON MICROSCOPY

INERT GAS BUBBLES IN METALS: A REVIEW

C. Templier

Laboratoire de Métallurgie-Physique, 40 Avenue du Recteur Pineau
UFR Sciences, 86022 Poitiers Cédex France

ABSTRACT

Heavy rare gas atoms implanted into metals precipitate and form highly pressurised bubbles, in solid or fluid form, depending on implantation conditions (temperature, fluence). Different analysis techniques like TEM, HRTEM, X–ray diffraction, EELS, XANES, etc. have been used in order to study solid bubbles and their epitaxial relationship with the host matrix. It has been found that solid bubbles have fcc and hcp structures in fcc and hcp matrices respectively. For all matrices a general relationship clearly appears at the bubble matrix interface, rare gas solid close–packed planes are parallel to the close–packed planes of the matrix. During further annealing of samples containing solid bubbles, the solid/fluid phase transformation occurs under approximately the same conditions (temperature, pressure) as in bulk solid rare gas.

1.   Short Historical Overview and Introductory Remarks

The behaviour of rare gases in materials have been investigated from approximately the middle of this century, in connection with the development of nuclear technology. Helium, which will be produced by large amounts in fusion reactors, has been the most studied and the general features of its evolution are now well-known. Because it is almost insoluble in metals, helium precipitates in a high concentration of small bubbles and then affects the mechanical properties of materials (see ref [1] for a brief review). Heavier rare gases are also of interest in this research field since they may be residues of nuclear reactions and some investigations have been performed, especially concerning krypton and xenon (100 fissions of Pu^{239} give respectively on an average 2.5 and 22.8 Kr and Xe atoms [2]). A combined implantation and sputtering technique was developed at Harwell in order to store the radioactive Kr^{85} in metallic matrices [3].

For all that, interest in the interactions of rare gases with materials not only results from the nuclear research field. Rare gas ions are mostly used in several surface treatment techniques like sputter cleaning, ion etching or ion beam mixing. With very few exceptions [4], users paid little attention to the behaviour of the introduced rare gas atoms as long as these later did not change dramatically the physical properties of their materials.

Since they are the simplest atomic solids, the heavier rare gas solids (RGS) are of great interest in testing the predictions of solid state theories and RGS properties have been widely investigated [5]. The pressure induced insulator-metal transition of solids is also a subject of absorbing interest and many experiments in diamond-anvil cells have been performed, especially with xenon which is expected to have a metallic state around 130 GPa [6].

Fundamental Aspects of Inert Gases in Solids
Edited by S.E. Donnelly and J.H. Evans, Plenum Press, New York, 1991

In 1984, the discovery of rare gas bubbles in the solid phase has given an impetus to investigations on heavier rare gas precipitation. By using two different techniques, a German research team found that after implantation in aluminium, argon and xenon (but not neon) are gathered into solid bubbles with a face centered cubic structure [7]. At the same time, identical results for xenon in aluminium were independently found by a French team [8, 9]. Diffraction patterns obtained by Transmission Electron Microscopy (TEM) showed two different sets of reflections in all directions of reciprocal lattice, one from the host metal and the other from rare gas bubbles in the solid phase. These results were consistent with an epitaxial alignment between the solid rare gas and the matrix.

A little later, the presence of solid krypton bubbles in copper, nickel and gold was found by Evans and Mazey [10]. The packing density of Kr inside the solid bubbles ($n_{Kr} \approx 2.85 \ 10^{22}$ at.cm^{-3}), deduced from the lattice parameter measurement, was not unlike those obtained from earlier experiments [3] by using macroscopic density measurements ($n_{Kr} \approx 3.2 \ 10^{22}$ at.cm^{-3}). This previous result is well over the threshold packing density for solid krypton at room temperature ($n_{Kr} \approx 2.32 \ 10^{22}$ at.cm^{-3}) and it is clear that, though authors did not conclude at the time to the presence of solid Kr, it was the first step on the way of RGS bubble discovery.

Compared to diamond anvil experiments, implantation is a simpler way to obtain RGS and it might facilitate their study. This was clearly demonstrated for the first time by Donnelly and Rossouw in 1985. Until then, due to the small and fragile environmental cells needed for bulk RGS studies, TEM experiments were not easy to perform [11] and high resolution microscopy had never been applied until Donnelly and Rossouw achieved their lattice imaging study of both Xe and Ar solid bubbles in aluminium [12, 13].

From 1984 to 1990, many investigations on RGS bubbles in materials (mostly metallic matrices) were performed using different techniques such as Positron Annihilation Technique (PAT), Glancing X-Ray Diffraction, Rutherford Back Scattering (RBS), Extended X-Ray Absorption Fine Structure (EXAFS), Mössbauer Spectroscopy and Soft X-Ray Emission. But since bubbles are easily detectable by TEM, it remains the most useful technique and it gives unambiguous results which are often needed for the interpretation of data from these other techniques. However, these techniques may give additional information, particularly when bubbles are undetectable by TEM.

Though some features remain unclear, many experimental results and some tentative modelling are now available. Thus it is now possible to give some general characteristics about rare gas precipitation in materials; this is the main aim of the present review.

A sample which contains rare gas bubbles may be simply characterized by C_g, the average concentration of rare gas and C_b the bubble density. Within a sample, each bubble is then defined by r_b the bubble radius (if a nearly spherical shape is assumed) and n_b the rare gas packing density. It is quite clear that all these parameters depend on implantation conditions (fluence, temperature and beam energy) but also on the implanted rare gas and the matrix nature. Thus, one must be very careful about experimental conditions when comparing experimental results. In addition we have to bear in mind that in most samples, bubble radius values are spread over a sometimes large range, a fact which consequently leads to a similar spreading of rare gas packing density. This bubble size distribution has to be taken into account for a straightforward interpretation of experimental results since all the used techniques, apart TEM imaging, only give average values over a large number of bubbles.

In section 2 the experimental results from as-implanted samples are compiled, and the influence of the different parameters such as implantation conditions will be described. The results are briefly discussed and compared to existing models in section 3 and for more information about these models, the reader is referred to original articles or to papers which appear elsewherein this volume. Section 4 deals with annealing and cooling results after implantation and the solid-fluid phase transformation will be discussed.

2.   General Results From As-Implanted Samples

2.1.   The structure of RGS bubbles

If room temperature is much lower than 0.5 T_M (T_M is the matrix melting temperature) and the fluence is not too high (lower or around 10^{16} ions.cm^{-2} but depending on T_M), Ar, Kr and Xe atoms implanted at RT precipitate in the form of solid bubbles. For a 3-5 at% rare gas concentration, the typical bubble mean radius is in the range 1-2 nm and the bubble concentration is around 5 10^{18} cm^{-3}.

The bubble structure is easily deduced from either TEM or X-Ray diffraction experiments by checking RGS reflections or RGS peaks.

In all fcc (Al, Cu, Au, Ag, etc.) and hcp metallic matrices (Zn, Zr and Ti), RGS bubbles have the same structure as the host metal [10, 14, 15]. It is quite different in bcc matrices (Mo and Fe) [15, 16] where the solid rare gas appears to probably have an fcc structure. Although annealings up to about 700°C are needed to obtain the characteristic additional reflections, xenon bubbles in silicon, a diamond structure material, exhibit also fcc crystallographic structure [17, 18].

Despite large lattice mismatches, between 1.2 and 1.6 (see Table 1), there always exists an epitaxial relationship between the solid rare gas and the host material, whatever the former and the later. The general relationship is the following: one of the RGS {111} close-packed planes is parallel to one of the matrix close-packed planes, i.e. respectively {111}, (0002) basal plane and {110} for fcc, hcp and bcc matrices. This general experimental result agrees well with a model elaborated by Finnis [19] which shows that, due to the high pressure inside bubbles, a RGS close-packed plane parallel to the metallic surface is energetically favourable compared to any other arrangement, whatever the lattice structure of the host.

In fcc matrices, any RGS (hkl) plane must be parallel to the same index matrix plane. At first glance, it seems that is really the case and that consequently the bubble structure is perfectly fcc. However, a thorough examination of several diffraction patterns obtained from a number of various Xe:Al samples shows that the lattice parameter, deduced from a simple ratio method by using xenon and Al reflections, is slightly dependent on the chosen reflection and the lower the xenon fluence, the higher the difference [20]. A tentative explanation was made by assuming the existence of four kinds of precipitates (variants), each of them associated with one of the <111> directions and having a rhombohedron structure resulting from a light distortion along the same <111> direction. However, due to the small bubble size and also to the scatter in bubble packing density, TEM results were not accurate enough for a confirmation of this hypothesis.

The formation of hcp RGS in hcp matrices is not really surprising. Although bulk RGS (except He) have an fcc structure and there is no transition to other structure for all temperatures and for not too high pressure (above 75 GPa, xenon might have an hcp structure [21]), the energy difference between the fcc and the hcp form is very weak, with $(G_F - G_H)/G_F \approx 10^{-4}$ [22] where G_F and G_H are the Gibbs free energies for both structures. Probably due to the interactions between the matrix atoms and the RG atoms, less energy is needed to reproduce the matrix structure compared to the fcc structure since it provides larger areas of matching planes [14]. However, we have to notice that whatever the host metal, for both hcp Xe and hcp Kr, the ratio c/a ($\approx 1.60 \pm .03$) is very close to $(8/3)^{1/2}$ (see Table 1), the ratio for an ideal hcp structure. Thus there is every sign that only the general relationship previously mentioned above is observed.

2.2.   The bubble pressure

The pressure measurement

It is not obvious that thermodynamical properties of bulk rare gas apply to RGS microscopic bubbles. However, a theoretical study of cluster size dependence on the surface tension [23]

Table 1 RGS packing density and pressure after room temperature implantation in various matrices. The average RG concentration is in the range 2–5 at%. P_R, P_B, P_E and P_{LP} are respectively the Ronchi pressure, the pressure deducted from P–V curves, the equilibrium pressure $2\Gamma/r$ and the pressure required for loop–punching. Ar, Kr and Xe results are from TEM experiments while the DSA thechnique was used for Ne results [44]. (* Result from 10^{16} Kr$^+$ in Ni) at RT [38].

Matrix Characteristics

Nature	Struc	Lattice Param. (nm)	Surface Tension (J.m^{-2})	Shear Modul. (GPa)	T_M (K) 300/T_M	Nature	Lattice Param. (nm)	Lattice Mism.	Molar Volume cm^3/mol	Pressure (GPa) P_R	P_B	P_E	P_{LP}
Al	fcc	0.405	1.1	27	933 0.32	Ne Ar Kr Xe	0.485 0.530 0.585	1.20 1.31 1.44	8.3 17.2 22.8 30.1	3.5 2.3 2.3	4.2 2.8 1.9 1.3	1.5	2.8
Ag	fcc	0.409	1.1	29	1234,.24	Xe	0.590	1.44	30.9	2.0	1.1	1.5	3.5
Au	fcc	0.408	1.4	28	1336 0.22	Kr Xe	0.550 0.600	1.35 1.47	25.1 32.5	1.2 1.6	1.0 0.8	1.9	3.8
Ni	fcc	0.352	1.82	76	1726 0.17	* Kr Kr Xe	0.515 0.500 0.560	1.46 1.42 1.59	20.6 18.8 26.4	3.7 6.2 4.8	2.9 4.5 2.9	2.4	7.6
Cu	fcc	0.361	1.43	48	1356 0.22	Ar Kr Xe	0.495 0.520 0.580	1.37 1.44 1.61	18.3 21.2 29.4	2.5 3.2 2.6	2.1 2.0 1.4	1.9	5.2
Fe	bcc	0.287	2.1	82	1809,.17	Xe	0.560		26.4	4.8	2.9	2.8	8.4
Mo	bcc	0.315	1.9	122	2883 0.10	Ne Kr	0.510		6.0 20.0	8.8 4.5	20.0 3.2	2.5	10.8
Ta	bcc	0.330	2.1		3269,.09	Ne			6.3	6.4	14.0	2.8	
Ti	hcp	a=0.295 c=0.468	2.1	41	1941 0.15	Kr Xe	a=0.375 c=0.610 a=0.415 c=0.680	1.27 1.30 1.41 1.45	22.4 30.5	2.3 2.2	2.1 1.4	2.8	5.6
Zn	hcp	a=0.266 c=0.495	0.3	42	693 0.43	Xe	a=0.440 c=0.705	1.65 1.42	35.6	1.0	0.5	0.4	3.2

Table 2   Required pressure and the corresponding molar volume for the obtention of the rare gas solid phase at room temperature.

Rare Gas	P_S (GPa)	V_S cm^3.mol^{-1}	Lat. Par. (nm)
Ne	4.83	8.15	0.378
Ar	1.43	19.40	0.505
Kr	0.81	25.98	0.557
Xe	0.41	35.81	0.620

shows that the bulk value is nearly obtained when an argon cluster contains at least 100 atoms.

When assuming a spherical shape, a 2 nm diameter solid argon bubble contains 130 atoms, allowing us to use bulk RGS results as soon as they are large enough to be seen on TEM images. The pressure inside bubbles is usually deduced from packing density measurements by using a pressure-volume equation of state, either Ronchi EOS [24] or experimental P-V curves at room temperature. The Ronchi EOS which is based on a perturbation hard-sphere model, is available for dense rare gases but the deduced pressure is very often overestimated compared to that from experimental curves.

Unfortunately, P-V experimental curves are often from investigations on tentative rare gas metallization and there are sparse results in the relatively low pressure range 0-5 GPa which corresponds to the bubble pressure range. Consequently, the inaccuracy on deduced bubble pressure is about ten times higher than those on experimental packing density (a few %). However, more accurate results are expected since the need for the bubble pressure knowledge has originated both experimental [25] and theoretical studies [26] in the lower pressure range. Results from diamond anvil measurements at room temperature have been used for neon [27], argon [27, 28], krypton [25] and xenon [29, 30, 31] and the resulting pressure P_B is given in Table 1, together with the Ronchi pressure P_R. The threshold pressure P_S needed to obtain solid rare gas at room temperature is given in Table 2 together with the molar volume and the lattice parameter.

<u>Dependence on nature of rare gas</u>

There is a general agreement that within the pressure measurement uncertainty, the bubble pressure is rather independent of the nature of the inert gas [14, 32]. However, very likely due to the higher damage level, it seems that for about the same average concentration, the heavier the rare gas, the lower the pressure [33]. But this effect, being second order compared to other parameters effects, can be ignored when we discuss the dependence upon matrix nature in the following section.

<u>Dependence nature of matrix</u>

The pressure inside bubbles after room temperature implantation ($0.1\,T_M < T < 0.43\,T_M$) is given in Table 1 and corresponds to samples which contain an average rare gas concentration in the range 2 to 5 at%.. Since the implantation conditions were not exactly the same, it can be used for only a rough evaluation of the connection between the pressure and the matrix nature.

As previously mentioned by Evans and Mazey [14], this later dependence can be evidenced by plotting the pressure versus the matrix shear modulus (Fig 1.a). For a better comparison, we have used only TEM experimental results and a crude evaluation of the fitting curve leads to a general linear law, whatever the rare gas nature: $P \approx A + \mu/B$ with $A \approx 0.6$ GPa and $B \approx 23$ when using the Ronchi pressure PR and $A \approx 0.5$ GPa and $B \approx 40$ when the pressure is deduced from the experimental P-V curves. Quite similar results have been reported by Deconninck and Lefevbre [34, 35]. By using Induced X-Ray emission from 2 at% neon implanted samples,

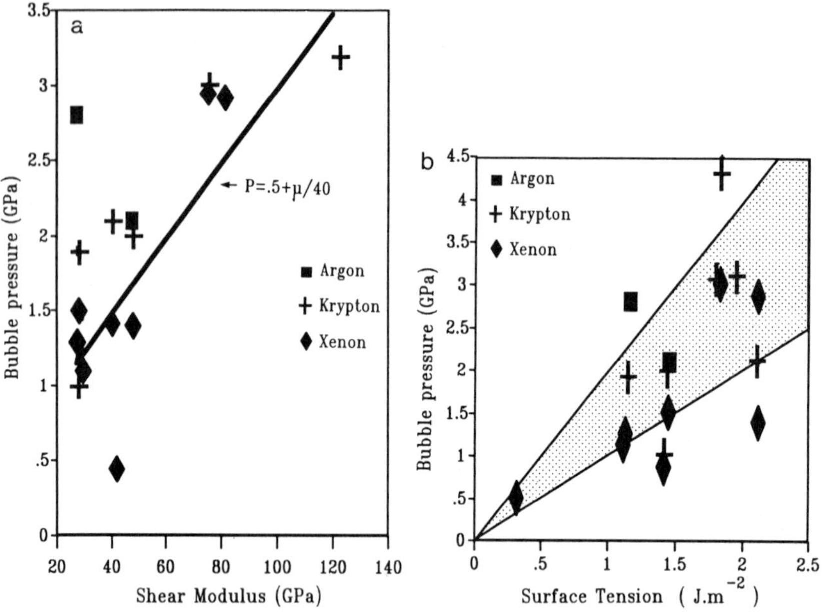

Fig. 1   The average P_B pressure (after RT implantation) inside Ar, Kr and Xe bubbles versus: a. the matrix shear modulus, b. the surface tension, the dashed area corresponds to $P=2\Gamma/r$ with 1<r<2nm.

they show that, except for Pb and Bi which have low melting temperature, the 2p width band W could be approximated by $W(eV) \approx 0.017\ \mu$ (GPa) + 1.3. Although there is not yet to my knowledge an experimental relationship between W and the neon bubble pressure, W broadening is probably an effect of pressure. Thus, it appears that μ is a useful parameter for describing the pressure dependence of matrix nature but it is not the only one and another matrix parameter such as the surface tension Γ could be used. When the pressure is plotted against Γ, there is no obvious fitting curve but one can see that, except for Kr in Ni, the bubble pressure is found close to the equilibrium pressure $P=2\Gamma/r$ when assuming a bubble radius in the range 1 to 2 nm. Interpretations of these results will be discussed in section 3.

2.3.   *Effects of implantation conditions on bubble characteristics*

<u>Fluence Dependence</u>

a) Bubble size r_b and bubble concentration C_b: As reported by the different authors who have used the TEM technique, there is a typical evolution of r_b and C_b upon fluence in Ni and Al.

   At low fluence ($\phi < 10^{14}$ ions.cm^{-2}) a dislocation network develops from the interaction of dislocation loops [36, 37] and it is only for higher fluences ($\phi > 10^{15}$ ions.cm^{-2}) that one can see a high concentration of small cavities. The presence of RGS inside these cavities is detected only after a threshold fluence which is around 6 10^{15} ions.cm^{-2} for Kr in Al [36, 37], 5 10^{15} ions.cm^{-2} for Kr in Ni [38] and 2 10^{15} ions.cm^{-2} for Xe in Al [39]. For these fluences, the typical mean bubble size and bubble concentration are 2r $\approx$ 2 nm and $C_b \approx$ 2 10^{18} cm^{-3}.

   Although there is a non-negligible width of the bubble size distribution, only one peak is observed for fluences lower than a few 10^{16} cm^{-2}. An increase of the Kr fluence results in a bubble size increase together with a decrease of the bubble concentration and an equivalent increase of the total observed cavity volume [38, 39]. As long as the fluence is lower than a few 10^{16} cm^{-2}, there is a total retention of implanted rare gas atoms [37]. It should be noticed that bubbles with a larger size than the average can be observed at grain boundaries.

122

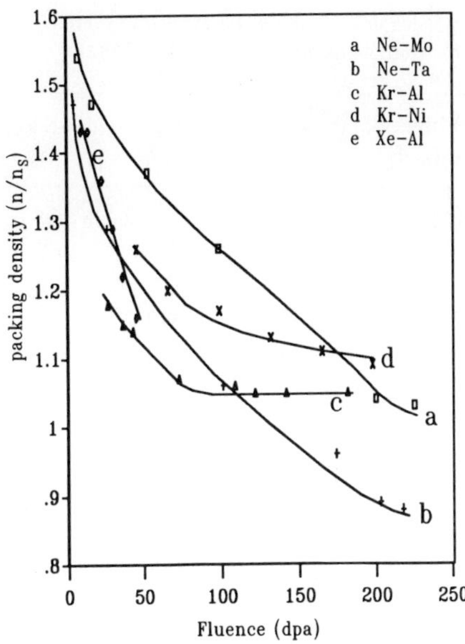

Fig. 2   Evolution of the rare gas packing density versus fluence.

For higher fluence, a double size distribution is seen in TEM observations; as an example, after a fluence of 2.5 10^{16} Kr$^+$.cm^{-2}, a Cu sample contained small bubbles (2r ≈ 4 nm) together with larger bubbles (2r ≈ 15 nm) [40]. The double size distribution after RT implantation has been also found by Andersen et al [41] for Kr in Al (the average concentration was around 9 at%), by using Glancing X-Ray Diffraction. They measured a mean diameter of 3.5 nm for the small bubbles and of 9 nm for the largest ones. When the bimodal size distribution occurs, a diffuse ring appears on diffraction patterns which may indicate a solid-fluid change in largest bubbles [38], a fact previously mentioned by Tyagi et al [42] for Ar bubbles in Al when the fluence reaches 10^{17} Kr$^+$.cm^{-2}.

Identical results have been recently reported for Xe in β-tin [43] but since the Sn melting temperature is very low, large fluid bubbles are observed for fluences as low as 5 10^{15} ions.cm^{-2} and for higher fluences, bubble sizes range from 1.5 nm up to 100 nm.

b) Bubble packing density and bubble pressure: There have been some systematic studies of bubble packing density n_b upon fluence, either by TEM for Xe-Al [37], Kr in Ni [38] and Xe in Al [39] or by Doppler Shift Attenuation (DSA) for Ne in Mo and Ta [44]. Since the RG packing density is deduced from lattice parameter measurements in TEM experiments, results are only related to solid bubbles while the DSA method gives average information on both solid and fluid bubbles.

Although in this later case the deduction of n_b is not straightforward, its evolution with fluence is quite similar to those obtained from TEM. Results are shown in Fig 2 and for a better comparison, we have plotted n_b/n_s (where n_s is the lowest packing density needed for bulk RGS at 300K) versus φ in dpa (for a 10^{16} ions.cm^{-2} and according to the used ion energies, the number of displacements per atoms is 12.5 for Ne-Mo, 14.5 for Ne-Ta, 33 for both Kr-Al and Kr-Ni and 45 for Xe-Al). One can observe that n_b decreases as φ increases and that in solid bubbles a saturation value close to n_s is reached for high fluence (Kr-Al).

As n_b decreases, the bubble pressure also decreases and this is connected with the bubble size increase. According to both packing density and bubble size measurements, it appears that, whatever the fluence, the pressure in solid bubbles is always close to the equilibrium value [38, 43] while large fluid bubbles seems to be underpressurized [43].

Table 3 The dependence of Xe bubble characteristics on implantation temperature. Lattice parameter measurements are performed at RT [39].

T_{imp} T_i / T_M	Fluence (cm^{-2})	C_b (cm^{-3})	$2r_b$ (nm)	a_{Xe} (nm)	P_b (GPa)
300K	10^{15}		no detectable RGS bubbles		
0.32	$5\ 10^{15}$	$2\ 10^{18}$	3–4	0.560	2.8
473K	10^{15}	$2\ 10^{16}$	4	0.570	2.1
0.51	$5\ 10^{15}$	$2\ 10^{16}$	4–5, 20	0.605	0.7
573K	10^{15}	10^{16}	4–5	0.580	1.6
0.61					

Temperature dependence

As at room temperature, the bubble concentration decreases while the average bubble size increases with increasing fluence (see Table 3) but the rate of increase of the total volume precipitates is higher (3.4 times for Kr-Ni at 500°C [38]). Compared to RT implantations, the threshold fluence for detectable RGS extra reflections is lowered and the distribution size becomes bimodal for lower fluence. After implantation at 200°C in Al (T/T_M =0.51), an inspection of electron diffraction patterns reveals weak additional spots for a fluence as low as 10^{15} Xe^{2+}.cm^{-2} and small bubbles ($2r \approx 5$ nm) are observed together with largest faceted bubbles ($2r \approx 20$ nm) for only $5\ 10^{15}$ Xe^{2+}.cm^{-2}.

These later observations are quite similar to those obtained after xenon RT implantation in tin (T_i/T_M =0.58) and it appears that T_i/T_M is the relevant parameter for the temperature dependence of bubble characteristics.

For a given fluence and connected with the bubble size increase, the RGS packing density (measured at RT after implantation) and the deduced pressure in solid bubbles decrease with increasing T_i (see Table 3).

The influence of implantation temperature on the rare gas packing density is only observed when T_i/T_M is around or higher than 0.45 (Ti > 500°C in Ni [38] and Ti > 200°C in Al [39]). Since the bubble pressure is lowered for the high values of T_i/T_M, it may explain why the pressure after RT implantations in Zn (T_i/T_M = 0.43), Pb (T_i/T_M = 0.50 and Bi (T_i/T_M = 0.55) do not agree with the fitting curve on Fig 1 or in ref [35].

2.4. *Bubble shape and bubble lattice*

Authors usually report that small bubbles (a few nanometers in diameter) are more or less round shaped whereas large bubbles are observed to be faceted. Generally, due to the interfacial tension anisotropy, the facets develop mainly on matrix close packed planes and bubbles are often truncated by facets on other densely packed planes [37, 43]. However, the bubble shape appears to be dependent on both the rare gas and the matrix nature. For example, only the smallest Xe bubbles are faceted in platinum while the largest ones are approximately round. In copper, the large faceted bubbles develop facets on {110} planes when they contain xenon, but mostly {100} planes for argon and {111} for krypton [40]. Since the matrix surface tension Γ is only slightly dependent on crystallographic directions, the influence of rare gas on interfacial surface tension Γ^* may be sufficiently high enough to give rise to rare gas dependent faceting.

Although helium bubble super-lattices have been often reported, there are only few results of bubble ordering for heavier rare gases. In titanium, Mazey and Evans [45] have observed that Kr bubbles order in a planar manner parallel to the hcp basal plane with random positions within the plane and with an interplanar distance in the range 5.4 to 5.8 nm. Similar results have been recently reported for Zr [46]. In nickel, precipitate ordering along <100> crystal directions is obtained as long as the temperature and ion flux remain extremely stable during implantation [38].

3. Nucleation and Growth of RGS Bubbles

Implantation of heavier rare gases in solids will result in a high damage level (typically a few dpa for a fluence of 10^{15} ions.cm^{-2}) and high supersaturation of vacancies and interstitials. For example, according to TRIM calculations [47], a 200 keV xenon ion product nearly 1850 Frenkel defects in aluminium, although only a few percent will survive the displacement cascade recombination [48]. Nevertheless, these defects must play an important role in bubble nucleation, particularly as the implanted rare gas atoms will interact strongly with the vacancies. The nucleation and growth process is then dependent on both the production rate and the migration rate of interstitials, vacancies and rare gas atom-vacancy complexes. For more detailed discussion on the clustering process, the reader is referred to a review by Van Veen [49].

3.1. Bubble nucleation

At very low fluences ($\phi < 10^{12}$ ions.cm^{-2}), the displacement cascades do not overlap and an implanted atom may stop in a substitutional configuration or may associate with irradiation vacancies. Gas atom-vacancy complexes are then formed which can thereafter act as nucleation centres. As soon as collision cascade overlapping occurs, the following implanted atoms may be trapped in already existing nuclei which leads to the beginning of the growth stage of the precipitation process since bubble growth starts when stable embryos are formed. This is partly confirmed by channelling measurements on Kr implanted aluminium [50–52] which show that after low fluence implantation as low as 10^{14} Kr^{+}.cm^{-2}, Kr atoms may be located at random sites (34%), ie inside rare gas bubbles, substitutional sites (34%), or may form vacancy complexes KrV_4 (17%) and KrV_6 (15%). In a similar study, Andersen et al [41] concluded that almost 10% of Kr atoms would be at interstitial sites, a quite surprising result because such large impurities as Kr atoms would be expected to adopt a substitutional configuration.

After a few 10^{15} ions.cm^{-2} implantation at room temperature, the typical bubble concentration is around 10^{18} cm^{-3}. If we assume that the number of nucleation centres had the same value, we are able to roughly evaluate the mean number of gas atoms within bubbles after a fluence of 10^{14} ions.cm^{-2}. Since almost one third of rare gas atoms are gathered into cavities [52], we find that small undermicroscopic bubbles contain less than 10 atoms. Since it is generally assumed that a gas bubble is nucleated when cavities contain two or more gas atoms, bubble growth processes appear to start around a fluence of 10^{13} ions.cm^{-2}, i.e. very soon after the implantation onset.

3.2. Bubble growth

Since TEM is not able to detect clusters of size below the resolution limit around 1 nm, there is no real confirmation concerning the first stages of bubble growth process. In addition, small bubbles contain very few atoms (≈ 10 atoms in a 1 nm diameter bubble) and they probably have not yet a well defined structure. If diffraction intensity was available from such small bubbles, it would give an undetectable weak diffuse ring and the necessary fluence (a few 10^{15} ions.cm^{-2}) for the extra spots detection on diffraction patterns (i.e. the solid bubble detection) is well above the fluence corresponding to the initiation of growth.

Detailed description of the various mechanisms of bubble growth can be found in the literature [53–55], so they will be briefly mentioned in the following. A bubble may grow by loop-punching mechanism as long as the pressure within bubbles is higher than a threshold P_{LP} given by:

$$P_{LP} = \frac{2\Gamma}{r_b} + \frac{\mu b}{2\pi r_b (1-v)} \ln(r/b) \qquad (1)$$

where r_b is the bubble radius, μ the matrix shear modulus, b the Burgers vector of the emitted dislocation loop, v the Poisson ratio and Γ is the interfacial tension. Since the surface tension, although negative for solid rare gas at room temperature, is negligible (around 10^{-2} J.m^{-2}), it is generally assumed that Γ is the matrix surface tension.

Bubbles may also grow by collection of vacancies, either irradiated vacancies or thermal vacancies if the implantation temperature is higher than ≈ 0.5 T_M. In this regime, if the bubble is overpressurized ($P > 2\Gamma/r_b$), then vacancy collection is driven by the excess pressure inside bubble. Bubbles can also trap gas atoms or gas atom-vacancy complexes but it has been suggested that the presence of a strain field around overpressurized bubbles may reduce the gas precipitation rate toward them [56]. Finally bubbles can grow by the migration and coalescence mechanism. Although resolution and ripening is virtually possible, it is very unlikely because of the low rare gas solubility and of the continuous rare gas supply as implantation proceeds.

From a general point of view, bubble growth mechanisms depend on many parameters: the number of interstitials and vacancies (either thermal or irradiation vacancies), the nature and concentration of sinks, the diffusity and the sink strength of mobile species.

By assuming that cavities (or bubbles) and dislocations are the dominant sinks for vacancies, interstitials and gas atoms, Evans [57, 58], concluded that in the temperature range 0.15 T_M–0.5 T_M, bubble growth during implantation occurs mainly by a pressure-driven mechanism (loop-punching) as long as the cavity sink strength is sufficiently larger than the dislocation sink strength. But loop-punching is no longer valid if the bubble pressure is lower than the threshold P_{LP}. Then the knowledge of the bubble pressure obtained after implantation may give some indications about the bubble growth process. Within the uncertainty in deduced pressure, it appears from experimental results collected in section 2 that the pressure is never far from the equilibrium value P_E. If bubbles are overpressurized, the excess pressure deduced from eqn. 1 is always under P_{LP} (see Table 1). In addition, Kamada claims in a recent work [59] that during implantation, two additional terms to P_{LP} must be taken into account, one from the local swelling effect due to the high bubble concentration and the other from the vacancy supersaturation as implantation proceeds. It leads to a threshold pressure for loop-punching a few times higher than the value given by eqn. 1. Eventually, it seems that loop punching mechanism is apparently not appropriate to describe the bubble growth during implantation. If need be, it might occur from isolated bubbles, either during implantation or further annealings.

In their bubble growth model in amorphous materials, Knuyt et al [60, 61] assume that bubbles are always overpressurized and that the excess pressure $P-2G/r_b$ remains constant. By using appropriate parameters and by taken into account the bubble coalescence, results from this model fit quite well the He bubble size evolution versus fluence in amorphous materials. As a characteristic feature, bubble growth appears to start only after an incubation dose, because rare gas atoms have to fill the amorphous alloy free volume. This is consistent with experimental results from electron microprobe analysis [62] on argon bubbles in sputtered amorphous Nb3Ge; a pinning strength increase was observed around 4 at% of Ar as a consequence of the bubble formation. The pressure (7 GPa) deduced from these experiments and from additional TEM observations is similar to those from simulation [63]. However, as

126

the fluence increases, the bubble pressure in crystalline materials appears to be very about equal to the equilibrium pressure (and sometimes lower for the large fluid bubbles) and the constant excess pressure hypothesis is no longer available. Thus, the later model cannot describe the entire growth process and it does not explain the double size distribution which is observed for high fluence or high implantation temperature.

Although the critical radius concept [64] may explain the bimodal size distribution, an alternative explanation arise from the more recent theory from Rest and Birtcher [32] which gives rather good results for the size distribution dependence on fluence and implantation temperature. By assuming gas driven process and by taken into account additional parameters like bubble diffusity changes, this model leads to the double size distribution for high fluence at high implantation temperature. A main parameter is the cut-off bubble radius which is the maximum radius at which bubbles are solid and then assumed to be immobile while the largest fluid bubbles may migrate by surface diffusion mechanism. Results from this model, when applied to Kr implantation in nickel, indicate that even at room temperature(T_i/T_M =0.17), bubbles are maintained close to equilibrium because of the high damage level and this is rather consistent with experimental results. Nevertheless, as emphasized by Evans [65], the whole system of defects and defects sinks has to be considered for a detailed description of bubbles growth and the validity of some assumptions used in the model from Rest and Birtcher remains open to discussion: no interstitials effects, surface diffusion of fluid bubbles and zero mobility of solid ones (see section 4.1).

Eventually, if we look upon both experimental and theoretical results, it appears that the relevant growth mechanisms involved during the implantation process is well described by a gas driven mechanism together with bubble coalescence when they are able to migrate. If the assumption that gas atom resolution from bubbles cannot operate is reliable, then the diffusion of implanted gas atoms to already existing bubbles is clearly demonstrated by the detection of RGS "alloys" when two different rare gases are successively implanted in the same Al sample, whatever the nature of the first implanted rare gas [66]. When compared to the theoretical study of Kr-Ar (or Xe-Kr) alloy parameter as a function of Kr (or Xe) concentration [67], the small discrepancy is easily explained by taking into account the higher damage production as the concentration of the heavier rare gases increases which leads to a lowering of the bubble packing density.

4.   Post-Implantation Annealing and Cooling

4.1.   Bubble growth during annealing

In most metallic matrices, bubbles begin noticeably to grow at temperatures around and above $0.5\,T_M$, when thermal vacancies are available. The gain of vacancies leads to a decrease of the rare gas packing density with an accompanying decrease of the bubble pressure. As they become larger, bubbles develop facets whose orientation depends on both rare gas nature and matrix nature (see 2.4). As already noticed by several authors, [14, 37, 68], rearrangement and coalescence of submicroscopic bubbles or gas atom-vacancy complexes occurs in the first time of annealing. This is consistent with the channelling result [52] which shows that in a 10^{15} Kr$^+$.cm^{-2} Al sample, the proportion of Kr atoms inside bubbles changes from 55% to 80% after annealing at 593K (T / T_M = 0.64) for 30 minutes.

Above $0.5\,T_M$, it is quite clear that bubble growth occurs mainly by absorption of thermal vacancies and by migration and coalescence process. It is mostly assumed that bubble migration is dominated by surface diffusion mechanisms [69] and inhibited by inert solid rare gas [37, 69, 70] and also that the diffusion rate is controlled by the nucleation of ledges when they become faceted. However, solid and fluid bubble migration mechanisms still remain a controversial topic because Yamaguchi *et al* [71] point out that solid Kr bubbles in Al randomly migrate and that the diffusion coefficient value is in agreement with a volume diffusion

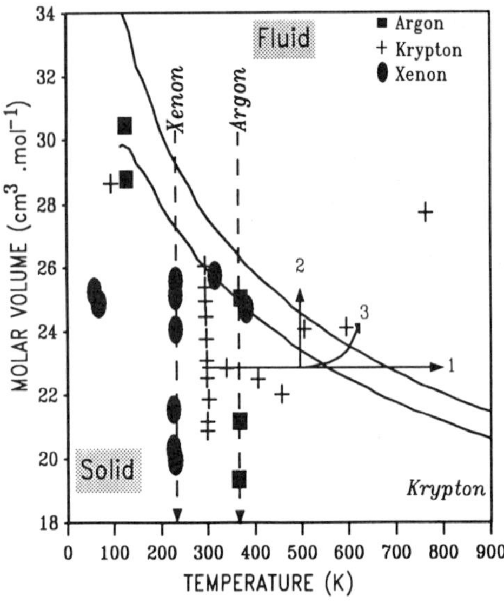

Fig. 3   V-T diagram of solid Kr. V_f and V_s are respectively the solid and the fluid molar volume.(1), (2) and (3) represent the possible routes to obtain fluid bubbles.

mechanism, whether the bubbles are solid or fluid [72]. In previous experiments, Ronchi and Matzke [73] also had concluded that the motion of fluid Xe bubbles in Pt is controlled by a volume diffusion process and that the shape of bubbles does not dramatically affect the bubble mobility: round shaped bubbles migrate approximately the same distance as faceted ones. Very recently, investigations on bubble migration mechanism have been performed by Cox *et al* [74] by introducing impurities which segregate at the Ne bubble-Al matrix interface. They shown that the presence of oxygen clearly reduces the bubble diffusivity up to 400–500 °C but not above this temperature. The working mechanism was discussed but no conclusions were reached.

Below 0.5 T_M, when the concentration of thermal vacancies is low, changes in either positron lifetime [75] (which is connected to the packing density by a simple linear law for Kr bubbles in Al or Ni) or in Doppler-broadening annihilation measurements [76] suggest that a different growth mechanism occurs. Since these changes are observed at 0.40 T_M in both Kr-Cu and Kr-Ni samples, a temperature which corresponds to the melting of Kr bubbles, it is assumed [70, 76] that migration and coalescence of fluid bubbles allow them to have an inner pressure above the threshold required for loop-punching growth process, an alternative mechanism when thermal vacancies are not available.

Since the effects of annealing at higher temperatures (surface fracture, blistering, etc.) are not discussed in this review, the reader is referred to the SEM study of Hashimoto *et al* [77].

4.2. Melting of bubbles

Fluid bubbles are observed after high fluence or high temperature implantation and it is of interest to know about the melting change (or freezing change when cooling) characteristics (pressure, temperature). During annealing experiments, the melting of RGS bubbles is observed either when quickly increasing the annealing temperature T_A or when increasing the annealing time. In this later case, the gain of thermal vacancies (if $T_A > 0.5\ T_M$) allows the bubble packing density to reach the maximum value above which bubbles are no longer solid at T_A (see Fig. 3).

From an experimental point of view, melting is defined in both TEM or X-Ray diffraction

128

studies by the reduction and the gradual disappearance of the rare gas reflection. Here, we have to take into account the size distribution of RGS bubbles. Thus the diffraction results are an average on a large number of bubbles, small ones with high pressure and larger ones with lower pressure.

If melting of individual bubble is observed by imaging it in dark field, then a typical result is that the melting temperature depends on the bubble radius (ie the bubble pressure): the lower the radius, the higher the melting temperature. From a thermodynamical point of view, if we assume a constant volume transformation, the melting of an individual bubble should takes place over a temperature range of about 100K (see Fig. 3), while its solid phase gradually transforms in the fluid phase. Nevertheless, it appears from dark field observations that bubbles switch-off in a very narrow temperature range (a few Kelvin) and that they sometimes switch-on again, even for higher temperature [33]. So it seems that bubbles may lose their epitaxial orientation before they melt [78] and this can be the interpretation of the diffraction ring which is observed on TEM diffraction patterns (with such a diameter that it coincides with 111 reflections of the solid rare gas) when the gradual reduction of reflection intensity takes place rather than a transitory amorphous state [79].

Molar volume values from the literature for different rare gas-matrix samples at several temperatures are plotted on Fig. 3. The V-T melting curves have been plotted for krypton by using empirical equations [80] and for a better comparison, we have used the corresponding state principle (in terms of critical parameters T_c and V_c) in order to plot argon and xenon results in the same figure. One can see that except for a Kr-Ni sample at 500°C, all the experimental dots are under the coexistence curves.

Eventually, if we ignore the Kr-Ni values which need confirmation, all the observations from the different rare gas-matrix samples are quite consistent with the bulk rare gas melting behaviour and there is apparently no detectable superheating.

Several authors have reported that even at annealing temperatures where almost all the bubbles are in the fluid phase, some small solid bubbles are still present. However, they are not numerous enough to give any diffraction reflections but they are easily detected by X-ray diffraction [41] and channelling [52].

Together with the diffuse ring, streaks in <111> directions are observed on diffraction patterns, for Kr in Al [79], Kr in Ni [38] and Xe in Al [39]. Three tentative explanations have been proposed : microcracks [79], oxide platelets [81] and rare gas adsorption on bubble facets [39]. This later hypothesis is consistent with a molecular dynamic simulation from Jensen et al [75] on the Kr-Cu system which shows that an ordering process takes place at the Kr-Cu interface, even when the rare gas is in the fluid phase.

As already quoted by Jensen et al [75], there could be an explanation why small bubbles, for which almost all the gas atoms are on the bubble surface, remain in a solid-like state. An effect of RG and matrix atoms at the interfaces is the lowering of thermal vibrational amplitude compared to bulk value which is connected with an increase of the Debye temperature [82, 83, 18].

4.3. Recrystallisation and adsorption phenomena

When cooling down to RT after annealing, a recrystallisation phenomena is observed if the rare gas volume at the end of the annealing is lower than the threshold for RGS at RT (see Table 2), otherwise a lower cooling temperature is required. Solidification of bubbles occurs with the same epitaxial relation as before annealing except when they are partly filled with solid. When samples at RT contain both small solid and large fluid bubbles, diffraction patterns at around 100K show clearly two sets of RGS reflections, one from the already solid bubbles and the other at smaller diffraction vectors from the solidified large bubbles [37]. So the mean value of the RGS lattice parameter increases [33, 41, 83, 84] from RT to 100K . However a decrease of Xe lattice parameter from 0.615 nm to 0.595 nm has been observed in a sample which contains both solid and fluid bubbles but with almost the same packing density (in this case, there is only one set of reflections at 100K). A tentative explanation is the contraction of the Al matrix upon cooling but it leads to only one third of the lattice parameter modification and the reason for this discrepancy is not yet understood.

For very large underpressurized fluid bubbles at RT (resulting from high temperature annealing), cooling to LN_2 temperature leads to only partial filling [37, 85] and rare gas atoms epitaxially condense on cavity facets. This is observed with Kr and Xe bubbles but not when bubbles contain Ar gas atoms [40, 84]. By assuming that bubble were in equilibrium at the end of the annealing experiment, it is found that bubbles with a mean diameter around 40 nm have an internal pressure as low as 10^7 Pa. Thus the temperature required for the solidification of the inner gas is very close to the triple point temperature which is respectively 162K, 110K and 80K for Xe, Kr and Ar. The uncertainty on the cold holder temperature suggests it is probably a little too high to form solid Ar while it is cold enough for Kr and Xe atoms to solidify on bubbles facets. This is observed in copper where Xe atoms are adsorbed on the bubble facets with {111} xenon planes parallel to {110} Cu planes with the preferential orientation $<110>_{Xe}$ // $<110>_{Cu}$ [86]. Such an orientation had been previously found as a result of Xe physisorption on copper at 77K at very low pressure [87]. Very similar results have been obtained by Mitchell *et al* [43]; xenon atoms condensed with their {111} planes parallel to the {100} and {110} close packed planes of β-tin (body centered tetragonal) and also with preferential orientations between these planes.

5.   Conclusions

Although some open questions remain, particularly concerning bubble nucleation and growth process and bubble migration, this review has allowed us the possibility of giving evidence of some general but sometimes rough characteristics of RGS bubbles in materials. As already quoted by Evans [88] in the early beginning of the RGS bubble discovery, highly pressurized bubbles in metals are a neat way to obtain solid rare gases; their studies may give new and interesting results on other research fields like melting and the surface [89], precipitation of insoluble species [78], or physisorption. Collaboration with scientists from these fields would be probably very fruitful in the future.

ACKNOWLEDGEMENTS

I should like to thank S.E. Donnelly and J.H. Evans for previous stimulating discussions and some help for the preparation of this paper and it is my pleasure to acknowledge H. Garem for help with electron microscopy, and J.C. Desoyer and J. Delafond for very fruitful discussion.

REFERENCES

1. H. Ullmaier, Rad. Effects **78**, 1 (1983).

2. Hj. Matzke, Ann. Chimie Fr. **14**, 133 (1989).

3. J.H. Evans, R. Williamson and D.S. Whitmell in *Effects of Radiation on Materials*: 12th Int. Symp., F.A. Garner and J.S. Perrin ed., ASTM proceedings, Philadelphia (1985).

4. J. Th. de Hosson, Phys. Stat. Sol. (a) **40**, 293 (1977).

5. M.L. Klein and J.A. Venables ed., *Rare Gas Solids*, Academic Press, London (1977) .

6. K.A. Goettel, J.H. Eggert, I.F. Silvera and W.C. Moss, Phys. Rev. Letters **62**, 665 (1989).

7. A. Vom Felde, J. Fink, Th; Müller-Heinzerling, J. Pflüger, B. Scheerer, G. Linker and D. Kaletta, Phys. Rev. Letters 53 922 (1984).

8. C. Templier, C. Jaouen, J.P. Rivière, J. Delafond and J. Grilhé, C.R. Acad. SCi. Paris **299**, 613 (1984).

9. C. Templier, R.J. Gaboriaud and H. Garem, Mat. Sci. and Eng. **69**, 63 (1985).

10. J.H. Evans and D.J. Mazey, J. Phys. F. **15**, L1 (1985).

11. R.J. Keyse and J.A. Venables, J. of Cryst. Growth **71**, 525 (1985).

12. S.E. Donnelly and C.J. Rossouw, Science **230**, 1272 (1985).

13. S.E. Donnelly and C.J. Rossouw, Nucl. Instr. Meth. in Phys. Res. **B13**, 485 (1986).

14. J.H. Evans and D.J. Mazey, J. Nucl. Mat. **138**, 176 (1986).

15. C. Templier, H. Garem and J.P. Rivière, Phil. Mag. **A53**, 667 (1986).

16. J.H. Evans and D.J. Mazey, Scripta Met. **19**, 621 (1985).

17. C. Templier, B. Boubeker, H. Garem, E.L. Mathé and J.C. Desoyer, Phys. Stat. Solid. (a) **92**, 511 (1985).

18. G. Faraci, A.R. Pennisi, A. Terrasi and S. Mobilio, Physica **B158**, 602 (1989).

19. M.W. Finnis, Acta Met. **35**, 2543 (1987).

20. C. Templier, J.C. Desoyer, H. Garem and J.P. Rivière, Acta Met. **37**, 393 (1989).

21. A.P. Jephcoat, H.K. Mao, L.W. Finger, D.E. Cox, R.J. Hemley and C.S. Zha, Phys. Rev. Letters **59**, 2670 (1987).

22. R.J. Keyse, J.A. Venables, J. of Physics **C23**, 4435 (1985).

23. K. Nishioka, Phys. Rev. **A16**, 2143 (1977).

24. C. Ronchi, J. Nucl. Mat. **96**, 134 (1981).

25. A. Polian, J.P. Itié, E. Dartyge, A. Fontaine and G. Tourillon, Phys. Rev. **B39**, 3369 (1989).

26. V. Karttunnen, J. Ignatius, J. Keinonen and R.M. Nieminen, J. Phys. Cond. Matter **1**, 4885 (1989).

27. L.W. Finger, R.M. Hazen, G. Zou, H.K. Mao and P.M. Bell, Appl. Phys. Lett. **39**, 892 (1981).

28. M. Ross, H.K. Mao, P.M. Bell and J.A. Xu, J. Chem. Phys. **85**, 1028 (1986).

29. K. Asaumi, Phys. Rev. **B29**, 7026 (1984).

30. A.N. .Zisman, I.V. Aleksandrov and S.M. Stishov, J.E.T.P. Letters **40**, 1029 (1984).

31. K. Syassen and W.B. Holzapfel, Phys. Rev. **B18**, 5826 (1978).

32. J. Rest and R.C. Birtcher, J. of Nucl. Mat. **168**, 312 (1989).

33. C. Templier, PhD Thesis, University of Poitiers 1987.

34. G. Deconninck and A. Lefebvre, Mat. Sci. and Engn. **90**, 167 (1987).

35. A. Lefebvre and G. Deconninck, Nucl. Instr. Meth. **B15**, 616 (1986).

36. R.C. Birtcher and W. Jäger, Nucl. Inst. Meth. Phys Res. **B15**, 435 (1986).

37. C. Birtcher and W. Jäger, Ultramicroscopy **22**, 267 (1987).

38. R.C. Birtcher and A.S. Liu, J. Nucl. Mat. **165**, 101 (1989).

39. C. Templier, H. Garem, J.P. Rivière and J. Delafond, Nucl. Inst. Methods **B18**, 24 (1986).

40. A. Raqi and C. Templier, Unpublished.

41. H.H. Andersen, J. Bohr, A. Johansen, E. Johnson, L. Sarholt-Kristensen and V. Surganov, Phys. Rev. Letters **59**, 1589 (1987).

42. A.K. Tyagi, R. Khanna and G.V.N. Rao, Scripta Met. **20**, 181 (1986).

43. D.R.G. Mitchell, S.E. Donnelly and J.H. Evans, Phil. Mag. **A61**, 531 (1990).

44. A. Luukkanein, J. Keinonen and M. Erola, Phys Rev. **B32**, 4814 (1985).

45. D.J. Mazey and J.H. Evans, J. Nucl. Mat. **138**, 16 (1986).

46. J.H. Evans, A.J.E. Foreman and R.G. McElroy, J. Nucl. Mat. **168**, 340 (1989).

47. J.P. Biersack and L.G. Haggmark, Nucl. Instr. Meth. **B174**, 257 (1980).

48. A.J.E. Foreman and W.J. Phythian, private communication.

49. A. van Veen, in *Erosion and Growth of solids stimulated by atom and ion beams*, G. Kiriakidis, G. Carter and J.L. Whitton ed., Martinus Nijhoff Publishers (1986).

50. E. Yagi, M. Iwaki, K. Tanaka, I. Hashimoto and H. Yamaguchi, Nucl. Instr. Meth. Phys. Res. **B33**, 724 (1988).

51. E. Yagi, Nucl. Instr. Meth. Phys. Res. **B39**, 68 (1989).

52. E. Yagi, I. Hashimoto and H. Yamaguchi, J. Nucl. Mat. **169**, 158 (1989).

53. B.M.U. Scherzer, p.222 of ref [49].

54. P.J. Goodhew, Scripta. Met.**18**, 1069 (1984).

55. H. Trinkaus, Rad. Effects **78**, 189 (1983).

56. C. Ronchi, J. Nucl. Mat. **148**, 316 (1987).

57. J.H. Evans, Nucl. Instr. Meth. **18**, 16 (1986).

58. J.H. Evans , Mat Res. Soc. Proc. **100**, 219 (1988).

59. K. Kamada, A.Sagara, H. Kinoshita and H. Takahashi, Rad. Eff. **106**, 219 (1988).

60. G. Knuyt, M. D'Olieslaeger, L. De Schepper and L.M. Stals, Mat. Sc. and Engineering **98**, 523 (1988).

61. G. Knuyt, M. D'Olieslaeger, L. De Schepper and L.M. Stals, Phil Mag **A57**, 277 (1988).

62. A. Pruymboom, P. Berghuis, P.H. Kes and H.W. Zandbergen, Appl. Phys. Letters **50**, 1645 (1987).

63. G. Knuyt, M. D'Olieslaeger, L. De Schepper and L.M. Stals, Phil Mag **A58**, 243 (1988).

64. L.K. Mansur and W.A. Coghlan, J. Nucl. Mat. **118**, 1 (1983).

65. J.H. Evans, Symposium on Noble Gases in Metals, TMS-AIME, Las Vegas, unpublished. (1989).

66. C. Templier, H. Garem, J.C. Desoyer and J. Delafond, Scripta Met. **20**, 1705 (1986).

67. H. M. Gilder, Phys. Rev. Letters **61** 2233 (1988) and private communication.

68. J.C. Desoyer, C. Templier, J. Delafond and H. Garem, Nucl. Instr. Meth. **B19/20**, 450 (1987).

69. N. Marochov, L.J. Perryman and P.J. Goodhew, J. Nucl. Mat. **149**, 296 (1987).

70. K.O. Jensen, M. Eldrup, N.J. Pedersen and J.H. Evans, J. Phys. **F18** 1703 (1988).

71. Y. Yamaguchi, I. Hashimoto, H. Mitsuya, K. Nakamura, E. Yagi and M.Iwaki, J. Nucl. Mat. **161**, 164 (1989).

72. K. Takaishi, T. Kikuchi, K. Furuya, I. Hashimoto, H. Yamaguchi, E. Yagi and M. Iwaki, Phys. Stat. Solidi a **95**, 135 (1988).

73. Hj. Matzke and C. Ronchi, European Applied Research Reports, **5**, 1105 (1984).

74. R.J. Cox, P.J. Goodhew and J.H. Evans, Nucl. Instr. and Meth. **B42**, 224 (1989).

75. K.O. Jensen and R.M. Nieminen, Phys. Rev. **B136**, 8219 (1987).

76. D. Britton, P.C. Rice Evans, J.H. Evans, Phil. Mag. **A55**, 347 (1987).

77. I. Hashimoto, H. Yorikawa, H. Mitsuya, H. Yamaguchi, K. Furuya, E. Yagi and M. Iwaki, J. Nucl. Mat.**150**, 100 (1987).

78. L. Gråbæk, J. Bohr, E. Johnson, H.H. Andersen, A. Johansen and L. Sarholt-Kristensen, Phys. Rev. Lett. **64**, 954 (1990).

79. I. Hashimoto, H. Yorikawa, H. Mitsuya, H. Yamaguchi, K. Takaishi, T. Kikuchi, K. Furuya, E. Yagi and M. Iwaki, J. Nucl. Mat.**149**, 69 (1987).

80. R.K. Crawford, as in [5].

81. R.J. Cox and P.J. Goodhew, Phil. Mag Letters **58**, 291 (1988).

82. C.J. Rossouw and S.E. Donnelly, Phys. Rev. Letters **55**, 2960 (1985).

83. G. Faraci, A.R. Pennisi, A. Terrasi and S. Mobilio, Phys. Rev. **B38**, 13468 (1988).

84. R.J. Cox, P.J. Goodhew and J.H. Evans, Acta Met **35**, 2497 (1987).

85. G.L. Zhang and L. Niesen, J. Phys. Cond. Matter **1**, 1145 (1989).

86. J. Guillot, M. Cartraud, H. Garem, C. Templier and J.C. Desoyer, CRAS Paris **30**, 161 (1987).

87. A. Glachant, M. Jaubert, M. Bienfait and G. Boato, Surface Science **115**, 219 (1981).

88. J.H. Evans, Phys. Bull. **36**, 59 (1985).

89. R.W. Cahn, Nature **323**, 668 (1986).

FORMATION AND ANNEALING OF Kr PRECIPITATES IN Ni THIN FILMS *

R. C. Birtcher

Materials Science Division
Argonne National Laboratory
Argonne, IL 60439

ABSTRACT

Precipitation of Kr implanted into Ni at temperatures between 25 and 560°C, and the annealing behavior up to 600°C of Kr precipitates produced at room temperature have been studied with TEM. At implantation temperatures below about 300°C, Kr precipitation is driven by stochastic (non-diffusional) processes, and the precipitate size-distribution is monomodal. Precipitates with radii less than about 30Å are solid Kr while larger precipitates are in the liquid or gas state. The crystal axis of the solid precipitates are aligned with the axis of the Ni. At higher implantation temperatures, the size-distribution evolves with dose from monomodal to bimodal. The small precipitates are solid Kr while large precipitates are faceted and nonsolid. The average Kr lattice parameter increases with increases in average precipitate size produced by either increasing Kr concentration or implantation temperature. During annealing of room temperature implanted Ni, little growth occurs until temperatures at which solid Kr melts. After Kr melting, precipitates become mobile leading to coalescence and rapid growth. Growth towards an equilibrium shape after coalescence frequently results in additional coalescence and catastrophic swelling. Rate theory modelling of both precipitation and annealing has suggested that solid Kr inhibits precipitate motion, and that Kr melting during implantation or thermal annealing is a precursor to rapid growth by coalescence due to precipitate migration.

1.   Introduction

In recent years rare-gas precipitates formed by implantation were found to be under high internal pressure to the extent that it is possible for precipitates to be in the solid phase [1,2]. The internal pressure arises from the work required to remove host atoms to make a cavity for the precipitate. Several studies have shown that heavy rare-gases (Ne, Ar, Kr, Xe) condense within high-pressure cavities as a liquid or as a crystalline solid aligned with the lattice of crystalline host material [1-8]. In this work, Kr precipitation in Ni has been studied with transmission electron microscopy (TEM) for implantations at temperatures between room temperature and 500°C. Thermal annealing of precipitates produced at room temperature has also been studied. When the Kr concentration is increased by injecting gas atoms into a crystal, lattice atoms are displaced from their normal sites. Each Kr ion used in this work produces a defect cascade containing about 1000 interstitials and vacancies. Most of these defects quickly recombine and disappear. Remaining vacancies provide the space required to accommodate the Kr atoms. Precipitate growth during implantation or thermal annealing is controlled by the phase of the precipitates as well as the concentration and mobility of defects

---

*Work supported by the U S Dept. of Energy, BES-Materials Sciences, under Contract W-31-109-Eng-38.

and Kr atoms. Similarities and differences between Kr implantation and annealing provide insight into the mechanisms involved in precipitate growth.

2.   Experimental

Thin single-crystal Ni films (70 nm thick) were prepared by evaporation of high purity Ni in a vacuum less than $1\ 10^{-7}$ Torr onto freshly cleaved NaCl at a temperature of 350 C. The thin film had a <100> surface normal. The coated NaCl was cleaved into small pieces, and specimens were floated on a water-methanol mixture onto Cu TEM grids. Implantations were performed with 180-keV Kr^+ ions at fluxes of less than $2\ 10^{16}\ m^{-2}sec^{-1}$. The implantation energy was selected on the basis of results from the TRIM computer code [9]. AT 180 keV, the Kr concentration depth profile within the Ni films peaks near the foil center and decreases to near zero at both foil surfaces. Estimates for a fluence of $1\ 10^{20}\ Kr^+m^{-2}$ yield a damage level of 33 displacements per atom at the location of the Kr concentration peak of 3 at% . For a fluence of $1\ 10^{20}\ Kr^+m^{-2}$, about 8 nm (or less, owing to reduced sputtering of the surface oxide) of material will be removed by sputtering [10]. Gradual removal of the surface layer results in a more uniform Kr concentration through the thickness of the Ni thin film and reduces the maximum gas concentration.

Post-implantation TEM observations were made with a JEOL JEM-100-CX at an operating voltage of 100 keV and at magnifications up to 100,000 times. Details larger than 1 nm were detectable within the specimens. In situ elevated-temperature observations were made with a Gatan heating stage; however, specimen oxidation restricted the observations to temperatures below about 650°C. Images were recorded at increasing temperatures after the specimen holder had equilibrated and the specimen position was stable. At most temperatures, this required about 10 minutes. Additional observations were made at room temperature after rapid cooling from 650°C.

3.   Microstructure After Kr Implantation

Evolution of the microstructure after Kr implantation at room temperature into separate single-crystal Ni thin films is shown in Fig. 1 for fluences between 1 and $4\ 10^{20}\ Kr^+m^{-2}$. The dominant features observed in the microstructure are Kr precipitates and a dislocation network. Evolution of the dislocation network has not been followed in this work. However, the dislocation network is fully developed by a fluence of $1\ 10^{18}\ Kr^+m^{-2}$ or about 0.3 dpa. For implantation temperatures below 300°C, the precipitates are initially small (<2nm) and, within microscope resolution, appear spherical in shape. Gas precipitates were detectable (resolution 1 nm) for fluences of $0.5\ 10^{20}\ Kr^+m^{-2}$ or greater at room temperature. With increasing fluence, precipitates increase in size and decrease in number. Large, irregularly shaped precipitates develop at fluences greater than about $3\ 10^{20}\ Kr^+m^{-2}$. Size distributions of Kr precipitates produced at room temperature are shown in Fig. 2 for two different Kr concentrations. The distribution at each fluence consist of a single peak, and the average size increases slowly with fluence from 15 Å at $2\ 10^{20}\ Kr^+m^{-2}$ to 20 Å at $4\ 10^{20}\ Kr^+m^{-2}$. This amounts to a doubling of the total precipitate volume upon a doubling of the Kr fluence. At higher fluences, specimen thinning by sputtering becomes important, and larger precipitates may disappear after rupturing through the foil surface. This alters the size distribution at large sizes.

The nature of the microstructure produced in Ni by Kr implantation changes drastically when implantation temperatures are above about 400°C. Evolution of the microstructure for Kr implantations at 500°C is shown in Fig. 3 for fluences between 0.7 and $4\ 10^{20}\ Kr^+m^{-2}$. Implantations of less than $1\ 10^{20}\ Kr^+m^{-2}$ result in small spherical precipitates similar to those observed after room temperature implantation. However, continued implantation at high temperatures, produces large precipitates which are faceted. Viewing along different crystallographic directions demonstrates that the faceted precipitates are three dimensional,

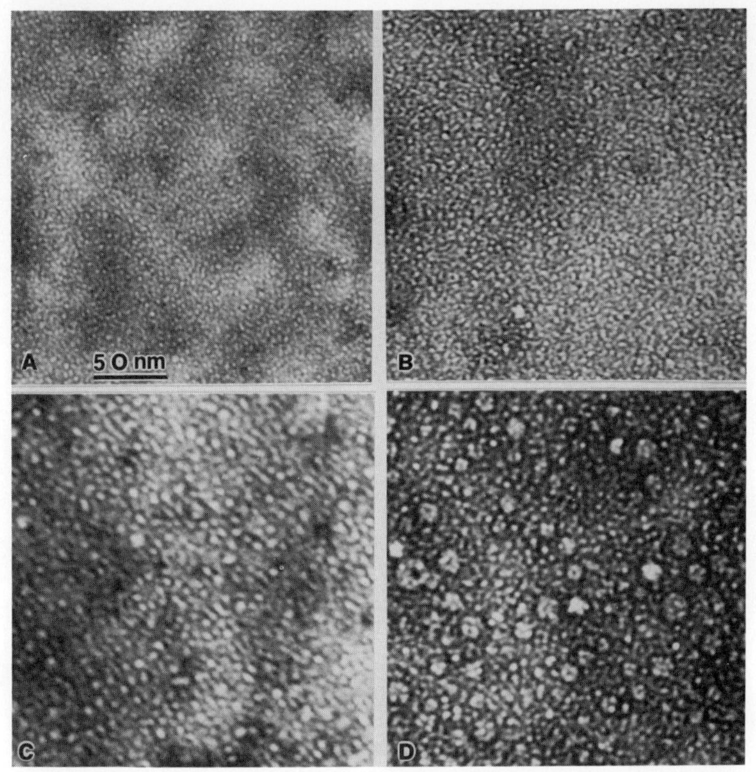

Fig. 1 Electron micrographs of Ni after Kr implantation at room temperature to doses of A.) $1 \cdot 10^{20}$ m^{-2}, B.) $2 \cdot 10^{20}$ m^{-2}, C.) $4 \cdot 10^{20}$ m^{-2} and D.) $6 \cdot 10^{20}$ m^{-2}.

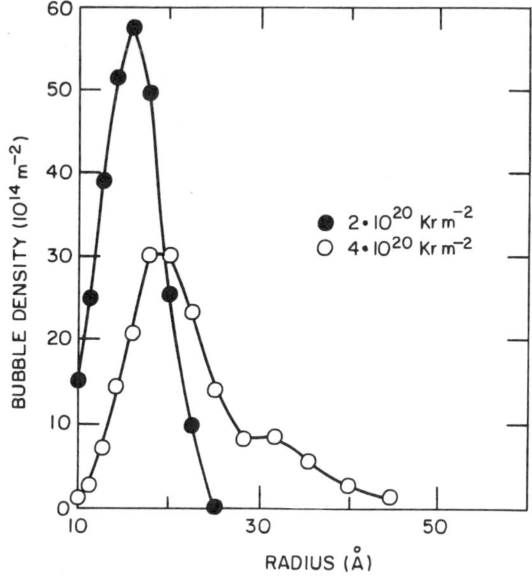

Fig. 2 Size variation of precipitates produced at room temperature.

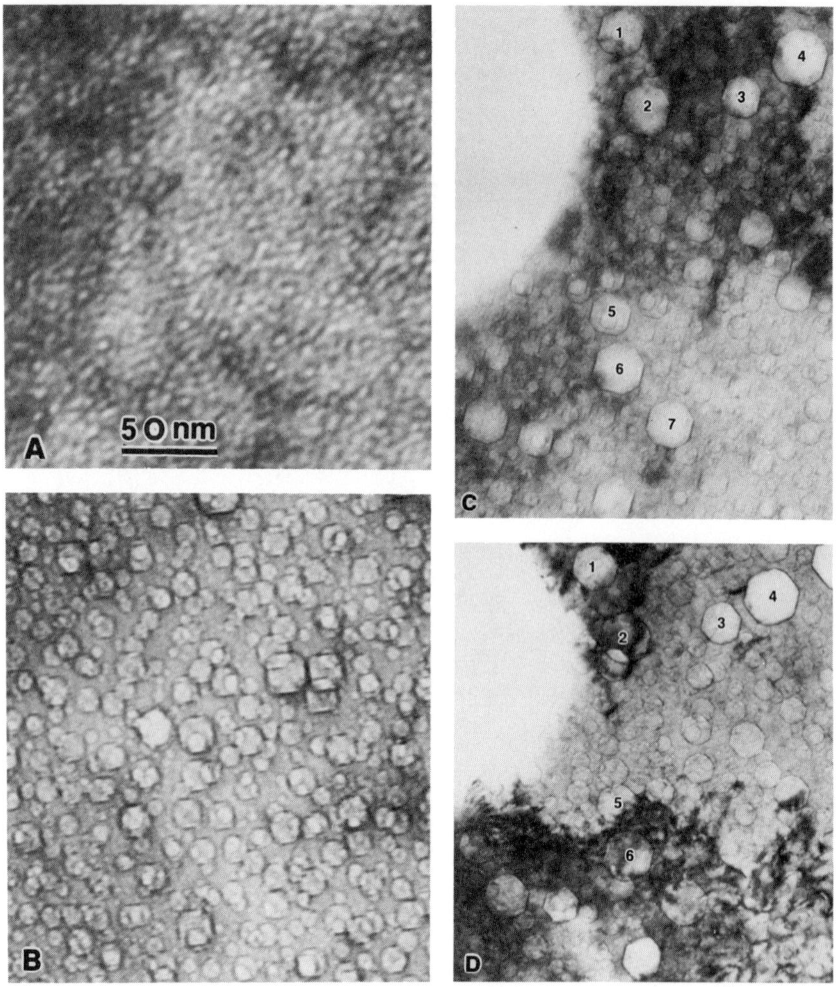

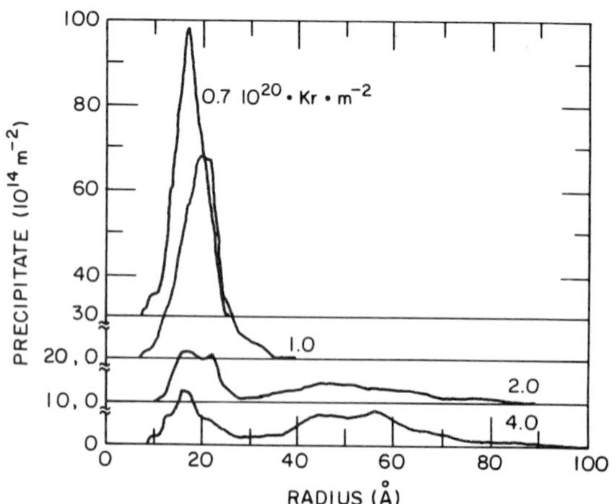

Fig. 3 Electron micrographs of Ni after Kr implantation at 500 °C to doses of A.) $1\,10^{20}\,m^2$, B.) $2\cdot10^{20}\,m^{-2}$, and $4\cdot10^{20}\,m^{-2}$ with foil normal oriented near C.) <100> & D.) <110>.

Fig. 4 Size variation of precipitates produced at 500 °C.

136

Fig. 3C and D. They have 14 faces and are formed from octahedra ({111} planes) truncated by cube faces ({100} planes). The facets generate streaks in diffraction patterns normal to habit planes of the facets. Such patterns indicate that the facets also include a small amount of other {h11} variants. These are found at precipitate corners. A similar behavior is found for He precipitates and voids in Ni [11-13].

Size distributions of the Kr precipitates produced at 500°C are shown in Fig. 4. As at lower implantation temperatures, the size distribution shifts to larger sizes with increasing Kr fluence. At temperatures above 400°C and fluences below $1\,10^{20}\,Kr^+m^{-2}$, the size distribution contains only small, spherical precipitates (radius less than 3 nm). However, with increasing fluence, the size distribution becomes bimodal and contains both small, spherical precipitates and larger, faceted ones. Large precipitates rapidly increase in size with fluence until their growth is halted by intersection with the specimen surfaces due to either growth or material removal due to sputtering. At all fluences, there is a population of small precipitates with radii less than 3 nm. Regardless of the Kr fluence, the distinction between spherical and faceted precipitates occurs at about a radius of 3 nm. This radius separates solid from nonsolid precipitates and will be discussed later. As at room temperature, the total precipitate volume increases linearly with fluence at 500 °C, at least up to $4\,10^{20}\,Kr^+m^{-2}$. However, the rate of increase at 500 °C is 3.4 times that observed for room temperature implantation.

Electron diffraction at room temperature from solid Kr was observed for Kr fluences greater than $1\,10^{20}\,Kr^+m^{-2}$ for implantation temperatures less than 300°C and at fluences greater than $0.3 \cdot 10^{20}\,Kr^+m^{-2}$ for implantation temperatures greater than 400°C. Differences in the lower fluence limit arises from the temperature dependence of precipitate nucleation and growth as exhibited by the resultant size distributions. For all implantation temperatures electron diffraction at room temperature from solid Kr is observed when precipitates are at least 1 nm in diameter and visible in TEM images. The kinetics of precipitate nucleation and growth determine the Kr fluence required to reach this size at any given implantation temperature. It appears that during 500 °C implantation, Kr precipitates are close to melting. A change in precipitate growth upon Kr melting at the irradiation temperature has been used explain the development of the bimodal size distribution and the constancy of solid precipitates sizes during high temperature implantations [15]. Slower cascade cooling at higher temperatures could increase the growth rate of small precipitates.

4. Microstructural Changes After Thermal Annealing

After high dose implantation at room temperature, thermal annealing results in motion of defects, Kr interstitials and Kr precipitates. Figure 5 shows TEM images taken after approximately 10 minute stabilization periods at increasing temperatures during thermal annealing up to 650 °C of a specimen initially implanted at room temperature to a dose of $4\,10^{20}\,Kr^+m^{-2}$. There was little change in the Kr precipitates during annealing at temperatures below 550°C. Above 100 °C, slight precipitate growth occurred during each heating pulse and the time required to equilibrate the specimen. There appears to be only a few coalescence events. At 550°C electron diffraction indicated that all solid Kr precipitates had melted. Visible precipitate growth began during heating to 590 °C; however additional growth occurred very slowly during continued annealing at 590 °C. As shown in Fig. 5, only minor changes occurred during an additional 10 minute annealing period. A continuous increase in the precipitate sizes occurred after heating to 650 °C. Precipitate migration occurred in concert with Kr melting. These observations are in agreement with those of Jensen *et al.* who observed Kr bubble migration at temperatures above 400 °C in Ni in correlation with Kr melting [14].

The rapid growth at 650 °C is displayed in Fig. 6 by a sequence of micrographs taken at 1 minute intervals after an initial 10 minute period of temperature and specimen stabilization. The events leading to the formation of any particular large bubble can be easily followed during this anneal. At this temperature, small precipitates were observed to migrate. Because

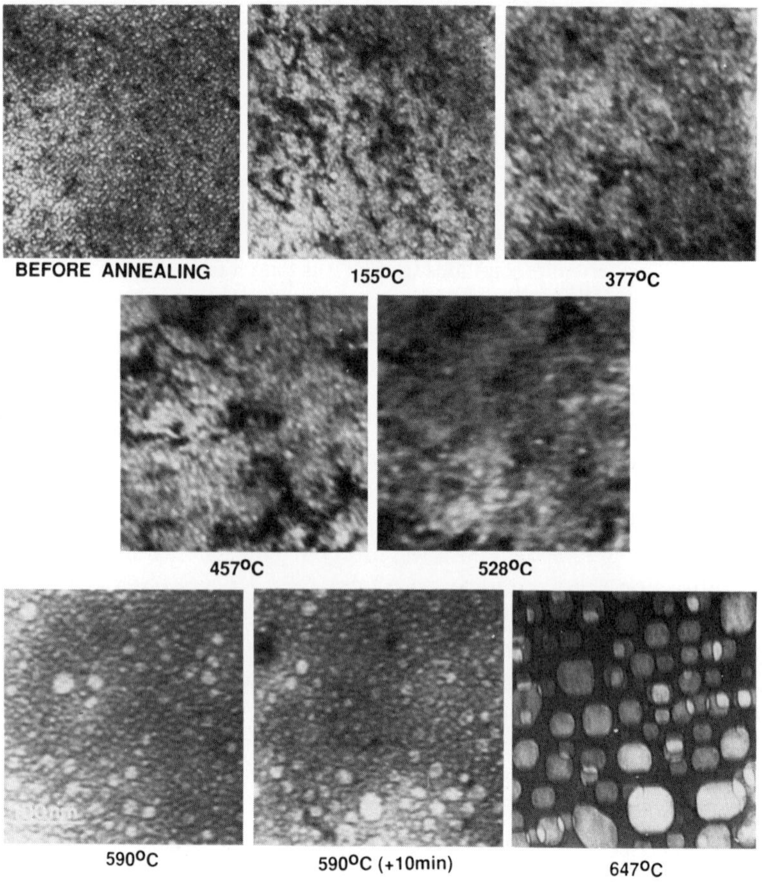

Fig. 5 Electron micrographs taken during heating of Ni after Kr implantation at room temperature to a dose of $4 \cdot 10^{20}$ m^{-2}.

the precipitates were initially separated by distances on the order of their diameter, this movement resulted in a large amount of coalescence and a rapid growth of Kr precipitates into bubbles. Even after bubbles grew to large sizes at which they became nearly immobile, growth continued by a secondary process of "round-out coalescence". This process is initiated by growth, or rounding-out, towards an equilibrium shape of an irregular-shaped, non-equilibrium bubble newly formed by coalescence. Coalescence between bubbles greater than about 10 nm occurred by formation of a thin neck through the material separating their corners. Examples of necking between large bubbles is shown in Fig. 7. The inverse curvature of the bubble surface in a neck is energetically very unfavorable and is quickly removed. The subsequent approach toward equilibrium shape of a given bubble was relative slow and not always discernable in-situ. However, during this growth phase, secondary coalescence or necking could occur between immobile large bubbles when the rounding-out process resulted in their close approach. Secondary coalescence resulted in a sudden increase in bubble sizes as evident in the nonuniform growth rates within each 1 minute time step shown in Fig. 6. Also evident in Figures 6 and 7 are denuded zones around a few large bubbles that achieved equilibrium shape without further secondary coalescence. After a secondary coalescence event, approach towards equilibrium resumed anew, and the process repeated. This recurring process lead to runaway precipitate growth. Very large bubbles occasionally intersected the specimen surface. When this occurred, the resulting steps on the specimen surface rapidly migrated away from the site of the bubble. Rapid bubble growth finally lead to failure of the thin film specimen.

138

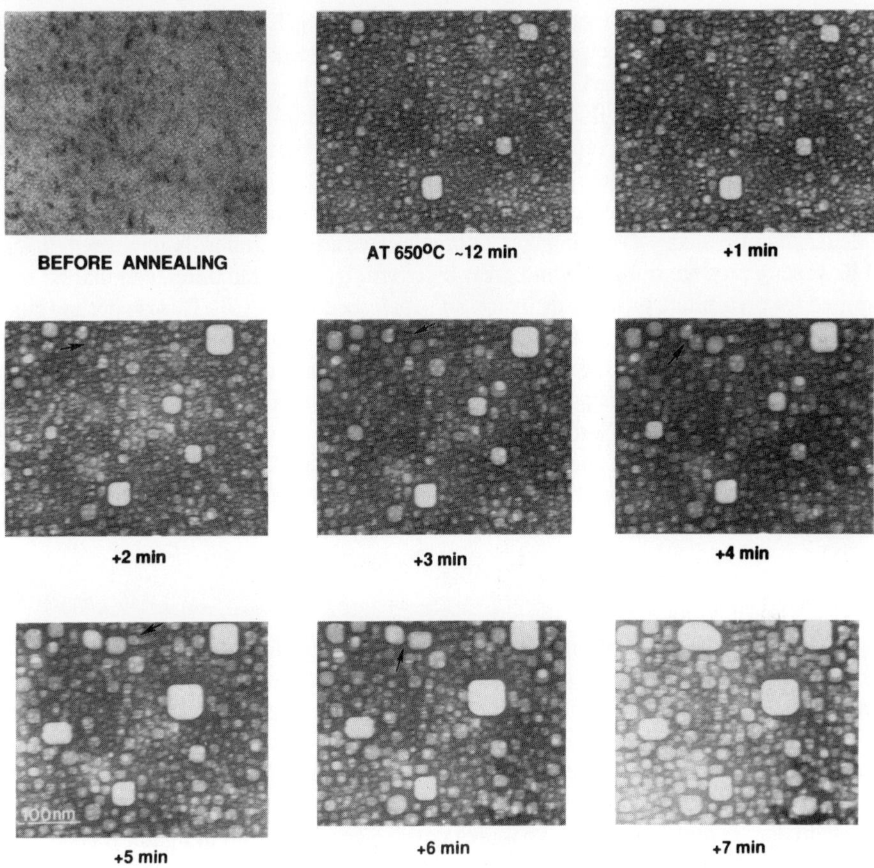

Fig. 6   Electron micrographs taken during thermal annealing at 650 °C of Ni after Kr implantation at room temperature to a dose of $4 \cdot 10^{20}$ m^{-2}.

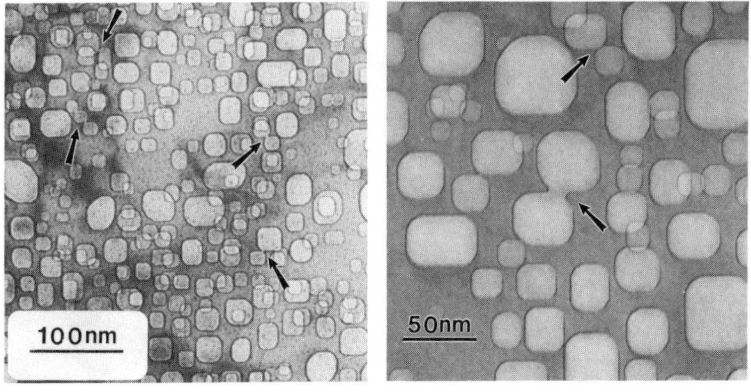

Fig. 7   Electron micrographs taken after thermal annealing at 650 °C for 20 minutes of Ni after Kr implantation at room temperature to a dose of $4 \cdot 10^{20}$ m^{-2}. Arrows mark coalescence events.

5. Influence of Kr Melting on Precipitate Growth

The transition between solid and liquid Kr plays an important role in the evolution of the precipitate sizes at elevated temperatures. With increasing Kr fluence at implantation temperatures above 400 °C, precipitates change from spherical to faceted at about the size required for melting. This marks the separation between the two classes in the bimodal size distribution. With increasing fluence, faceted precipitates rapidly grow while the distribution of smaller, solid precipitates remain nearly constant. Since the Kr lattice parameter at the 500 °C implantation temperature is near the melting limit, it appears that the Kr phase controls the growth kinetics of Kr precipitates. In order to investigate this, precipitate growth was modeled using rate theory assuming that Kr precipitates are continuously nucleated, that small solid Kr precipitates are immobile and grow by atomic gas accumulation, and that Kr melting is required for precipitate grow by diffusion and coalescence [15,16]. This modelling suggests that solid Kr prevents surface diffusion of the precipitates by preventing vacancy formation at the Kr-Ni interface. When the Kr density in a precipitate decreases because of melting, precipitate migration becomes possible leading to coalescence and to the rapid formation of larger precipitates. At 500 °C, this occurs at fluences above about $1 \ 10^{20} \ Kr^{+}m^{-2}$, and is responsible for development of the bimodal size distribution. During thermal annealing after room temperature implantation this occurs in a few precipitates during temperature pulses and in general at temperatures above 550 °C.

6. Conclusions

Many features of rare-gas precipitation in metals are universal. At temperatures below those required for atomic diffusion, precipitation occurs on a stochastic basis, and precipitates sizes increase slowly with implantation dose. The defect cascades induced by rare-gas implantation are capable of melting volumes larger than the precipitates and generating a thermal spike in an ever larger volume. This repeated melting provides a mechanism for rare-gas atom mobility to precipitates and allows annealing of defects in precipitates. This melting is more complete in denser metals which have more confined cascades than in lighter metals. Denser cascades should lead to a lower concentration of rare gas in solution, precipitate nucleation at lower doses and more rapid precipitate growth. The defect cascade is more intense for the heaver rare gases. Thus it might be expected that a heavy rare gas is dispersed in solution at lower concentrations. At higher temperatures, precipitate growth could be controlled by the phase of the rare gas precipitate and mobile defects. Solid precipitates are incapable of diffusive motion, while nonsolid precipitates diffuse and coalesce. This leads with increasing ion dose to an evolution of the precipitate size distribution from a monomodal shape to a bimodal shape. Likewise, thermal annealing of rare-gas precipitates is controlled by rare-gas melting, and growth by coalescence occurs above a temperature threshold determined by precipitate melting. Additional coalescence occurs as merged precipitates move towards equilibrium shapes, and breakaway swelling follows.

REFERENCES

1. A. vom Felde, J. Fink, Th. Müller-Heinzerling, J. Pflüger, B. Scheerer and G. Linker, Phys. Rev. Lett. **53**, 922 (1984).
2. J. H. Evans and D. J. Mazey, J. Phys. F:Met.Phys. **15**, L1 (1985).
3. R. C. Birtcher and W. Jäger, Ultramicroscopy, **22**, 267 (1987).
4. R. C. Birtcher and A. S. Liu, in *Beam-Solid Interactions and Transient Processes*, edited by M.O. Thompson, S.T. Picraux and J.S. Williams (Mater. Res. Soc. Proc. **74**, Boston, Ma. 1986) pp. 345.
5. C. Templier, H. Garem and J. P. Riviere, Phil Mag. **A53**, 667 (1986).

6. D. J. Mazey and J. H. Evans, J. Nucl. Mater. **138**, 16 (1986).

7. R. C. Birtcher and A. S. Liu, in *Beam-Solid Interactions and Transient Processes*, edited by M. J. Aziz, L. E. Rehn and B. Stritzker (Mater. Res. Soc. Proc. **100**, Boston, Ma. 1987) pp. 345-350.

8. R. C. Birtcher and A. S. Liu, J. Nucl. Mater **165**, 101 (1989).

9. J. Biersack and L. G. Haggmark, Nucl. Instr. and Meth. **174**, 257 (1980).

10. H. H. Andersen and H. L. Bay, in Sputtering by Particle Bombardment, R. Behrisch ed. Topic Appl. Phys., vol. **47** (Springer, Berlin 1981), page 145.

11. W. Jäger and J. Roth, J. Nucl. Mat. **93 & 94**, 756 (1980).

12. P.B. Johnson, D.J. Mazey and J.H. Evans, Rad. Effects **78**, 147 (1983).

13. M.D'Olieslaeger, L.deSchepper, G.Knuyt and L.M.Stals, J. Nucl. Mat. **138,** 27 (1986).

14. K. O. Jensen, M. Eldrup, N. J. Pedersen and J. H. Evans, J. Phys. **F18**, 1703, 1988.

15. J. Rest and R. C. Birtcher, 14 Int. Symp. Effects Radiat. on Mater., June 27 -29,1988, Andover, Ma, USA.

16. R. C. Birtcher, J. Rest and B. S. Bergstrom, in *Beam-Solid Interactions and Transient Processes*, edited by J. A. Knapp, P. Børgesen and R. A. Zuhr (Mater. Res. Soc. Proc. **74**, Boston, Ma. 1989) pp. 277.

CROSS SECTION TRANSMISSION ELECTRON MICROSCOPY (XTEM) ON INERT GAS IMPLANTED METALS

E. Gerritsen[1] and J.Th.M. De Hosson[2]

[1] *Philips Research Laboratories, P.O. Box 80000*
5600 JA Eindhoven, The Netherlands

[2] *Department of Applied Physics, Nijenborgh 18*
University of Groningen, The Netherlands

ABSTRACT

TEM specimen foils prepared perpendicular to the implanted surface allow a direct observation of the entire region modified by the injected ions. In this way depth resolved microstructural information can be obtained. The preparation technique developed for this purpose is extensively described here and applications are shown for copper and stainless steel implanted with various inert gases.

1. Introduction

Transmission electron microscopy (TEM) has been extensively used for studying microstructural effects in ion implanted metals. However, nearly all of this microscopy has made use of planar foils prepared parallel to the implanted surface. In this way no direct information can be obtained about the depth distribution of the microstructural features. This is for instance of interest in relation to the observation that the increased wear resistance of ion implanted metals often extends to depths far exceeding the projected range of the implanted ions [1].

For depth resolved microstructural observations, cross-section samples prepared perpendicular to the implanted surface, are required. Whereas the use of this cross-section TEM (XTEM) technique is common practice for use on implanted or layered semiconductor structures [2] its application to metals has been very limited. This is probably due to the specimen preparation technique for metals, which, although simple in principle, is quite tedious in practice. We have developed a preparation technique which has been applied to ion implanted copper and stainless steel.

The technique has been used as one of the microstructural tools in a larger study on the modification of metal surface properties by ion implantation [3]. Additional techniques included RBS/Channeling, Mössbauer spectroscopy, glancing angle X-ray diffraction as well as computer simulation.

The observation by XTEM of deep implantation damage, at depths far in excess of the projected ion range, has been reported previously [4]. In the present paper an extensive description of the XTEM preparation procedure is given with a summary of results obtained on inert gas implanted copper and stainless steel.

Fundamental Aspects of Inert Gases in Solids
Edited by S.E. Donnelly and J.H. Evans, Plenum Press, New York, 1991

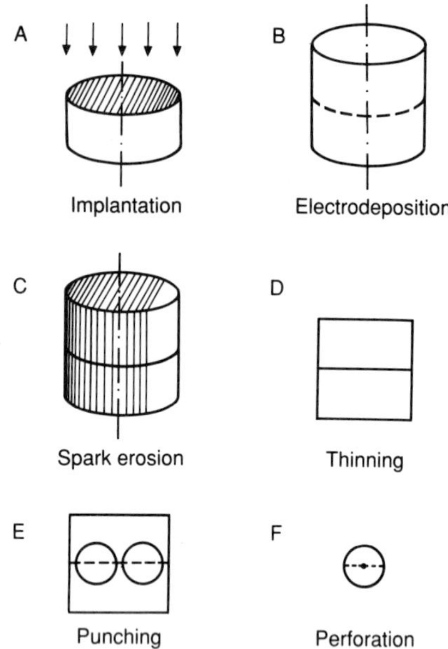

Fig. 1   General outline of our XTEM specimen preparation procedure.

2.   Experimental procedure

The basics of our preparation technique are outlined in Fig. 1. Generally a TEM specimen can be made transparent to electrons either by ion milling or by electrochemical thinning. Our experience was no exception to the common observation [5] that ion milling of implanted metals gives rise to fine defects which interfere with the implantation effects to be studied. Therefore we have thinned our samples by electrochemical jet polishing. This however, precludes the use of a protective layer which is glued on top of the surface to be studied. This is usually done for XTEM specimens of semiconductors which are made by ion milling. This thin layer of glue however, disturbs the flow of electrolyte during chemical thinning.

We, therefore, applied a thick electroplated layer to protect the implanted surface during cross-section preparation (Fig. 1b). This electroplating is the most crucial step in the XTEM preparation procedure. We have developed two specific preparation techniques for copper and stainless steel (AISI 316: 17%Cr,13%Ni) but, in fact, the technique described for copper can be directly transposed to other elementary metals which can be easily electrodeposited. These metals include silver, gold, zinc, cadmium, chromium, cobalt and nickel [6].

A strict requirement for the electrodeposit is that it should adhere strongly to the implanted surface. For this purpose specific adhesion layers can be applied [6]. For copper this layer is grown from a cyanide copper bath containing 40 g/l copper cyanide, 50 g/l sodium cyanide, 30 g/l potassium sodium tartrate ($KNaC_4H_4O_6.4H_2O$) and 15 g/l sodium carbonate with a current density of 2 A/dm^2 for 5 minutes.

The stainless steel was electroplated with nickel. Because this steel is covered with a passive layer of chromium oxide, adhesion can only be obtained by selectively removing this layer, taking care not to remove the thin implanted layer beneath. This process is carried out by immersion in a Woods nickel bath (240 g/l $NiCl_2$, 124 ml HCl, 1 l H_2O) for 1 minute. The amount of material removed was found from the RBS profile of 200 keV gold ions implanted in a stainless steel foil. After treatment this profile was found to have been shifted 10 nm towards the surface, which is much less than the typical range of 100-200 keV ions. After 1 minute of activation the sample is connected to the power supply at a current density of 3 A/dm^2 for 5 minutes to grow a thin adhesion layer of Woods nickel.

144

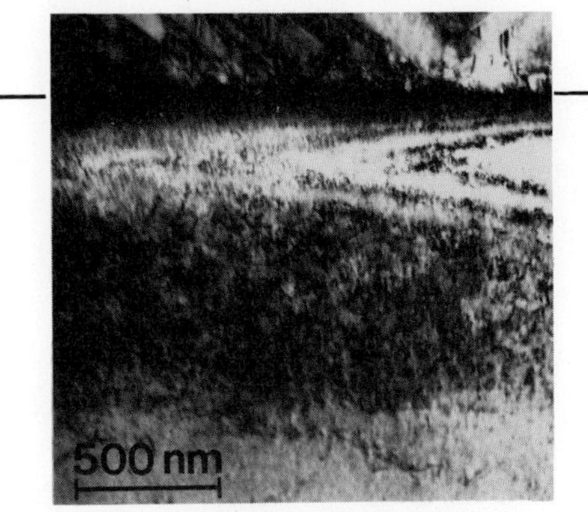

Fig. 2 XTEM micrograph of a copper (110) crystal implanted with 230 keV argon ions to a dose of 10^{17} Ar/cm^2. The position of the surface is indicated by the horizontal line.

Having applied the thin adhesive layers, plating has to be continued to give a thick deposit of 3 to 4 mm. To avoid stress build-up in the layer, plating currents have to be kept low, which implies plating times of several days. Gas bubbles formed during plating are removed from the surface by rigorous stirring, in order to avoid a porous deposit. The thick copper plating is obtained from an acid copper bath containing 220 g/l copper sulphate and 60 g/l sulphuric acid at a current density of 3 A/dm^2. The steel is plated with nickel in a Watts nickel bath at 50°C containing 300 g/l nickel sulphate, 55 g/l nickel chloride and 43 g/l boric acid at a current density of 5 A/dm^2.

After electrodeposition cross-sectional slices with a thickness of about 0.3 mm are obtained by wire spark erosion (Fig.1c). The copper slices are electrochemically thinned to a final thickness of 0.1 mm (Fig.1d). In order to prevent preferential thinning of the edges these must be protected by lacquer or by a Teflon sample holder. The polishing is performed at 2.35 V in an electrolyte containing 50% phosphoric acid (85%) and 25% glacial acetic acid.

Using a GATAN specimen punch, discs of 3 mm are cut from these slices (fig.1e), with the target-electrodeposit interface along the centre of the disc. Finally the discs are electro-chemically thinned to perforation in a Struers TENUPOL double jet polishing apparatus.

For the copper cross-sections, Struers D2 electrolyte is used. Because of the diffcrent polishing rates of steel and nickel the steel cross-sections are thinned using two electrolytes. Electrolyte 1 works faster on the nickel and electrolyte 2, which is used till perforation, works faster on the steel. A perforation near the steel/nickel interface is then obtained by balancing the polishing times. Electrolyte 1 contains 30% nitric acid and 70% methanol applied at 10V and a temperature of 0°C for 1 minute. Electrolyte 2 contains 10% perchloric acid and 90% glacial acetic acid, 20 g/l chromium oxide and 10 g/l nickel chloride applied at 45V and room temperature.

3.   Results

3.1.   Inert gas implanted copper

The XTEM micrograph in Fig. 2 shows a copper (110) crystal implanted with 230 keV argon ions along a <110> axis at a dose of 10^{17} Ar/cm^2. The position of the surface is indicated by the horizontal line. On top of it the electrodeposited layer can be observed. Beneath the surface a wedge shaped area has been etched away preferentially during XTEM thinning due to the presence of argon bubbles. The implantation induced defects extend to a depth of 850 nm. This damage range largely exceeds the LSS projected range which is tabulated to be $R_p = 90$ nm with $\Delta R_p = 40$ nm [7]. Deep damage has been extensively studied for copper implanted

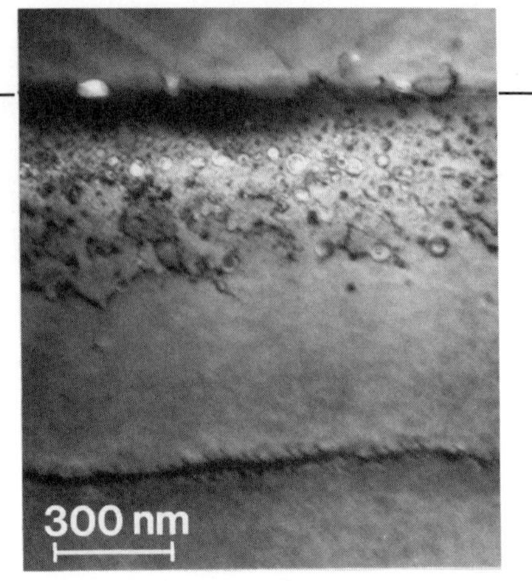

Fig. 3   XTEM micrograph showing a deep dislocation wall in a (110) copper crystal implanted with 230 keV argon ions after heating for 1 minute at 600°C.

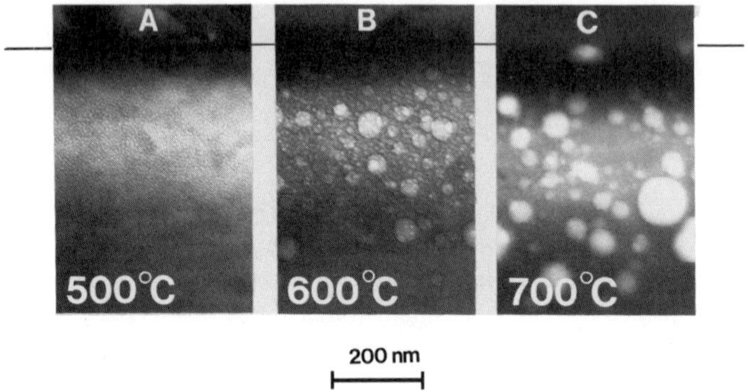

Fig. 4   XTEM micrograph showing bubble coarsening in a copper (110) crystal implanted with 230 keV argon ions after heating for 1 minute to 500, 600 and 700°C.

with aluminium ions [3,4]. In that study it was found that the damage range was determined by the maximum penetration of a small fraction of channelled ions. These ions, with initial energy E_0, slow down mainly by electronic energy losses $(dE/dx=S_{el})$ and their maximum range R_{max} can be calculated from:

$$R_{max} = \int (S_{el})^{-1} dE = \int (k\sqrt{E})^{-1} dE = (2/k)\sqrt{E_0} \tag{1}$$

The electronic stopping constant k is determined by the ZBL-formalism [8] which gives R_{max} = 700 nm for this 230 keV argon implant. By EDX analysis (detection limit $\approx$1%) in the transmission electron microscope, argon was detected up to a depth of 600 nm.

A striking observation in the SADP patterns of this specimen was the significant shift of the Kikuchi line pattern in going from the modified top layer to the underlying copper crystal. This indicates that there is an orientational difference between the two regions. From the size and direction of the pattern shifts it was found that the lattice planes in the surface layer were tilted with respect to the bulk over 1.2 ± 0.1 degrees, with the tilt axis being parallel to the implanted surface. This effect has been observed previously by RBS/Channeling on As-implanted vanadium [9] as well as by XTEM on Al-implanted copper [10]. It is attributed to dislocation glide driven by the strain fields around the implanted atoms and at the interface between the implanted and the unimplanted bulk region.

Fig. 3 shows the damage recovery after in-situ heating the specimen for 1 minute at 600°C. The deeper part of the modified layer seems to have been fully recovered whereas the recovery of the top layer has been hindered by the presence of the argon bubbles. Furthermore a fairly straight low angle boundary is observed at a depth of 850 nm, which is formed by the line-up of dislocations. This recovery process, polygonisation, is commonly observed during the anneal of plastically bent crystals [11]. The resulting microstructure might be particularly effective in inhibiting fatigue crack initiation by blocking the surface penetration of slipbands originating in the bulk [12].

The series of micrographs in Fig. 4 focusses on the growth of the bubbles after heating to 500, 600 and 700°C. In Fig. 4a a dense distribution of bubbles can be observed in the depth region from 50 to 400 nm. Heating to 600 and 700°C results in a further coarsening of the bubble size.

Fig. 5 shows two preliminary examples of the application of XTEM on the sputter induced topography on implanted surfaces. This phenomenon has been studied for many years [13,14] but the mechanism of the development of the two major features, etch pits and pyramids, is still not clear. XTEM has some potential to clarify the underlying, i.e. subsurface, microstructural details accompanying the formation and evolution of this topography. Fig. 5a is obtained from a copper crystal implanted with krypton, whereas the sample in Fig. 5b was implanted with lead, which is also insoluble in copper.

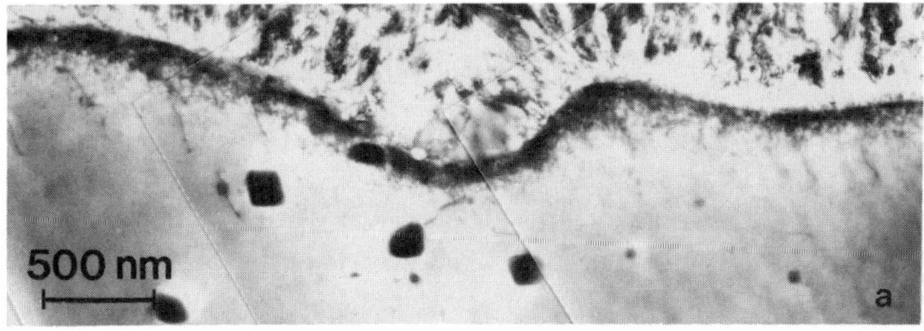

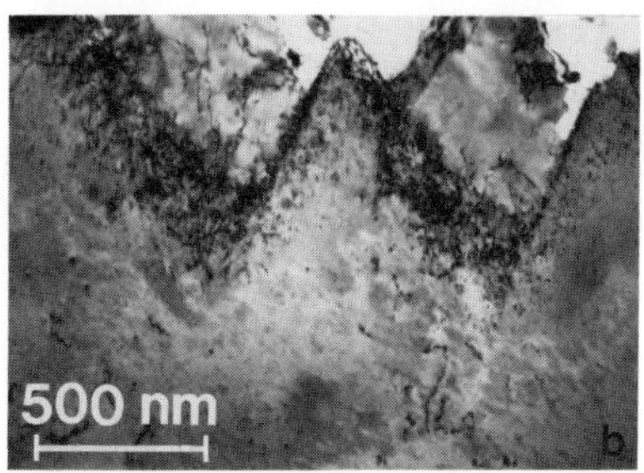

Fig. 5 XTEM micrographs showing surface topography on implanted copper crystals.
a) Cu(151): 35 keV Kr, $2.5 \cdot 10^{18}$ Kr/cm^2.
b) Cu(111): 150 keV Pb, $2 \cdot 10^{18}$ Pb/cm^2.

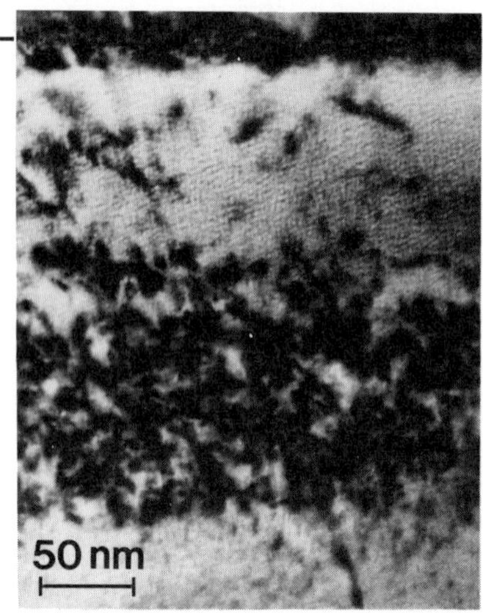

Fig. 6   XTEM micrograph of AISI316 stainless steel implanted with 200 keV argon to a dose of 10^{16} Ar/cm^2.

3.2. *Inert gas implanted stainless steel*

Fig. 6 shows an XTEM micrograph of a 200 keV Ar implanted foil of austenitic (f.c.c.) stainless steel. Implantation of inert gases above a certain threshold dose has been found to induce a martensitic transformation from f.c.c. to b.c.c. in this type of steel [15]. These implantations might therefore be of practical importance since the conventional work-hardening of this steel is due to the stress-induced formation of martensite. In Fig. 6, a fairly regular areray of argon bubbles can be observed up to a depth of 130 nm, which compares well with the LSS ion range plus straggling calculated to be $R_p = 80$ nm and $\Delta R_p = 40$ nm. Implantation damage can be observed up to a depth of about 300 nm.

Another 200 keV argon implantation is shown in Fig. 7. Here the target is a single crystal of similar composition as the AISI 316 steel of Fig. 6 (17%Cr:13%Ni). The crystal was implanted along the <321> surface normal, which gives rise to major channeling effects along the (111) planes perpendicular to the surface. The damage range is now extended to about 500 nm. According to equation (1) the maximum penetration of channeled argon ions is calculated to be 510 nm, corresponding to the observed damage range.

The specimen of Fig. 7 is also used to obtain Fig. 8, where defocussed phase contrast is applied to show the argon bubbles more clearly. It can be recognized that there is a certain size distribution of the bubbles. The smallest bubbles are found up to a depth of about 200 nm and the largest ones are seen at about 70 nm. The latter depth corresponds to the projected ion range of 80 nm. (Note that 10 nm of the surface is removed during XTEM specimen preparation).

Fig. 9 shows the bright field and dark field XTEM micrographs of a 17Cr:13Ni steel single crystal of (321) surface orientation implanted along the surface normal with 230 keV xenon to a dose of 10^{16} Xe/cm^2. The top layer of about 50 nm consists of a dense array of xenon inclusions. This thickness coincides with the LSS range (36 nm) plus straggling (12 nm) for the xenon ions. In the dark field micrograph of Fig. 9b small defects induced by individual collision cascades can clearly be observed up to about 300 nm. For this implantation the maximum penetration depth of channeled ions is calculated to be 280 nm.

For the implantation of 230 keV Xe in a 17Cr:13Ni steel single crystal the threshold dose for martensite formation was found to be 2 10^{16} Xe/cm^2 [15]. The XTEM micrograph of Fig. 10 was obtained from a crystal implanted to a dose of 5 10^{16} Xe/cm^2. The martensite can be

148

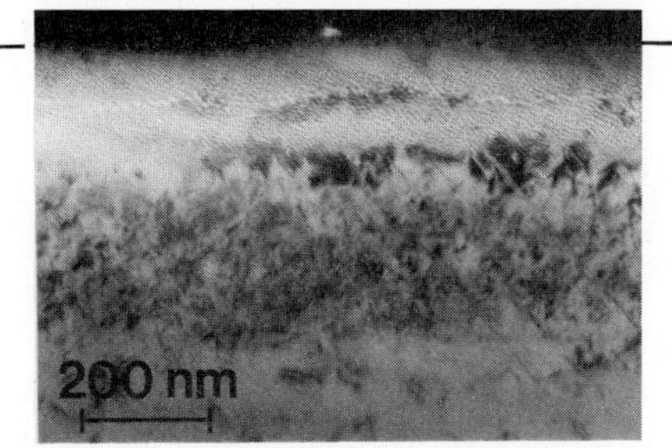

Fig. 7   XTEM micrograph of a 17Cr:13Ni (321) single crystal implanted with 200 keV argon to a dose of 10^{17} Ar/cm^2.

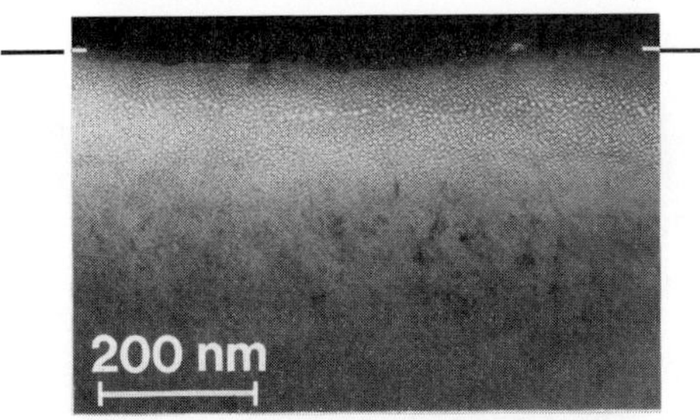

Fig. 8   XTEM micrograph of the sample of Fig. 7, showing an argon bubble layer just under the surface.

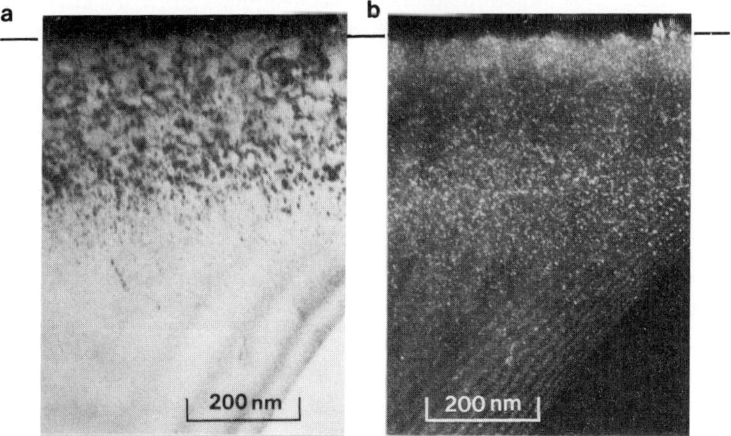

Fig. 9   XTEM micrographs showing a top layer of xenon bubbles and small defects up to a depth of 300 nm in a 17Cr:13Ni (321) single crystal implanted with 230 keV xenon to a dose of 10^{16} Xe/cm^2.   a) Bright field   b) Weak beam dark field.

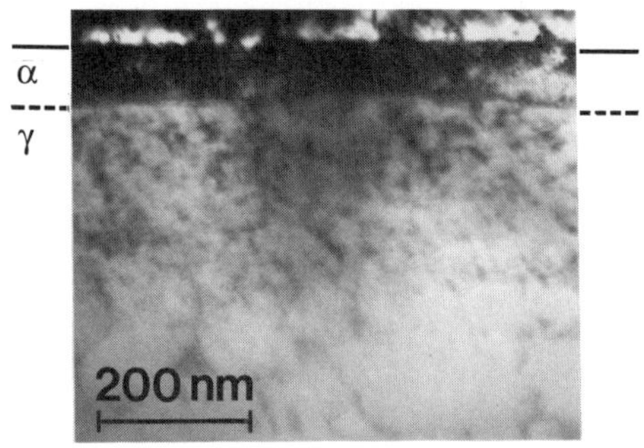

Fig. 10 XTEM micrograph showing a top layer of martensite (α) in a 17Cr:13Ni (321) single crystal implanted at random incidence with 230 keV xenon ions to a dose of 5.10^{16} Xe/cm^2.

seen as a homogeneous layer extending from the surface to a depth of 90 nm, where it ends in a rather sharp martensite(α)-austenite(γ) interface. This, as well as other experiments [16,17] suggests that the martensite nucleates at the surface and grows to larger depths with increasing dose. Recent experiments [18] however, give evidence that nucleation occurs in the vicinity of the largest gas inclusions, located at the projected ion range, where there is a maximum shear stress component. Beneath the martensite layer in Fig. 10 a region of 200 nm is observed having a high density of dislocation tangles.

The microbeam (diameter 100 nm) diffraction pattern taken from the martensite top layer is given in Fig. 11a. It shows a clear b.c.c. pattern with (131) zone axis. The mixed pattern of Fig. 11b is obtained from the martensite/austenite interface region. It consists of a superposition of a (131) and a (112) pattern (Fig.11c). This orientation relationship, with the closest packed planes ((111) and (110)) in each phase being parallel, is close to the Nishiyama-Wassermann relation, which has been observed for martensitic transformations in many other systems [19]. The clear pattern in Fig. 11a indicates that the top layer contains a large fraction, presumable 100%, of martensite. This transformation efficiency is much larger than that during conventional formation of martensite by severe deformation. There the efficiency is at most 40% at room temperature [20,21].

4.   Conclusions

Cross-section TEM (XTEM) on inert-gas-implanted metals gives additional depth-selective information which is not obtained from conventional planar TEM specimens.Using XTEM it is found that the depth distribution of gas bubbles is in agreement with depth profiles as calculated using standard range theories (such as L.S.S.). However, implantation damage is observed up to depths far in excess of the calculated ion range. This deep damage is induced by a small fraction of channelled ions. In copper, the stress gradient in the surface layer gives rise to a tilting with respect to the unimplanted bulk region. In austenitic stainless steel these high stresses induce a martensitic phase transformation in the surface layer.

ACKNOWLEDGEMENTS

The effort in applying XTEM on implanted metals has been very much stimulated by the many discussions with E. Johnson (Univ. Copenhagen).The assistance of H.A.A. Keetels in specimen preparation is gratefully acknowledged. Implantations were performed by J. Politiek and by A. Johansen (Univ. Copenhagen). The samples in Figs. 5a and 5b were provided by G. Carter (University of Salford) and L.Sarholt-Kristensen (University of Copenhagen). Part of this work has been funded by the EEC programme Basic Research for Industrial Technology in Europe (BRITE) under contract number 1357.

150

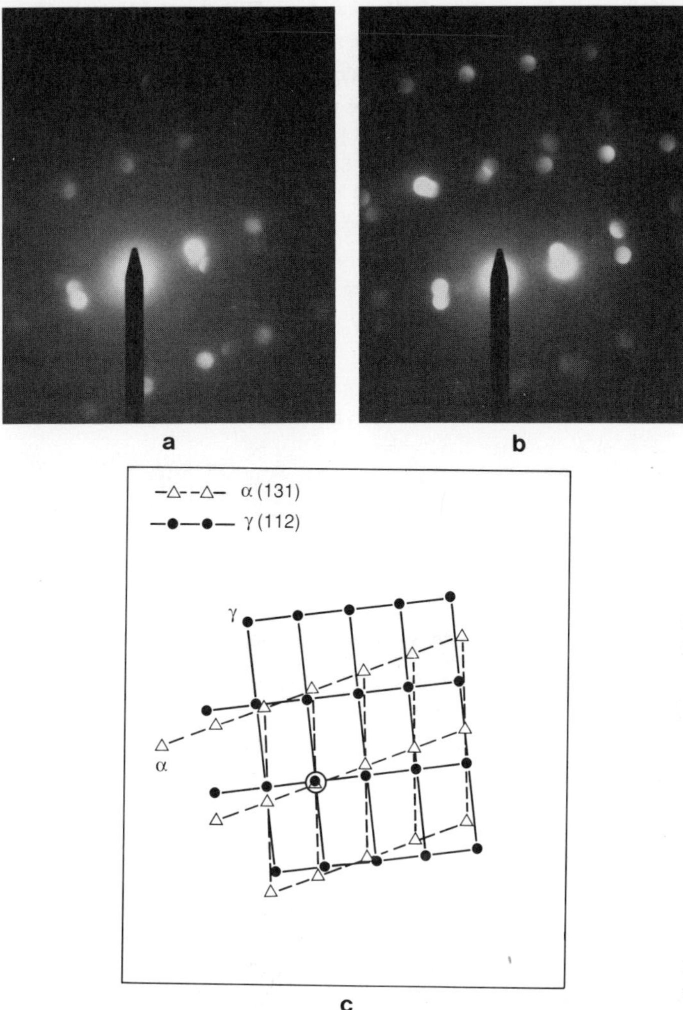

Fig. 11 Selected area microprobe diffraction patterns of the specimen in Fig. 10. a) Taken from the top layer. b) Taken from the α-γ interface region. c) Schematic indexing of pattern b.

REFERENCES

1. H. Dimigen, K. Kobs, R. Leutenecker, H. Ryssel, P. Eichinger, Mat. Sci. Eng. **69**, 181 (1985).

2. C. Bulle-Lieuwma, P.C. Zalm, Surf. Int. Analysis **10**, 210 (1987).

3. E. Gerritsen, *Surface Modification of Metals by Ion Implantation*, Ph.D thesis, University of Groningen, The Netherlands (1990).

4. E. Gerritsen, H.A.A. Keetels, H.J. Ligthart, Nucl. Instr. Meth. **B39**, 614 (1989).

5. J.Th.M. de Hosson, Phys. Stat. Sol. **A40**, 293 (1977).

6. T. Rodgers, *Handbook of Practical Electroplating*, Mc. Millan, New York (1959).

7. A. Burenkov, F. Komarov, M. Kumakov, M. Temkin, *Tables of Ion Implantation Spatial Distributions*, Gordon and Breach Publ. (1986).

8. J. Ziegler, J. Biersack, U. Littmark, *The Stopping and Range of Ions in Solids*, vol. 1, Pergamom Press (1984).

9. O. Meyer, A. Azzam, Phys. Rev. Lett. **52**, 1629 (1984).

10. E. Gerritsen, H.J. Ligthart, T.E.G. Daenen, Mat. Res. Soc. Proc. **93**, 329 (1987).

11. D. Hull, *Introduction to Dislocations*, p.201, Pergamom Press (1975).

12. S. Spooner, K. Legg, in: *Ion Implantation Metallurgy*, eds C. Preece, J. Hirvonen, AIME (1980).

13. J.L. Whitton, G. Carter, M.J. Nobes, J.S. Williams, Radiat. Eff. **32**, 129 (1977).

14. G. Carter, I.V. Katardjiev, M.J. Nobes, L. Tanovic, N. Tanovic, Nucl. Instr. Meth. **B48**, 576 (1990).

15. E. Johnson, E. Gerritsen, N.G. Chechenin, A. Johansen, L. Sarholt-Kristensen, H.A.A. Keetels, L. Grabaek, J. Bohr, Nucl. Instr. Meth. **B39**, 573 (1989).

16. A. Johansen, E. Johnson, C. Sarholt-Kristensen, S. Steenstrup, E. Gerritsen, C.J.M. Denissen, H. Keetels, J. Politiek, N. Hayashi, I. Sakamoto, Nucl. Instr. Meth. **B50**, 119 (1990).

17. J. Klostermann, W.G. Burgers, Acta Met. **12**, 355 (1964).

18. J. Noordhuis, J.Th.M. de Hosson, Acta Met., **38**, 2067 (1990).

19. C. Barret, T. Massalski, *Structure of Metals*, Pergamom, (1987).

20. T. Angel, J. Iron Steel Inst. London **177**, 165 (1954).

21. D. Cook, Metall. Trans. **A18**, 201 (1987).

FUNDAMENTAL AND APPLIED ASPECTS OF NOBLE GAS BUBBLES IN STEEL

J. Noordhuis and J.Th.M. De Hosson

Department of Applied Physics, Materials Science Centre
University of Groningen, Nijenborgh 18, 9747 AG Groningen, The Netherlands

ABSTRACT

This paper reports a study aimed at improving the wear performance of common steels by noble gas ion implantation and laser melting. It turns out that the shear stress field of a noble gas bubble in a metastable austenitic steel (SS 304) induces a martensitic transformation, provided that bubble size and pressure are large enough. The wear resistance decreases with increasing dose. In laser melted stable austenitic steel (RCC) subsequently implanted with noble gas ions, transmission electron micrographs has provided clear evidence of the formation of a large number of bubbles, Orowan looping and pinning of dislocations by these bubbles. The wear rate decreases with increasing dose. However, this is certainly not a general observation of laser melted–ion implanted steels. The low carbon steel CK22 for instance shows a martensitic structure after laser melting with a high dislocation density. In contrast to RCC, the wear rate increases with increasing dose suggesting that moving dislocations are interacting with forest dislocations rather than with the noble gas bubbles.

1. Introduction

Despite the advantages of laser processing for the production of wear resistant materials, laser melting results in tensile stresses. These tensile stresses may lead to severe cracking and to deleterious effects on the wear behaviour. Here, our basic idea is to convert these tensile stresses into compressive ones by noble gas implantation. When pressurized bubbles are nucleated, one can imagine that the corresponding stress field could annihilate the tensile stress field of the laser treated material. Furthermore, the surface layer might also be strengthened during wear, because of the interaction between moving dislocations and the noble gas bubbles.

Further, the technique of ion implantation has been applied successfully in the past to improve the wear performance of metals and ceramics. Usually the creation of second phase particles is the underlying mechanism. Because the penetration depth is inversely related to the ion mass, light elements are favourable. Therefore boron, carbon and nitrogen are frequently used as ion species [1]. Other mechanisms are an increase in hardness due to radiation damage or an increased toughness due to amorphization. Pinning of dislocations can also play an important role [2-4].

This paper reports of our current research on the effects of neon implantations in different steels and we will show that in a metastable austenitic steel a pressurized neon bubble can induce even a martensitic transformation by means of its shear stress field.

Wear and hardness measurements are performed in order to obtain information of the mechanical properties, whereas TEM and X–ray diffraction are used to detect microstructural changes induced.

2. Theoretical Concepts

2.1. Nucleation of martensite

When austenitic stainless steel (fcc) is implanted with certain kinds of atoms, the top layer transforms to the stable phase (bcc) [5–7]. Because of the low temperature and the step–like nature [5, 6], the transformation must have a military (martensitic) character. Basically there are two driving forces for the transformation:

- the build–up of stress in the top layer [8].
- a compositional change that favours the bcc phase. In this case the reverse transformation is also possible [4].

By using inert gases one can exclude the latter one; doing so, the efficiency is not lowered, but actually increased [9-11]. It is even possible to cause a transformation in SS 304 implanted with helium [7], an element which causes hardly any implantation damage, and thus a low level of damage induced stress. It can be stated that the martensitic transformation is characterized by its shear–like nature [12,13]. An appropriate shear–stress will thus favour a martensitic transformation. By considering the shear stress, the chemical free energy, the interfacial energy and the resultant strain energy, the total energy curve can be calculated as a function of the nucleus size [14].

Consider a lens–shaped nucleus with radius r and thickness t. The mean shear strain will be:

$$e_0 = \frac{\theta t}{2r} \tag{1}$$

where θ is the martensite shear angle. When there is an appropriate stress τ already present with a corresponding strain e_1, the net strain is:

$$e = e_0 + \frac{\theta t}{2r} - \frac{\tau}{G} \tag{2}$$

The negative sign is only valid if the shear stress on a plane favours the formation of martensite on that particular plane.

The energy per unit volume is given by:

$$E_v = \frac{1}{2} G e^2 = \frac{1}{2} G \left(\frac{\theta t}{2r} - \frac{\tau}{G} \right)^2 \tag{3}$$

where G represents the shear modulus. The total energy of formation of the nucleus can be written as:

$$E = 2\pi r^2 \gamma + \frac{1}{2} \pi r^2 t \left(Q - \frac{4}{3} \tau \theta \right) + \frac{1}{6} \pi r t^2 G \theta^2 \tag{4}$$

where Q is the change in chemical free energy. Hydrostatic and normal components are not taken into account. We assume that we have a shear–field in the plane of the nucleus. It is possible however, to tilt the nucleus over a small angle, so that the shear–component equals $\tau_s = \tau \cos(2\phi)$ and a normal component $\tau_n = \tau \sin(2\phi)$. The normal component favours the volume expansion. It is found that the angle is about five degrees and the energy gain is only two

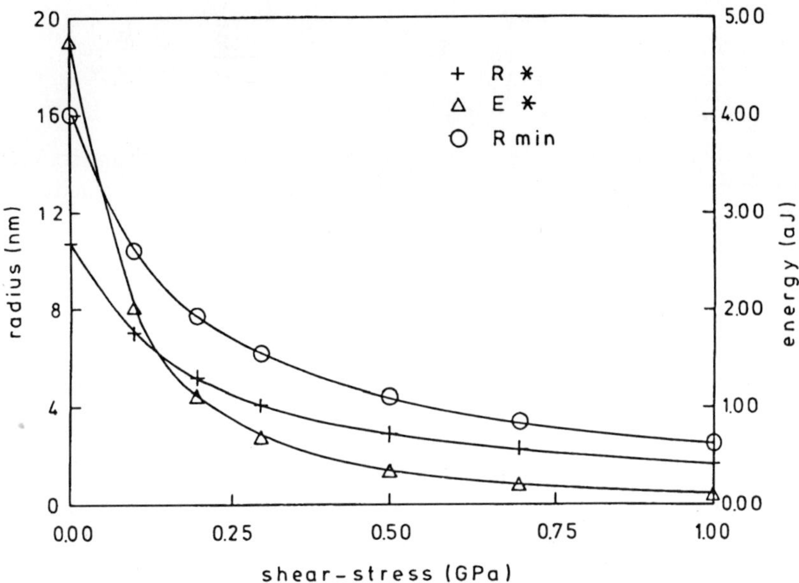

Fig. 1 Shear stress dependence of minimum size, critical nucleus size and activation energy, as given by the theory for homogeneous nucleation. (See Eq. 4).

percent. Therefore, it seems reasonable to neglect this effect and to look in the following only at shear stresses.

In the description of the process of nucleation there exist three parameters: activation energy E^*, critical size and minimum size of the nucleus. They are calculated with the aforementioned equations for various stresses and are depicted in Fig. 1 with $G = 70$ GPa, $\theta = 11.5°$, $Q = 150$ MJ/m^3, $\gamma = 20$ mJ/m^2 and $t = r/10$. It must be emphasized that the value for is the one for coherent nuclei [15]. From Fig. 1 it is clear that nucleation is more likely to occur in a strong shear field. In particular, the activation energy is drastically lowered, and comes in the range below 1 aJ when a stress of about 300 MPa is applied. On the other hand, one has to consider that the shear stress at yielding of SS 304 is only about 150 MPa. Consequentially such large stresses are not possible under normal conditions in annealed SS 304.

As the nucleus thickens or the shear–field disappears due to the transformation, the growth of the martensite has to be assisted by dislocations or by twinning.

2.2. Stress fields in the implanted layer

During implantation Frenkel pairs are produced; about 450/neon atom at 50 keV. Most of them annihilate, but neon will be trapped by the vacancies left. As a consequence the interstitials forming loops or trapped at dislocations will contribute to a compressive stress. This stress field can be in the order of 1 GPa [8,9]. When a bubble absorbs an interstitial neon or coagulates with another bubble, the pressure in the new bubble may exceed the equilibrium pressure ($P_{eq} = 2\Gamma/r$, Γ is the surface free energy and a value of 2.5 J/m^2 is often used for fcc iron [9]). This excess pressure can be reduced by emitting interstitials or dislocation loops, thereby contributing to a compressive stress. The bubbles have a stress field consisting of a hydrostatic and shear components [16]:

$$\sigma_H = \frac{\bar{p}\,S}{1-S} + \frac{(p-\bar{p})\,\alpha S}{1-\alpha S} \tag{5}$$

155

$$\sigma_{ij} = \frac{a^3}{2r^3} \left(\delta_{ij} - 3\frac{x_i x_j}{r^2} \right) \left(\frac{\bar{p}}{1-S} + \frac{p-\bar{p}}{1-\alpha S} \right) \tag{6}$$

where S is the volume fraction of the bubbles. r and x_i represent the length and components of the radius vector with the origin in the bubble centre, and δ_{ij} is the unit tensor, and α is defined by:

$$\alpha = \frac{(1-\bar{G}/G)}{(1+4\bar{G}/3K)} \tag{7}$$

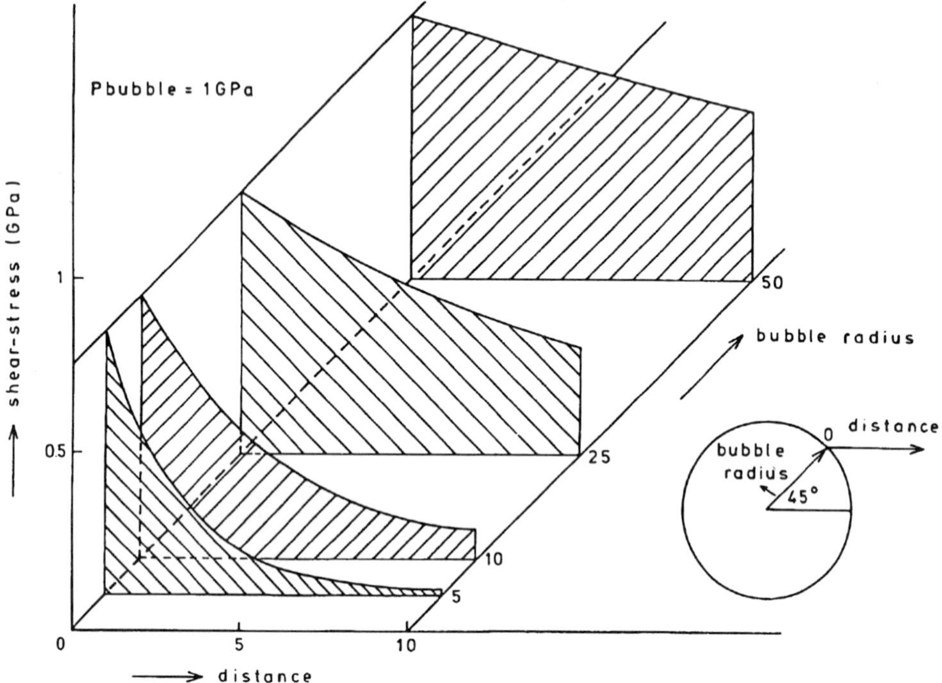

Fig. 2   Shear field of a bubble as a function of distance and size of the bubble.

It is easy to see that σ_H plays a minor role because it is scaled with S. When we take a realistic bubble pressure of 1 GPa, the shear stress can be calculated as a function of bubble size at various distances from the edge of the bubble; this is illustrated in Fig. 2. When there are two bubbles close to each other the fields will accumulate. This is the reason for inter–bubble fracture when the inter–bubble distance becomes too small. When on the other hand the inter–bubble distance becomes too large, there is no repulsive force on emitted dislocation loops, and the pressure can more easily be lowered to equilibrium pressure.

3. Experiments

3.1. *Ne implanted stainless steel*

The material investigated is SS 304 (19 wt% Cr, 10 wt% Ni, 2 wt% Mn, < 0.08 wt% C and bal Fe), because of its low austenite stability, i.e. its low position in the Schaeffler diagram [17]. Samples were made fully austenitic by annealing at 1065 C followed by water quenching. TEM–samples were implanted at 50 keV Neon before as well as after thinning. Samples for X–ray measurements were implanted with 110 keV in order get a penetration (0.1 μm), closer to that of the X–rays (Copper K–α: 6–10 μm for the used angles). The fluences used range up to $3\ 10^{17}$/cm^2; the implantation temperature was kept below 150 °C.

Fig. 3a shows a picture of a sample that was implanted with $3\ 10^{17}$/cm^2, and prepared following the bulkside–thinning method. Here bubble radii up to 40 nm can be observed within a broad distribution. From the diffraction pattern and the dark field micrograph it is evident that the dark areas represent a bcc phase. In contrast to other researchers [10,11] the bcc phase has hardly any orientation relationship with the matrix. It is also noteworthy that the size of the

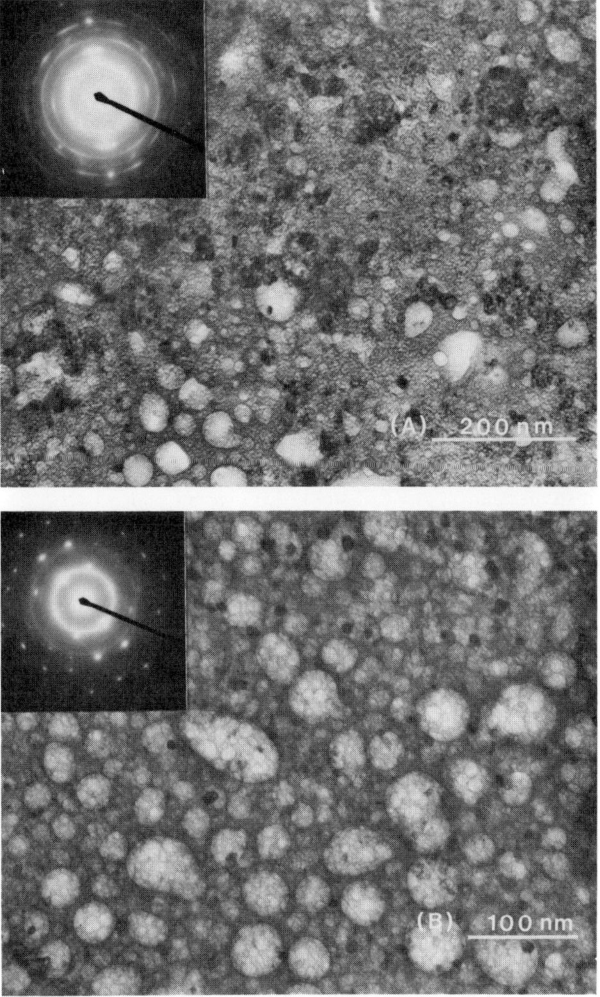

Fig. 3   Bright field TEM micrograph of an implanted bulkside thinned sample (a) and an implanted thin foil (b), showing bubbles and bcc particles (dose $3\ 10^{17}$/cm^2).

particles is smaller than observed by [10]: 50 nm instead of 120 nm.

In a thin foil, implanted with 1 10^{17}ions/cm^2, the bubble radii vary from 1 to 5 nm, and no bcc phase can be detected. When a thin foil is implanted with 3 10^{17}/cm^2 the bubble size distribution is the same as in the bulkside–thinned sample, but the bcc particles look more spherical and are even smaller: about 8 nm diameter (Fig. 3b). In this case there is no orientation relationship at all (see inset Fig. 3). The critical dose for transformation in thin foil samples amounts to about 1.5 10^{17}/cm^2. To investigate the effect of the bubble size an implanted thin foil, dose: 1 10^{17}/cm^2, has been annealed at 400 C for 144 hours. After this anneal the maximum bubble size was increased from 5 to about 25 nm, and in the vicinity of a group bubbles some bcc particles, have grown. In annealed thin foils twinning can often be observed. Fig. 4 shows a bcc particle, which is larger than average and internally twinned in a sample annealed for 67 hours. The twins are 1–4 nm in size, and are also observed in the bulkside thinned samples.

In the annealing experiments there was however a drastic decrease of the amount of the bcc phase as the thickness of the foil increased. In order to exclude the effects of foil thickness, the same experiment was carried out for X–ray samples with doses of 0, 1, 2, 3 10^{17}/cm^2. From other experiments [18], it was known that the critical dose for transformation for unannealed samples is about 2.3 10^{17}/cm^2. After annealing for 67 hours, the number of X–ray diffraction peaks of the sample implanted with 3 10^{17}/cm^2 remained unchanged. From the sample implanted with 2 10^{17}/cm^2 a small bcc peak could be detected. The unimplanted sample did not exhibit a bcc phase after annealing.

In order to characterize the mechanical properties of the implanted layer, hardness and wear measurements have been performed. Hardness measurements were done with a Knoop indenter because of the large length to depth ratio of the indentation. Wear measurements were performed on a standard pin–on–disk apparatus, and analyzed with an interference microscope [3].

The hardness of 304 steel implanted with 3 10^{17}/cm^2 neon increases substantially; 202 HK$_{0.015}$ before and 251 HK$_{0.015}$ after implantation. From the plot in Fig. 5, it is clear that the wear resistance decreases with increasing neon dose, in spite of the increased hardness at the highest dose.

3.2. Ne/He implanted laser melted steels

The material under investigation is RCC steel (2.05 wt% C, 11.05 wt% Cr, 0.62 wt% W and bal Fe). It has been chosen because it shows a constant hardness profile after laser melting. Furthermore, this material doesn't have a martensitic phase after laser melting and compression due to martensite is not possible.

In this study a transverse flow Spectra Physics 820 CO$_2$ laser system has been used. The parameters of the treatment are: a measured power on the surface of 1350 W, a focal length of 127 mm, a focus point at 15 mm above the surface, resulting in a melted track of 2 mm. Tracks are made adjacent to each other at a distance of 1 mm apart under a protective argon atmosphere. After laser melting the microstructure consists of a dendritic structure of retained austenite, which is surrounded by M$_3$C carbide. The surface roughness created by laser melting is smoothened by grinding with SiC paper and finally polished using diamond paste.

Implantations are carried out with doses of 3 10^{16}, 1 10^{17} and 3 10^{17}/cm^2 at an energy of 50 keV per Ne atom. TEM observations of laser melted-Ne implanted material showed bubbles with a radius up to 34 nm. To determine the implantation depth the bulkside thinning method applied has a disadvantage since there is no guarantee that the front surface near the polished hole will not be affected by the polishing solution. Therefore thin foils have been implanted as well. Fig. 6 shows a stereo TEM picture of an implanted foil exhibiting the depth distribution of the bubbles throughout an austenitic cell. The bubble density is 5.7 10^{20} bubbles/m^3.

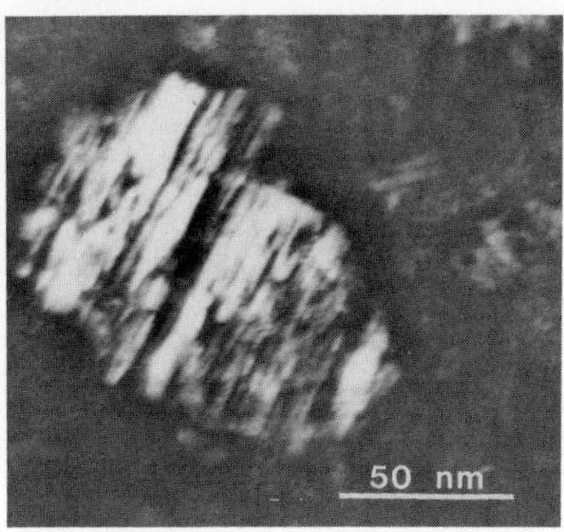

Fig. 4 Dark field micrograph of a large twinned bcc particle.

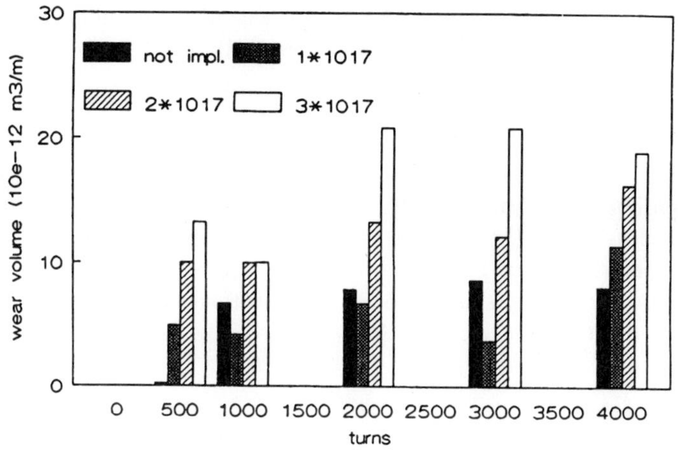

Fig. 5 Histogram showing the wear characteristic SS 304 for various doses.

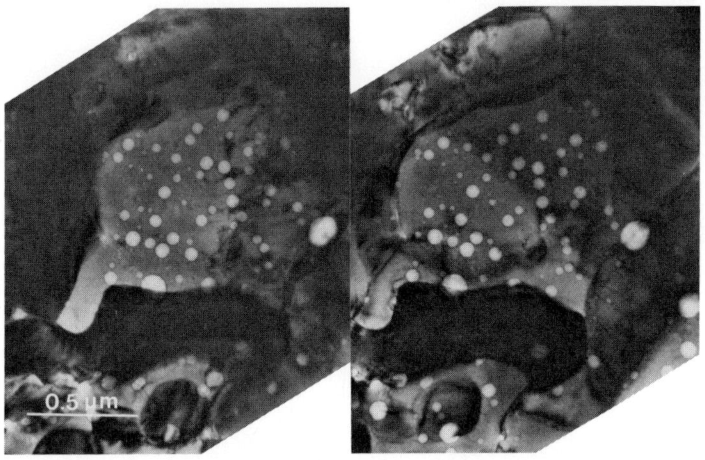

Fig. 6 Stereo TEM picture of Ne bubbles in laser treated RCC steel (stereo angle 20°, dose 3 10^{17} Ne/cm^2).

A distribution of bubble radii is shown in Fig. 7. The mean bubble radius is 20 nm and the mean volume is 4.5 10^4 nm^3.

The wear performance is depicted in Fig. 8. Clearly, the wear rate decreases with increasing Ne dose. At a dose of 3 10^{16} Ne ions/cm^2 the wear rate increases after 6000 turns. The depth of the wear track is then about 160 nm, which is two times the calculated implantation depth. This increased wear rate is due to the interfacial layer causing both compressive and tensile stresses to be present in the wear track. Such a transition is also found with the highest dose implantation. Extended measurements done at lower loads show, after 15000 turns, when the depth of the wear track is about the same as compared to the lowest dose implantation, an increase in wear rate. The wear rate increases gradually to a value of the unimplanted laser melted steel.

The measured hardness profiles show a hardness varying from 650 $HV_{0.025}$ up to 950 $HV_{0.025}$. This means an extra hardening due to Ne of 350 $HV_{0.025}$. It has to be emphasized that the real hardness of the implanted layer is difficult to measure, because the indenter penetrates the layer. We may only conclude that there is a significant hardening which is at least up to 900 $HV_{0.025}$. This extra hardening might be due to the compressive stress which increases the yield stress. Moreover, hardening is to be expected from noble gas bubbles which impede dislocation movements.

From Fig. 8 it was concluded that the wear rate decreases with increasing dose. This is certainly not a general observation of laser melted–ion implanted steels. CK22 (0.2 wt% C, 0.25 wt% Si, 0.5 wt% Mn, bal Fe) for instance shows a martensitic structure after laser melting, with a high dislocation density. After noble gas implantation, (3 10^{17} Ne/cm^2 at 28 keV followed by 8 10^{17} He/cm^2 at 110 keV) the increase in hardness is only negligibly small. Wear experiments were performed by giving different parts of the sample a different implantation dose to study the wear coefficient as a function of dose. In contrast to RCC, from Fig. 9, it is clear that the wear rate increases with increasing dose.

4. Discussion

The analysis of the experimental results suggests that a shear stress produced at the Ne–bubble is a plausible nucleation site for the martensitic transformation. Although the magnitude of the stress field is large, its range strongly depends on the bubble size.

The critical dose of about 2.3 10^{17}/cm^2 (in the bulk samples), is exactly in the range where there is a strong increase in bubble size.

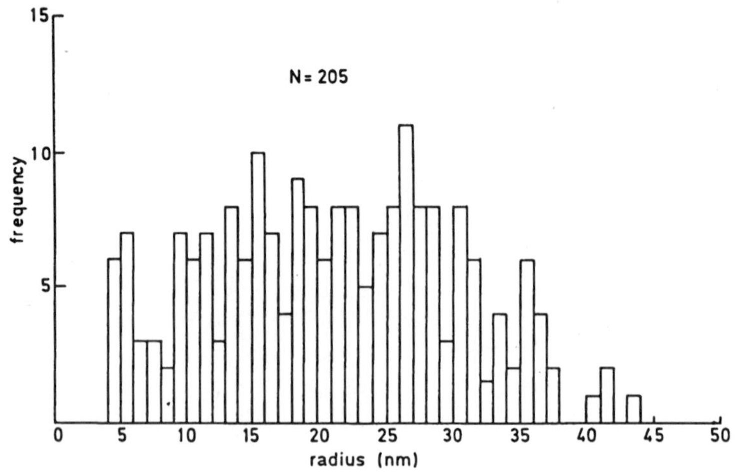

Fig. 7 Bubble size distribution after implantation of 3 10^{17} Ne/cm^2 into laser melted RCC steel.

160

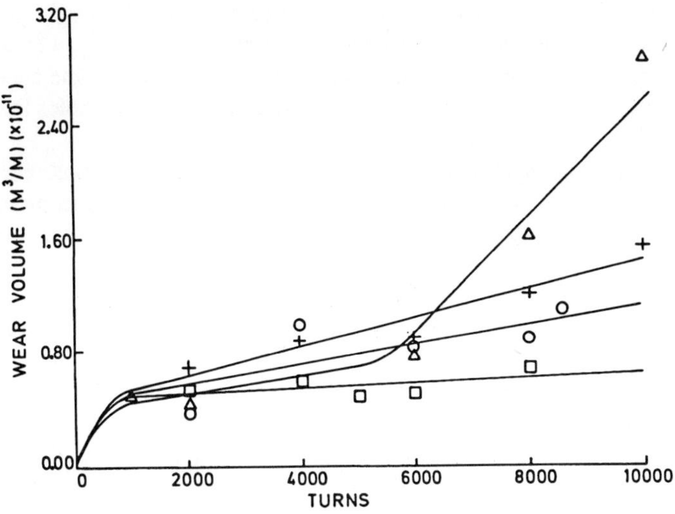

Fig. 8 Measured wear volume of laser-melted RCC steel: $+$ unimplanted, $\triangle$ 3 10^{16}, $\bigcirc$ 1 10^{17}, $\square$ 3 10^{17}/cm^2.

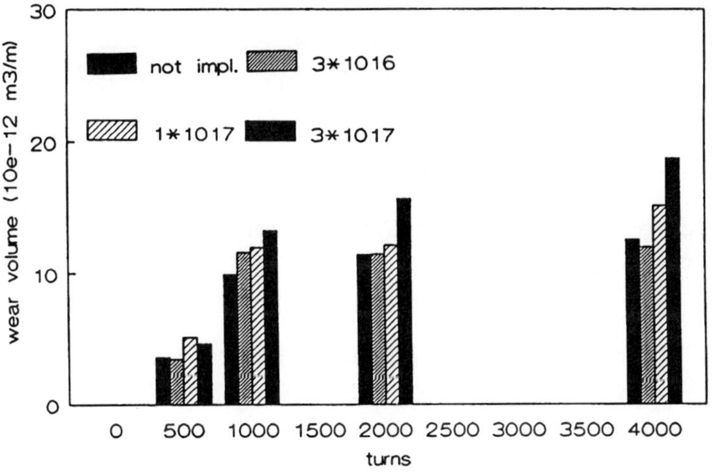

Fig. 9 Wear characteristic of laser treated CK22 steel for various doses.

On the other hand we have to note that besides the stress fields of the bubbles there is also a radiation damage induced stress field, which is a driving force for the martensitic transformation. This, however, does not explain the effect in the thin foil samples.

By inspecting stereo–micrographs it was not possible to obtain a 3–D impression of the bcc particles because of the random orientation and thus random Bragg–conditions. Therefore it is not possible to conclude whether the phase transformation occurs preferentially at the bubble surface, or in between two bubbles because of the accumulation of the stress–fields, or even at the sample surface. The edge of the bulk–side thinned samples however, representing the top of the implanted layer, shows no bubbles or bcc particles. This supports the conclusion that the transformation does not start at the surface, but more deeper where there are more bubbles.

From the annealed specimens we prove that the bubble size is a crucial factor in the martensite nucleation. It is important that the bubbles in the annealing experiments maintain their pressure. This means that it is not allowed to rise the temperature to a value where

161

thermal vacancies become available, because in that case the bubbles might grow by absorption. According to [19], one should keep the temperature below 0.45 T_m, which is satisfied in our experiment. So it is reasonable to assume that the large bubbles in the annealed samples still are at least pressurized to the equilibrium pressure of the small bubbles from which they are grown.

If the scale of a bubble precipitate is to coarsen, then there are two possible explanations. Firstly, the gas in the bubbles may go back into solution, re–emerging preferentially into larger bubbles where the pressure might be lower. Secondly the small bubbles themselves may migrate as entities, collision between bubbles leading to coalescence and hence coarsening.

There is no evidence of resolution, the first alternative. The difficulty of resolution lies in the high energy of solution (about 5eV for Ne in Fe) rather than the migration energy. Taking the second alternative, Ne bubble migration can be expected in a temperature gradient, and due to interaction with dislocations during annealing. Since bubble migration requires the transfer of solid material from one side of the bubble to the other, it seems likely that surface diffusion around the bubble periphery occurs upon annealing at 400 °C. Further, when dislocations move by climb during annealing one may expect them to sweep up the bubbles in their path, and hence cause a coarsening.

In addition, Goodhew [20] pointed out that around the annealing temperature of 0.45 T_m surface diffusion is the predominant mechanism of coarsening. The growth rate appears to increase with increasing bubble concentration. In the present study the bubble concentration is very high, i.e. 10^{24} bubbles /m^3.

The main difference between our result and the ones obtained before [10,11] is the lack of a distinct orientation relation. However, it has to be emphasized that [10] refers to Sb implantations, implying that the stress fields are not localized around bubbles, and in [11] Kr has been taken as implant. Since Ne possesses a much higher compressibility compared to Kr it is quite obvious that the critical dose of Ne implantations should be higher.

Consequently the nucleation behaviour is altered in such a way that the volume fraction and bubble sizes are much larger in our case. This, together with high pressure in the bubbles causes a deformation around the bubbles, and thereby a reduction of the maximum size of the nuclei. The small size and the deformation cause the random orientation. This idea is in agreement with the increase in orientation relationship when we compare the unannealed thin foil samples ($3\ 10^{17}/cm^2$), with the bulk–side thinned ($3\ 10^{17}/cm^2$) or annealed thin foil samples ($1\ 10^{17}/cm^2$). This can be seen in Table 1 and in Fig. 3. In the case of the bulkside–thinned samples there is a restriction to the deformation because there is only one free surface. In the case of the annealed samples the volume fraction of the bubbles is much lower, causing less deformation. The sizes of the bcc particles also fit in this idea.

The experiments also show that the hardness increases when there is a high concentration of neon gas bubbles present. It is likely that the bubble induced martensitic phase is responsible for the improvement. Looking at the wear characteristic, we have to conclude that a large bubble density has detrimental effects on the wear behavior. This might be explained by the formation of blisters, which will inevitably occur when a large bubble will reach the surface as some material above it is removed during the wear process.

In laser melted RCC the wear performance improved after noble gas implantations. This improvement is likely to be due to the dislocation–bubble interaction. Dislocations may bypass the bubbles either by pure glide (Orowan process) and cross slip or by shearing the bubbles. The stress field around a bubble determines whether a dislocation will reach the bubble and penetrate or will bypass by bowing around the bubble. Usually in aged–hardened alloys the contribution of modulus hardening to the shearing force, required to cut a particle by glide dislocations, is small. Since the elastic energy of a dislocation is proportional to the shear modulus, a change in energy and hence a force will be associated with a dislocation interacting with a particle whose shear modulus differs from that of the matrix. The extreme case is the

Table 1 Details of samples

Sample: Dose	Preparation Method	Average bcc Size	Twinning Observed	Orientation Relation
1 10^{17}	Foil, Annealed	25 nm	Yes	Hardly
3 10^{17}	Foil	8 nm	No	None
3 10^{17}	Bulkside thin	50 nm	Yes	Hardly

Fig. 10   Interaction between Ne bubbles and dislocations.

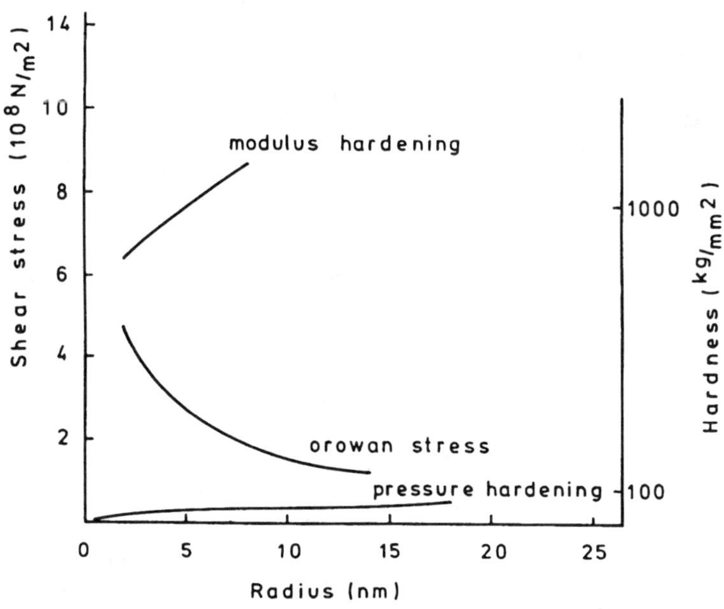

Fig. 11   Critical shear–stress as function of bubble size (volume fraction 0.025).

interaction with a void or a bubble, in which the elastic energy is reduced to zero and the system can be described by image forces. The modulus hardening in the case of Ne bubbles may be a crucial contribution to the hardening since the dislocations experience in the neighbourhood of the bubbles always an attractive force component towards the bubbles, which is confirmed by TEM observations (Fig. 10). In Fig. 11 the critical shear stress is depicted as a function of bubble size, given a volume fraction of 0.025. In this graph, the Orowan stress is calculated using the traditional Orowan expression, rewritten for spherical obstacles of radius R and a fixed volume fraction [21]. Pressure hardening is calculated, combining an expression for shearable objects [22], and an expression for the average force between an edge dislocation and a pressurized bubble [23,24]. Finally modulus hardening is calculated using an expression quoted by [25]. It is easy to see that modulus hardening contributes most compared to Orowan and pressure hardening.

As a last reason we mention that the bubbles will aid the formation of martensite during wear or indentation tests, thereby increasing the wear resistance and hardness. It is quite feasible that the interaction between bubbles and dislocations is not a predominant mechanism in a martensitic microstructure like laser melted CK22 steel. Due to the high dislocation density present in this microstructure, moving dislocations are interacting with these forest dislocations rather than with the noble gas bubbles.

5.   Conclusions

In conclusion we can state:
- The shear stress field of a noble gas bubble can induce a local martensitic transformation in metastable austenite, provided that bubble size and pressure are large enough.
- The sizes of the smallest bcc particles observed are just above the calculated minimum size.
- The martensitic transformation in thin foils is more effective than near the free surface of bulk samples implanted.
- There is no orientation relationship with the matrix when the sizes of the bcc particles are small.
- The bcc particles are allowed to grow to a larger size when the volume fraction of the bubbles is smaller, or when there is a restriction to the deformation around the bubble. In that case twinning is observed and an orientation relationship comes into being.
- In all cases studied implantation leads to an increased hardness.
- The implantation induced compressive stress leads to a decreasing cracking probability.

ACKNOWLEDGEMENTS

This work is part of the research program of the Foundation for Fundamental Research on Matter (F.O.M. Utrecht) and has been made possible by financial support from the Netherlands Organization for Research (N.W.O.–The Hague).

REFERENCES

1. N.E.W. Hartley, in J.K. Hirvonen, Ed., *Treatise on Materials Science and Technology*, Vol. 18, Ion implantation, Academic Press, New York (1980).
2. L.J. Cuddy, H.E. Knechtel and W.C. Leslie, Metallurgical Transactions **5**, 1999 (1974).
3. H. De Beurs and J.Th.M. De Hosson, Applied Physics Letters **53**, 663 (1988).
4. R.G. Vardiman and I.L. Singer, Materials Letters **2**, 150 (1983).
5. I. Sakamoto, N.Hayahi, B. Furubayashi and H. Tanoue, Hyperfine Interactions **42**, 1005 (1988).
6. E. Johnson, E. Gerritsen, N.G. Chechenin, A. Johansen, L. Sarholt–Kristensen, H.A.A. Grabaek and J.Bohr, Nucl. Instr. and Meth. **B39**, 573 (1989).
7. N. Hayashi and T. Takahashi, Applied Physics Letters **41**, 1100 (1982).
8. N.E.W. Hartley, J. Vac. Sci. Technol. **12**, 485 (1975).

9. E. Gerritsen, Thesis, Univ. of Groningen (1990).

10. E. Johnson, U. Littmark, Johansen and C. Christodoulides, Phil.Mag. **A45**, 803 (1982).

11. E. Johnson, A. Johansen, L. Sarholt–Kristensen, L. Grabaek, N. Hayashi and I. Sakamoto, Nucl. Instr. and Meth. **B19/20**, 171 (1987).

12. S. Kajiwara, Metall. Trans. A, **17A**, 1693 (1986).

13. J.R. Patel and M. Cohen, Acta Metall. **1**, 531 (1953).

14. J.C. Fisher and D. Turnbull, Acta Metall. **1**, 310 (1953)

15. K.E. Easterling and A.R. Tholen, Acta Metall. **24**, 333 (1976).

16. W.G. Wolfer, Phil. Mag. **A59**, 87 (1989).

17. D.A. Porter and K.E. Easterling, *Phase Transformations in Metals and Alloys*, 257 (1981).

18. Y.J. Xu, B. van Brussel, J. Noordhuis, P.M. Bronsveld and J.Th.M. De Hosson, Surface Engineering, Elsevier, 167 (1990).

19. J.H. Evans, Nucl. Instr. and Meth. **B18**, 16 (1986).

20. P.J. Goodhew, Scripta Metall. **18**, 1069 (1984).

21. U.F. Kocks, Mat. Sci. Eng. **27**, 291 (1977).

22. J.W. Martin, *Micromechanisms in Particle–Hardened Alloys*, Cambridge University Press (1980).

23. V. Gerhold and H. Haberkorn, Phys. stat. sol. **16**, 675 (1966).

24. A.J. Ardell, Met. Trans. A **16a**, 2131 (1985).

25. P.M. Kelly, Int. Met. Rev. **18**, 31 (1973).

GAS BUBBLE LATTICES IN METALS

P.B. Johnson

*Department of Physics, Victoria University of Wellington
P.O. Box 600, Wellington, New Zealand*

ABSTRACT

The ordering of small (~ 2 nm diameter) helium bubbles in metals to form a gas bubble superlattice is a striking phenomenon. Studies contribute understanding of the mechanisms underlying the development of microstructure in metals during ion irradiation. The earliest investigations were for the bcc metal molybdenum. Since then extensive studies have been made of the structural nature of the gas bubble superlattice in copper. In this metal, and the other fcc metals including nickel, stainless steel, and gold the bubble array contains structural variants— that is regions where the ordered bubble array has an orientation which is rational with, but different from, that of the crystal lattice of the host metal. Previous work is reviewed and the knowledge of the structural nature of bubble lattices gained from studies based on transmission electron microscopy is summarised. The variation of bubble structure with depth and dose for monoenergetic (30 keV) helium implantation in copper at normal incidence is outlined. Recent experiments at high helium doses have led to the discovery of several new types of bubble structure. In vanadium and copper, the breakup of the superlattice leads to bubble arrays ordered on a coarser scale. These arrays in turn break up into a remarkable "honeycomb" structure involving a dense random array of large cavities separated by thin metal walls of uniform thickness. It is thought that it is cracking of the metal walls, of typical thickness ~ 1.0 nm, that eventually leads to radiation blistering of the surface. The honeycomb structure explains the high swellings postulated earlier on the basis of surface displacement measurements. As recently reported, a spectacular gas bubble macrolattice has been discovered in gold. The ordered array has the same symmetry as the atomic lattice of the metal and a spacing ~ 20 times greater than that of the superlattice found in the same specimen.

A.   REVIEW OF PREVIOUS WORK

1.   The BCC Metal Molybdenum

The formation of the helium gas bubble superlattice is an important stage in the development of radiation damage structures in many metals implanted with helium at low temperature. Early studies based on transmission electron microscopy (TEM), such as those by Barnes and Mazey [1] in 1960, showed that ion irradiation of metals with He isotopes could result in the formation of microscopic gas bubbles. Originally it was assumed that the thermal production and mobility of vacancies was a necessary condition for bubble growth [2]. The phemomenon of He gas bubbles ordered on a space lattice—a gas bubble superlattice—was first observed by Sass and Eyre [3] in the bcc metal Mo implanted at a temperature ~ 300 K. The thermal production and mobility of vacancies are low at this temperature in Mo [4] and yet bubble growth had occurred. Electron diffraction from the specimens exhibited splitting with satellite spots evident around the matrix reflections. The spacing of bubbles in the ordered array was calculated from the separation of these satellite reflections in diffraction and from the spacing of bubble rows in bright-field micrographs.

Fundamental Aspects of Inert Gases in Solids
Edited by S.E. Donnelly and J.H. Evans, Plenum Press, New York, 1991

Bubble ordering in Mo was subsequently (1976) investigated in considerable detail by Mazey *et al.* [6]. The observations of superlattice formation were extended to He implantation temperatures up to 970 K. In addition a superlattice was observed following neon implantation. The conclusion was reached that the superlattice had a bcc structure and was aligned parallel to the host metal matrix. In looking back to this foundation work on bubble ordering in a bcc metal several points should be noted: (i) The spacings of the dense-packed (110) bubble planes, d_{110}, were deduced from densitometer measurements made on diffraction patterns of the position and distance of the matrix and lattice spot maxima from the pattern centre. However, the interplanar distances found were quoted as "(110) bubble lattice spacings". This has subsequently led to a misinterpretation that the values quoted are for the lattice constant, a_1. See, for example, the review article by Krishan [7]. (For Mo the values listed by Krishan as bubble lattice parameters should be increased by a factor $\sqrt{2}$). (ii) The typical lattice parameter found was ~5 nm. The corresponding bubble concentration (i.e. number density) is 1.8×10^{25} bubbles m^{-3}. No systematic change in lattice parameter was observed with change in implantation temperature. (iii) Superlattice formation occurred for both implantation into thick targets and into pre-thinned targets. (iv) Some pre-thinned specimens were irradiated in-situ in the electron microscope and the damage monitored throughout the course of implantation. (v) Whereas the diffraction patterns were generally consistent with three dimensional bubble ordering, it was in one case only that conclusive evidence was obtained. (The evidence required to demonstrate full three-dimensional ordering is discussed in B2.1(ii)). (vi) The ordered bubble structure was found to develop from an earlier stage in which the bubbles were random.

2. The Fcc Metals

2.1. Bubble Ordering

(i) Introduction

A close parallel was found between the bubble lattice in Mo and the void lattice produced by neutron or ion bombardment as first observed by Evans [5] and by Wiffen [6]. There were obvious, although not necessarily significant, differences in scale: the bubble lattice spacing, ~5 nm, was a factor ~5 smaller than the void lattice spacing, while the ratio of cavity lattice spacing to cavity radius was ~4 for bubbles and ~10 for voids. The theoretical description of cavity lattice formation had concentrated almost exclusively on the void case [7] and even in that case the basic processes involved were still the subject of debate [8]. Of the other observations of void lattices, most had been for bcc metals (see, for example, Ref.[8]). For fcc metals, void lattices had been seen in Ni [9] and Al [10, 11], in Ni-Al solid solution alloys [12] and in stainless steel [13]. Void ordering for the fcc case seemed less clear-cut than for the bcc metals. For example, good void ordering had been found in Ni-Al alloys but not in pure Ni irradiated under similar conditions. Against this background it was not at all certain whether bubble lattices should be expected to form in metals having structures other than bcc or indeed, in metals other than Mo.

(ii) Helium in copper

In 1978 Johnson and Mazey [14] commenced a series of TEM studies to investigate the bubble structures in fcc metals. The He implantations were performed at room temperature into polished polycrystalline foils having a high degree of crystalline perfection within grains. The dose levels were close to the critical dose for radiaiton blistering of the surface. Thin sections were prepared for electron microscopy by jet electropolishing to perforation from the unirradiated side of the foil. The micrograph of gas bubbles in Cu shown in Fig.1 provides an example of the results obtained. Considerable bubble alignment can be seen with dense-packed rows parallel to matrix {111} trace directions. Diffraction from the bubble array

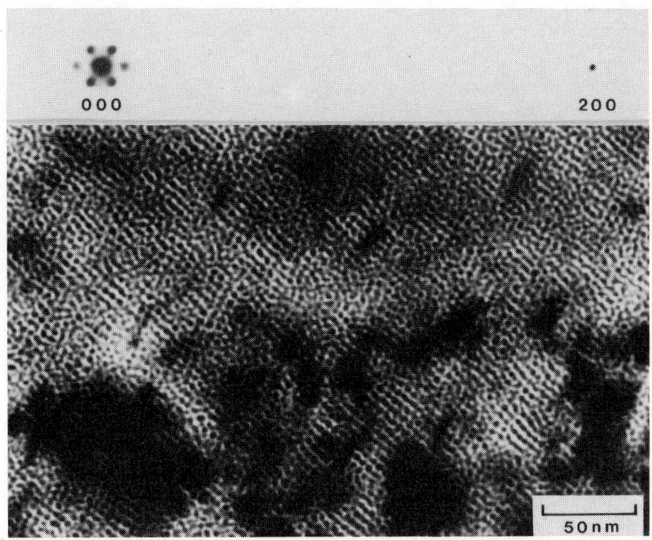

Fig. 1 A transmission electron micrograph (overfocus image) of helium gas bubbles in copper following implantation to 4×10^{21} 30 keV He^+ m^{-2} (approx.) at 300 K. The electron beam is along [110] in the matrix and is normal to the plane of the bubble layer. The bubble rows evident in the figure are in the directions of the traces of (111) matrix planes in the (110) plane. Shown at the top of the figure is a section of an electron diffraction pattern (reverse contrast print) taken from the same area of the specimen. (The pattern is in approximately the correct relative orientation). The two pairs of strong satellite reflections around the 000 transmitted beam are assigned 111 indices.

caused the splitting around matrix reflections in electron diffraction patterns as shown in the superlattice pattern at the top of the figure. The strongest reflections in this pattern were assigned (111) indices as they were in line with the (111) matrix spots and were in the positions expected for diffraction from the dense-packed planes seen edge-on as bubble rows in the micrograph. The results were interpreted as showing that the bubble lattice was, at least predominantly, an fcc structure aligned parallel with the Cu matrix. The bubble diffraction patterns for the matrix orientations [100], [111] and [112] supported this conclusion. Also alignment conformed to crystallographic features in the Cu, bubble rows ran parallel to matrix slip traces and assumed the twin-direction on crossing twin boundaries.

The lattice constant found from diffraction from $\{111\}$ bubble planes, $a_1 = 7.6 \pm 0.3$ nm, corresponded to a bubble concentration $\sim 10^{25}$ bubbles m^{-3}. The bubble spacings found from measuring the separation of the bubble rows in direct micrographs were in good agreement with the results from diffraction. There was also good agreement between the results both for samples from the same irradiaition and for the several different Cu targets that were used.

The average bubble diameter found using a through-focus technique was 2.0 ± 0.5 nm with a rms spread in bubble size of only 0.3 nm. This small spread in bubble size is a characteristic of ordered bubble structures. Micrographs taken at [110]—see for example Fig.2—showed domains of good ordering, typically of five interbubble spacings across, separated by regions of comparable extent in which the ordering was less obvious. It is interesting to note that even in this first study diffraction anomalies were found. The lattice constant based on diffraction from a given set of planes was consistent for the electron beam directed down [100], [110], [111] and [112] in the metal matrix. There were, however, small but apparently significant differences between the different planes. For example, diffraction from the bubble planes parallel with the $\{200\}$ matrix planes gave a consistent value for a_1 of 6.8 ± 0.4 nm. From the

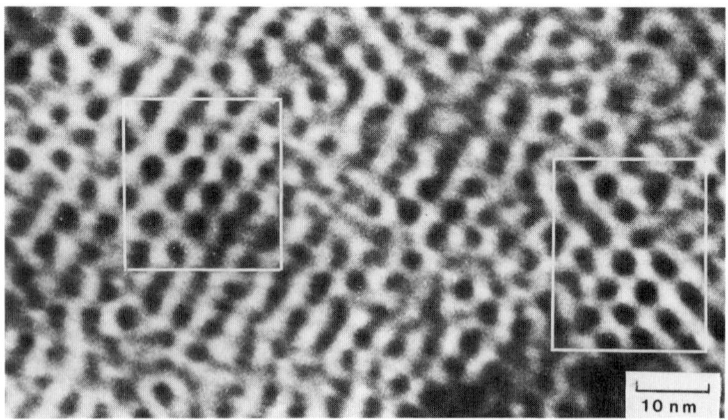

Fig. 2 A section of the micrograph of Fig.1 at high magnification to highlight the domain structure in the ordered bubble array. Electron micrographs taken at [110] show domains of good bubble ordering (see boxed areas), typically of five to ten inter-bubble spacings across, separated by regions where the ordering is less obvious. The projected bubble images have confirmed the presence of structural variant lattices in the bubble array (see A2.4(ii)).

bubble volume fraction calculated from the measured superlattice parameters and estimates of the local helium concentration (based on the He dose and calculations of the depth distribution) it was concluded that much of the helium resided in the matrix rather than in bubble cavities.

(iii) Other Early Studies

Similar helium-bubble lattices were seen after irradiation at 300 K in the two other fcc metals studied, Ni and AISI 321 austenitic stainless steel. The lattice constants based on diffraction from (111) superlattice planes were 6.6 ± 0.5 nm and 6.4 ± 0.5 nm respectively. By 1979 the observations of superlattice formation had been extended to include the fcc metal AISI 316 stainless steel ($a_1 = 6.4 \pm 0.5$ nm) [15]. As in Mo, vacancy mobility did not seem to play a major role in lattice formation in the fcc metals, at least in the limited range of low mobility covered in these experiments. Bubble ordering had also been found at 300 K in the hcp metal titanium [15]. The bubble sizes and spacings were approximately 50% greater than those found in the fcc metals. Because of the similarities with the fcc case the assumption was made that the ordering was three dimensional. (Evans [16] has recently found evidence to suggest that the bubble ordering may have been confined to the basal planes.)

A study was also made of hydrogen implantation into Cu [15,17]. Hydrogen is largely insoluble in Cu and in this sense it is similar to He. The implantation conditions, such as damage rate and gas deposition depth profile were chosen to match those in the He experiments. Considerable bubble alignment was observed. The bubble array exhibited a much wider range of bubble sizes and many more structural defects than were evident in the helium case. The degree of ordering was not sufficient to cause splitting of the diffraction pattern. In (110) grains bubble rows (resulting from the projections of dense-packed bubble planes) were parallel to matrix trace {111} directions. For bubbles of diameter 2 nm, the {111} interplanar spacing was found to be ~7 nm by measuring row spacings in direct micrographs. The damage rate per implanted gas atom was a factor of 10 lower than for the helium implantations but because the final dose was higher the total damage in the two cases was similar. It seemed that the nucleation density of bubbles was lower in the case of hydrogen implantation.

170

2.2. *Stability of an existing superlattice*

(i) Collisions

To investigate the role of displacement damage collisions on bubble growth (in the absence of further gas deposition) specimens containing a superlattice of He bubbles formed during prior He implantation were irradiated at 300 K with 1 MeV electrons to create a specific collision regime. In both Cu and Ni [18] the result was to induce bubble growth and an increase in the lattice constant of the superlattice. Similar results were obtained using sub-threshold electron energies (~400 keV). The conclusion was reached that collision or displacement events acting alone could give rise to both bubble growth and changes in bubble structure. Of the possible interpretations the simplest was perhaps that the bubbles had gained further He atoms and possibly vacancies, that were displaced from the matrix in collision events. Nelson [19] suggested a possible mechanism for athermal vacancy mobility. An implication of this interpretation was that a substantial amount of helium was held outside the bubbles. It was recognised that other interpretations were possible.

(ii) Heating

The behaviour of the gas bubble lattice in Cu during thermal annealing up to 920 K following He implantation at 300 K was investigated. The superlattice remained intact, with unchanged lattice constant, during heating to 470 K (0.35 T_m) in the hot-stage of the electron microscope. On heating to 600 K (0.44 T_m) bubble growth and coalescence became evident but with little obvious change in lattice constant. At higher temperatures complicated processes of rapid bubble growth and coalescence occurred. Dynamic sequences were recorded on video tape and ciné film [20, 21].

2.3. *Low energy helium implantation*

In the first studies the He energy was ~30 keV with the ion beam incident along the normal to the target surface. Under these conditions, the He concentration initially increases rapidly with increasing depth to reach a maximum at a depth ~150 nm; thereafter the concentration falls rapidly—see Fig.3.

Thin sections typically ~60 nm in thickness, are prepared from near the front (implanted) surface. Jäger and Roth [22] investigated the microstructural modifications of Ni surfaces after multiple energy helium implantation at low energies in the range from 250 eV to 8000 eV. Under these conditions the He concentration increased from low values at a depth of 140 nm to reach a maximum concentration at the surface. A bubble lattice was observed for implantation temperatures below 570 K. It was concluded that thermally activated vacancy migration seemed to have no influence on the formation of the bubble lattice. A given superlattice spot in the TEM electron diffraction pattern was assigned the indices of the matrix reflection in the same direction. From the separation of superlattice spots the average spacing between bubble planes were determined to be 4.7 nm for {111} planes and 3.8 nm for {220} planes. These spacings were in good agreement with spacings measured for these planes directly on TEM micrographs. The spacing of the {220} planes relative to that of the {111} planes was clearly not correct for a complete fcc structure and this anomaly was not resolved. The bubble ordering was associated with a shallow buried layer. Between this layer and the surface lay an interconnected layer of random bubbles. In a later work [23] similar bubble ordering was observed for shallow He implants in AISI 316 stainless steel.

Van Swijgenhoven et al [24] investigated bubble growth in Ni for 5 keV He implantation. Bubble ordering was found. The diffraction patterns were interpreted as indicating that the bubbles were ordered in the {220}, {111} and {331} planes of the host lattice but not on the {200} and {131} planes. Under the assumption that the bubble lattice had the same orientation as the host lattice, the bubble lattice constant derived from the {111} reflections was 6.9 ± 0.5 nm a value in good agreement with that found for deeper implants by Johnson and Mazey [14].

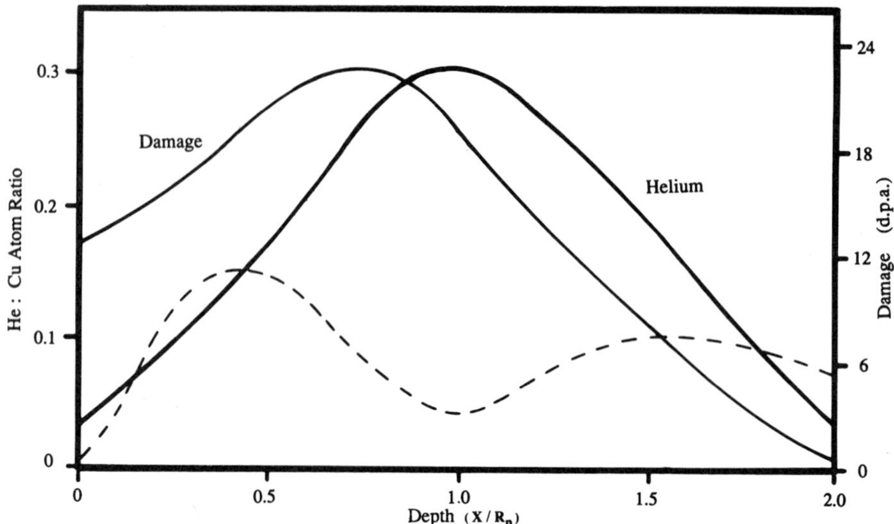

Fig. 3 The helium and damage depth profiles in copper, calculated using the Harwell version of the computer code E-DEP-1 [52] for 4×10^{21} 30 keV $He^+ m^{-2}$. The depths are normal to the surface and expressed as a fraction of the mean projected range. The thickness of specimens prepared for TEM examination typically corresponds to 0.4 R_p, where $R_p \sim 150$ nm. The lower curve (dashed) is the depth profile of the retained helium in Cu at 300 K found by Terreault and Veilleux for a dose $\sim 5 \times 10^{21} He^+ m^{-2}$ (scaled to suit the depth scale).

The lattice constant derived from the {220} and {331} reflections were respectively 11 ± 0.8 nm and 16 ± 1 nm. Apart from noting that the interplanar spacings calculated from bubble reflections were always about 4 ± 0.3 nm again the anomaly was not resolved. It was concluded that much of the He resided in the matrix rather than in bubble cavities, at least in the early stages of implantation.

2.4. Structural variants

(i) Diffraction Studies

It had seemed that the bubble lattice in the fcc metals had the same symmetry as the host metal (i.e. fcc bubble lattice in an fcc host) with the principal axes of the bubble lattice along the corresponding axes of the matrix. (In the following discussion the orientation of a bubble lattice which is parallel with the matrix is referred to as the *m* orientation). However, electron diffraction exhibited unusual features which indicated that the structure of the bubble lattice was more complicated than suggested by this simple model. In Cu the spacing of diffraction satellites gave only two interplanar spacings 3.4 nm and 4.3 nm [21, 25]. If the superlattice reflections were indexed in the same way as the matrix, the lattice constants obtained varied widely. For the dense-packed bubble planes parallel with {111} matrix planes, the interplanar spacing was 4.3 nm and the lattice constant 7.5 nm. The interplanar spacings and lattice constants found for other orientations were respectively: {200}, 3.4 nm, 6.8 nm; {220}, 4.3 nm, 12 nm; {311}, 4.3 nm, 15 nm; and {331}, 4.4 nm, 19 nm. (Uncertainties in spacings ranged from 4% to 8%.)

It was noted that a lattice constant of 7.5 nm was obtained if 200 indices were replaced by 210 and 111 indices were used for all other superlattice reflections. An explanation was sought in terms of the presence of structural variants in the ordered bubble array, [21, 25]. Bright-field micrographs (see Fig.2) showed that the bubble structure was not homogeneous but contained domains of good ordering. It was proposed that within any given domain the

172

bubbles lay on an fcc lattice with lattice constant ~7.5 nm in one of several possible orientations relative to the Cu matrix. For the electron beam along a given zone in the matrix, the bubble diffraction pattern was regarded as a composite with contributions from several variant bubble structures having zones with low indices close to the electron beam direction, **B**.

In the polycrystalline Cu targets the most common grain orientation was (011). To explain the diffraction results for (011) grains, two structural variants Δ and Ω were proposed in addition to m-ordered lattices [25]. The Δ, Ω and m are identical fcc lattices with the same lattice parameter, 7.5 nm, which differ only in orientation. Referred to coordinate systems based on the variant structures, the Δ and Ω each have a <112> direction coincident with the [011] surface normal. An m oriented lattice has two sets of {111} planes normal to the grain surface. A set of {111} planes in Δ is parallel to one of these sets, and a set of {111} planes in Ω is parallel to the other set. The relative orientations of the Δ, Ω and m are summarised in the superimposed stereograms of Fig.4. The [$\bar{1}$12] directions in both Ω and Δ coincide with [011] in m; [$\bar{1}$1$\bar{1}$] in Δ coincides with [$\bar{1}$1$\bar{1}$] in m; and [$\bar{1}$1$\bar{1}$] in Ω coincides with [11$\bar{1}$] in m. All three lattice types contribute to 111 superlattice reflections: m contributes to both sets whereas Δ and Ω each contribute to one set only. In addition, Δ and Ω give rise to 210 type reflections near both 011_m and 100_m (only those near 011_m are shown in the stereogram).

The use of indices such as 210 (indices which are normally disallowed for an fcc lattice) is accounted for in terms of the imperfections in the bubble lattice and small domain size [25]. Six variant structures each having an orientation different from that of the matrix, were proposed to account for the diffraction with the electron beam directed down other principal zones in the matrix [25]. Further evidence of the existence of these variants in the bubble array has been provided by tilt experiments (P.B. Johnson and D.J. Mazey - unpublished work). In these experiments the diffraction from the bubble array was observed as the specimen was systematically tilted from one major crystallographic zone in the matrix to others. Good agreement was found between the results obtained and the behaviour expected on the basis of superimposed stereograms of the type shown in Fig.2. Although detailed studies of projected bubble images in direct micrographs were not done the broad features were found to be consistent with the structural variants proposed.

(ii) Projected bubble images

The first detailed examination of projected bubble images was by Diprose (née Malcolm) [26] and by Johnson, Malcolm and Mazey [27-29]. The method used was to first determine the pattern of bubble images to be expected in areas where a single structural variant only occurred. These individual patterns were combined to find the patterns that would be observed for various combinations of the structural variant lattices. The patterns obtained experimentally were then compared with those theoretically predicted patterns to find how the various lattice types were distributed through the array. It was found that in <110> grains over half of the ordered bubble array consisted of the variants Δ and Ω, bubble lattices having orientations that differed from that of the Cu matrix. This, taken together with the diffraction results, contradicted the widely held assumption that the superlattice was in the matrix orientation only. The analysis of projected bubble images also supported the existence of the other structural variants that had been proposed earlier on the basis of diffraction evidence.

2.5. Implications

(i) Bubble movement

The theories of bubble ordering, taken over from the void case, had assumed a bubble lattice in the m orientation only. The symmetry and alignment of the superlattice was explained in terms of some anisotropic property of the crystal lattice of the host metal. Anisotropies had been included through the elastic constants and through surface energies via cavity facetting. In this

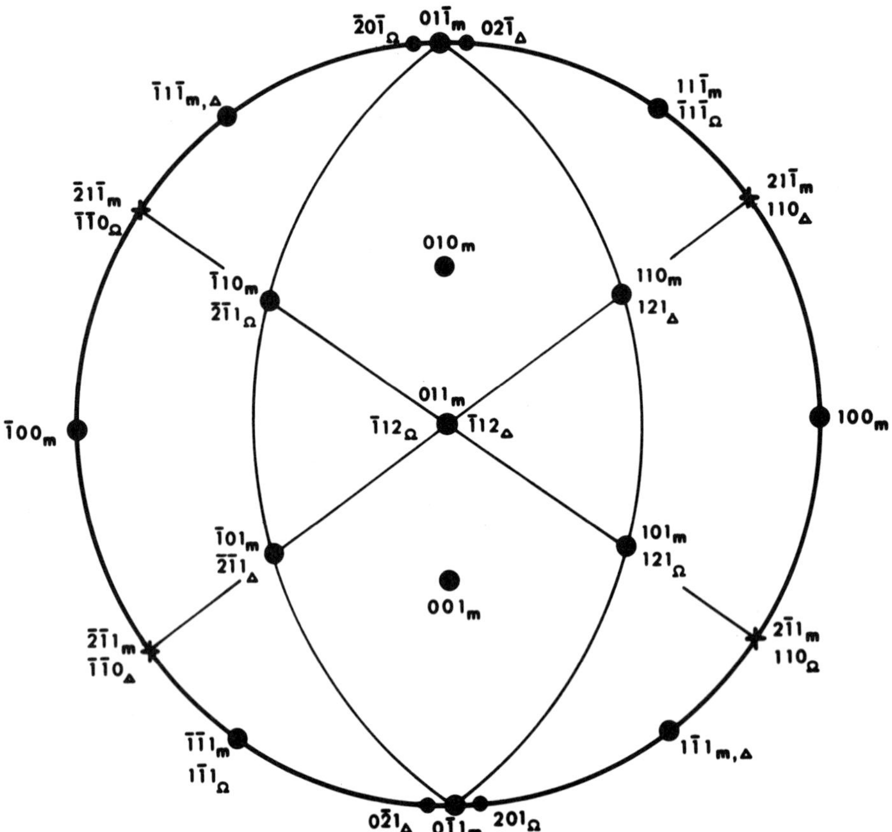

Fig. 4 Stereograms for the two structural variant bubble lattices, Δ and W, superimposed on the matrix (m) stereogram centred on $\mathbf{B} = [011]_m$. Notice that Δ and Ω have a common <112> direction which coincides with $[011]_m$. Δ and Ω each have a set of {111} planes parallel to a set of {111} planes in m, and each have <210> directions near both $[01\bar{1}]_m$ and $[100]_m$. (Only those near $[01\bar{1}]_m$ are shown).

way the structure of the bubble lattice was intimately linked to the crystallographic properties of the host metal. In terms of these theories it was difficult to understand the formation of structural variants having an orientation different from that of the matrix. Furthermore, these variants had the same symmetry and lattice parameter as the variant *m*. This compounded the difficulty because it suggested that these two important characteristics of the bubble lattice were at least to some degree independent of the directional properties of the metal lattice.

It seemed certain that the symmetry and lattice spacing of an ordered bubble array must be determined by the directional properties of the host metal. That the Δ, Ω and *m* had the same lattice parameter and same fcc symmetry as the Cu, strongly suggested that all the bubbles in the ordered array must have ordered initially parallel with the host metal. Consequently, it was proposed that the bubbles ordered first into the *m* orientation and it was at that stage that the symmetry and lattice parameter were determined by the directional properties of the metal lattice. At a later stage the bubbles in some domains moved to take up a structural variant orientation.

(ii) Dislocations

The bubbles produced by implantation at low temperature were expected to be overpressured [30–32] a property that has been confirmed by recent work [33, 34]. The matrix surrounding an overpressured bubble is in radial compression and tangential tension. The microstress in the region between two neighbouring bubbles involves compressive forces along the line joining them and tensile forces normal to this line [31, 32]. Bubble growth at low temperatures was expected to be by an athermal process such as dislocation punching [30-32]. An isolated bubble growing in a stress-free matrix was expected to punch out prismatic loops which moved along glide cylinders in <110> directions in the matrix [35-37]. Any impediment to the glide of these dislocation loops, such as the presence of another bubble on the same glide cylinder, would result in a back-stress on the bubble [31, 32]. It was recognised that this back-stress would not only inhibit bubble growth but could lead to a strong repulsion between bubbles in an array ordered parallel with the matrix [31].

The line joining the centres of two neighbouring bubbles in an fcc lattice in the m orientation lies along $<110>_{Cu}$. The two bubbles are linked by a common glide cylinder. Because there is a strong tendency for each bubble to punch out dislocations on this cylinder in the direction of the other, the Cu matrix between the bubbles will be highly compressed. The resulting back-stresses were expected to give rise to a repulsion between the bubbles along the common $<110>_{Cu}$ direction [31], [27–29]. This repulsion was in addition to the attractions and the repulsions proposed in the early theories of cavity ordering [38–40]. Originally [31, 41], it was proposed that the repulsion was due to dislocations actually present on the common glide cylinder. Several considerations caused this proposal to be modified. For any given bubble in an fcc metal there are twelve possible <110> directions available for dislocation punching. The ordered bubbles are only ~2 nm in diameter and the atoms displaced from a bubble cavity of this size can be accommodated in ~12 interstitial loops of diameter comparable with that of the bubble. This implies that a given bubble would need to contribute only one dislocation to each <110> glide cylinder. Furthermore, it was thought improbable that loops would remain intact in the presence of the high damage rates in the metal during implantation (see also discussion in B 2.1(iii)). Although dislocations on a fine scale were present at high density in the implanted layer (e.g. refs [14], [15], [18] and [20]) dislocation loops had never been observed in the metal between the ordered bubbles.

(iii) <u>Stress-stress interactions</u>

It was for the above reasons that Johnson et al [27–29] attributed the repulsion to the strong compression of the metal within the common glide cylinder due to the *tendency* of overpressured bubbles to punch dislocations, rather than to the *presence* of dislocations. However, it seemed reasonable to *model* the repulsion as an interaction between a dislocation at the surface of one bubble and a dislocation on the same glide cylinder at the surface of the neighbouring bubble. A similar approach seems to be implicit in the work of Wolfer [42]. On this basis it was suggested [27] that the repulsion would increase rapidly with increasing bubble size and that this could eventually lead to the lattice becoming unstable against bubble movement. In particular it was argued [27–29] that it was this repulsion that drove the bubbles in some domains out of the m orientation and into one of the several possible variant orientations.

Analysis of projected bubble images had shown that the Δ (Ω) frequently lay immediately above or below m in terms of depth from the specimen surface with the domains having a common set of dense-packed {111} bubble planes which were orthogonal to the grain surface. This sharing of a common plane was attributed [27–29] to the effects of lateral stress—a biaxial compression in planes parallel with the specimen surface (see for example, refs [43, 44]). It was argued that in the presence of lateral stress, a set of dense-packed bubble planes lying normal to the surface was energetically favoured because then neighbouring bubbles could cooperate fully to produce strain in the copper matrix against the component of the lateral stress that acted normal to the common set of planes. It was proposed that the bubble movements that carried sections of lattice from the m orientation to the Δ (Ω) configuration took place within these common planes [28, 29]. Because the domains were small the bubble movements required were not large (< 2.7 nm). The movement of any individual bubble would involve a series of small displacements which were correlated with the displacements of neighbouring bubbles so as to preserve the local bubble ordering. The relative orientations of the Δ, Ω and m, see for example, Fig.4, suggested that these variants were complementary structures in accommodating the lateral stress.

It was argued that in the presence of a strong lateral stress the energy was minimised when all three orientations m, Δ and Ω were represented in the bubble array and it was for this reason that the bubble lattice in some domains remained in the m orientation. The conclusion was reached that stress-stress interactions played a dominant role in the later stages of bubble ordering in the fcc metals. The initial ordering of bubbles into the m orientation is the normal case treated by theories of cavity ordering.

3.   The Hcp Metals

The theories of bubble ordering divide into three classes according to the following mechanisms (i) an elastic interaction between bubbles [38–40], (ii) the planar diffusion of host interstitial atoms [45-48] and (iii) an interaction between dislocations punched out from overpressured bubbles [27, 31, 41, 42]. Interest in bubble ordering in the hcp metals has been stimulated recently by the recognition that this case may allow a critical evaluation to be made of competing theories [47, 48]. With the discovery of bubble ordering in Ti [15], the phenomenon had been demonstrated to occur in metals representing the three main structures: bcc, fcc and hcp. Recent work has re-examined the case of He in Ti and found that the bubbles are ordered only on planes parallel with the basal plane of the metal with little apparent ordering within the planes themselves [49]. A similar result has been found for krypton bubbles in both Ti [49] and Zr [50]. Planar ordering of this type finds a ready explanation in terms of the theories based on the planar diffusion of intersititial atoms and no easy explanation in terms of the other theories [50].

B.   METAL STRUCTURES AT HIGH DOSE

1.   Introduction

1.1.   General

Measurements of the displacement of the surface of metals following He implantation indicated values of local swelling ~40%—see, for example, ref. [51]. For ordered bubble structures, the maximum swelling found from TEM studies of the bubble lattice, was only ~10% [4, 14]. The metal "wall" separating nearest-neighbour bubbles was never less than ~2 nm in thickness. Under the assumption that any reduction in this wall thickness would lead to bubble coalescence it seemed that the bubble structure must coarsen beyond the superlattice stage for the swelling to rise above ~10%. This was demonstrated in 1982 for Cu [52] when it was observed that continued He implantation into an existing superlattice caused the bubble lattice to break up to form a structure involving large bubbles with no discernible order. The purpose of the present work was to investigate further the coarse structures that form beyond the superlattice stage. As an introduction the results of the 1982 experiment are summarised below.

1.2.   Variation of bubble structure with depth

Many of the superlattice studies have been based on 30 keV He implantation with the ion beam normal to the metal surface. The calculated He and damage depth profiles for 30 keV He implantation in Cu are shown in Fig.3. The dose level of ~5 10^{21} He$^+$ m^{-2} at 320 K was close to the critical dose for radiation blistering of the Cu surface. The bubble structure was investigated by removing varying amounts of metal from different specimens from the same irradiation by argon ion milling of the irradiated surface at 300 K. Thin sections, for TEM

176

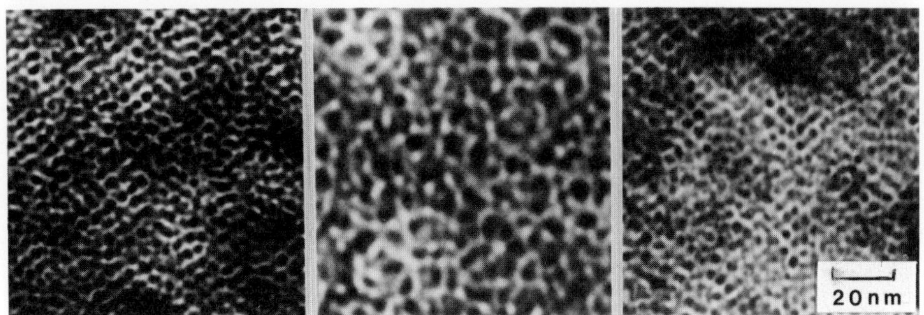

Fig. 5 Direct-image transmission electron micrographs of helium gas bubbles at various depths in Cu following 30 keV He$^+$ irradiation to a level ~ 5 10^{21} He$^+$ m^{-2} at 320 K. The electron beam is along [110] in the matrix and is normal to the plane of the bubble layer. The micrographs in order from left to right correspond to depths below the surface (expressed as fractions of R$_p$ - see fig.3) of 0.7, 1.1 and 1.5 respectively.

examination, corresponding to the different depths were then prepared by chemical back-thinning [14]. The bubble structure at three selected depths is shown in Fig.5. At depths shallower and deeper than the mean projected range, R$_p$, the bubbles were ordered on a superlattice. However, at depths near R$_p$, which correspond to the maximum He concentration, the bubble structure had a coarse porous character. Random cavities, ~4 to 10 nm across, appeared to be separated by metal walls ~2 nm thickness. At lower doses, at this depth near R$_p$, a gas-bubble superlattice with characteristics similar to those of Fig. 5 is found (Johnson, Mazey and Diprose, to be published). The conclusion can be drawn that the coarse structure observed, has evolved from a previous stage in which gas bubbles were ordered on a superlattice. In the present work an investigation is made of the evolution with He dose of bubble structures in the bcc metal vanadium.

2.   Helium in Vanadium

2.1. Superlattice

(i) Structure

For the bcc metals, bubble ordering had been studied previously only in the case of molybdenum. As part of a programme to extend the observations to include a range of bcc metals Johnson and Mazey have recently made a study of superlattice formation in vanadium [15]. Vanadium is of particular interest for comparison with Cu because the two metals have an almost identical shear modulus. The mean range (R$_p$) of 30 keV He ions in vanadium is 130 nm. Foils were implanted to a dose level ~4 10^{21} He$^+$ m^{-2} at 320 K and thin sections typically 60 nm in thickness were prepared as in previous studies. He bubbles typical of those observed in the many specimens investigated are shown in Fig.6. Considerable bubble alignment is seen with bubble rows parallel to each of the three matrix trace {110} directions. Diffraction of electrons from the bubble array caused the splitting which is evident around the 000 transmitted beam in the section of the electron diffraction pattern shown as an inset in the figure. The diffraction spots in the superlattice pattern are assigned 110 indices because they are in line with the 110 matrix spots and in the positions expected for diffraction from the dense-packed {110} planes seen edge on as bubble rows in Fig.1b. The interplanar spacing of the 110 bubble planes deduced from the radial separation of the corresponding superlattice spots, has the same value, (2.7 ± 0.1) nm, for each of the three sets of 110 planes which lie perpendicular to the specimen surface. Bubble row separations measured on direct micrographs give interplanar spacings in good agreement with those from diffraction.

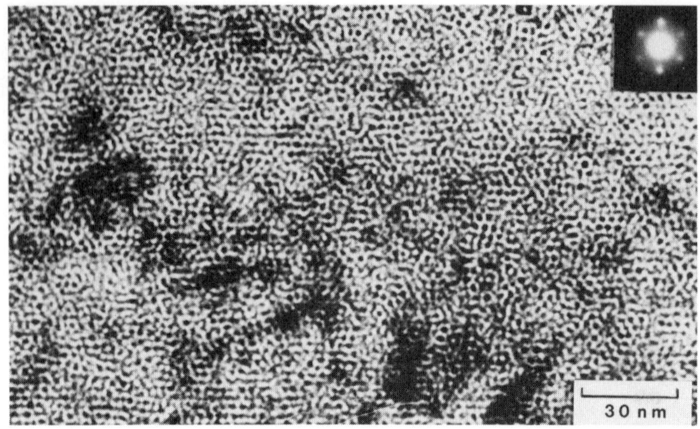

Fig. 6   A direct image transmission electron micrograph (over-focus contrast) of helium gas bubbles in the bcc metal vanadium following He$^+$ irradiation to a level ~ 5 10^{21} He$^+$ m^{-2} at 320 K. The electron beam is along [111] in the matrix and is normal to the specimen surface. There are three sets of dense-packed {110} planes normal to the surface. The trace of one of these sets in the (111) surface plane, runs parallel with the lower edge of the micrograph. Each of the other two sets is at 60° to this direction. Bubble rows in each of the three directions are an obvious feature. A section of the superlattice diffraction pattern from the ordered bubble array is shown as an inset.

(ii) <u>Degree of Ordering</u>

It should be emphasised that it is not possible to conclude simply on the basis of the above diffraction evidence that the lattice is fully ordered in three dimensions. The observed diffraction pattern could be explained, for example, in terms of the ordering of bubbles into {110} planes without there necessarily being bubble ordering within the planes themselves [53]. Suppose that in any given region the bubbles were ordered into one set of {110} planes only and that there were three types of region corresponding to the three sets of {110} crystal planes which are perpendicular to the specimen surface. The diffraction pattern of Fig.6 could then simply result from a sum of the contributions from regions of the three types that lie within the selected-area diffraction aperture. There is a three-dimensional character to the ordering only when there are two or more intersecting sets of planes within the *same* region. For this reason satellite reflections from dense-packed bubble planes alone are not sufficient to demonstrate three-dimensional ordering. Only if, in addition, there are reflections from other planes can it be concluded that the ordering has a three-dimensional character. Recently, for example, Kalin et al [54] have reported superlattice formation in type 18-10 austenitic stainless steel and 20-45 high-nickel base alloy. It has not been demonstrated, however, that the ordering is complete in three dimensions.

Two observations in particular show that the bubble ordering in vanadium is genuinely three-dimensional. (i) Many areas can be found where bubble rows in each of the three trace {110} directions intersect (see, for example, Fig.6). In such areas the pattern of projected bubble images is commonly that expected for a fully ordered bcc structure. (ii) The diffraction pattern of Fig.7 is from an ordered bubble array in an (011) grain. The spacing of the 011 bubble planes calculated from the radial separation of the 011 satellite spots associated with the 000 transmitted beam is again (2.7 ± 0.1) nm. For a bcc structure the associated bubble lattice constant is (3.8± 0.2) nm. The two satellites of the 011 matrix reflection, indicated by arrows, are in <112> directions relative to the (011) matrix spot. On the basis of their angular and radial positions these satellites are assigned indices of 1/2{112}—indices which are normally excluded for a bcc lattice. The bubble lattice constant calculated from the radial position of the satellites using 1/2{112} indices is (3.8 ± 0.2)nm. That this value is in excellent

Fig. 7   A section of an electron diffraction pattern from He gas bubbles in an (001) grain in vanadium. The electron beam is directed approximately along $[011]_m$. The two satellite spots associated with the 000_m transmitted beam are the allowed $\{110\}_b$ reflections from the ordered bubble array. The two satellite spots associated with the $0\bar{1}1_m$ matrix reflection are from bubble planes parallel to $\{112\}_m$. These satellites are assigned indices of $1/2\{112\}$ — indices which are normally disallowed for a bcc lattice.

agreement with that based on diffraction from 011 bubble planes provides further strong evidence that the ordering is fully three-dimensional over small regions at least.

A study based on computer modelling of the diffraction (Cook and Johnson, to be published) explains the main features observed here (including the presence of disallowed reflections) in terms of three-dimensional ordering of bubbles in small domains. Within any given domain bubble positions are represented by a perfect bcc lattice with individual bubbles shifted from lattice sites by a small random displacement within a {110} plane that is perpendicular to the surface of the grain. The results confirm the conclusion reached in earlier work [25] that the presence of normally excluded reflections can be attributed to imperfections in the bubble lattice and small domain size.

(iii) <u>Discussion</u>

Bubble images can be up to 1.5 nm across, are often of an irregular shape and commonly appear to be facetted. The lattice constant of 3.8 ± 0.2 nm corresponds to an ultra-high bubble concentration of $3.6\ 10^{25}\ m^{-3}$. This concentration is unprecedented for metals. For a bcc lattice nearest-neighbour bubbles lie along <111> directions. The bubble concentration is so high that the thickness of the metal wall separating nearest-neighbour bubbles corresponds to the spacing of 6 vanadium atoms only, measured along a <111> atom row. The results for the superlattice in vanadium raises some important points. The swelling is estimated to be ~6%, maximum. It is interesting that small bubbles, despite high concentrations, do not give rise to a very high swelling. In earlier work [51], high values of swelling were accounted for by proposing the existence of a large population of gas-filled defects and small bubbles with sizes lying below the TEM resolution limit. The existence of such a population seems unlikely. Large values of swelling are almost certainly associated with large bubbles.

As discussed in A 2.5 (ii) and A 2.5 (iii) some theories of ordering require the retention of interstitial dislocation loops, with diameters comparable with the bubble size, on the common glide cylinders linking nearest-neighbour bubbles. The results for the vanadium superlattice highlight a difficulty with these theories. It seems improbable that the metal in the thin wall between bubbles is sufficiently strong to accommodate the two or more dislocation loops required by these theories, especially under the conditions of continual displacement of host atoms from matrix lattice sites that occur during the He implantation. Cotterill [55] has considered the limiting strain energy that a metal can withstand and yet still retain a crystalline character. The limiting dislocation density for Cu is estimated to be 10^{16} to $10^{17}\ m^{-2}$. A similar value might be expected for vanadium. It is interesting to apply this concept of a limiting dislocation density to the present case. For two dislocation loops, having radii comparable with those of the bubbles, to be accommodated in the thin wall between bubbles would require a dislocation density exceeding the Cotterill limit by a factor of thirty or more. This is a conservative estimate because the strain energy associated with both the overpressured bubbles and the presence in the metal of He and defects has been neglected. It is recognised that the strength of a metal on a microscopic scale can greatly exceed that measured for bulk specimens. Nevertheless it seems certain that such a high level of dislocation could not be sustained, particularly in the presence of damage exceeding 10 d.p.a. in the implanted layer. Suppose the loops were to move out of the <111> glide cylinders to become uniformly distributed through the metal locally in the implanted layer. If the loops remained at a size comparable with that of the bubbles, the average dislocation density would still exceed the Cotterill limit. The combination of high stress levels and large damage rates suggest that any dislocation loops punched from bubbles must evolve towards a coarser structure with lowered energy. One could speculate that these coarser dislocations are in the walls separating bubble domains. TEM studies show that the dislocations associated with bubble lattices are on such a fine scale and at such high density that they appear as a single dense dark image rather than as a series of discrete dislocation images. With the electron beam directed precisely down a major zone axis in the matrix, strain constrast from the dislocation structure is sufficient to completely obscure the bubble images. Direct micrographs and selected area diffraction patterns from bubble lattices are, for this reason, usually taken with the specimen tilted a few degrees off a major axis. (Patches of strain contrast from dislocations are particularly obvious in Fig.1 for example.)

2.2. Coarse bubble structures

(i) Experimental

Previous experiments have indicated that high values of swelling imply the existence of coarse structures involving large bubbles. These structures have been observed following 30 keV He$^+$ implantation at normal incidence (see B 1.2). However, this low energy implantation leads to He and damage profiles that change rapidly with depth below the surface (see Fig.3) and the large bubbles are in a shallow layer that spans only a narrow range of depths. To obtain a more uniform He concentration, extending deeper into the target, a different method of implantation has been adopted. The He is implanted at higher energy (160 keV) and at an oblique angle (60° to the surface normal) to produce a broad distribution of He with a maximum He concentration at a depth ~250 nm. The thin sections for TEM examination have been prepared by backthinning from the unimplanted side of the foils using argon ion-beam milling rather than chemical jet electropolishing. An ion-beam technique has been preferred because of the need to preserve the porous structure exposed by the thinning process.

The targets were prepared from high purity vanadium foils, 25 μm in thickness, which were heat treated in vacuo. The foils were electropolished to a high surface finish immediately prior to implantation at a temperature of 430 K to a dose ~1.2 10^{22} He$^+$ m^{-2}. High resolution electron microscopy on the back-thinned specimens was performed in a Philips EM 420ST 120 keV electron microscope.

(ii) Results

Structures covering a range of bubble sizes were found. In local areas typically 100—200 nm in extent, bubbles were generally of uniform size. Near the thin edge of the specimens the bubbles were typically smaller than in the thicker regions. These small bubbles were often ordered on a superlattice of lattice constant ~4.4 nm, having general features similar to the superlattice of Fig.6. (This lattice constant corresponds to an interplanar spacing for the {200} bubble planes of 2.2 nm.) Away from the edge in the thicker regions, larger bubbles dominated the structure. These bubbles were also spatially ordered but on a coarser scale—see the upper micrograph of Fig. 8. The degree of ordering was sufficient to give rise to satellite reflections in electron diffraction patterns taken from the bubble layer (see inset Fig.8). From

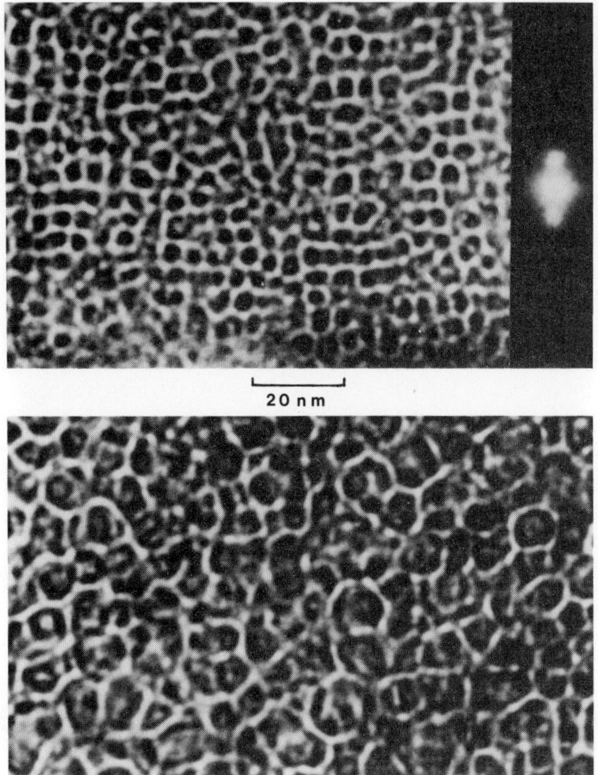

Fig. 8   Bright-field TEM micrographs (over-focus contrast) of bubble structures in vanadium following implantation at a temperature of 430 K to a dose ~ 1.2 10^{22} 160 keV He$^+$ m^{-2} with the ion beam at 60° to the surface normal. The electron beam is directed approximately along [200]$_m$ with the specimen tilted slightly off-zone to clear obscuring dislocation images. The thin metal walls of uniform thickness are a striking feature. In the thinnest regions of the specimen the bubbles are ordered on a superlattice similar to that shown in Fig.6. Away from the thin edge the bubble structure is ordered on a coarser scale (see upper micrograph). The ordering is sufficient to give rise to satellites around the 000$_m$ transmitted beam in the electron diffraction pattern shown (in approximately the correct orientation) as an inset. (The $g = [020]$ direction is parallel with the right hand edge of the figure.) Further away from the edge the coarse porous structure shown in the lower micrograph is found. The degree of local swelling associated with these coarse structures is estimated to be in the range of 50% to 80%.

the radial separation of the satellites along a <200> direction, the spacing of the corresponding {200} bubble planes was calculated to be 5.4 nm. This value is in agreement with the spacing of bubble rows measured on direct micrographs. (In other areas of the specimen diffraction satellites corresponded to interplanar spacings as large as 8 nm.) In the thickest regions of the specimen a remarkable "honeycomb" structure was observed—see the lower micrograph of Fig.8. This structure consisted of a dense random array of large cavities, 8 to 10 nm across, separated by thin continuous metal walls having a uniform thickness in the range ~1.0 to 1.4 nm. (It is not yet certain what focus conditions should be used to image these walls so that accurate thickness measurements can be made.)

(iii) <u>Discussion</u>

The formation of this porous honeycomb structure will result in very high values of swelling in the implanted layer. There is some evidence that the structure may form rapidly in terms of dose at a late stage in the implantation. The rapid formation of such a structure could explain, for example, the non-linear volume expansion observed at high dose in He implanted erbium by Blewer and Beezhold, [56]. Such thin metal walls could be expected to have very low electrical and thermal conductance. One could speculate that a high temperature rise in the implanted layer would accompany the formation of the structure and that this may contribute to the onset of radiation blistering of the surface. The coarsest structure is associated with depths near R_p. Using the same implantation and thinning procedures we have observed a similar structure in Cu. Terreault et al [57] have observed the development of double peaked He depth profiles in Cu at a He dose close to the critical dose for radiation blistering (see the dashed curve in Fig.3). The double peaking was attributed to a loss of He from depths near R_p. At depths both closer to the surface and beyond R_p the He continued to be retained. These are the depths where the superlattice structure is likely to remain intact—see B.1.2. On the basis of the observations here the following development of bubble structure with increasing dose at depths near R_p is proposed. In the first stage bubbles are nucleated randomly. As the bubbles grow patches of local order develop. This ordering is almost certainly not three-dimensional and is characterised by small bubble sizes and small inter-bubble distances. With further bubble growth the superlattice stage is reached. The breakup of the superlattice leads to larger bubbles ordered on a coarser scale. The final stage is a random array of large cavities separated by a mesh of thin walls of metal of uniform thickness.

3.   The Gas Bubble Macrolattice

Recently an investigation has been made [58] of the bubble structures in He implanted Au. The same techniques as above were used for implantation and thinning. A striking new phenomenon was discovered—a macrolattice—in which large ~60 nm diameter artifacts, thought to be He bubbles, form an ordered array. The macrolattice—see Fig. 9—had the same symmetry as the gold atomic lattice but had a spacing some 400 times larger. A gas-bubble

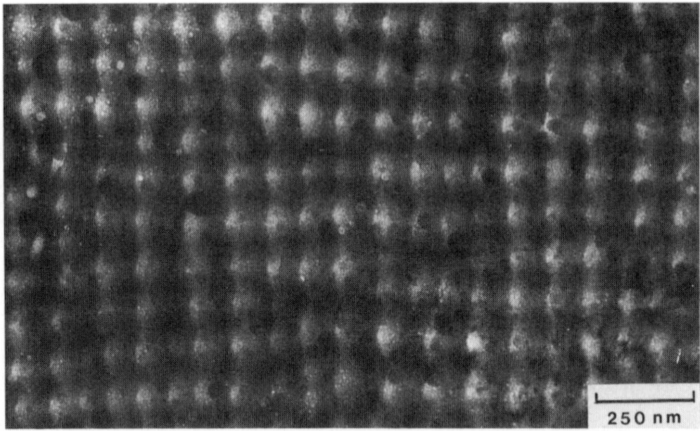

Fig. 9   A TEM micrograph of artifacts thought to be helium gas bubbles forming a gas-bubble macrolattice, in a (100) gold grain following implantation with $1\ 10^{22}$ 160 keV He$^+$ m^{-2} at 280 K. The helium ion beam was incident at an angle of 60° to the surface normal to give a broad distribution of helium with depth below the target surface. The specimen was back thinned using argon ion beam milling to preserve the coarse structure. The specimen is tilted to bring the electron beam approximately along [112] in the fcc gold lattice. The size and spacing of artifacts is remarkably uniform throughout the specimen with typical values of ~ 70 nm diameter and ~ 105 nm nearest-neighbour separation (measured along <220>). At higher magnification the images of small bubbles (diameter ~ 2 nm) ordered on a superlattice of lattice constant a_l~ 8 nm can also be identified.

superlattice of spacing ~20 times that of the atomic lattice was also present. The coexistence in a single specimen of three separate lattices of very similar structure which differ so markedly in scale is providing a new perspective on ordering. It is thought that long range elastic interactions between bubbles may play an important part in forming the macrolattice.

4. Conclusion

Previous work on the bubble lattice formation in metals has been reviewed. In addition, more recent results have been included, These results were obtained using a new way of implanting metal foils with He, based on higher He energy (160 keV) and oblique incidence of the ion beam. Implanted foils have been back-thinned using argon ion beam thinning in preference to electro-chemical erosion. The combination of these two techniques has allowed the coarse bubble structures that develop at high dose to be preserved in thin sections suitable for subsequent TEM examination. The results for vanadium and copper have enabled the evolution with dose of the bubble structure at depths near the maximum in He concentration to be determined. The final stage, near the critical dose for radiation blistering of the surface, is a random array of large cavities separated by a remarkable mesh of very thin metal walls of apparently uniform thickness. In He implanted gold a spectacular array of large bubble-like features ordered on a macrolattice has recently been discovered.

ACKNOWLEDGEMENTS

It is a pleasure to acknowledge the stimulating contribution of Dr D.J. Mazey, my coauthor on many publications. I thank Dr J.H. Evans for many interesting discussions and the research students who have participated in this work. I am grateful for the technical assistance of Mr T. Young and Mr R.W. Thomson, and for financial assistance from the Internal Grants Committee of Victoria University of Wellington.

REFERENCES

1. R.S. Barnes and D.J. Mazey, UKAEA Unclassified Report, AERE R-3348, (1960) and Phil.Mag **5**, 290 (1960).
2. B.M.U. Scherzer, p 271 in *Topics in Applied Physics*, Vol.52—*Sputtering by Particle Bombardment II*, R. Behrisch, ed.,Springer-Verlag, Berlin (1983).
3. S.L. Sass and B.L. Eyre, Phil.Mag **27**, 1447 (1973).
4. D.J. Mazey, B.L. Eyre, J.H. Evans, S.K. Erents and G.M. McCracken, J. Nucl. Mater. **64**, 145 (1977).
5. J.H. Evans, Radiat. Effects **10**, 55 (1971) and ibid **17** 69 (1972).
6. F.W. Wiffen in *Radiation Induced Voids in Metals*, J.W. Corbett and L.C. Ianniello, eds., Albany, USAEC-CONF-71061, p386 (1972).
7. K. Krishan, Radiat.Effects **66**, 121 (1982).
8. L.T. Chadderton, E. Johnson and T. Wohlenberg, Comments Solid State Phys **7**, (5) 105 (1976).
9. G.L. Kulcinski, J.L. Brimall and H.E. Kissinger in *Radiation Induced Voids in Metals* W. Corbett and L.C. Ianniello, eds., Albany, USAEC-CONF-71061, p449 (1972); J. Nucl.Mater **40**, 166 (1971).
10. D.J. Mazey, S. Francis and J.A. Hudson, J. Nucl.Mater **47**, 137 (1973).
11. A. Risbet and V. Levy, J. Nucl.Mater **50**, 116 (1974).
12. L.J. Chen and A.J. Ardell, J.Nucl.Mater **75**, 177 (1978).
13. S.B. Fisher and K.R. Williams, Radiat.Effects **32**, 123 (1977).
14. P.B. Johnson and D.J. Mazey, Nature **276**, 595 (1978); Harwell Report AERE R-9290, (1978).
15. P.B. Johnson and D.J. Mazey, J. Nucl.Mater **93/94**, 721 (1980).
16. J.H. Evans and D.J. Mazey, J. Nucl.Mater **138**, 176 (1986).
17. P.B. Johnson and D.J. Mazey, Harwell Report AERE R-9685 (1980).
18. P.B. Johnson and D.J. Mazey, Nature **281**, 359 (1979).
19. R.S. Nelson, Harwell Report AERE R-9380 (1979); J. Nucl.Mater **88**, 322 (1980).
20. P.B. Johnson and D.J. Mazey, Radiat.Effects **53**, 195 (1980).
21. P.B. Johnson, D.J. Mazey and J.H. Evans, AERE R-10874 (1983); Radiat.Effects **78**, 147 (1983).
22. W. Jäger and J. Roth, J.Nucl.Mater **93/94**, 756 (1980).

23. W. Jäger and J. Roth, Nucl.Instr.& Meths **182/183**, 975 (1981).

24. H. Van Swijgenhoven, G. Knuyt, J. Vanoppen and L.M. Stals, J. Nucl.Mater **114**, 157 (1983).

25. P.B. Johnson and D.J. Mazey, AERE -R10644 (1984); J. Nucl.Mater **127**, 30 (1985).

26. A.L. Diprose, *The Helium Gas Bubble Superlattice in Copper* Ph D Thesis, unpublished, Victoria University of Wellington,Wellington, New Zealand (1988).

27. P.B. Johnson, A.L. Malcolm and D.J. Mazey, Nature **329**, 316 (1987).

28. P.B. Johnson, A.L. Malcolm and D.J. Mazey, J. Nucl.Mater **152**, 69 (1988).

29. P.B. Johnson, A.L. Diprose and D.J. Mazey, J.Nucl.Mater **158**, 108 (1988).

30. J.H. Evans, J. Nucl.Mater **68**, 129 (1977); **76/77**, 228 (1978); **79**, 241 (1978).

31. P.B. Johnson, 4th Ann. Solid State Phys. Mtg Aust./N.Z. Inst. Phys. 1980 (Phys. Rpt 1/82, Victoria University, Wellington 1982).

32. P.B. Johnson and W.R. Jones, J. Nucl.Mater **120**, 125 (1984).

33. S.E. Donnelly, Radiat. Effects **90**, 1 (1985).

34. J.H. Evans and D.J. Mazey, J. Nucl.Mater **132**, 176 (1986).

35. R.S. Barnes and D.J. Mazey, Acta Metall **11**, 281 (1963).

36. K. Shiraishi, A. Hishinuma and Y. Katano, Radiat Effects **21**, 161 (1974).

37. W.R. Wampler, T. Schober and B. Lengeler, Phil.Mag **34(1)**, 129 (1976).

38. K. Malen and R.Bullough, R. in Proc. Conf. on *Voids Formed by Irradiation of Reactor Materials*, (Brit.Nucl.Energy Soc.) S.F. Pugh, M.H. Loretto and D.I.R. Norris, eds., AERE Harwell, U.K., 109 (1971).

39. A.M. Stoneham, in Proc, Harwell Consultants Symposium on the Physics of Irradiation Produced Voids, R.S. Nelson, ed., AERE Report R-7934, 319 (1974).

40. A.G. Khachaturyan and V.M. Airapetyan, Phys.Status Solidi **A26**, 61 (1974).

41. V.I. Dubinko, V.V. Slezov, A.V. Tur and V.V. Yanosky, Radiat.Effects **100**, 85 (1986).

42. W.G. Wolfer, Phil.Mag **A59**, 87 (1986).

43. E.P. EerNisse and S.T. Picraux, J. Appl.Phys **48**, 9 (1977).

44. S.E. Donnelly, M. Renier, A.A. Lucas and H.J. Whitlow, J. Nucl.Mater **115**, 347 (1983).

45. J.H. Evans, Mater.Sci.Forum **15-18**, 869 (1987).

46. C.H. Woo and W. Frank, Mater.Sci.Forum **15-18**, 875 (1987).

47. J.H. Evans in Encyclopedia of Materials Science and Engineering, R.W. Cahn, ed., Pergamon Press, Oxford, **Suppl.Vol.1**, 547 (1988).

48. J.H. Evans in Proc. of NATO Advanced Study Institute on *Patterns, Defects and Instabilities*, Cargese, Corsica, Sept (1989). D. Walgraef and N.M. Ghoniem, eds., Kluwer Academic Publishers, 1990 p.347.

49. D.J. Mazey and J.H. Evans, J. Nucl.Mater **138**, 16 (1986).

50. J.H. Evans, A.J.E. Foreman and R.J. McElroy, J. Nucl.Mater **168**, 340 (1989).

51. R.G. Saint-Jacques, G. Veilleux and B. Terreault, Nucl.Instr. & Meths **170**, 461 (1980).

52. P.B. Johnson and D.J. Mazey, J.Nucl.Mater **111 & 112**, 681 (1982).

53. P.B. Johnson and D.J. Mazey, J. Nucl.Mater **170**, 290 (1990).

54. B.A. Kalin, S.N. Korshunov, I.I. Chernov, A.V. Markin, I.V. Reutov, V.N. Chernikov, A.P. Zakharov, V.M. Gureev and M.I. Guseva, J.Nucl. Mater **161**, 228 (1989).

55. R.M.J. Cotterill, Phys.Lett. **60A**, 61 (1977).

56. R.S. Blewer and Wendland Beezhold., Radiat.Effects **19**, 49 (1973).

57. B. Terreault and G. Veilleux, J. Nucl.Mater **89**, 392 (1980).

58. P.B. Johnson, R.W. Thomson and D.J. Mazey, Nature, **347**, 265 (1990).

MACROSCOPIC PHENOMENA INDUCED BY HIGH DOSE MeV ENERGY IMPLANTATION OF He, Ne AND Ar IONS

F. Pászti

Central Research Institute for Physics
H–1525 Budapest P.O.Box 49, Hungary

ABSTRACT

The nuclear fusion product He ions that will be implanted into the first wall of an operating thermonuclear reactor will cause serious problems. One of them is the macroscopic surface deformation on the first wall components. Sputtering and lattice damage caused by the particles of different kinds originated from the active zone may considerably influence the processes. The present paper reviews our most important results on macroscopic surface deformations caused by 0.8 - 4 MeV energy high dose He ion implantation. To simulate the effect of sputtering and damage, heavier noble gas ion implantations were also applied.

1. Introduction

If one wants fusion to take place at the necessary rate in a thermonuclear reactor, the temperature in the active zone has to be very high, corresponding to several tens of keV, where the deuterium and tritium components are in the plasma state [1]. The reaction product He ions that will be implanted into the first wall of the device will cause serious problems. One of these is the macroscopic surface deformation. This will be very dangerous because the affected layer is soon removed from the surface (e.g. simply falls down or, because this layer is in poor heat contact with the bulk, evaporates) and reaches the active zone of the reactor. The heavy impurities cool down the fusioning plasma and the economical self sustaining nuclear fusion fails. The cold He gas that will be reemitted during the surface deformations may have similar effects. The material loss itself may also be a problem.

Other phenomena, such as sputtering by plasma ions of several keV energy or unipolar arcing under high electromagnetic fields will also have their effect on the first wall and plasma purity, so it may become necessary to replace regularly the most affected first wall components. It is evident that the interaction of the fusioning plasma and the reactor wall is a rather complex problem, especially when the synergism between the different effects and processes is also taken into account. If one wants to design a fusion reactor, the data about the different processes and their interactions are crucial. This is why wide scale model experiments are carried out in different laboratories all over the world.

As reviewed by others [e.g. 2-5], many studies have been made over the last 15 years on the formation and evolution of surface effects—blistering, flaking—as a result of helium implants in the 1-200 keV ion energy range. Our "Plasma–Wall Interaction" group in CRIP, Budapest formed and joined this research in 1978. However, our work is mainly based on a 5 MeV Van de Graaff accelerator and this focussed our investigations on the effects of MeV energy He ions. This research is relevant, because the 3.52 MeV He ions produced in the fusion reaction slow down only when penetrating the plasma of the active zone. The fraction hitting the wall

Fundamental Aspects of Inert Gases in Solids
Edited by S.E. Donnelly and J.H. Evans, Plenum Press, New York, 1991

with MeV energies strongly depends on the type and operating regime of the reactor. It can be just a few percent in the reactors of magnetic confinement (i.e. tokamaks) and much more in devices of inertial confinement.

The aim of the present paper is to review our most important results in the light of other investigations done mainly at 1–200 keV ion energies.

2. The Mechanism of Surface Deformation

It is well established that if the implanted He gas is unable to escape, it will coalesce and precipitate to form gas filled bubbles in the host material. These bubbles may grow by capturing additional gas atoms and vacancies or by punching out interstitial dislocation–loops. If the temperature is low and the radiation damage is not too serious they are overpressurized [6]. The pressure inside these bubbles may lead to inter–bubble cracking [7], or the bubbles of equilibrium pressure will coalesce to form a crack parallel to the sample surface. The layer separated from the substrate by this crack may contain a huge compressive lateral stress. This stress and/or the pressure of the gas accumulated in the cavity formed by the crack lifts the covering layer up, deforms it plastically and may even tear it off from the bulk [8–10]. If this process takes place on well defined, usually circular spots, it is called blistering, otherwise exfoliation. If the detached layer falls off, the process is referred to as flaking [11]. These processes are accompanied with enhanced reemission of He [12–13].

3. Summary of Results With MeV Energy He Implantation

We will refer to the model investigations via ion implantation at mainly 1 – 200 keV ion energies at different laboratories in a general way rather than with individual references. The mentioned results and theoretical considerations are mainly reviewed in references [2–4].

3.1. Appearance

At these relatively low energies, it was shown that the mentioned processes can take place in all the solid materials including metals, semiconductors, ceramics, glasses, metallic glasses, sintered powders and carbonic materials of different structure, etc. We have also investigated materials of different crystal structure (e.g. polycrystalline Au [14–15], Al [16–19], Inconel and stainless steel [20], single crystalline Si [21–22] and amorphous metallic glasses [23–26]). It was found that instead of blistering which is frequently observed in the low energy range, exfoliation and flaking takes place. We supposed that there is a "transition energy" between the two processes and it is related to the width of "suppressed zones" surrounding the blisters [15]. From these enough He migrates into the blisters to prevent the appearance of a new blister. If the radius of the blisters is smaller than the thickness of the suppressed zone, the blisters will never coalesce. Increasing the implantation energy the blister radius increases, so by overcoming the thickness of the suppressed zone, blister coalescence is then possible. At even higher energies one cannot observe separated blisters [15] but instead the exfoliation will expand in jumps when new blisters appear and join it [14]. It was also shown that the transition energy sharply increases with the implantation temperature. In the case of pure Al targets it was below 600 keV at room temperature, but it increased to 2000 keV at a temperature of 360 K [18]. This fact is in accordance with the speculation that the width of the suppressed zone sharply increases together with the diffusion coefficient.

Hoping to get information about the formation mechanism of exfoliations, we opened them. The traces of the primary blisters and of the expansion of the exfoliation in sudden jumps were clearly seen as contour lines consisting of small lamellae on both the inner surface of the exfoliated layer and the surface left behind on the bulk [14]. This observation confirms our speculation on the mechanism lying behind the transition energy. The mentioned surfaces frequently contained small secondary blisters or suffered secondary flaking [14,15,20]. It was suggested that this was due to the high He concentration in the vicinity of these surfaces. To make more direct experiments, the so called "quasi–multiple energy implantation" technique was developed [27]. This technique is based on a periodically tilted absorber foil placed right in front of the target. By this technique it was confirmed that surface deformations may take place even when the implanted concentration is high at the sample surface [17,22]. This is in contradiction with the observations at smaller energies where a channel network formed in the He containing layer. This network was open to the surface and through it the implanted gas continuously escaped. At high energies, however, this mechanism sometimes may fail.

To get information on the appearance of the channel network, a special experiment was performed [19]. Here, a wide gas depth profile was implanted into Al by multiple energy implantation, then sharp monoenergetic peaks were added to it. The He distribution was regularly checked using in–situ proton RBS. No gas redistribution was observed before the onset of exfoliation. Also, the gas escaped through the holes in the exfoliation lid only from layers of highest gas concentration. From this experiment one can draw the conclusion that the channel network develops more or less simultaneously with the surface deformation. These may be competing processes. Near to the surface, channel network formation can be favoured and surface deformations do not develop. If the implanted profile is wide enough and decreases only slowly with depth, the gas may escape from the near surface region by diffusion, so channel formation may not prevent the appearance of surface deformations.

3.2. *Critical dose*

In the work at lower implant energies, it was shown that when the implanted dose reaches a critical Φ_c value, macroscopic surface deformations appear suddenly. This critical dose has special interest because of its relation to the lifetime of the reactor wall. It soon became obvious that Φ_c strongly depends on experimental parameters such as the target material, its lattice and macroscopic structure, temperature and surface roughness, as well as the implantation energy or energy distribution and the angle of incidence. It was generally supposed that the ion energy, the angle of incidence and their distributions have their effect on Φ_c only via the critical peak concentration n_c: Φ_c will be reached if the He concentration at the maximum of its depth distribution reaches n_c. This approximation can fail at low implantation energies, where the surface is close enough to disturb the profile, and the lattice damage changes fast with the depth and implantation energy. In our case, however, it was valid as clearly proved in experiments on widened implantation profiles [17, 19, 22, 26] or on different implantation energies [18, 20, 28].

The effect of target material on n_c is moderate. We found it to be 22, 24, 33, 38 and 45 at.% for Au [14], Al [16–19], stainless steel [20], metallic glass [23] and Si [21–22] respectively. The effect of alloying was demonstrated on an AlMgSi sample (n_c = 32 at.% instead of 24 at.% for pure Al) [18].

In case of pure and alloyed Al we found that increasing the temperature n_c decreases in accordance to the yield or tensile strength of the material [18, 28]. This change can be dramatic. For example n_c decreases from 24 at.% at room temperature to 3 at.% at 530 °C for pure Al. Also, the experiments showed that in this case a post implantation annealing is equivalent to an elevated temperature during implantation. If we implanted the sample at room temperature with a dose ϕ, that was equal to the critical one at a given higher temperature T_0, and performed a post implantation annealing by slowly increasing the temperature, the surface deformations suddenly appeared when reaching T . This way we can define the critical dose Φ_c (T) and critical temperature T_c (Φ) functions and conclude that they coincide in the investigated case.

The same experiment [18] clearly demonstrated that surface deformation (namely blistering) can occur even at 90 % of the melting temperature of the material T_m. This is in marked contrast to the rule of thumb found for lower implant energies that: "surface deformations do not occur above 0.5 T_m ". The reason is obvious: at high temperatures channels may form by inter–bubble coalescence. At low energies the implanted profile is shallow, and these channels intercept the surface; as a result the gas escapes from this porous structure without any macroscopic surface deformation. If the implantation energy is high enough, the channels cannot intercept the surface, and the surface deformation is inevitable.

3.3. The effect of lateral stress

As it was pointed out in the section concerning the mechanism of surface deformations, some theories state that the effect of the huge compressive lateral stress generated by the extra material and damage introduced during the implantation is crucial in these processes. Some of our experiments clearly demonstrated that this stress can also cause new types of phenomena that were unknown from experiments using lower implantation energies.

If our sample was homogeneous (i.e. single crystalline Si or amorphous metallic glass), we sometimes observed a rather regular wave pattern of protrusions with triangle shaped cross section on the crack made surfaces, both on the remaining bulk and the detached layer (see Fig. 1) [21, 23–26]. This phenomenon was explained by a stress model, assuming that the lateral stress in the zone around the projected range of the impinging ions (R_p) can reach a critical limit (approximately a tenth of the Young's modulus). At this stress value, the layer will be subjected to mechanical instability and become rippled with a wave–length of $\lambda \approx 4.3$ h, where h is the thickness of the implanted profile (or the projected range straggling in FWHM sense) [21, 29–30]. When the crack is propagating in a layer of periodically changing stress caused by the rippling process, it should be disturbed, resulting in wave–like structures.

The rippling was also directly observed on the sample surface if the implanted profile was close to the surface, i.e. in the case of He implantation of wide energy distribution when the implanted profile extended from 0 to 3 µm depths in Si [22] and after implantations into Si and metallic glass of Ne and Ar ions of 1.4 and 0.7 MeV energies, respectively [31]. It is worth noting that this new type of surface deformation develops without the appearance of any crack in the material.

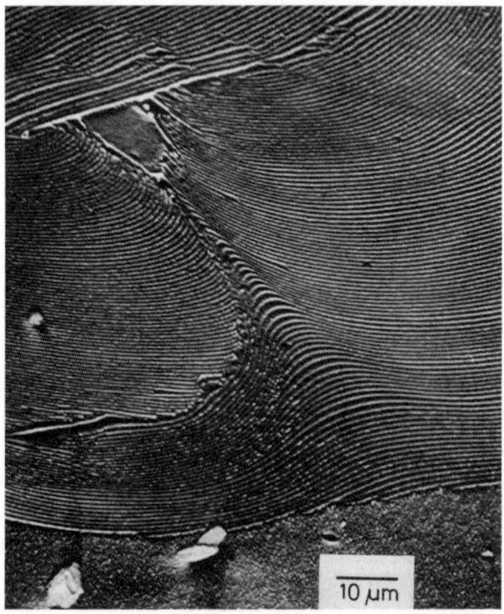

Fig. 1   Wave pattern on the surface left behind after flaking on a 2 MeV ^{4}He$^+$ implanted amorphous METGLASS-2826A sample.

The already mentioned Si sample with a wide He profile [22] also showed another phenomenon related to the lateral stress. The He containing layer was unstable: it flaked off from big spots under the weakest mechanical influence resulting in shell shaped craters.

3.4. The thickness of the detached layer

In early work with helium implants in the lower energy range, there was considerable discussion on the relation between this thickness and the position of the helium peak [3]. Although some discussion ensued on the role of lateral stress, displacement damage and other factors, it was eventually recognised that while bubble swelling [32] could probably account for earlier discrepancies, some aspects were not totally resolved. In our experiments the thickness of the detached layer always coincided with the depth of peak He concentration within the experimental error. This was valid even at implantations of wide energy distribution [17, 19, 22, 26]. Moreover, if the depth distribution had two well separated equivalent peaks, the surface layer detached alternatively at both of them [19]. This observation excludes the validity of speculations based on macroscopic effects.

Microscopically, however, our results seem to be plausible as follows. If the He energy is higher than ≈ 10 keV, the caused lattice damage is almost negligible. So, the MeV energy ions cause lattice damage mainly at the end of their path (i.e. at the final ≈ 0.4 μm in Fe). As a consequence, the shift in the damage profile as compared to the depth distribution of implanted atoms is significant only if the profile is not too wide (≤ 1 μm in Fe). This means that at energy distributed implantations or at implantations of energies exceeding a limit (≈ 30 keV in Fe), the shift will be negligible. In this case there is no big difference in the damage to implanted atom density ratio along the depth and the conditions that are necessary for crack formation will be fulfilled first around the depth of peak He concentration.

4. Ne and Ar Implantations

The main goal of these experiments was to compare the effect of He and other noble gases in high dose MeV energy implantations. We hoped that implantation of heavier noble gas ions can simulate the fusion reactor environment in a more complex way: the effect of the lattice damage of fast neutrons, the sputtering caused by the deuterium–tritium fuel and other impurities, as well as the bubble formation and surface deformations caused by high energy He fusion products can be modelled simultaneously [28, 31]. Different types of samples were used in the experiments: amorphous metallic glass, Single crystal Si and polycrystalline Al.

The main experimental conditions and results of these experiments are presented in Table 1. For comparison some experiments with He ions are also included in the table.

The critical concentrations are lower for Ne and Ar ions than for He. This fact is partly connected to the different atomic volumes. The higher levels of radiation damage (see ξ in Table 1.) can also play some role by enhancing gas atom and vacancy migration. In metallic glasses the amorphous state of the material makes a further increase in n [23]. The relatively low values for Ne and Ar ions in this case can also be connected to recrystallisation under ion irradiation [35–37]. The lower critical concentration of Ne implanted into Si at a higher dose rate can be partly due to the elevated sample temperature. In the weak and plastic pure Al the critical concentration for He is low. For Ar ions, the depth distribution is close to the sample surface, and the surface roughness increases the critical concentration.

In case of He implantation the projected range was big and we observed flaking or exfoliation only. For Ne and Ar, because of shallower implantation depths blisters became more preferred. Sometimes, however, the "blister" density was so high that it was more precise to speak about ripples on the surface instead of distinct blisters (Fig. 2). At doses only slightly higher than the critical one the ripples were ordered into a wave pattern in accordance to the stress theory. Their sizes are also included in Table 1. The mean wavelengths of these

Table 1 Comparison of different experiments. The different columns are as follows: target material; ion species, energy and dose rate of implantation; critical dose; diameter or wave length of the observed blisters, rippling or waves, respectively; appearance of flaking or exfoliation; mean projected range and its straggling in FWHM sense as calculated by TRIM code [33]; thickness of the sputtered layer determined from the critical dose and the sputtering coefficients from ref. [34]; peak damage by TRIM; critical concentration that was calculated from the critical dose, range straggling and the thickness of the sputtered layer; references of the experiments.

mat.	ion	E (MeV)	dose rate (10^{13} ion/cm²s)	Φ_c (10^{18} ion/cm²)	d blist (µm)	d ripple (µm)	λ wave (µm)	flak.	exf.	R_p (µm)	ΔR_p (µm)	D_s (µm)	ξ (dpa)	n_c (at%)	Ref.
metglass	He	2.0	9	1.5	–	–	1.2±.1	+	–	2.98	.33	.00	100	32	24
	He	1.0	29	1.5	–	–	0.8±.1	+	–	1.51	.26	.00	100	38	24
	Ne	1.4	8	0.44	3–4	1–2	1.5±.1	–	–	0.74	.33	.01	300	12	31
	Ar	0.7	110	0.30	≤5	.7–1.5	1.2±.1	–	–	0.29	.19	.05	500	12	31
Si	He	2.0	20	1.8	–	–	4.0±.3	+	–	6.95	.59	.00	100	41	21
	Ne	1.4	13	0.84	6–13	2–3	4.3±.3	+	+	1.55	.42	.02	500	27	31
			220	0.68	15–36		–					.01	400	23	
	Ar	0.7	70	0.57	8–20	2–5	–	+	+	0.64	.28	.06	700	25	31
Al	He	2.0	10	0.95	–	–	–	–	+	6.73	.40	.00	50	27	18
	Ar	0.7	55	1.0	4–14	–	–	–	–	0.61	.28	.13	1000	27	31

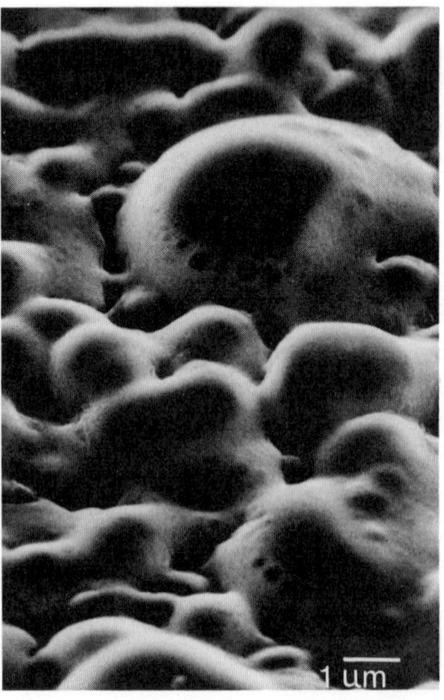

Fig. 2 Rippling on a 0.7 MeV Ar implanted amorphous METGLASS-2826A.

patterns were in good accordance with the stress model (see in the table), so it was reasonable to suppose that this phenomenon was also caused by the same stress driven mechanism.

There remain, however, two important questions to answer: how can the amplitude of these ripples become big enough to be observed and why does not the stress relax before reaching the critical value. The possible explanations are as follows. Diffusion of host atoms, vacancies and implanted noble gas atoms as well as orientation and gliding of interstitial dislocation loops are also influenced by stress gradients. This leads , however, not only to stress release, but it can decrease the effective Young's coefficient for long term processes (creep). Thus ripples can form far below the critical stress calculated from unimplanted conditions. The same mechanism can lead to the enhanced amplitude of ripples via transporting material to places where tensile stress arises under the layer moving away from the bulk. Gas migration from bubbles of smaller size onto larger ones (Ostwald ripening) can also play an important role. It is worth noting that there is no crack involved in this mechanism.

Cylindrical or mushroom shaped blisters were also observed on Si samples. They may also be connected to the processes mentioned above. Here the bottom of the blister walls was the place where tensile stress arose, and continuous material transport built up the observed cylindrical walls. Similar formations were observed on the inner surface of a Cf^{252} neutron source capsule [38]. A detailed explanation of the phenomenon is also presented there.

5. Relevance For Fusion Reactors

One can predict that surface deformations caused by *thermalised* He ions will not appear in fusion reactors, because:

1. the erosion rate from other processes (i.e. sputtering) will be too high, and the critical He concentration will be never reached;
2. the maximum gas concentration will be reached at the surface, and the gas escapes through channel network that develops under such conditions;
3. the wall temperature will be above half of the melting point of the material, and the gas escapes similarly to case 2.

Our results show, that the last two points alone cannot guarantee against surface deformations by MeV energy He ions in a fusion reactor. If the competitive processes are suppressed somehow, and they do not erode the surface with a rate that is high enough, then the surface deformations will cause serious problems. The effect of radiation damage and the elevated wall temperature may decrease the critical gas concentration by a factor of ten, as compared to the values from classical experiments at room temperature and low ion energies.

REFERENCES

1. T.J. Dolan, Fusion Research, Pergamon, New York (1982).
2. S.K. Das and M. Kaminsky in *Radiation Effects on Solid Surfaces*, Advances in Chemistry Series **158**, M. Kaminsky ed., American Chemical Society, Washington 112 (1976).
3. B.M.U. Scherzer in *Sputtering by Particle Bombardment*, Topics in Applied Physics **52**, R. Behrisch ed., Springer, Berlin 271 (1983).
4. J.H. Evans in *Encyclopedia of Materials Science and Engineering*, M. B. Bever ed., Pergamon, Oxford, 2123 (1983).
5. M.I. Guseva and Yu. V. Martynenko in *Physics of Radiation Effects in Crystals*, R.A. Johnson and A.N. Orlov ed., Elsevier, Amsterdam 621 (1986).
6. H. Trinkaus, Rad. Eff. **78**, 189 (1983).
7. J.H. Evans, J. Nucl. Mat. **68**, 129 (1977).
8. M. Risch, J. Roth and B.M.U. Scherzer in *proc. of the Int. Symp. on Plasma Wall Interaction*, Pergamon Press, Oxford, 391 (1977).
9. G.K. Erents and G.M. McCracken Rad. Eff. **18**, 245 (1973).
10. O. Auciello in Beam Modification of Materials, Vol. 1.: *Ion Bombardment Modification of Surfaces*, O. Auciello and R. Kelly ed., Elsevier, Amsterdam, 1 (1984).

11. G. Mezey, F. Pászti, M. Fried, A. Manuaba, Gy. Vizkelethy, Cs. Hajdu and E. Kòtai in *Twenty Years of Plasma Physics*, B. McNamara ed., World Scientific, Philadelphia, 95 (1984).

12. J. Ehrenberg, R. Behrisch and B.M.U. Scherzer, Nucl. Instr. Meth. **194**, 501 (1982).

13. B.M.U. Scherzer, P. Borgesen J. Ehrenberg and W. Möller, Nucl. Instr. Meth. **B7/8**, 71 (1985).

14. F. Pászti, L. Pogány, G. Mezey, E. Kòtai, A. Manuaba, L. Pòcs, J. Gyulai and T. Lohner, J. Nucl. Mater. **98**, 11 (1981).

15. G. Mezey, F. Pászti, L. Pogány, A. Manuaba, M. Fried, E. Kòtai, T. Lohner, L. Pòcs and J. Gyulai in *Ion Implantation Into Metals* V. Ashworth ed., Pergamon Press, Oxford, (1982) 293.

16. N.T. My, G. Mezey, A. Manuaba, F. Pászti, L. Pogány, E. Kòtai, M. Fried and L. Pòcs in *proc. of the 12 European conf. on Controlled Fusion and Plasma Phys.*, L. Pòcs and A. Montvai ed., European Physical Society, Budapest 639 (1985).

17. N.T. My, A. Manuaba, G. Mezey, F. Pászti, Kòtai, L. Pòcs, E. Klopfer, P. Kostka and M. Fried in *proc. of the 12 European conf. on Controlled Fusion and Plasma Phys.*, L. Pòcs and A. Montvai ed., European Physical Society, Budapest 642 (1985).

18. N.T. My, F. Pászti, G. Mezey, A. Manuaba, E. Kòtai and J. Gyulai, J. Nucl Mat **165**, 222 (1989).

19. A. Manuaba, F. Pászti and E. Kòtai, J. Nucl. Mater. **75**, 58 (1990).

20. F. Pászti, G. Mezey, L. Pogány, M. Fried, A. Manuaba, E. Kòtai, T. Lohner and L. Pòcs, Nucl. Instr. Meth. **209/210**, 1001 (1983).

21. F. Pászti, Cs. Hajdu, A. Manuaba, N.T. My, E. Kòtai, L. Pogány, G. Mezey, M. Fried, Gy. Vizkelethy and J. Gyulai, Nucl. Instr. Meth. **B7/8**, 371 (1985).

22. F. Pászti, A. Manuaba, L. Pogány, Gy. Vizkelethy, M. Fried, E. Kòtai, H.V. Suu, T. Lohner, L. Pòcs and G. Mezey, J. Nucl. Mater. **119**, 26 (1983).

23. A. Manuaba, F. Pászti, L. Pogány, M. Fried, E. Kòtai, G. Mezey, T. Lohner, I. Lovas, L. Pòcs and J. Gyulai, Nucl. Instr. Meth. **199**, 409 (1982).

24. F. Pászti, M. Fried, L. Pogány, A. Manuaba, G. Mezey, E. Kòtai, I. Lovas, T. Lohner and L. Pòcs, Nucl. Instr. Meth. **209/210**, 273 (1983).

25. F. Pászti, M. Fried, L. Pogány, A. Manuaba, G. Mezey, E. Kòtai, I. Lovas, T. Lohner and L. Pòcs, Phys. Rev. **B28**, 5688 (1983).

26. M. Fried, L. Pogány, A. Manuaba, F. Pászti and C. Hajdu, Phys. Rev. **B41**, 3923 (1990).

27. F. Pászti, M. Fried, A. Manuaba, G. Mezey, E. Kòtai and T. Lohner, J. Nucl. Mater. **114**, 330 (1983).

28. N.T. My, F. Pászti, A. Manuaba, G. Mezey and E. Kòtai, J. Nucl. Mater. **168**, 76 (1989).

29. C. Hajdu, F. Pászti, M. Fried and I. Lovas, Nucl. Instr. Meth. **B19/20**, 607 (1987).

30. C. Hajdu, F. Pászti, I. Lovas and M. Fried, Phys. Rev. **B41**, 3920 (1990).

31. F. Pászti, Mat. Sci. Engin. **A115**, 57 (1989).

32. B. Emmoth, Radiat. Effects **78**, 365 (1983).

33. J.P. Biersac and L.G. Haggmark, Nucl. Instr. Meth. **174**, 257 (1980).

34. N. Matsunami, Y. Yamamura, Y. Itikawa, N. Itoh, Y. Kazumata, S. Miyagawa, K. Morita and R. Shimizu, Report of Institute of Plasma Physics, Nagoya University, IPPJ–AM–14 (1980).

35. R.V. Nandekar and A.K. Tyagi, Rad. Eff. Letters **58/3**, 91 (1981).

36. A.K. Tyagi, R.V. Nandekar and K. Krishan, J. Nucl. Mater. **114**, 181 (1983).

37. N. Nayashi and I. Sakamoto, Phys. Letters **88A/6**, 299 (1982).

38. W.R. McDonell, J. Nucl. Mater. **85/86**, 1117 (1979).

INERT GAS BUBBLES IN METALS FOLLOWING ROOM-TEMPERATURE IMPLANTATION: OTHER TECHNIQUES

POSITRON STUDIES OF INERT GASES IN METALS

K.O. Jensen

School of Physics, University of East Anglia
Norwich NR4 7TJ, United Kingdom

ABSTRACT
A review is given of recent advances in the application of positron techniques to study metals implanted with inert gas atoms. The article focuses on how to use positron experimental methods to determine properties of defects, such as the density of inert gas inside inert gas bubbles. Recent theoretical models are described and a number of examples of experimental studies with both the 'conventional' positron techniques, such as positron lifetime spectroscopy, and positron beam techniques are given.

1.   Introduction

Techniques relying on the interaction of positrons with condensed matter are finding increasing use in studies of metallic systems [1–4]. The sensitivity of positrons to defects, such as vacancies, vacancy clusters, voids, and dislocations, has been particularly useful. In this article I will examine the application of positron techniques to the defects created when inert gases are introduced into metals.

The key property governing the behaviour of inert gases in metals is their extremely low solubility which implies a strong tendency to precipitation. Thus, defects, such as small vacancy-gas agglomerates (clusters) and larger gas-filled cavities (bubbles), are usually found in metals containing inert gases. As impurity concentrations increase these defects become larger and eventually lead to detrimental effects on the mechanical properties of the metal as well as blistering of surfaces. These can be major technological problems, in particular in future fusion reactors since neutrons escaping the fusion plasma will produce He by (n,α) reactions in the surrounding reactor walls [5]. Bombardment of metals with heavier inert gas ions is a process frequently used in, e.g, sputtering and ion-beam mixing. Accordingly, there has been a growing interest in the study of inert gases and gas-defect interactions in metals, not only from a technological viewpoint, but also in terms of achieving a better fundamental understanding of the behaviour of inert gas in metals [6].

The most direct determination of defect sizes and concentrations can be obtained by transmission electron microscopy (TEM) which, however, is only sensitive to cavities with radii larger than ≈ 10 Å. Further, TEM normally provides no direct information on the density of gas inside defects which is of particular interest since gas atomic density inside bubbles can be comparable to the surrounding metallic density. From a macroscopic point of view this implies enormous pressures of the order of several GPa inside bubbles. A direct consequence of the high pressures is that the heavier inert gases can be found in a solid phase inside bubbles even at room temperature and above [7]. (Despite the atoms being in a solid phase, it is customary still to refer to these defects as bubbles).

For solid inert-gas precipitates the density can be directly obtained from the lattice parameter measured in a diffraction experiment. The situation is much more difficult if the confined gas is in a disordered phase or if the precipitate is small. A number of techniques have been applied for density determination, in particular for He [8]. In addition to the positron techniques, these include electron energy-loss spectroscopy (EELS) [9] and vacuum ultra-violet absorption spectroscopy (VUVAS) [10].

The emphasis in the present article will be on how to relate measurable parameters in positron experiments to properties of the defects, such as the gas density inside bubbles. This means that the discussion will be focused on the fundamentals of the interaction of positrons with inert gas defects but illustrative examples of applications will also be given. I will also outline the background for the present research by describing important earlier work and review some of the more recent investigations.

The structure of the article is as follows: In section 2 the basic principles of positron studies of condensed matter are described. Section 3 discusses the interaction of positrons with inert gas defects in metals. Experimental positron studies of inert gases in metals are described in section 4, with the special techniques employing variable-energy positron beams being discussed separately in section 5. Finally, section 6 summarises and concludes.

2.   Positron Studies of Solids. Basic Principles

The field of positron studies of condensed matter can generally be separated into two subfields. The so-called 'conventional' positron techniques rely on radioactive isotopes as the source of positrons and are primarily suited for studies of the bulk of materials. More recently variable energy positron beams have increasingly been employed to study surfaces and the near-surface region of solids. In the following, the two subfields will be described.

2.1.   Conventional positron techniques

The basic process underlying most positron studies of matter is the annihilation of an electron with its anti-particle, the positron, resulting in the emission of γ-rays, in most cases two. A typical positron annihilation experiment consists of the introduction of positrons into the sample material followed by detection of the resulting γ-quanta by nuclear detection techniques. The source of positrons in conventional positron measurements is a β^+-emitting isotope such as ^{22}Na. Positrons from this type of source are implanted up to depths of the order of 100 μm in metallic samples.The experimental methods available for positron studies include the lifetime, angular correlation of annihilation radiation, and Doppler broadening techniques.

In lifetime measurements [11, 12] one expoits the fact that when the positron is emitted from a ^{22}Na nucleus there is almost simultaneous emission of a 1.28 MeV γ-ray. The detection of this γ-quantum provides a 'start' signal to a time-to-amplitude converter, while the detection by a second detector of one of the 511 keV annihilation quanta gives the 'stop' pulse. Thus, the time difference between the two pulses measures the lifetime of the positron. A lifetime spectrum is measured by collecting the distribution of time-differences for a large number of such events, typically 10^6. The geometry usually employed is to sandwich the positron source between two identical samples which are then placed between two scintillation detectors detecting the γ-rays.

The annihilation rate, which is equal to the inverse of the (mean) positron lifetime, is proportional to the overlap of electron and positron densities. In metals positron lifetimes are normally in the range 100–500 picoseconds (ps).

The total momentum of an annihilating electron-positron pair in the laboratory frame causes a small angular deviation from co-linearity of the two annihilation photons, of the order of milliradians. The distribution of these deviations are measured in angular correlation of annihilation radiation (ACAR) experiments. The momentum of the annihilating particle pair also leads to a Doppler shift of the measured energy of the 511 keV γ-quanta. This brings about a Doppler broadening (DB) of the annihilation line which can be measured by a high-resolution solid state detector [11, 12]. Both the ACAR and DB thus measure the momentum distribution of the annihilating particles. Since positrons rapidly (within a few ps or less) slow down to thermal energies after implantation into a metal [13,14] this distribution is primarily governed by the distribution of electron momenta sampled by the positron.

The positron state after thermalisation in a defect-free metal can be described by a delocalised Bloch state. However, due to the strong repulsion between the positive ion cores and a positron, any region with lower than average ion density, such as vacancies and other open-volume defects, attracts the positron and thus forms a possible trapping site. Trapped in such a defect the positron will experience a lower electron density than in the defect-free bulk and its lifetime is therefore increased. Similar, the ACAR and DB spectra become narrower. These changes in the annihilation characteristics for defect-trapped positrons form the basis of the now well-established use of positron techniques in metal defect studies.

When defect trapping is present, the lifetime spectrum is composed of several exponential decay components due to untrapped positrons and positrons trapped in different types of defects. To extract maximum information from the spectra a least-squares fitting procedure [15] is therefore applied to decompose the spectra into the various components.

2.2. *Positron beam techniques*

A new development in positron physics in recent years has been the availability of variable-energy mono-energetic positron beams. A recent review of the whole of this area is found in ref. [16]. A main advantage of positron beams is that they allow depth profiling of defects, as will be described below. The current status of this application of the technique is reviewed in ref. [17].

The range of the implanted positrons in the sample depends on the injection energy. It is thus possible to probe different depths by varying the positron beam energy. Typically, for implantation energies of the order of 10 keV, beam ranges are of the order of 5000 Å and varies with energy approximately as $E^{1.6}$ [16]. After a positron has slowed down it can undergo a number of different processes. It may annihilate while diffusing in the solid or get trapped in a defect with subsequent annihilation from a trapped state. It may also reach the surface, where it can be ejected into the vacuum either as a free positron or bound to an electron in a positronium (Ps) atom, or it can get trapped in a surface state [16]. The measurable parameters in positron beam experiments include the probability of the positron returning to the surface (the back diffusion probability) as well as the annihilation characteristics of the injected positrons, such as the Doppler broadening of the annihilation line shape. How these techniques can be used in defect profiling will be discussed in section 5.

An important feature of both conventional and beam positron techniques is that they are essentially non-destructive for metallic systems.

3. Positron Interaction with Defects Containing Inert Gases

3.1. *Background*

It has been realised for some time that the lifetime of a positron trapped in a nominally empty vacancy cluster increases with the size of the cluster [18, 19]. The first theoretical work on positron states in gas-vacancy agglomerates was directed at vacancies or small clusters of vacancies occupied by single gas atoms [20-23] and it was shown that the presence of inert gas atoms reduces the lifetime from the empty cavity value.

This effect was expected also for large cavities [24, 25] but Hansen and co-workers [22, 26, 27] made what appear to be the first realistic attempts to correlate lifetimes and gas densities inside bubbles. They argued that the annihilation rate of a trapped positron, λ, is a sum of two contributions, one from annihilation with gas atom electrons, λ_{gas} and one from annihilation with electrons from the surrounding metal, λ_{metal}

$$\lambda = \lambda_{gas} + \lambda_{metal}.$$ (1)

The gas contribution at a given density inside the bubble was assumed to be identical to the annihilation rate of a positron in a uniform gas at the same density, which for He is known experimentally to be nearly proportional to the density [28]. The metal contribution was subject to considerable uncertainty since it was not known how the presence of the gas modifies the positron state in the cavity. If the positron is in a free state inside the bubbles, λ_{metal} will be close to 0. However, if the positron state is similar to the state localised at the surface in empty voids, λ_{metal} will be similar to λ_{void}, the annihilation rate in an empty void, but it may not then be realistic to use the annihilation rate in a uniform gas for λ_{gas}. These uncertainties affect the deduced bubble parameters to a very large extent [26, 27] which made it clear that a more detailed description of the positron state in bubbles was required. This was the motivation for the theoretical work described in the next subsection.

3.2. *Positron states in inert gas bubbles*

The calculations for large bubbles were done as a combination of molecular dynamics (MD) simulations of the atomic structure of gas bubbles and detailed calculations of positron states at the metal-gas interface. The description given here is very brief but more details can be found in refs. [29] and [30]. Since large bubbles generally have three-dimensional faceted shapes, it is a reasonable approximation to consider planar interfaces representing the facets. Consequently, the MD simulations were performed for a sandwich geometry with several layers of metal atoms on both sides of a collection of up to 700 gas atoms. In MD simulation the motion of the particles in the system is followed by solving the classical equation of motion numerically with forces determined by the interatomic interactions.

Examples of simulated gas density profiles perpendicular to the interface are shown in Fig.1. The profiles were obtained by averaging over a large number of integration time steps after the system had been equilibrated at the required temperature. The peaks in the profiles are not a result of the weak Van der Waals attraction between the metal and gas but originate from the statistical packing of gas atoms next to the repulsive wall resulting from the strong close-range repulsion between metal and gas atoms [31]. It is seen that the peaks become more pronounced as the average gas density is increased.

The calculations of the positron states at the interface were done through an intermediate step in which the positron model was applied to metal surfaces with ordered overlayers of inert gas atoms. The results from these calculations were then combined with the density profiles from the MD simulations to obtain positron binding energies and annihilation rates at the interface. The model used in the positron calculations is based on the corrugated-mirror model [32] for the image potential induced surface state.

The results showed that the ground state for the positron in all cases was localised at the metal-gas interface and the wave function was similar to that for a positron trapped at a clean surface apart from a depletion around the ion cores of the inert gas atoms. This implies that the annihilation rate with metal electrons is very similar to that for a positron at a clean surface, i.e. λ_{void}, which is about 2 ns^{-1}. The contribution to the annihilation rate from inert gas electrons was found to be somewhat lower than for a uniform gas at the same density. Adding the two contributions, cf. eqn.(1), gives the positron lifetime, which is the inverse of λ, at the interface. The lifetime results for the Cu-Kr system are shown in Fig.2. The straight line approximating the theoretical data correspond to the relationship:

$$Cu\text{-}Kr: \quad \tau = 500 \, \text{ps} - 9.2 \text{ps/nm}^{-3} \times \bar{n}_{Kr}$$ (2)

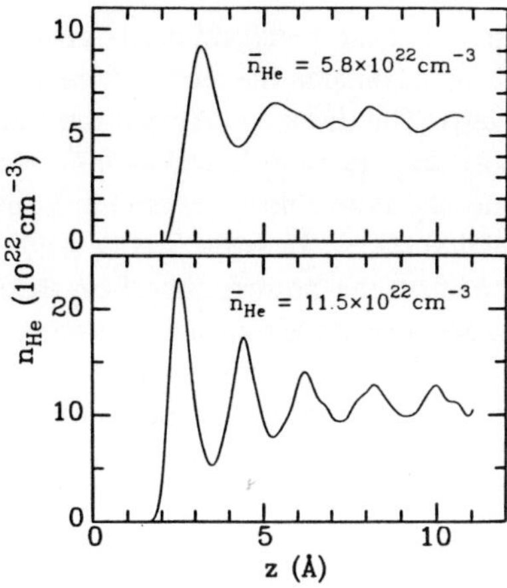

Fig. 1   He density profiles perpendicular to an Al-He interface at 300 K obtained from molecular dynamics simulations [30]. The topmost Al atoms are positioned at z=0. The densities $\bar{n}_{He}$ are the average densities for large z.

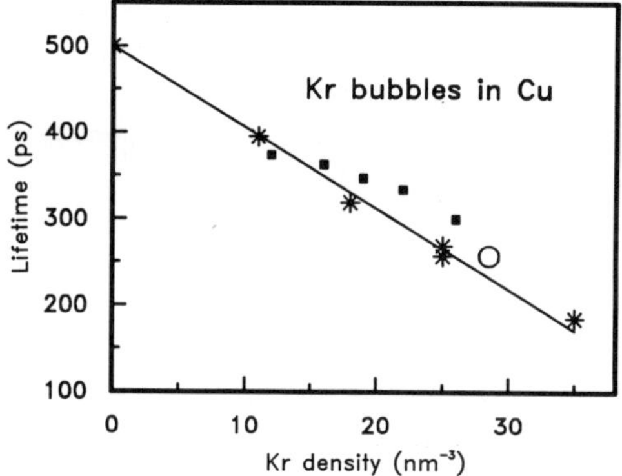

Fig. 2   The positron lifetime as a function of Kr density in Kr bubbles in Cu. The stars are the theoretical results [30] and the line is a fit to these data. The circle and the squares are from an experimental study of Cu containing ≈ 3 at.% Kr [33] with the Kr density determined from electron diffraction (circle) or from TEM by assuming the bubbles to be in thermal equilibrium (squares).

where τ is the positron lifetime and $\bar{n}_{Kr}$ is the Kr density in the bubbles. Fig.2 also shows experimental results obtained from the study described in section 4.2 and it is seen that the agreement between theory and experiment is quite good.

The theoretical results for the Al-He system produced a relationship [29,30]:

$$Al\text{-}He: \qquad \tau = 500 \text{ ps} - 2.35 \text{ ps/nm}^{-3} \times \bar{n}_{He}, \tag{3}$$

confirmed by experimental results on 600 MeV proton-irradiated Al (see section 4.1).

Although detailed calculations were made only for the Al-He and Cu-Kr systems (with a single calculation done for Cu-He) there are reasons to expect that equations (2) and (3) can be used also for other metallic hosts as a first approximation. The metal influences the lifetime-gas density relationship in a number of ways. First of all, the gas density profile near the interface is determined by the interaction potential between the He and metal atoms. This interaction is mainly repulsive and the metal primarily acts as a rigid wall. The effective position of this 'wall' is the point outside the surface where the electron density becomes appreciably different from zero. The distance between the 'wall' and the topmost layer of atoms therefore corresponds to the 'size' of the metal atoms. Since the 'size' of different metal atoms is more or less the same (the atomic densities are fairly similar) the gas density profiles will look similar for all metal surfaces. How the gas density profile influences a positron trapped at the surface depends on the positron interface-state wave function. The image potential outside different metals is nearly identical for different surfaces [34], and the binding energy of positrons to different surfaces is found experimentally to be fairly constant, in the range 2-3 eV [16]. This means that λ_{gas} is probably similar for different surfaces. Finally, there is the contribution of annihilation with metal electrons to the annihilation rate, given approximately by λ_{void}. Since the lifetimes measured for empty voids in different materials, including Al [35], Cu [36], and Ni [33], all seem to be about 500 ps, this contribution also appears to be similar for different materials. For these reasons, the relationship between lifetime and gas density is expected to be approximately the same for most metals. This notion is supported by the close agreement between calculated results for the Cu-He and Al-He systems [30].

Recently Dunn et al [37] have examined this question using a theory which is simpler (but no less ingenious) than the one outlined here. Their results substantiate the general arguments given above. The same authors have also done calculations for all of the heavier noble gases in bubbles in Cu [38] with results similar to those for He and Kr.

Experimentally ACAR (and thus Doppler Broadening) spectra are known, like the positron lifetime, to depend on gas density [33,39]. However, ACAR spectra are considerably more difficult to calculate theoretically than lifetimes. The attempt by Dunn et al [40] to do ACAR calculations for gas bubbles is probably unrealistic since it ignores annihilation with metal electrons. Thus, no theoretical analysis of ACAR spectra for gas bubbles is yet available.

3.3. Positron states im smaller vacancy gas clusters

In the limit of small bubble sizes the positron-interface model must inevitably break down, since the long-range image interaction, which attracts the positron to the interface, will be replaced by a shorter-range correlation with the metallic electrons and the positron state will extend over the whole of the defect volume. For this reason a separate set of calculations was performed [30] for small defect complexes using the method of Puska and Nieminen [21] which relies on a local approximation to describe the positron-electron correlation. It was found that the presence of gas reduces the lifetime from the empty cluster value, and that for a given gas atom/vacancy ratio the lifetime increases with cluster size. However, the increase with size for gas-filled clusters is found to be less steep than for empty clusters.

The cluster calculations were restricted to clusters with 13 vacancies or less since the image effects, responsible for the evolution from a volume to an interface/surface state for increasing cavity size, were not considered. Results were thus only obtained for the two extremes of the cavity size range of small clusters and large bubbles because no adequate model existed for the intermediate size range where non-local effects are important, yet cannot be represented by a simple image potential. However, since these results were obtained a non-local approximation to the positron electron correlation within the framework of density-functional theory has been constructed which gives realistic results for the positron surface state [41] and is equivalent to the local-density approximation in nearly homogeneous systems. This approximation has recently been applied to calculate positron states and annihilation rates in cavities in Al as a function of cavity size [42]. The results for the lifetime are shown in Fig.3.

It is seen in Fig.3 that the saturation radius, above which the lifetime does not depend on void size, is predicted to be about 5 Å. The positron wave function in voids with larger radii becomes increasingly similar to the wave function at a planar surface. Although calculations

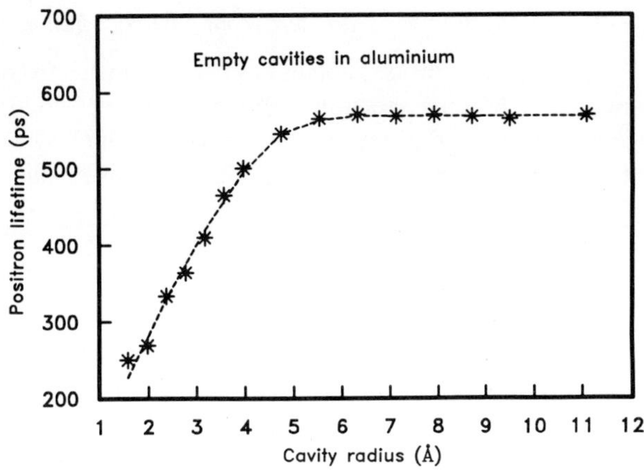

Fig. 3 Lifetime of a positron trapped in cavities in Al as a function of cavity radius calculated with non-local positron-electron density-functional theory using a jellium model for the cavities [42].

have not yet been made for gas-filled cavities, this result vindicates the use of the relationships between density of inert gas inside cavities and positron lifetime (equations 2 and 3), derived for nominally large cavities, for cavities with radii as small as 5 Å.

3.4. Positron trapping rates into cavities

The trapping rate of positrons from the freely diffusing state into a trapped state in a bubble can, as the lifetime in the bubbles, be extracted from the fitting analysis of the lifetime spectra (unless the trapping rate is so high that essentially all positrons are trapped before annihilation). It is usually reasonable to assume that the trapping rate κ is proportional to the defect concentration C:

$$\kappa = \mu C \tag{4}$$

where μ is the specific trapping rate, although this is only an approximation to a full treatment of the positron diffusion to the defect and the transition into the trapped state [43]. The specific trapping rate depends on temperature T [43, 35, 44] and the cavity radius R [45] and is predicted [30] to depend, albeit weakly, on the gas density inside. The R dependence for Al at T=300 K has been determined empirically from experimental results for trapping rates into bubbles and voids combined with TEM determinations of cavity sizes and concentrations in a range of samples [45]. The results correspond approximately to a relation:

$$\mu(R) = (1/AR + 1/BR^2)^{-1} \tag{5}$$

where A and B are constants. Above about 50 Å, μ is found to be proportional to R which indicates that it is the diffusion to the cavities that limits the trapping. Since the positron diffusion constants in different metals are quite similar [73] the $\mu(R)$ relationship obtained for Al can therefore be used as a reasonable first approximation also for other metals. Since the diffusion does not depend on the gas density (if one can neglect the influence of strain fields in the metallic matrix around the cavity) this also means that μ will not depend on the gas density for the larger bubbles.

3.5. Determination of inert gas bubble parameters from positron lifetime data

After the (average) gas density n_{gas} inside bubbles has been estimated from the lifetime of the gas bubble component in a lifetime spectrum using equation (2) or (3), it is possible to estimate the average bubble size and concentration in two different ways.

If one assumes that the bubbles are in thermal equilibrium, the pressure p is approximateiy related to R as:

$$R = 2\gamma/p \qquad\qquad (6)$$

where γ is the specific surface energy of the metal [46]. Since the pressure is related to the gas density through the equation-of-state, R can be determined directly from n_{gas}. This requires no knowledge of the trapping rate but if κ is determined the bubble concentration can then be obtained from eqn. (4) with $\mu(R)$ from eqn. (5).

Alternatively, if the trapping rate has been determined and the total amount of gas in the sample is known and is assumed all to be contained in the bubbles, R and C can be estimated without having to assume equilibrium bubbles [39], since the equation expressing the gas content in terms of n_{gas}, R, and C and eq.(4) provide two coupled equations in the two unknowns R and C.

4. Conventional Positron Studies of Inert Gas Defects

4.1. Helium implanted materials

In the first positron lifetime study of a He-irradiated material [47] 0.6 ppm He was injected into Al by irradiating with 50 MeV He ions. Defects with lifetimes longer than that expected for monovacancies were detected in a subsequent isochronal annealing study. These defects could be identified as He-vacancy clusters.

In subsequent studies of He-implanted materials the amount of He has generally been higher, typically of the order of 100–1000 ppm. With the higher He concentrations it has been easier to identify defect components due to He-vacancy clusters and He bubbles. It was found that defect trapping persisted in He-irradiated samples at temperatures above which defects are annealed out in deformed samples or samples irradiated by other types of particles, such as neutrons or deuterons [25, 48]. A common observation was that positron lifetimes in defects got longer as annealing temperatures were increased [25, 27, 48, 49, 50, 51] which could signify either a decrease in He density in the defects or an increase in their size. Although attempts at quantitative interpretations of the results were made [25, 27, 49, 50, 51] these were rather uncertain since no detailed understanding of the behaviour of positrons in inert gas defects existed.

With the development of the model described in the previous section, a basis for the interpretation of the positron results was provided. The theoretical work described above was made in tandem with a series of positron measurements on 600 MeV proton-irradiated Al which contains He as a result of nuclear transmutation reactions (but contains no hydrogen) [39]. A large number of samples irradiated at different temperatures and with different He contents were investigated and some of the samples were subjected to isochronal annealing studies.

The effect of annealing on the lifetime of positrons trapped in He bubbles is illustrated in Fig. 4. It is seen that when the annealing temperature exceeds about 600 K the lifetime starts to increase for the three samples with the lowest irradiation temperatures, T_{irr}, initially containing bubbles with radii ~10 Å. The lifetimes indicate, cf. eqn. (3), He densities of about 70 nm^{-3} for the as-prepared samples, decreasing to ~40 nm^{-3} at the highest annealing temperatures. For the sample with higher T_{irr}, initially containing bubbles with radii ~25 Å, the increase in the lifetime starts at a higher temperature showing that the onset of bubble growth, which

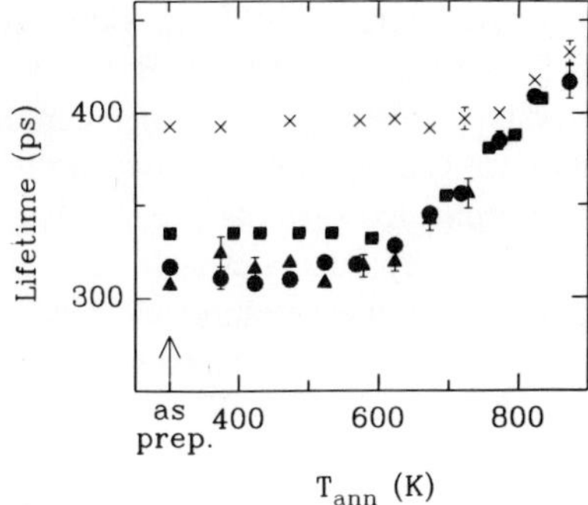

Fig. 4 Lifetime of a He bubble component in positron lifetime spectra measured on 600 MeV, proton-irradiated Al as a function of annealing temperature [39]. All spectra were recorded at room temperature after 30 minutes annealing. The filled symbols denote results for three samples with irradiation temperatures T_{irr} in the range 375–404 K, and the crosses give results for a T_{irr} = 471 K sample.

leads to the decrease in n_{He}, occurs at a higher temperature for the sample with the larger bubbles, in accordance with theories of bubble growth [52,53].

The positron data was analysed in the way described at the end of section 3.5 to obtain estimates of bubble radii and concentrations. Some of the results obtained are shown in Fig. 5. It is seen that R grows from ~10 Å to ~50 Å at the highest annealing temperatures. At the same time C drops by several orders of magnitude. The values of R and C at the final annealing steps are in good agreement with TEM examinations of the same samples [39]. Fig. 5 also includes TEM results for annealed He-implanted Al. Despite large differences in implantation energies and total He contents, there is good general agreement between TEM and positron results.

Although an indirect technique like positron lifetime spectroscopy will probably never compete with TEM for determination of bubble sizes and concentrations for large bubbles, an analysis like the one presented above allows direct comparison with TEM and therefore gives a valuable confirmation of the whole interpretation of the positron results, including the determination of gas densities inside the bubbles.

A number of other groups have recently presented detailed analyses of positron lifetime

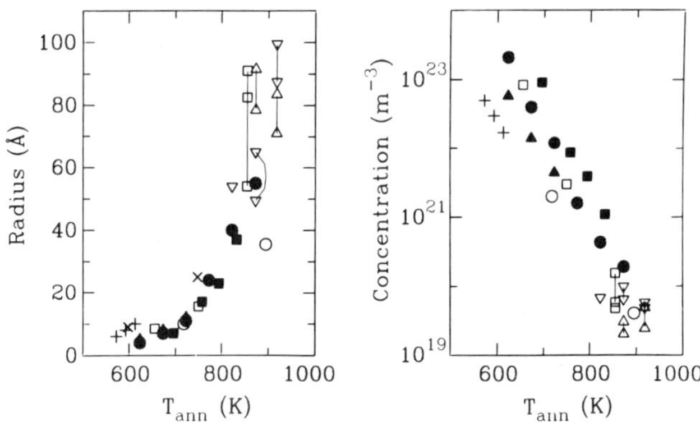

Fig. 5 Radii and concentrations of He bubbles in Al as a function of annealing temperature. Filled symbols denote positron lifetime results for samples irradiated with 600 MeV protons at temperatures in the range 375-404 K with He contents from 32 to 215 ppm [39]. Other symbols are TEM results for samples implanted with Al with a wide range of implantation energies and He contents [54-58].

203

results using the methods described here: The Kalpakkam group has studied a range of materials, including stainless steel [59], Cu [60]; Cu-B alloys [60,61], Ni and Ni-Ti alloys [59,62], in which He was introduced either by direct implantation or, in the case of Cu-B, by neutron-induced nuclear reactions. The agreement between the positron analysis and TEM was found to be good even in complicated systems like stainless steel [59].

He-irradiated Ni was also studied by Segers et al [63]. Results were similar to those of ref.[62] but no attempt at a quantitative interpretation was made.

Rajainmaki et al [64] irradiated Al at a very low temperature (20 K) followed by isochronal annealing at temperatures ranging from 20 to 900 K. The low irradiation temperature meant that bubbles were not formed in the as-irradiated samples and it was possible to observe the nucleation and initial growth during the subsequent annealing using the theoretical results for small clusters (section 3.3) to estimate the size of the defect complexes. After annealing above 600 K large bubbles had formed and the analysis procedure outlined in section 3.5 could be applied.

The analysis by Sen *et al* [65] for He-irradiated Al probably needs revision, since they used the proposed gas density-lifetime relation of ref.[27] which, as shown by the detailed model presented here, is probably inaccurate. Also the application of the ideal gas equation-of-state at the high pressures found in He bubbles is unrealistic.

In ref. [66] He was introduced in refractory metals by He-irradiation or by decay of tritium in tritiated samples. The absence of long lifetime components in the tritiated materials indicated that without accompanying radiation damage He preferentially precipitates in small defects and not in large bubbles.

Other recent studies of He-induced defects can be found in refs. [67,68]. However, a reliable interpretation of these results is difficult since He and defect concentrations are highly inhomogeneous in the depth region sampled by the positrons.

It is appropriate at this point to stress that quantitative interpretation of positron lifetime experiments relies on the ability to separate the lifetime components due to trapping into defects in the fitting analysis of the lifetime spectra. If the component interpreted as being due to, for example, bubbles includes contributions from other unresolved components with lifetimes in the vicinity of that of the bubbles, the extracted information will be erroneous. This emphasises the importance of, whenever possible, making independent examinations of samples by other methods such as TEM, which may test some of the predictions made based on the positron data. This difficulty should always be borne in mind when reading positron papers in the literature.

4.2. Heavier inert gases

Very few studies of samples containing inert gases other than He by the conventional positron techniques have been made, mainly because it is difficult to implant the heavier inert gases deep enough to make a significant fraction of the positrons sample the region where the gas is. One way of getting around this problem is to deposit the metal atoms as the gas is being implanted in which case thick homogeneous samples can be obtained [69].

Cu containing ~3 at. % Kr prepared in this way was first studied by Eldrup and Evans [24] and later by Britton *et al.* [70] and Jensen *et al.* [33] who also studied Ni with high Kr content. In as-prepared samples the Kr is contained in solid precipitates (bubbles) [71]. The lattice parameter of the Kr provides a direct determination of the Kr density which can be correlated with the lifetime of positrons trapped in the bubbles [33]. The circle in Fig.2 was obtained this way. Upon annealing above ~600 K an increase in the lifetime is observed signifying a decrease in Kr density, and the Kr is no longer solid at room temperature. The densities obtained from the positron lifetimes are consistent with the assumption that bubbles are in thermal equilibrium (cf. the agreement between the squares and the theory in Fig.2). At higher temperatures extensive swelling is observed prior to and simultaneous with the release of the majority of the Kr, leaving a structure dominated by μm-sized pores. It was possible to detect the presence of Kr inside these pores, despite the fact that the Kr density is very low, by observing the reduction in positron lifetime when Kr condensed on the surfaces of the pores at low temperatures [72]. Positron lifetime spectroscopy was thus able to provide unique information about the gas inside bubbles throughout the range of annealing temperatures.

5. Positron Beam Studies of Samples Implanted with Inert Gases

5.1. Defect depth profiles immediately below the surfaces

The probability of the positron diffusing back to the surface after implantation is sensitive to the presence of defects close to the surface, since defect trapping reduces the effective diffusion length of the positron. This effect can be utilised to estimate the defect depth profile by varying the positron energy and hence the implantation depth and measuring the back diffusion probability. Since this technique relies on the diffusion of positrons to the surface it can only be used to study defects at depths less than typical positron diffusion lengths which are of the order of 1000 Å [73] but very high resolution, $\approx$20 Å, can be obtained just below the surface [17]. This method has been applied to Al and Mo surfaces irradiated with low energy ($\approx$1 keV) Ar or N_2 ions [74,75]. In all cases it was found that the defect profile consisted of a highly damaged region close to surface (~25 Å) and a tail of lower defect concentration extending to depths of the order of 100 Å. These findings were in qualitative agreement with computer simulations of the irradiation process.

5.2. Defect depth profiles at larger depths

The Doppler broadening (DB) technique can easily be employed in positron beam as well as in conventional positron studies, with the additional advantage in the beam case that the implantation energy and hence the sampling depth of the positrons can be varied. Since the positron implantation profile is fairly wide, it is generally necessary to perform a fitting analysis of the DB-vs.-energy data with an assumed shape for the depth profile depending on a set of parameters. Using beam energies of up to 50 keV it is then possible to obtain approximate depth profiles for depths as high as a few μm [17].

Typical studies of He-implanted metal with the positron beam DB method can be found in refs. [76–79]. It has been possible to follow how the depth profiles of vacancy-type defects vary with incident He energy and total He dose and how the profiles change upon annealing. The analysis of Lynn et al. [78] yielded defect profiles in Ni implanted with 30-180 keV He that agreed well with the damage profiles obtained by Monte Carlo simulations using the TRIM code [80], although the detailed shape of the profiles could not be determined from the experimental data. Uedono et al. [79] demonstrated that defects can be detected for He doses as low as 10^{13} He/cm^2 for 70 keV He implantation.

For the heavier gases studies have been made on Mo and Ti implanted with 100 keV Kr [81, 82] and Cu implanted with 30-100 keV Ne [83]. Both in the Ti-Kr and Cu-Ne cases, the data indicate positron trapping into defects well beyond the predicted range of implanted ions. TRIM calculations [80] for Kr implantation in Ti indicates that the deep damage can be due to recoil atoms penetrating beyond the Kr profile [82]. These results emphasise the fact that positrons detect the *damage* profiles, not the impurity profiles. The positron information is therefore complementary to that which can be obtained by other depth profiling techniques such as Rutherford Backscattering.

Although, in principle, it should be possible to deduce properties of the defects in the damaged regions, such as the gas density inside bubbles, from the Doppler broadening measurements using an approach similar to the one outlined in section 3, the data has not allowed such a detailed interpretation. It will be interesting to see if this becomes possible when positron lifetime spectroscopy in conjunction with variable energy positron beams [84] is applied to study subsurface defects.

6. Conclusions

In the previous sections I have described how positron techniques can be applied to study defects in metallic systems in which inert gas impurities have been introduced. It was indicated how detailed information, such as the gas density inside bubbles, could be obtained using positron lifetime spectroscopy and how the sensitivity to submicroscopic defects allowed bubble nucleation and initial stages of growth to be studied. The depth profiling of defects near surfaces using variable energy positron beams was also briefly discussed.

All the positron techniques have the unique property of being sensitive to the *defects* in the metals rather than the impurities, and the positron is one of the few probes that is sensitive to defects throughout the size range from single vacancies to μm-sized cavities. The fact that the techniques are non-destructive also means that consecutive measurements can be made on the same samples after, for instance, annealing treatments between measurements, and the samples can be examined by other techniques as well.

The high defect sensitivity means that inert gas defects can be detected when gas concentrations are as low as 1 appm. The He production rates in fusion reactors are estimated to be about 100 appm/year [5]. This means that the He concentration range where the positron method is best suited is also the most important in relation to fusion reactor materials problems. The requirement of bulk samples for the conventional positron techniques, such as lifetime spectroscopy, will also be an advantage in this context, since the He is created by 14 MeV neutrons from the fusion plasma which penetrate deep into the reactor wall materials. The use of bulk samples also reduces the influence of extraneous effects due to the sample surfaces. This is in marked contrast to techniques like EELS and VUVAS which require thin films and at. % concentrations of gas.

If, on the other hand, one is interested in defect structures close to surfaces and in thin films, the positron beam method is unique in providing information about the vacancy-type defects to complement the information about impurity profiles that can be obtained by other methods.

These unique properties mean that both the conventional positron and positron beam techniques are likely to develop into powerful tools for studying metal-inert gas systems.

ACKNOWLEDGEMENTS

I am pleased to acknowledge the support of my collaborators on the work described in this article, especially Morten Eldrup, Risto Nieminen, John Evans, Bachu Singh, Alison Walker and Geoff Dunn, who is also thanked for sending me results prior to publication. I would also like to thank the Commission of the European Communities for a Research Grant, contract no. SCI-0163, under the Stimulation Action scheme.

REFERENCES

1. W. Brandt and A. Dupasquier, eds., *Positron Solid-State Physics*, North-Holland, Amsterdam (1983).

2. P. Hautojärvi, ed., *Positrons in Solids*, Springer, Berlin (1979).

3. P.C. Jain, R.M. Singru, and K.P. Gopinathan, eds., *Positron Annihilation*, World Scientific, Singapore (1985).

4. L. Dorikens-Vanpraet, M. Dorikens, and D. Segers, eds., *Positron Annihilation*, World Scientific, Singapore (1989).

5. H. Ullmaier, Nuclear Fusion **24**, 1039 (1984).

6. H. Ullmaier, ed., Proc. Int. Symp. on Fundamental Aspects of Helium in Metals, Radiat.Eff. **78**, 1 (1983).

7. A. vom Felde, J. Fink, Th.Müller-Heinzerling, J. Pflüger, B, Scheerer, G. Linker, and D. Kaletta, Phys. Rev. Lett. **53**, 922 (1984).

8. S.E. Donnelly, Radiat.Eff. **90**, 1 (1985).

9. R. Manzke and M. Campagna, Solid. Stat. Commun. **39**, 313 (1981).

10. J.C. Rife, S.E. Donnelly, A.A. Lucas, J.M. Gilles, and J.J. Ritsko, Phys. Rev. Lett. **46**, 1220 (1981).

11. I.K. Mackenzie, in *Positron Solid-State Physics*, W. Brandt and A. Dupasquier, eds., North Holland, Amsterdam, (1983), p.196.

12. L.C. Smedskjaer and M.J. Fluss, in *Methods of Experimental Physics* vol. 21 , J.N. Mundy et al, eds., Academic Press, New York (1983), p.77.

13. R.M. Nieminen and J. Oliva, Phys. Rev. **B22**, 2226 (1980).

14. K.O. Jensen and A.B. Walker, J. Phys. Condens. Matter **2**, 9757 (1990).

15. P. Kirkegaard, M. Eldrup, O.E. Mogensen, and N.J. Pedersen, Comput. Phys. Commun. **23**, 307 (1981).

16. P.J. Schultz and K.G. Lynn, Rev. Mod. Phys. **60**, 701 (1988).

17. A. Vehanen, in *Positron Annihilation*, L. Dorikens-Vanpraet et al, eds., World Scientific, Singapore (1989), p.39.

18. M. Eldrup, O.E. Mogensen, and J.H. Evans, J. Phys. **F6**, 499 (1976).

19. P. Hautojärvi, J. Heiniö, M. Manninen, and R.M. Nieminen, Phil. Mag. **35**, 973 (1977).

20. P. Jena and M.J. Ponnambalam, Phys. Rev. **B26**, 5264 (1982).

21. M.J. Puska and R.M. Nieminen, J. Phys. **F13**, 333 (1983).

22. H.E. Hansen, R.M. Nieminen, and M.J. Puska, J. Phys. **F14**, 1299 (1984).

23. P. Jena and B.K. Rao, Phys. Rev. **B31**, 5634 (1985).

24. M. Eldrup and J.H. Evans, J. Phys. **F12**, 1265 (1982).

25. B. Viswanathan, W. Triftshäuser and G. Koegel, Radiat. Eff. **78**, 231 (1983).

26. H.E. Hansen, H. Rajainmäki, R. Talja, M.D. Bentzon, R.M. Nieminen, and K. Petersen, in *Microstructural Characterization of Materials by Non-Microscopical Techniques*, N. Hessel Andersen et al, eds., Risø National Laboratory, Roskilde, Denmark (1984), p.261.

27. H.E. Hansen, H. Rajainmäki, R. Talja, M.D. Bentzon, R.M. Nieminen, and K. Petersen, J. Phys. **F15**, 1 (1985).

28. P. Hautojärvi, K. Rytsölä, P. Tuovinen, A. Vehanen, and P. Jauho, Phys. Rev. Lett. **38**, 842 (1977).

29. K.O. Jensen and R.M. Nieminen, Phys. Rev. **B35**, 2087 (1987).

30. K.O. Jensen and R.M. Nieminen, Phys. Rev. **B36**, 8219 (1987).

31. M. Manninen, J.K. Nørskov, M.J. Puska, and C. Umrigar, Phys. Rev. **B29**, 2314 (1984).

32. R.M. Nieminen and M.J. Puska, Phys. Rev. Lett. **50**, 281 (1983).

33. K.O. Jensen, M. Eldrup, N.J. Pedersen, and J.H. Evans, J. Phys. **F18**, 1703 (1988).

34. N.D. Lang and W. Kohn, Phys. Rev. **B7**, 3541 (1973).

35. S. Linderoth, M.D. Bentzon, H.E. Hansen, and K. Petersen, in *Positron Annihilation*, P.C. Jain et al, eds., World Scientific, Singapore (1985) p.494.

36. G. Kögel, J. Winter, and W. Triftshäuser, in *Positron Annihilation*, R.R. Hasiguti and K. Fujiwara, eds., Japanese Institute of Metals, Sendai (1979) p.707.

37. G.M. Dunn, P. Rice-Evans, and J.H. Evans, Phil. Mag., to be published (1990); also Radiat. Eff. and Defects in Solids **115**, 19 (1990).

38. G.M. Dunn, P. Rice Evans, and J.H. Evans, J. Phys. Condens. Matter **2**, 10529 (1990).

39. K.O. Jensen, M. Eldrup, B.N. Singh, and M. Victoria, J. Phys. **F18**, 1069 (1988).

40. G.M. Dunn, P. Rice-Evans, and J.H. Evans, in *Positron Annihilation*, L. Dorikens-Vanpraet et al, eds., World Scientific, Singapore (1989), p. 360.

41. K.O. Jensen and A.B. Walker, J. Phys. **F18**, L277 (1988).

42. G.M.Dunn, K.O. Jensen, A.B. Walker, and P. Rice-Evans, in the Proceedings of the Workshop on *Slow-Positron Beam Techniques for Solids and Surfaces*, to be published by American Institute of Physics (1990).

43. R.M. Nieminen, J. Laakkonen, P. Hautojärvi and A Vehanen, Phys. Rev. **B19**, 1397 (1979).

44. M. Eldrup and K.O. Jensen, Phys. Stat. Sol.**A102**, 145 (1987).

45. K.O. Jensen, M. Eldrup, B.N. Singh, A. Horsewell, M. Victoria, and W.F. Sommer, Mater. Sci. Forum **15-18**, 913 (1987).

46. H. Trinkaus, Radiat. Eff. **78**, 189 (1983).

47. C.L. Snead, A.N. Goland, and F.W. Wiffen, J. Nucl. Mater. **64**, 195 (1977).

48. M. Doyama, K. Hinode, S. Tanigawa, and K. Shiraishi, J.Nucl Mater. **85-86**, 781 (1979).

49. K.O. Jensen, B.N. Singh, M.Eldrup, M. Victoria and W.V. Green, in *Microstructural Characterisation of Materials by Non-Microscopical Techniques*, N. Hessel Andersen *et al.*, eds., Risø National Laboratory, Roskilde, Denmark (1984), p.333.

50. G. Kögel, Q.-M. Fan, P. Sperr, W. Triftshauser, and B. Viswanathan, J. Nucl. Mater. **127**, 125 (1985).

51. B. Viswanathan, G. Amarendra and K.P. Gopinatham, in *Positron Annihilation*, P.C. Jain *et al*, eds., World Scientific, Singapore (1985) p.862.

52. E.E. Gruber, J. Appl. Phys. **38**, 243 (1967).

53. A.J. Markworth, Metall. Transact. **4**, 2651 (1973).

54. F.A. Smidt and A.G. Pieper, ASTM STP **570**, 352 (1975).

55. H. Shiraishi, H. Sakairi, E. Yagi, T, Karasawa, R.R. Hasiguti, and R. Watanabe, Trans. Jap. Inst. Metals **17**, 749 (1976).

56. K. Farrell, R.W. Chickering, and L.K. Mansur, Phil. Mag. **A53**, 1 (1986).

57. K. Ono, M. Inoue, T. Kino, S. Furono, and K. Izui, J. Nucl. Mater. **133-134**, 477 (1985).

58. R. Manzke, G. Crecelius, W. Jäger, H. Trinkaus, and R. Zeller, Radiat. Eff. **78**, 327 (1983).

59. B. Viswanathan, J. Phys. Condens. Matter **1**, SA71 (1989).

60. G. Amarendra, Ph.D. Thesis, University of Madras (1990).

61. B. Viswanathan, G. Amarendra, and K.P. Gopinathan, Radiat. Eff. **107**, 121 (1989).

62. G. Amarendra and B. Viswanathan, in *Positron Annihilation*, L.Dorikens-Vanpraet et al, eds., World Scientific, Singapore (1989), p.494.

63. D. Segers, I. Lemahieu, L. Deschepper, M. Dorikens, D. Geshef, L. Dorikens-Vanpraet, L.M. Stals, G. Severne, and A, Hermanne, in *Positron Annihilation*, L. Dorikens-Vanpraet et al, eds., World Scientific, Singapore (1989), p.416.

64. H. Rajainmäki, S. Linderoth, H.E. Hansen, R.M. Nieminen, and M.D. Bentzon, Phys. Rev. **B38**, 1087 (1988).

65. P. Sen, P.M.G. Nambissan, S.V. Naidu, and S C. Sharma, Phil. Mag. **A59**, 455 (1989).

66. P. Sperr, G. Kögel, J. Störmer, W. Triftshäuser, and R. Lässer, in *Positron Annihilation*, L. Dorikens-Vanpraet et al, eds., World Scientific, Singapore (1989), p.437.

67. P.M.G. Nambissan and P. Sen, Sol. Stat. Commun. **71**, 1165 (1989).

68. P.M.G. Nambissan and P. Sen, Phil. Mag. **A62**, 173 (1990).

69. DS. Whitmell, Radiat. Eff. **53**, 209 (1981).

70. D.T. Britton, P.C. Rice-Evans, and J.H. Evans, Phil. Mag. **A55**, 347 (1987).

71. J.H. Evans and D.J. Mazey, J. Phys. **F15**, L1 (1985).

72. K.O. Jensen, M. Eldrup, S. Linderoth, and J.H. Evans, J. Phys. Condens. Matter **2**, 2081 (1990).

73. E. Soininen, H. Huomo, P.A. Huttunen, J. Mäkinen, A.Vehanen, and P. Hautojärvi, Phys. Rev. **B41**, 6227 (1990).

74. J. Mäkinen, A. Vehanen, P. Hautojarvi, H. Huomo, J. Lahtinen, R.M. Nieminen, and S. Valkealahti, Surf. Sci. **175**, 385 (1986).

75. M.D. Bentzon, H, Huomo, A. Vehanen, P. Hautojärvi, J. Lahtinen, and M. Hautala, J. Phys. **F17**, 1477 (1987).

76. W. Triftshäuser and G. Kögel, Phys. Rev. Lett. **48**, 1741 (1982).

77. S. Tanigawa, Y. Iwase, A. Uedono, and H. Sakairi, J. Nucl. Mater. **133-134**, 463 (1985).

78. K.G. Lynn, D.M. Chen, B. Nielsen, R. Pareja, and S. Myers, Phys. Rev. **B34**, 1 449 (1986).

79. A. Uedono, S. Tanigawa, and H. Sakairi, Phys. Lett. **A129**, 249 (1988).

80. J.P. Biersack and L.G. Haggmark, Nucl. Instrum. Methods **174**, 257 (1980).

81. D.T. Britton, P.C. Rice-Evans, and J.H. Evans, J. Phys. **F18**, 1433 (1988).

82. D.L. Smith, P.C. Rice-Evans, D.T. Britton, J.H. Evans and A. Allen, Phil. Mag. **A61**, 839 (1990).

83. R.S. Brusa, A. Dupasquier, R. Grisenti, S. Liu, S. Oss, A.Zecca, J. Phys. Condens. Matter **1**, 5411 (1989).

84. G. Kögel, D. Schödlbauer, W. Triftshäuser, and J. Winter, Phys. Rev. Lett. **60**, 1550 (1988).

POSITRON STUDIES OF HELIUM IN Ni, Ni–Ti AND Ti–STABILISED STEEL

B. Viswanathan and G. Amarendra

Materials Science Division, Indira Gandhi Centre for Atomic Research
Kalpakkam-603102, India

ABSTRACT

Positron lifetime and Doppler lineshape studies have been made to obtain detailed information on bubble structures in Ni, Ni–Ti and Ti–stabilised stainless steel, implanted with 100 appm helium and annealed at temperatures up to 1200 K. The temperature of formation of stable bubble embryos, which retain helium, has been identified in Ni as 750 K. On the basis of computed positron lifetimes in helium decorated clusters, positron sensitivity to the size, helium-to-vacancy ratio and configurational structure of gas-defect complexes is shown. PAT analysis of the bubble coarsening stage in Ni indicates overpressure in bubbles, in excess of what is expected from the behaviour of equilibrium bubbles. The implication of this behaviour is discussed. Studies in dilute alloys of Ni–Ti show clear trends for resistance to bubble coarsening with increasing Ti–concentration. Finally, the influence of TiC precipitates on bubble structures in austenitic matrix is qualitatively discussed from PAT measurements on helium irradiated steel containing Ti.

1.   Introduction

The behaviour of inert gases in metals has evoked considerable interest in recent years. Apart from the technological interest, more fundamental issues on the gas behaviour are being increasingly addressed by a variety of experimental techniques in the area of helium in metals [1, 2]. In addition to the conventional determination of bubble size and concentration, a quantity of significant interest is the density of He in bubbles, especially since He densities comparable to metallic densities can develop implying high pressures. The determination of He densities in bubbles has been recently reviewed in detail by Donnelly [3]. Our present understanding of the effect of gas pressure on bubble mechanisms seems to raise several questions.

Positron annihilation technique (PAT) is an established method to provide defect-specific information in metals and alloys [4]. The high sensitivity of PAT for the study of helium-vacancy interaction and bubble evolution in post-irradiation annealing studies has been demonstrated earlier [5–7]. More recent theoretical developments in our understanding of positron states in gas bubbles have shown the utility and potential of PAT in providing structural and density information of various gas-defect complexes [8–10]. PAT offers good scope for the study of helium bubbles over a wide parameter range. In this article, we discuss results of our recent studies on the helium behaviour in Ni, dilute alloys of <u>Ni</u>–Ti and Ti–stabilised austenitic stainless steel. The questions addressed by the present study are: (a) positron lifetime sensitivity to helium decoration of vacancy-type defects; (b) Identification of the formation of stable helium clusters and the effect of helium pressure on bubble coarsening in Ni; (c) Effect of Ti-alloying on bubble structures in Ni–Ti and (d) Influence of extended

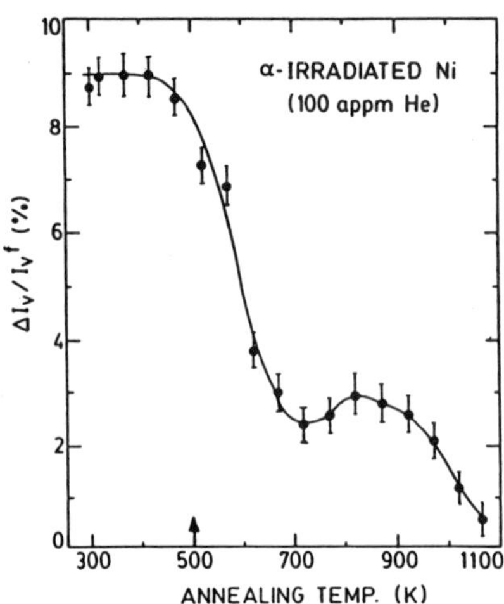

Fig. 1 Variation of Doppler lineshape parameter I_v, in reduced units, as a function of annealing temperature for helium irradiated Ni (100 appm He); Irradiation temperature is 500 K, as shown by arrow on temperature axis.

defects such as TiC precipitates on bubble properties in steel. The main results of the study are discussed here and complete details of the work are presently being published.

2. Experimental

High purity Ni and dilute alloys of Ni–Ti were vacuum annealed at high temperatures to produce defect-free reference samples for helium irradiation. Homogeneous helium implantation was made at the Variable Energy Cyclotron, Calcutta (India) on all samples to a dose of 100 appm He. The implantation procedure employed in the present study has been reported earlier [11]. Helium-free alpha irradiation to a comparable dose was also made on Ni by allowing 40 MeV α-particles to pass through the samples, thereby leaving behind only damage. The details of special thermomechanical treatment given to steel samples prior to irradiation are given in section 4. Post-irradiation isochronal annealing treatments on the samples were given in high vacuum from 300 to 1200 K for an annealing time of 1hr. Positron lifetime and Doppler lineshape measurements were made at room temperature after each annealing step. The lifetime spectrometer had a time resolution of 200 ps (FWHM) and the energy resolution of the Doppler broadening spectrometer was 1.1 keV at 514 keV gamma ray. Doppler lineshape was analysed in terms of the usual normalised peak parameter I_v [5]. Lifetime spectra were analysed for different components using the established programs: RESOLUTION and POSITRONFIT.

3. Helium in Nickel

3.1. PAT annealing results

Figs. 1 and 2 show the variation of Doppler lineshape I_v, expressed in reduced units, as a function of post-irradiation annealing temperature for Ni with 100 ppm He and helium-free irradiated Ni respectively. A comparison of both Figs. 1 and 2 shows that the broad shoulder seen between 700 and 900 K in the I_v-curve for Ni (100 ppm He) is absent in the case of helium-free irradiated Ni. Furthermore, defect recovery is almost complete around 900 K in Fig. 2, whereas I_v does not obtain pre-irradiation value even above 1000 K in Fig. 1. The behaviour observed at 700 K and above, in Fig. 1, is therefore clearly due to the presence of helium. Fig. 3 shows the variation of resolved lifetime parameters in Ni containing 100 ppm He. In the as-irradiated state, two lifetimes are resolved: $\tau_1 = 140$ ps of 70% intensity and

210

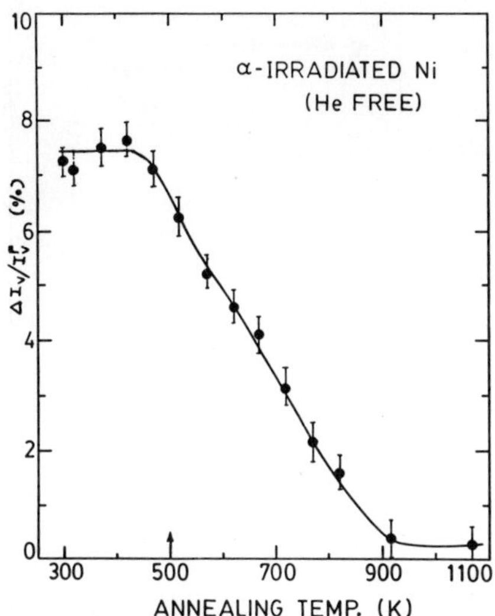

Fig. 2 Variation of the Doppler lineshape parameter I_v, in reduced units, as a function of annealing temperature for helium-free alpha irradiated Ni (Dose: $1.2\ 10^{17}\alpha/cm^2$). Irradiation temperature is marked by arrow on the temperature axis.

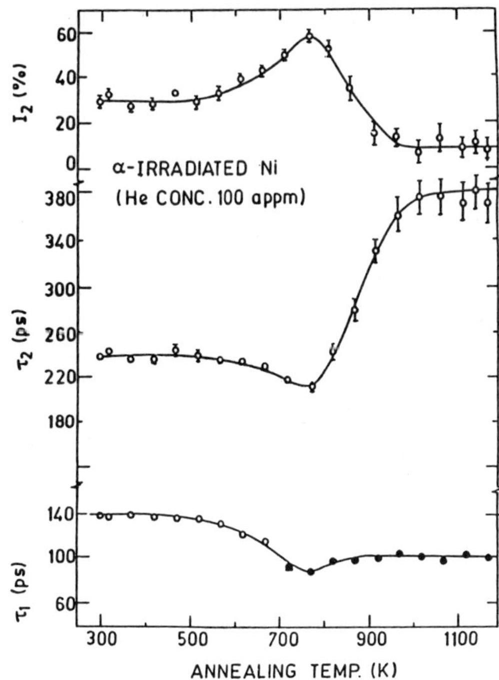

Fig. 3 Variation of the resolved lifetime parameters. τ_1, τ_2, and I_2 as a function of annealing temperature for helium irradiated Ni (100 appm He); closed circles for τ_1, correspond to the bulk state behaviour.

$\tau_2 = 240$ ps of 30% intensity. Between 500 and 750 K, τ_1 decreases, τ_2 shows a decrease before increasing again and I_2 increases before decreasing again. The onset of minimum in τ_2 and maximum in I_2 match with the temperature of occurrence of the shoulder in the I_v-curve (Fig. 1). Above 750 K, τ_2 exhibits a sharp increase before attaining a constant value of 380 ps

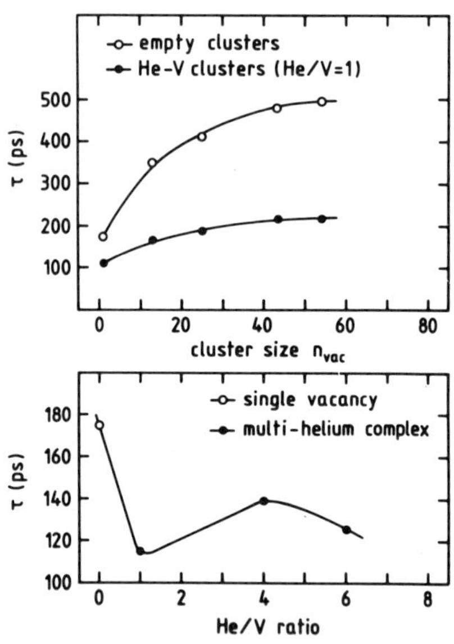

Fig. 4   (Top figure). Variation of computed positron lifetime as a function of cluster size for pure vacancy clusters (open circles) and He-V clusters (closed circles) in Ni; (Bottom figure). Variation of computed positron lifetime as a function of He/V ratio for multiple helium-single vacancy complexes in Ni.

while I_2 shows a marked decrease before levelling off. The present observations on Ni are qualitatively similar to those observed earlier by us on Cu [12] and C̲u̲–B [13]. We take up the interpretation of Fig. 3 after discussing the theoretical results on the effect of helium decoration on positron lifetime in the following section.

3.2.   Effect of helium decoration on positron lifetime in Ni

Calculations of annihilation characteristics in different helium-vacancy complexes have been made, based on the atomic superposition model of Puska and Nieminen [14]. The total positron potential for the defect complex is written following the earlier reported prescription [14, 15] and the Schrödinger equation solved in real space to obtain the positron wave function. The annihilation rate (inverse of the lifetime) is then computed from the overlap of the positron density with core and valence electron densities. The complete details of the study are given elsewhere [16, 17]. We report here the essential results. Fig. 4 (top) shows the variation of computed lifetime τ as a function of cluster size for empty vacancy clusters as well as for He–V clusters. For the latter, the ratio He/V is kept at unity as the cluster size is varied. The marked reduction in τ for He–V clusters, as compared to that for helium-free clusters, arises from the enhanced core electron density brought about by the presence of helium at the positron site. In the next set of calculations, the positron response to multiple helium decoration of a single vacancy has been studied. Here, the relaxed positions of He atoms with respect to the vacancy, as deduced from lattice statics calculations, have been used to define the internal structure of the defect complex. Fig. 4 (bottom) shows the variation of τ as a function of He/V ratio for the case of multiple helium-single vacancy complex. The sharp decrease of τ for He_1V and subsequent increase for He_nV arise from the fact that helium atoms occupy off-centre positions with respect to the vacancy for He_nV, while helium fills the vacancy centre for He_1V. The results in Fig. 4 show how lifetime is sensitive to the size as well as helium-to-vacancy ratio in helium decorated structures in Ni. These are in agreement with the results of Jensen and Nieminen [8] who first discussed these effects for small clusters in Al.

212

3.3. Interpretation of experimental results in Ni

Based on our understanding of positron states in helium-vacancy complexes discussed above, we now interpret the annealing curve for the measured lifetimes in Fig. 3. The measured shorter lifetime ($\tau_1 = 140$ ps) which is lower than that for monovacancy in Ni, matches well with the value calculated for He_4V complex. It seems more appropriate, however, to identify the experimental τ_1 with positron trapping at He_nV complex with $n \leq 4$, considering the model uncertainty of the calculations. The longer lifetime $\tau_2 = 240$ ps, measured in the as-irradiated state of Ni, may be explained as due to small vacancy clusters formed during irradiation (the present irradiation temperature of 500 K is just above the vacancy migration stage in Ni). Such a lifetime would correspond to a cluster of about 4 vacancies, as seen from the result in Fig. 4 (top). The existence of larger vacancy clusters ($n_v > 4$) containing one helium atom or so might also result in similar lifetime. In either case, τ_2 may be understood as due to positron states in multivacancy clusters. As the annealing temperature is increased to 700 K, τ_1 decreases to-wards a bulk-state value as seen in Fig. 3. This decrease of τ_1 indicates the dissociation of thermally unstable multihelium-single vacancy complexes. This observation is in agreement with the results of the recent thermal desorption study by van Veen [18] who reports a dissociation temperature of 700 K for He_nV complexes with $n = 2$ to 4. The formation of stable helium clusters is then brought about by the interplay between two processes, viz: (a) helium supply from the dissociating He_nV complexes and (b) vacancy supply from the annealing structures of irradiation induced defects. The reduction of τ_2 from 240 ps to 210 ps at 750 K with enhanced intensity I_2 (Fig. 3) is indicative of such a formation of stable helium clusters (bubble embryos), which is a precursor to growth. The lifetime τ_2 (210 ps) observed at 750 K is larger than the lifetime values typical of single vacancy complexes discussed earlier in Fig. 4. This would suggest that the embryo stabilises with at least two vacancies per cluster. At 750 K and above, bubbles are the dominant positron traps as seen from the bulk state behaviour of τ_1 in Fig. 3. The increase of τ_2 with a concomitant decrease of I_2 at higher annealing temperatures is in accordance with the behaviour of bubble coarsening, where there is an increase in average radius and decrease in bubble concentration.

3.4. PAT analysis of bubble coarsening in Ni

At $T_A = 820$ K and above, helium atom density n_{He}, bubble radius r_B, and bubble concentration C_B have been determined from an analysis of PAT data, using the procedure reported earlier [19, 20]. n_{He} is obtained directly from the experimental τ_2 using the relation proposed by Jensen and Nieminen [8]. The positron trapping rate in bubbles K_B is known from the lifetime parameters fitted to the well-known two state trapping model [4]. To relate C_B and K_B, the specific trapping rate μ_B should be known. At the present level of understanding, μ_B is assumed to depend only on r_B. If the total helium concentration N_{He}, which is known, is assumed to be contained in spherical bubbles, then the helium inventory equation can be written connecting N_{He}, n_{He}, r_B, K_B and μ_B. Using the functional form for $\mu_B(r)$ as known from the semi-empirical relationship given by Jensen et al [19], the helium inventory equation is solved to obtain r_B. Once r_B is known, C_B can be determined. Fig. 5 (top) shows the plot of n_{He} vs r_B, as deduced from the above PAT analysis. The dashed line shown for comparison, represents the densities corresponding to bubbles in thermodynamic equilibrium. Fig. 5 (bottom) shows the plot of r_B vs the corresponding pressure p_{He} obtained from the experimental n_{He} in Fig. 5 (top) via the Trinkaus equation of state [21]. The equilibrium pressure ($p = 2\gamma/r_B$) is also shown as dashed line for comparison in Fig. 5 (bottom). As seen from both curves in Fig. 5, there is a clear derivation from the behaviour expected for equilibrium bubbles. This deviation, which corresponds to an overpressure of nearly 3 to 4 Gpa is outside the uncertainty of the deduced parameters. Also, n_{He} or p_{He} tends to constant values independent of r_B at temperatures above 1000 K. The features seen in Fig. 5 raise an important question [22], which is

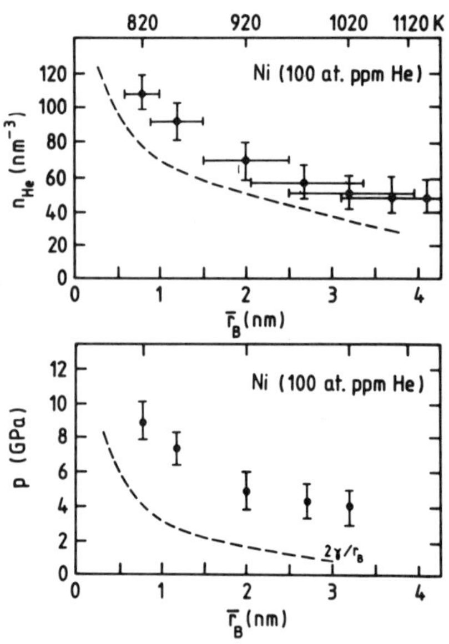

Fig. 5 (Top) Plot of helium density n_{He} vs mean bubble radius r_B (horizontal and vertical error bars indicate the uncertainty of the deduced parameters); Dashed curve corresponds to the helium density of equilibrium bubbles, assuming a surface energy $\gamma = 1.8\ Nm^{-1}$
(Bottom) Helium pressure p as a function of r_B, as obtained via the Trinkaus equation of state; dashed curve corresponds to equilibrium pressure.

contrary to our conventional understanding, namely, that annealing of bubble structures at $T_A \geq 0.6\ T_m$ results in equilibrium bubbles aided by the condensation of thermal vacancies. It is also perhaps worth pointing out from the present study that the Arrhenius plot of $\ln r_B$ vs $1/T$ yields a slope of $\approx 0.36\ eV$. However, no attempt is made here to discuss quantitatively the bubble coarsening mechanisms on the basis of the present isochronal annealing data alone. The following conclusions may be drawn from the present study on Ni with 100 appm He: (a) Bubble coarsening seems to occur under an overpressure which is in excess of equilibrium bubble pressure; (b) Under such conditions of vacancy deficit, strong pressure relaxation mechanisms of coarsening are perhaps not operating. Instead, a weakly activated coarsening process such as bubble migration and coalescence seems to operate, where the bubbles can be expected to maintain their overpressure. These conclusions are in accordance with the recent theoretical model of Trinkaus [22] and also in agreement with the conclusions reached from independent TEM [23] and SANS [24] studies reported on Ni containing higher helium concentrations (1000–1200 appm).

4. Helium in Ni-Ti Alloys

4.1. Experimental results

No detailed studies have been reported in literature, to the best of our knowledge, on the effect of impurity doping on bubble structures in metals. We discuss in this section results of a controlled study addressed to the effect of Ti addition in Ni matrix. Samples used are from Ni–Ti substitutional alloys over the range of Ti–concentration from 0.07 at% to 5 at%. In this range of concentration, Ti acts as oversized substitutional impurity with smaller electron density in the host Ni matrix. Fig. 6 shows the variation of measured lifetime parameters as a function of annealing temperature for Ni–0.5 at% Ti with 100 appm He. The distinct features in Fig. 6 can be summarized as follows: (a) Only a single lifetime of 160 ps is resolved in the as-irradiated state of the alloy and this persists till 700 K. This observation of single lifetime in

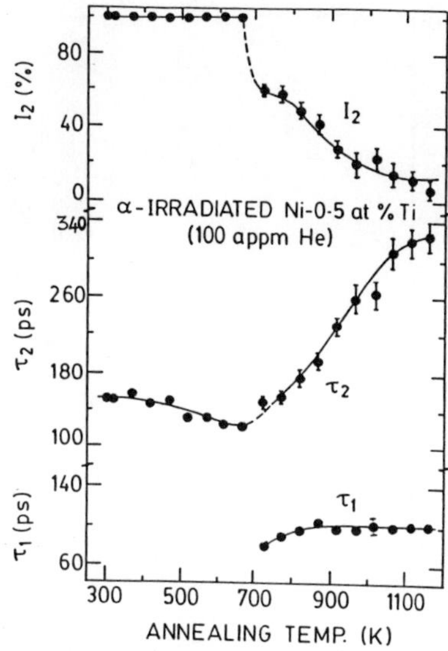

Fig. 6   Variation of the resolved lifetime parameters τ_1, τ_2 and I_2 as a function of annealing temperature for helium irradiated Ni–0.5 at% Ti (100 appm He); Note that the single lifetime ou to 700 K is shown as τ_2 with 100 % intensity.

the initial annealing range is in contrast to the behaviour seen earlier in Ni. (b) A clear stage is revealed at 700 K, where two lifetimes are resolved, with the intensity exhibiting an abrupt drop. The shorter lifetime τ_1 here corresponds to the bulk. Above 700 K the behaviour of τ_2 and I_2 are qualitatively similar to that seen for Ni. The observations on the lifetime behaviour of other alloys, Ni–0.07 at% Ti and Ni–5 at% Ti are similar to that of Ni–0.5 at% Ti. However, the near-saturation value of τ_2 at higher annealing temperatures shows a progressive decrease with increasing Ti concentration and these values in turn are lower than that for pure Ni.

4.2. *Effect of Ti on helium behaviour*

With a view to get some insight into the plausible helium-defect complexes in Ni–Ti, helium binding energy trends for different traps have been calculated. These calculations are based on the well known lattice statics method of calculating the configurational energies of the complex using a pair potential approach. The details are given elsewhere [16, 25]. To summarise the main results, the binding energy of Ti atom to vacancy is found to be ≈ 0.5 eV. For a He_1V_1 complex, the binding energy is about 2.66 eV, which decreases towards a value of 1.7 eV as more helium atoms decorate the vacancy. The Ti–V complex is found to be a competitive trap for helium, in addition to pure vacancy. For a Ti–V complex, the helium binding energy is 2.12 eV which decreases towards a value of 1.5 eV with increasing helium atoms per site. A single substitutional Ti atom does not seem to bind helium. On the basis of these trends, it may well be argued that upon irradiation, three types of complexes, viz, Ti–V, He–V and He–Ti–V are formed in the alloy. Positron trapping in all these complexes results in lifetimes which are too close to be resolved. This seems to be consistent with the observation of saturation positron trapping corresponding to the single lifetime of 160 ps which is lower than that for a pure vacancy. As the annealing proceeds, Ti–V complexes (undecorated by He) dissociate, as indicated by the slight decrease of τ till 700 K. Vacancy release from this process possibly stabilises helium clustering in the alloy. This can explain the appearance of the second lifetime τ_2 with τ_1 corresponding to the bulk state behaviour (Fig. 6). The increase of τ_2 accompanied by a decrease of I_2 is in accordance with the behaviour expected for bubble growth at higher temperatures.

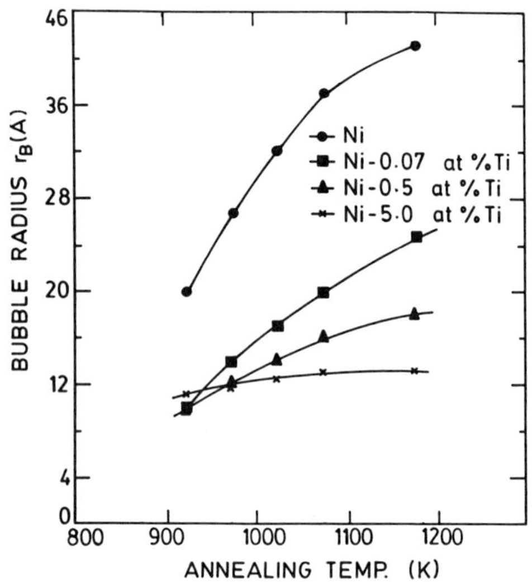

Fig. 7 Variation of the bubble radius r_B (in units of Å), as deduced from PAT analysis, as a function of annealing temperature for helium irradiated Ni (closed circle), Ni–0.07 at% (closed square), Ni–0.5 at% Ti (closed triangle) and Ni–5 at% Ti (crosses); All samples have the same helium concentration of 100 appm.

The bubble parameters have been determined from the analysis of PAT data, as discussed in section 3.4. Fig. 7 shows the plot of deduced bubble radius r_B vs annealing temperature for different Ni–Ti alloys. The variation of r_B for pure Ni is also Shown for comparison. The progressive reduction of r_B with increasing Ti–concentration for a given annealing temperature indicates a resistance to bubble coarsening upon Ti–addition. Also, the deduced bubble concentration shows a progressive increase as a function of Ti–concentration. These results may be understood as due to the effect of Ti atoms attached to the metal-gas interface, which impede the movement of bubble surface and thereby the bubble coalescence is suppressed. The resistance to coarsening and consequent reduction of bubble swelling is found maximum for Ni–5at% Ti.

5. Helium in Ti–Stabilised Austenitic Steel

Finally, we make a brief discussion of our first qualitative PAT study made on helium irradiated Ti–stabilised austenitic stainless steel. The influence of fine dispersive TiC particles on helium trapping has evoked a lot of interest in recent times [26, 27]. The present stainless steel samples (Ni 15%, Cr 15%, Ti/C = 4) whose chemical composition is close to the steel DIN 1.4970 reported in literature, has been given a thermomechanical treatment, following the earlier reported [27] prescription, to stabilise fine TiC precipitates in the austenitic matrix prior to helium irradiation. This sample treatment, consisted of solution annealing at 1373 K, followed by 20%, cold work and subsequent aging at 1073 K for 12 hr. These samples (called Ti SS$_2$) were irradiated at 373 ± 50 K to a helium dose of 100 appm. Fig. 8 shows the variation of lineshape I_v as a function of annealing temperature. The I_v-curve exhibits two clear annealing stages, one at 673 K and the other at 1000 K. Figs. 9 and 10 show the variation of resolved lifetimes and their intensities respectively for TiSS$_2$. The observation of two annihilation modes in the as-implanted TiSS$_2$, with one corresponding to a positron trapped state and the other to the bulk state, is in contrast to the single lifetime behaviour seen earlier in Ni–Ti alloy. This seems to suggest that upon irradiation of TiSS$_2$, a major fraction of positron traps are associated with extended defects such as TiC precipitates whose concentration is such that it does not result in saturation positron trapping. TiC is expected to be a highly biased

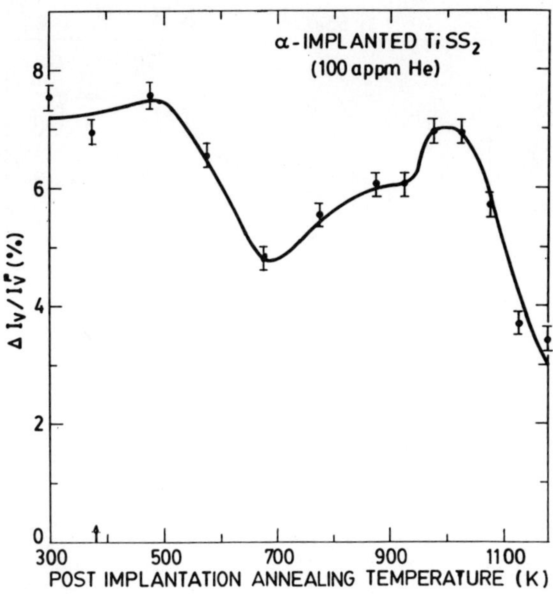

Fig. 8 Variation of the lineshape parameters I_v, in reduced units as a function of annealing temperature for helium irradiated Ti–SS$_2$ (arrow marks the irradiation temperature).

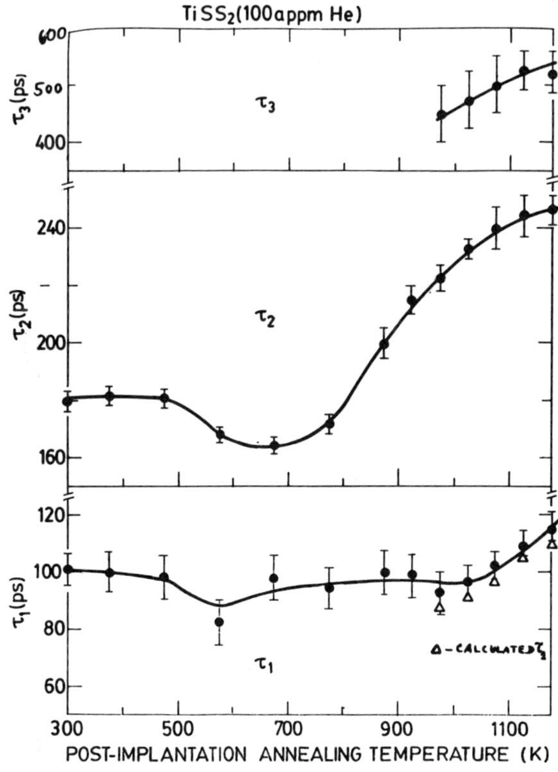

Fig. 9 Variation of resolved lifetimes τ_1, τ_2 and τ_3 as functions of annealing temperature for TiSS$_2$ (100 appm He); open triangles in the bottom figure represent the calculated τ_1 shown for comparison.

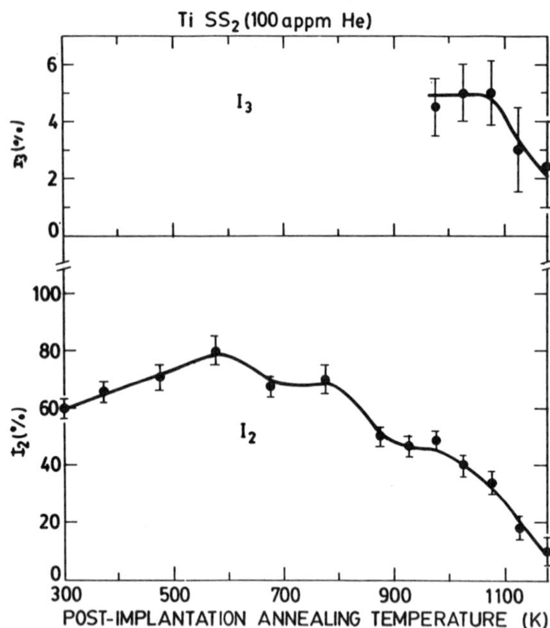

Fig. 10 Variation of the resolved intensities I_2 and I_3 as functions of annealing temperature for TiSS$_2$ (100 appm He).

vacancy absorber as a result of large volume increase due to the precipitation of |Ti, C| in austenitic matrix [26, 27]. Helium can therefore preferentially accumulate at or close to TiC–matrix interfaces. Accordingly, the observed lifetime τ_2 is qualitatively assigned to positron states at helium trapped to TiC. The decrease of τ_2 with somewhat enhanced I_2 between 500 and 700 K indicates the formation of stable helium clusters prior to bubble growth. In addition to the expected behaviour of τ_2 and I_2, a third component, τ_3 (400–600 ps) of weak intensity, develops at higher annealing temperatures ($\geq$ 1000 K) as seen in Fig. 9 and 10. This long lifetime possibly arises from large cavities formed in the vicinity of grain boundaries or surfaces where bubbles can relax in size due to strong vacancy sources. Alternatively, the origin of τ_3 as due to cavities of low density stabilised by constituent impurities in the steel alloy cannot be ruled out. This is a matter for further investigation.

A preliminary analysis of the PAT parameters τ_2 and I_2 corresponding to matrix bubbles, indicates a strong refinement of bubble structure in TiSS$_2$, as compared to that in Ti-free SS 316 [20]. Bubble radius r_B is found reduced by nearly a factor of 5 at any given temperature and bubble concentration C_B higher by nearly an order of magnitude as compared to those in SS316. The present observation on changes in bubble structure in TiSS$_2$, is in agreement with the recent microstructural results reported on similar steels [28]. Further plans of study on the material include correlation of PAT data with electron microscope data on the same samples and systematic PAT studies as a function of Ti/C ratio.

6.   Conclusions

Detailed understanding of helium behaviour in pure Ni has been possible through PAT measurements and analysis. Positron lifetimes are shown to be sensitive to the size and helium content as well as the configurational structure of helium decorated defect complexes in Ni. Analysis of the bubble coarsening stage in Ni indicates overpressure, in excess of that expected for equilibrium bubbles, thereby implying conditions of strong vacancy deficit. Clear trends of resistance to bubble coarsening with increasing Ti-concentration are seen in

218

Ni–Ti alloys. Finally, the influence of extended defects such as TiC precipitates on helium trapping is qualitatively seen in Ti-stabilised steel.

ACKNOWLEDGEMENT

We would like to thank K.P. Gopinathan for his keen interest in the present work, A. Bharathi for her help in the computations of annihilation characteristics and R. Rajaraman for his assistance in some of the measurements. Several interesting discussions with H. Trinkaus and H. Ullmaier are gratefully acknowledged.

REFERENCES

1. H. Ullmaier (ed.). Proc. Inter. Symp. *Fundamental Aspects of Helium in Metals* (Gordon and Breach, 1983) see Rad. Effects **78**, (1983).

2. G.J. Thomas, Rad. Effects **78**, 37 (1983).

3. S.E. Donnelly, Rad. Effects **90**, 1 (1985).

4. W. Brandt and A. Dupasquier (ed.) *Positron solid State Physics* (Amsterdam, North Holland) 1983.

5. B. Viswanathan, W. Triftshäuser and G. Kögel, Rad. Effects **78**, 231 (1983).

6. G. Kögel, Qin-min Fan, P. Sperr, W. Triftshäuser and B. Viswanathan J. Nucl. Mater. **127**, 125 (1985).

7. H.E. Hansen, H. Rajainmaki, R. Talja, M.D. Bentzon, R.M. Nieminen and K. Petersen, J. Phys. F **15**, 1 (1985).

8. K.O. Jensen and R.M. Nieminen, Phys. Rev. **B36**, 8219 (1987).

9. R.M. Nieminen in *Positron Annihilation*, L. Dorikens-Vanpraet, M. Dorikens and D. Segers (eds.) (World Scientific, Singapore 1989) p. 60.

10. K.O. Jensen, NATO ASI Series. This volume.

11. G. Amarendra, B. Viswanathan and K.P. Gopinathan, Cryst. Res. Technol. **22**, 1563(1987).

12. G. Amarendra, B. Viswanathan and K.P. Gopinathan, in *Effects of Radiation on Materials*. ASTM STP 1046, Ed. N.H. Pachan, R.E. Stoller and A.S. Kumar (ASTM, Philadelphia, 1989) Vol. 1, p. 639.

13. B. Viswanathan, G. Amarendra and K.P. Gopinathan, Rad. Effects **107**, 121 (1989).

14. M.J. Puska and R.M. Nieminen, J. Phys. F **13**, 333 (1983).

15. A. Bharathi and. B. Chakraborthy, J. Phys. F **18**, 363 (1988).

16. G. Amarendra, *Positron Annihilation Studies of Helium in Metals and Alloys*, Ph.D. Thesis (University of Madras, Madras 1990) unpublished.

17. G. Amarendra, B. Viswanathan and K.P. Gopinathan, paper under communication.

18. A. van Veen, Mater. Sci. Forum **15–18**, 3 (1987).

19. K.O. Jensen, M. Eldrup, B.N. Singh and M. Victoria, J. Phys. F **18**, 1069 (1988).

20. B. Viswanathan, J. Phys.: Condensed Matter 1 SA71–76 (1989).

21. H. Trinkaus, Rad. Effects **78**, 189 (1983).

22. H. Trinkaus, Scripta. Metal. **23**, 1773 (1989).

23. V.N. Chernikov, H. Trinkaus, P. Jung and H. Ullmaier, J. Nucl. Mater. **170**, 31(1990).

24. Li Qiang, W. Kesternich, H. Schroeder, D. Schwahn and H. Ullmaier, Acta. Metall. Mater. **38**, 2383 (1990); also see H. Ullmaier, NATO ASI Series, this volume.

25. G. Amarendra, B. Viswanathan and K.P. Gopinathan, paper under communication..

26. P.J. Maziasz, Proc. Symp. on Irradiation Effects on phase stability, TMS–AIME1981, p. 477.

27. W. Kesternich, Materials Research Society Symposium proceedings 62 229 (1986).

28. W. Kesternich, J. Nucl. Mater. **155–157**, 1025 (1988).

COMPARISON OF RESULTS FROM DIFFERENT EXPERIMENTAL TECHNIQUES (SANS, TEM, PAT, SEM) APPLIED TO BULK Cu AND Ni CONTAINING KRYPTON

M. Eldrup,[1] J. Skov Pedersen,[1] A. Horsewell,[1] K.O. Jensen[2] and J.H. Evans[3]

[1] *Risø National Laboratory, DK-4000 Roskilde, Denmark*
[2] *University of East Anglia, Norwich NR4 7TJ, U.K.*
[3] *Harwell Laboratory, Oxon OX11 ORA, U.K.*

ABSTRACT

The isochronal annealing behaviour of bulk samples of Cu and Ni containing 3-5 at% krypton is discussed on the basis of results from different experimental techniques. These include those mentioned in the title as well as sample weight and linear dimension measurements. Although the techniques give the same qualitative picture of bubble growth and increasing swelling, quantitative differences exist.

1.   Introduction

It is generally recognized that in studies of gases in solids no one single experimental technique can provide a full picture of the microstructural parameters (e.g. defect concentrations, size distributions, gas densities etc.) [1]. In the present contribution we shall discuss investigations carried out on bulk samples of Cu and Ni containing 3 to 5 at% Kr using several different experimental techniques. Fuller details of part of the work, in which results from the positron annihilation techniques (PAT), transmission and scanning electron microscopy (TEM and SEM) and length and weight measurements were compared, have been published elsewhere [2].

More recently we have extended the range of experimental techniques to include small angle neutron scattering (SANS). This technique is particularly sensitive to precipitates in the size range of about 1-100 nm [3], i.e. the same range as TEM. However, although being more indirect than TEM, SANS has the advantage that it averages over bulk dimensions of the sample, and—like PAT—is nondestructive.

The comparisons of TEM, SEM and PAT results in [2] were made on samples which had been prepared in similar, but not necessarily identical ways. For comparison with the SANS data, new PAT and TEM investigations were therefore carried out on samples which were identical to those used for SANS.

The material used in this work was obtained from Whitmell [4,5] who devised a combined sputtering/implantation process to make unique centimetre thick layers of copper and nickel containing high concentrations of krypton, a material which is very suitable for conventional PAT and SANS measurements that require bulk samples.

In previous work on this material it was established that, in the as-prepared condition, the substructure in both Cu-Kr and Ni-Kr consists of a very small grain size with a high density of small Kr bubbles within grains. Electron diffraction studies [6] showed that the Kr in the

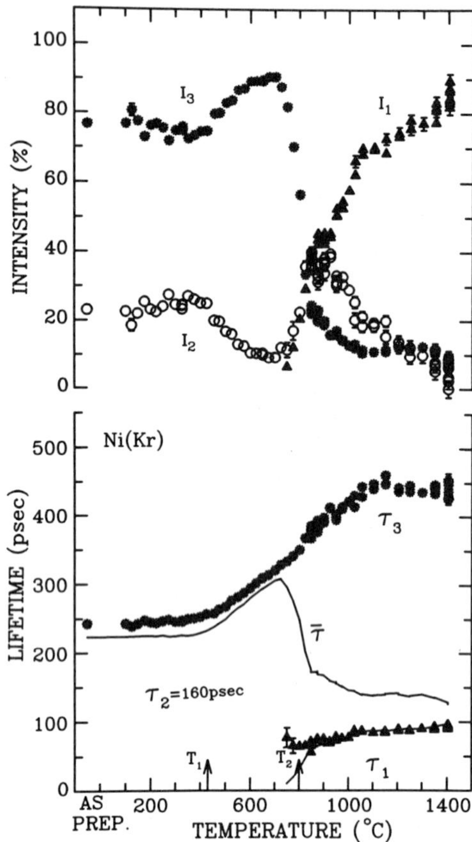

Fig. 1 Positron lifetimes and relative intensities for Ni-Kr as functions of annealing temperature. The curve marked $\bar{\tau}$ $(= \Sigma\, \tau_i\, I_i\,)$ is the mean lifetime. The temperature at which τ_3 starts to increase is defined as T_1. The lifetime component $\tau_3\, I_3$ is due to positrons trapped in Kr bubbles. T_2 is the temperature where I_3 decreases most steeply.

bubbles is precipitated in the solid phase with the high bubble pressures being associated with the bubble growth processes [7,8]. It was argued [9,10] that only the larger bubbles were seen in TEM, while a considerable fraction (about 75%) of Kr is present in submicroscopic bubbles or vacancy-Kr clusters.

Previous work also included annealing studies on Cu-Kr using both TEM and SEM. It was established that bubble migration leads to coalescence and to the growth of a grain boundary bubble population which eventually forms an extensive pore structure leading to the loss of gas to specimen surfaces.

In the next Section we shall summarize the results of reference [2], and in Section 3 we present the SANS results and compare them with the complementary TEM and PAT data.

2.   Annealing of Cu-Kr and Ni-Kr specimens

Figure 1 shows positron lifetime data for a Ni-Kr sample, isochronally annealed for 30 minutes after each 25°C step. As discussed in e.g [11,12] positrons become trapped at open volume defects such as vacancy clusters and gas bubbles. The positron lifetime increases with cluster size and decreases with gas density [12]. Up to about 700°C all positrons become trapped in defects (Fig. 1), and two representative lifetimes can be resolved. The shorter one

222

(τ_2) scattered around a mean value of 160 ps in the whole annealing range; for the analysis it was fixed at this value and only its intensity (I_2) is plotted in Fig. 1. The magnitude of τ_2 makes it reasonable to associate the τ_2,I_2 component with small Kr-vacancy clusters and/or dislocations, while the τ_3,I_3 component has to be due to positron trapping into Kr bubbles.

As discussed by Jensen [12], the lifetime of positrons trapped in voids or bubbles increases with cavity size up to a diameter of about 1 nm where it saturates. For voids the saturation lifetime is about 500 ps, but in bubbles this saturation value decreases with increasing gas density. The theoretical relationship between krypton density and positron lifetime is in a good approximation:

$$\tau = 500 - 9.2 n_{Kr} \qquad (1)$$

where τ is in ps and n_{Kr} in nm^{-3}. The Kr density derived from the τ_3 value of 243 ps at low annealing temperatures (Fig. 1) is 28 nm^{-3}. This is in fair agreement with Kr densities obtained from measurements of the sample mass density [10] on similar samples, and—as discussed in [12]— with the value determined from electron diffraction measurements [6].

On annealing, the lifetime τ_3 starts to increase at the temperature denoted T_1 (Fig. 1) while its intensity shows a sharp drop at T_2. The rise of τ_3 above T_1 signifies a decrease of the Kr density in the bubbles (Eq. (1)). It is suggested that above T_1, after melting of the krypton, bubble migration takes place, leading to bubble coalescence, a process which is also observed in TEM [2]. However, the coalescence in itself does not lead to a decrease of n_{Kr}. Simulations [8] suggest that thermal vacancies are too few at T_1 ($\approx 0.4T_m$) to effectively release the overpressure created by the coalescence, although at somewhat higher temperatures ($\approx 0.5T_m$) this process may take place. One source of vacancies may be the small clusters that contribute to the τ_2,I_2 component, since their vacancy to krypton ratio might be higher than in the bubbles. The decrease above T_1 of I_2 compared to I_3 is consistent with the disappearance of small Kr-V clusters which may coalesce with the mobile bubbles. Another source of vacancies may be the loop punching process. There is no loss of Kr from the sample around T_1. This is seen from Fig. 2 which shows the relative linear dimensions and the relative weight of the samples used for the PAT measurements as functions of annealing temperature.

In the temperature range from T_1 to T_2 the bubble lifetime τ_3 increases steadily, a sign of continued decrease of the Kr density in the bubbles due to coalescence and growth. This bubble coarsening was also observed by TEM [10], and by assuming the bubbles to be in thermal equilibrium, the Kr pressure P can be determined via the equilibrium condition:

$$P = 2\gamma/r \qquad (2)$$

where r is the observed mean bubble radius and γ the surface energy. From P the Kr density is determined via the equation of state [13]. The fairly good agreement between Kr densities derived from TEM and PAT on similar Cu-Kr samples is demonstrated in Fig. 2 in [12]. The bubble coarsening gives rise to a small swelling (linear dimensional increase of 1-2%), but no loss of Kr (Fig. 2), until dramatic parameter changes take place at around T_2. The strong drop of I_3 reflects an enhanced coalescence of bubbles that reduces the positron trapping rate into the bubbles, gives rise to a strong enhancement of the swelling and eventually provides pathways for the Kr to the grain boundaries and subsequently to the surface. The swelling and the loss of Kr is seen in Fig. 2, and SEM micrographs of sample surfaces in this temperature region clearly show the appearance of holes into the surface [2].

Above T_2, although I_3 drops to $\approx 10\%$, τ_3 continues to rise as the surviving bubble population coarsens. Eventually τ_3 reaches a value of ≈ 450 ps, in the range expected for large, near-empty cavities. After the final anneal at 1400°C, the linear dimensions are still more than 10% above the starting value, and a density measurement showed a swelling of 46% relative to pure nickel. This indication of a rather porous structure was confirmed by SEM as seen in Fig. 3. The structure is now dominated by 1-10 μm large cavities. Finally, the specimens were melted in a desorption equipment and the Kr concentration determined to be 200±100 appm. It may seem surprising that such a small amount of Kr is able to stabilize the observed massive swelling. However, once large cavities are formed, only a low Kr density is required to maintain them in equilibrium. From the total Kr concentration and swelling mentioned above the average Kr density inside the cavities can be calculated to $4\pm2\ 10^{25}$ Kr/m^3. This is equivalent to a pressure of 0.92±0.46 MPa (using the van der Waals equation with $T = 1400°C$). With a surface energy value for Ni of 1.9 J/m^2 [14] this predicts cavity radii for equilibrium bubbles between 2 and 8 μm, in fair agreement with the observed sizes (Fig. 3).

3.   Comparison of SANS, TEM and PAT

As mentioned above, the comparisons discussed in Section 2 between PAT, TEM and SEM results were made for similar, but not identical specimens. A renewed study was therefore undertaken on Cu-Kr specimens cut side by side from the same piece of material with the aim of comparing as far as possible identical samples and, not least, in order to utilize the potential of SANS measurements. One 1.5 x 14 x 17 mm^3 piece was used for SANS, with the main surface cut perpendicular to the direction of deposition of the Cu-Kr material. Two 1 x 6 x 6 mm^3 pieces were cut in the same way for the positron lifetime measurements, and a number of 3 mm discs were prepared with different orientations for TEM. Measurements with all three techniques were carried out on the as-prepared material and after annealing for 30 min. at the temperatures 275°C, 425°C, and 575°C.

3.1. SANS

The small angle neutron scattering measurements were carried out at the newly installed SANS facility at Risø [15]. The principles of SANS can be found in e.g. [3]. The neutrons scattered by the sample are detected by a two-dimensional position-sensitive detector. Scattering of neutrons with scattering vectors in the detectable range ($\approx 10^{-2}$ –10 nm^{-1}) is due to regions in the sample of dimensions of roughly 0.3-300 nm inside which the scattering length density deviates from that of the matrix. Inhomogeneities in this size range, like gas bubbles and other precipitates can therefore be investigated by SANS.

Fig. 4 shows the radial intensity distribution as derived from the two-dimensional raw data, after subtraction of two background contributions, a constant one due to incoherent scattering from the Cu matrix, and one with q^{-3} dependence, probably due to dislocations, that dominates at low q values (q is the scattering vector). In the present paper we shall briefly describe the analysis of the data in terms of a simple model which assumes that the Kr-bubbles are spheres of equal size. A detailed treatment of the data will be published elsewhere [16].

For a collection of particles randomly distributed, the scattered intensity is given by:

$$I(q) = 4\pi\ N \int p(r)\ \frac{\sin(qr)}{qr}\ dr \tag{3}$$

where N is the bubble density and p(r) the distance distribution function for the particles given by:

$$p(r) = r^2 \int \rho\,(\bar{r}')\,\rho\,(\bar{r}' + \bar{r})\ d\bar{r}', \tag{4}$$

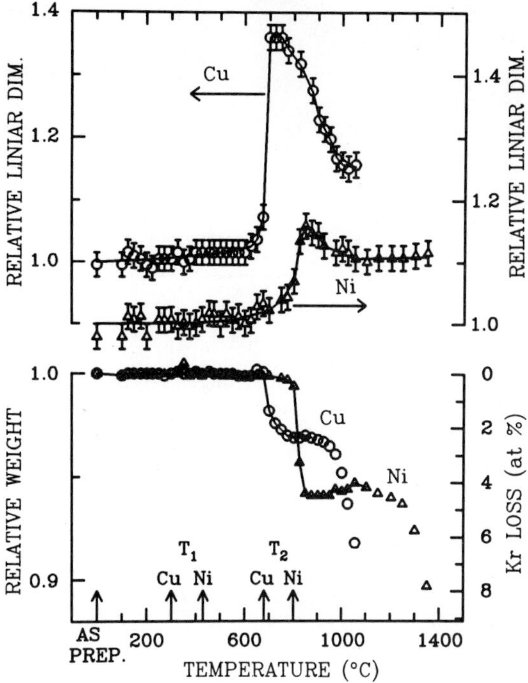

Fig. 2 Relative weights and linear dimensions of both Ni-Kr and Cu-Kr samples as functions of annealing temperature.

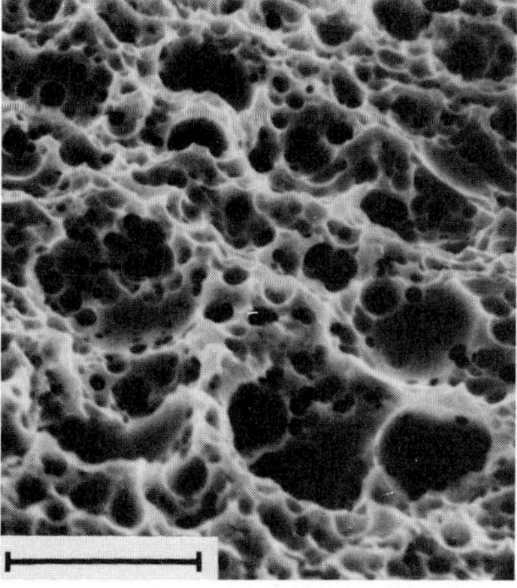

Fig. 3 Scanning electron micrograph of the surface of a Ni-Kr sample after annealing to 1400°C. The marker length is 10 μm.

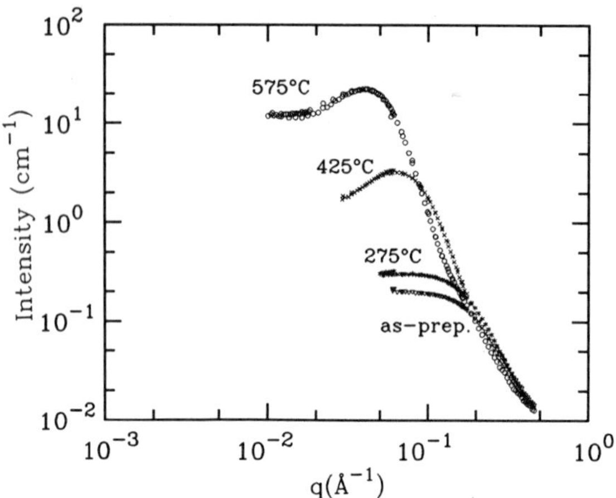

Fig. 4   SANS measurements for Cu-Kr. Scattering curves for the four annealing temperatures after background subtraction. The annealing temperature is indicated for each curve.

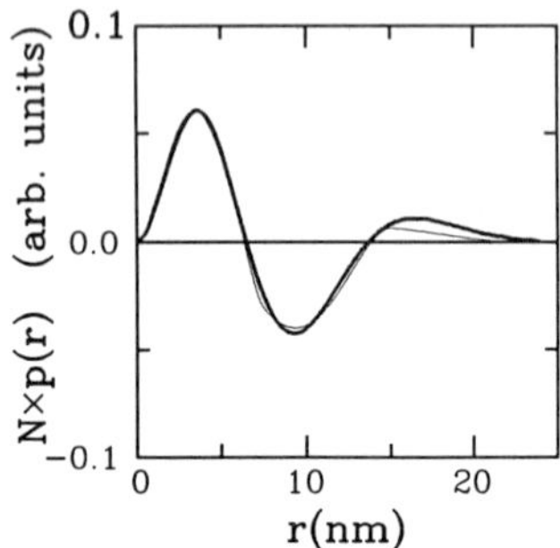

Fig. 5   The distance distribution function, p(r) multiplied by the bubble density N after the 575°C anneal, calculated as described in the text. The thin curve is obtained by fitting the "shell model" to N p(r).

where $\rho = \rho_o - \rho_{av}$ is the difference between the particle scattering length density and the average scattering length density in the sample. By fitting Eq. (3) to the measured spectra (Fig. 4), the distance distribution function can be determined. In practice p(r) has to be parameterized, and in the present work this was done by using the method proposed by Glatter [16,17] in which p(r) is described as a linear combination of a proper set of spline functions:

$$p(r) = \sum_n a_n B_n(r) \qquad (4a)$$

In the fitting, the coefficients a_n are determined. The p(r) resulting from the 575°C spectrum is shown in Fig. 5. Qualitatively, the peak at the lowest r (≈ 3 nm) represents the self-correlation of bubbles and the peak position is close to the mean bubble radius. The negative part may be associated with the regions around the bubbles which (because no bubbles are present in this region) contribute less than average to the scattering. Finally, the broad peak at around 17 nm may be associated with correlation between neighbouring bubbles; its position thus represents an average inter-bubble distance.

226

Table 1 Bubble parameters derived from 3 different techniques for identical Cu-Kr samples. r_1 and r_2 are the bubble mean radius and approximate inter-bubble distance (see text), S_i the volume swelling, r and C the bubble mean radius and concentration, n_{Kr} the krypton density in the bubble, and r_{eq} the mean bubble radius assuming equilibrium bubbles.

	SANS			TEM			PAT		
T_{ann} (°C)	r_1 (nm)	r_2 (nm)	S_s (%)	$\bar{r}$ (nm)	C (10^{24} m^{-3})	S_T (%)	n_{Kr} (10^{28} m^{-3})	r_{eq} (nm)	S_p (%)
as prep	1.1	3.5	2.6				2.66	1.5	8.2
275	1.1	3.7	3.1	0.6	15.0	1.7	2.62	1.1	8.3
425	2.3	7.1	5.1	1.2	3.9	4.9	2.07	2.6	10.5
575	3.8	11.4	7.4	2.5	0.3	4.1	1.59	5.3	13.7

This qualitative picture calls for a "shell model", in which the scattering length density equals ρ_1 inside the (bubble) radius r_1, $-\rho_2$ between r_1 and an outer radius r_2, and zero for larger distances. This "shell model" was inserted into Eq. (4) and p(r) then fitted to the distance distribution function derived from the experimental data (Fig. 5), r_1, r_2 and $N^{1/2}\rho_1$, $N^{1/2}\rho_2$ being the fitting parameters. The fitted curve is also shown in Fig. 5, and the bubble mean radii (r_1) as well as r_2 (the approximate distance to the nearest bubbles) are given in Table 1 for the four annealing temperatures. It is interesting that the ratio r_1/r_2 decreases slowly with increasing annealing temperature, reflecting increasing swelling. The swelling at a given temperature can also be estimated using the estimated scattering length density ρ_1 inside the bubbles and the fitted value obtained for $N^{1/2}\rho_1$. Using N determined this way, the swelling in Table 1 was calculated as $S_S = N(4\pi/3)r_1^3$.

3.2. TEM and PAT

TEM showed that at the two highest annealing temperatures the bubbles observed were facetted while at the lowest temperature no facetting was seen. Quantitative estimates of bubble sizes and concentrations were very difficult for the 275°C sample because of the small sizes and high concentrations, the latter giving rise to overlap of the bubble images. The bubble mean radii and concentrations are given in Table 1 together with the swelling calculated from the bubble size distributions.

The results of the positron lifetime measurements were in close agreement with the results in ref. [2] for Cu-Kr (qualitatively the same behaviour as for Ni-Kr, Fig. 1). Applying Eq. (1), the krypton density in the bubbles is obtained from the positron lifetime, and, assuming the bubbles to be in equilibrium, the bubble radius, r_{eq}, can be derived from Eq. (2), the pressure being obtained from the equation of state [13]. The results are given in Table 1.

The swelling, S_p of the specimen can be derived from n_{Kr}:

$$S_P = (n/n_{Kr})G \qquad (5)$$

where n is the Cu-matrix density of atoms ($8.36 \ 10^{28}$ m^{-3}) and G the total concentration of Kr in the sample (measured to be 2.6 at.%). The derived swellings are given in Table 1.

3.3. Comparison of results

A comparison of the results in Table 1 shows that there are rather large discrepancies between the values obtained for the mean bubble radii and the swellings by the three techniques. Some of the apparent differences between the r-values arise from the experimental uncertainties associated with the techniques. In particular the uncertainty on r_{eq} obtained by PAT is quite large, mainly because small uncertainties in n_{Kr} give rise to large errors in the krypton pressure and thereby in r_{eq}. The assumptions that the bubbles are in equilibrium and that the Ronchi EOS [13] is valid may also be questioned. (It should be mentioned that in other cases where saturation trapping of positrons does not take place, fewer assumptions are involved in the determination of r and C [12,18,19]). As mentioned above, the high density of bubbles, especially at low annealing temperatures, gives rise to uncertainties in the $\bar{r}$-values determined by TEM. They are estimated to ≈50% at 275°C and ≈20% at the higher temperatures. Also the SANS bubble radius (r_1) was derived from a rather simple, monodisperse, model which may also lead to some systematic errors in the values.

Apart from the experimental uncertainties, another factor is important when comparing the data from the different techniques. This is in the weighting of the different bubble sizes when "average" radii are derived from the experiments. In SANS the scattering is proportional to r^6, while in TEM all bubbles are in principle observed with the same probability (above the resolution limit of ≈1 nm). With PAT, the detection probability is roughly proportional to r^a ($1<a<2$) for bubbles in the present size range [12,20]. For a more detailed comparison, it is therefore important to derive the bubble size distribution from the SANS data. Such work is in progress [16].

The swelling values calculated from the PAT data are based only on the krypton density, n_{Kr}, and are therefore expected to have only a rather small uncertainty (≈20%). This is in good agreement with the ≈8% swelling derived from sample-density measurements on other Cu-Kr specimens in the as-prepared state [10]. On the other hand, the swellings calculated from the TEM data are considerably lower than the PAT values, in agreement with earlier estimates [10]. Although very conservative estimates of the errors on r and C could bring e.g. PAT and TEM swellings to agree within the uncertainties, a more likely explanation of the difference between S_P and S_T seems to be that even after the 575°C anneal an appreciable fraction of the Kr is trapped in submicroscopic vacancy clusters and/or other defects (grain- or twin-boundaries) [21].

Similarly, the SANS values for the swelling are lower than those determined by PAT, but higher than the TEM-values. As discussed for TEM, some of the sample swelling may be associated with Kr trapped in defects which are not resolved by SANS. However, some of the discrepancy may also arise from the different weighting of the bubble sizes as discussed above.

4.   Conclusion

The information obtained on the annealing of Cu (and Ni) containing ≈3 at% Kr shows that although the same qualitative picture emerges from PAT, TEM and SANS, quantitative differences emerge from the various techniques for mean bubble radii and for the volume swellings. The differences may in part be attributed to experimental uncertainties, to different ranges of sensitivity of the techniques and to different variations of the sensitivity inside these ranges.

The different techniques, however, also give complementary information, e.g. PAT the Kr density, TEM bubble size distributions and other microstructural information, and SANS also bubble size distributions when more refined analysis methods are applied [16]. It was shown also that very useful information can be obtained from simple macroscopic measurements (weight and dimension). A comparison between SANS, TEM and PAT also on Ni-Kr samples is in progress.

REFERENCES

1. S.E. Donnelly, Rad. Effects **90**, 1 (1985).
2. K.O. Jensen, M. Eldrup, N.J. Pedersen, and J.H. Evans, J. Phys. F: Met. Phys. **18**, 1703 (1988).
3. G. Kostorz, Treatise on Met.Sci. and Techn. **15**, 227 (1979).
4. D.S. Whitmell, Radiat. Effects **53** 209 (1981).
5. D.S. Whitmell, R.S. Nelson, K.J.S. Smith and G.J. Bauer, Eur. Appl. Res. Reports - Nucl. Sci. Techn. **5** 513 (1983).
6. J.H. Evans and D.J. Mazey, J. Phys. F: Met. Phys. **15** L1 (1985).
7. J.H. Evans and D.J. Mazey, J. Nucl. Mater. **138** 176 (1986).
8. J.H. Evans, Nucl. Instr. Methods **B18** 16 (1986).
9. M. Eldrup and J.H. Evans, J. Phys. F: Met. Phys. **12**, 1265 (1982).
10. J.H. Evans, R. Williamson and D.S. Whitmell in *12th Int. Symp. on Effects of Radiation on Materials*, F.A. Garner and J.S. Perrin, eds., ASTM STP 870 (1985) p. 1225.
11. M. Eldrup in *Defects in Solids*, A.V. Chadwick and M. Terenzi, eds., Plenum, New York (1986).
12. K.O. Jensen, this volume.
13. C. Ronchi, J. Nucl. Mat. **96**, 314 (1981).
14. V.K. Kumikov and Kh.B. Khokonov, J. Appl. Phys. **54**, 1346 (1983).
15. B. Lebech, Neutron News **1**, 7 (1990).
16. J. Skov Pedersen et al., in preparation.
17. O. Glatter, J. Appl. Cryst. **10**, 415 (1977).
18. K.O. Jensen, M. Eldrup, B.N. Singh and M. Victoria, J. Phys. F: Met. Phys. **18**, 1069 (1988).
19. B. Viswanathan et al., this volume.
20. M. Eldrup and K.O. Jensen, Phys. Stat. Sol. **a102**, 145 (1987).
21. R. Williamson, Harwell Report, AERE-R 11273 (1985).

^{83}KR MÖSSBAUER SPECTROSCOPY AND INERT GAS INCLUSIONS IN ALUMINIUM AND SILICON

M.J.W. Greuter,[1] G.L. Zhang,[1][*] L. Niesen,[1] F.J.M. Buters,[2] and A. van Veen[2]

[1] *Materials Science Centre, Laboratorium voor Algemene Natuurkunde, Westersingel 34, 9718 CM Groningen, The Netherlands*

[2] *Interfakultair Reactor Instituut, Technical University Delft, Mekelweg 15, 2629 JB Delft, The Netherlands*

ABSTRACT

^{83}Kr Mössbauer spectroscopy has been used to obtain information about inert gas precipitates in Al. Krypton inclusions in Si are studied with TEM, RBS and X-ray diffraction.

1. Introduction

This paper consists of three parts. First, the technique of Mössbauer spectroscopy and its relevance for the study of inclusions of noble gases in solids is discussed. The second part gives a summary of results from Mössbauer spectroscopy on the 9.4 keV transition in ^{83}Kr, applied to the study of Kr bubbles in Al. The last part is devoted to the study of Kr inclusions in Si, using a variety of techniques.

2. Mössbauer Spectroscopy

The Mössbauer effect, resonant recoilless emission and absorption of γ-rays, provides a means of probing the local environment of particular atoms in a solid. Detailed information can be obtained on the electronic and geometric structure as deduced from the hyperfine interactions, detected by Mössbauer spectroscopy as a shift or a splitting of the nuclear energy levels. The quantities that determine the position of these levels are the isomer shift, the magnetic hyperfine splitting and the electric quadrupole splitting. The second quantity is only present in a ferromagnetic or paramagnetic environment and will not be discussed here.

In a Mössbauer experiment the nuclear energy levels of an atom emitting a γ-ray (source) are compared with those of an atom absorbing the γ-ray (absorber). In most cases the nuclear energy levels in the source are not split (single line source). The source energy can be varied by moving it with respect to the absorber, producing a Doppler shift

$$\Delta E = E_0 \, v/c \tag{1}$$

where E_0 is the energy difference between nuclear excited state and ground state, v is the relative velocity between source and absorber and c is the velocity of light. Resonance

---

* Permanent address: Institute of Nuclear Research, Academia Sinica, Shanghai, People's Republic of China

Fundamental Aspects of Inert Gases in Solids
Edited by S.E. Donnelly and J.H. Evans, Plenum Press, New York, 1991

absorption occurs if the photon emitted by the source matches a transition in the absorber nucleus. This produces a typical Mössbauer spectrum consisting of dips in the gamma ray intensity behind the absorber.

The isomer shift S is the shift of the centre of gravity of the Mössbauer spectrum. It is proportional to the difference in total electron density at the Mössbauer nucleus in the source, $\rho_s(0)$, and in the absorber, $\rho_a(0)$:

$$S = \alpha_0 \, [\rho_a(0) - \rho_s(0)] \tag{2}$$

where α_0 is proportional to the difference in size of the nucleus in the excited state and in the ground state. This quantity monitors changes in the atomic density (or pressure) of the solid, although the precise relation can only be found via experimental calibration or electronic structure calculations.

If the Mössbauer nucleus is in a site with point group symmetry less than cubic, it will experience an electric field gradient with largest component $V_{zz} = \partial^2 V/\partial z^2$. The interaction with the nuclear quadrupole moment Q gives rise to the electrical quadrupole interaction that within the nuclear manifold with spin I can be expressed as

$$H_Q = eQV_{zz} \, [3I_z^2 - I(I+1) + \eta(I_x^2 - I_y^2)]/4I(2I-1) \tag{3}$$

This interaction leads to a splitting of the nuclear levels. If the site has a threefold or higher axis of symmetry, the asymmetry parameter is zero. In this case the energy levels are given by

$$E_m = E_{-m} = eQV_{zz} \, [3m^2 - I(I+1)]/4I(2I-1) \tag{4}$$

where $m = -I, ...+I$

Differences in E_m between excited and ground state, coupled with matrix elements for gamma emission and absorption, yield the allowed transition energies and amplitudes. Hence the Mössbauer spectrum is a fingerprint for the symmetry of the nearest environment of the Mössbauer atom.

The probability for zero-phonon emission or absorption of γ-radiation is known as the recoilless fraction. It is given by

$$f = \exp(-k^2 <x^2>) \tag{5}$$

where k is the magnitude of the wave vector of the γ-ray and $<x^2>$ is the mean square displacement of the emitting or absorbing nucleus along the γ-ray direction. Various moments of the phonon density of states can be obtained by measuring the area under the Mössbauer spectrum as a function of temperature of source or absorber. A more crude approach is to fit these measurements to a Debye model, with the Debye temperature as parameter. This works well as long as values for the different moments of the phonon density of states are not too different.

As a special case we will treat the 9.4 keV transition in ^{83}Kr. The decay scheme is given in Fig 1. The excited state has a half life of 147 ns, making it very suitable for Mössbauer measurements. Still, the isotope has been used only rarely, because of the limited number of solid materials in which Kr can be incorporated. The spins of the excited state and ground state are 7/2 and 9/2 respectively, leading to 11 lines in the case of an axial quadrupole interaction. The quadrupole moment Q for the ground state is +0.260(7) barn [1] while the ratio of quadrupole moments for excited and ground state is given by Q*/Q = 1.958 [2]. Three parent isotopes can be used to populate the excited state, but only one of them, ^{83}Rb, has a convenient half life (86 d).

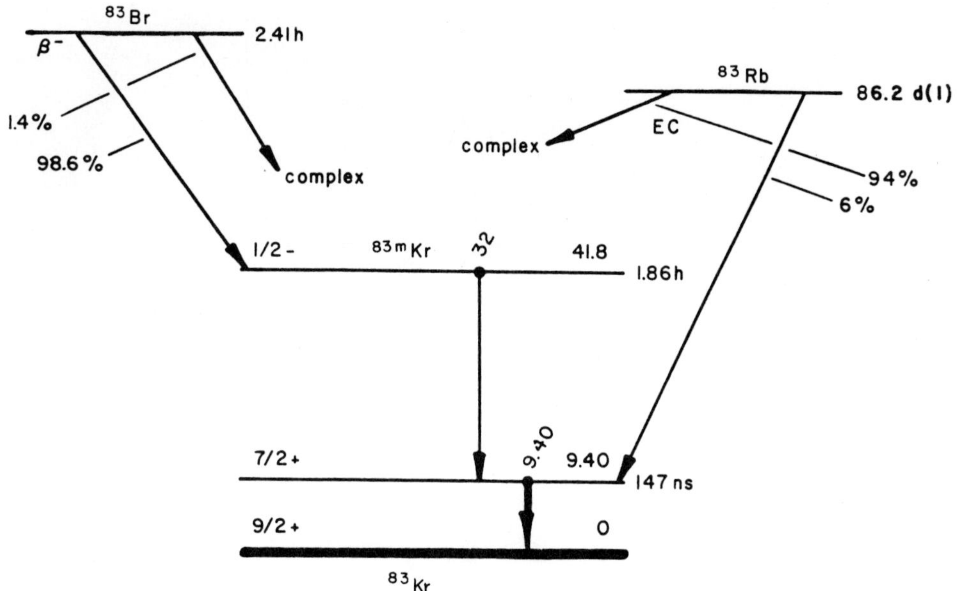

Fig. 1 Decay scheme of ^{83}Kr.

3. ^{83}Kr Inclusions in Al

^{83}Kr was implanted at room temperature in Al at an energy of 110 keV to a total dose of 1.7 10^{16} Kr/cm^2, corresponding to a maximum atomic concentration of $\approx$ 4%. TEM experiments show that under these conditions small solid precipitates are formed which are epitaxially aligned with the Al matrix [3-5]. The molar volume derived from the extra Kr reflections is 22.8 cm^3, i.e. 15% lower than the molar volume at 0 K and 1 bar. The corresponding pressure is 2.1 GPa at 300 K [6] and 1.1 GPa at 4.2 K [7]. For the Mössbauer experiments [8] a total area of 70 cm^2 was implanted at both sides, from which an absorber of 1.5 10^{18} Kr/cm^2 was made.

As a source we used 0.5 mCi of ^{83}RbCl, produced via the ^{85}Rb(p,3n)^{83}Sr reaction. A target of 750 mg cm^{-2} natural RbCO$_3$ was irradiated with 45 MeV protons to a total dose of 60 μAh, using the KVI cyclotron in Groningen. The resulting ^{83}Sr activity was chemically separated and allowed to decay to ^{83}Rb (86 d), after which a thin ^{83}RbCl source was made. Mössbauer spectroscopy was performed in transmission geometry, using a Si(Li) detector. During the measurements the source was always kept as 4.2 K, while the absorber temperature could be varied from 4.2 to 230 K. In addition measurements were performed at 4.2 K on a 2.4 mg cm^{-2} solid Kr layer adsorbed on the outer side of the Be window of the liquid helium cryostat.

The Mössbauer spectrum obtained on the as-implanted sample is displayed in Fig. 2. It is fitted with two components: a single line and a component split by quadrupole interaction. It was assumed that the electric field gradient V_{zz} is axially symmetric. Consequently three position parameters are necessary to fit the whole spectrum, the isomer shifts of both components and the quadrupole coupling $\Delta = eQV_{zz}$. For the single line we obtained S = +0.029(5) mm s^{-1}, while for the other component we found S = +0.017(7) mm s^{-1} and Δ = +1.83(6) mm s^{-1}.

Some spectra obtained at 4.2 K after vacuum annealing for 20 min at various temperatures are also displayed in Fig. 2. The line widths did not change significantly on annealing and thereafter were kept constant at their average values 0.49(5) mm s^{-1} (single line) and 0.69 mm s^{-1} (quadrupole component).

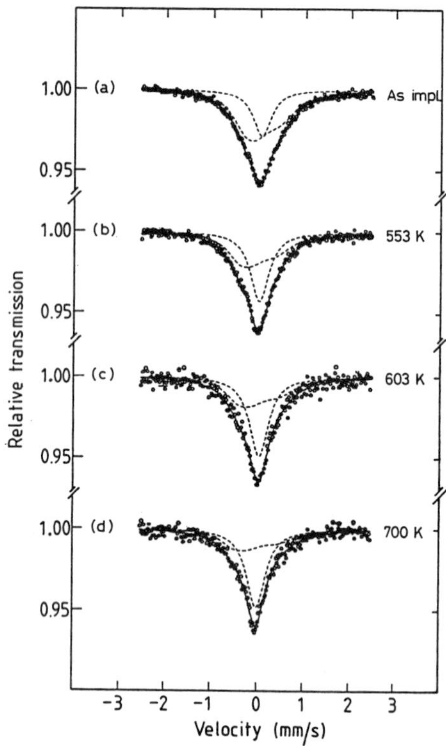

Fig. 2 Mössbauer spectra of ^{83}Kr implanted in Al, taken at 4.2 K with a source of ^{83}RbCl. Dose 1.7 10^{16} Kr/cm^2, implantation energy 110 keV. Spectra were taken directly after implantation (top) and after annealing during 20 min at the indicated temperatures.

The original analysis of the data implicitly assumed that the ^{83}RbCl source gives rise to a single line, as expected for a cubic environment. However, later experiments have shown that this is not the case. To demonstrate this, several spectra are compared in Fig. 3. Fig. 3(a) shows the spectrum taken on solid Kr, (b) shows the spectrum taken on the ^{83}Kr<u>Al</u> absorber which was annealed at 750 K for two hours. Since there is hardly any difference between the two spectra, this strongly suggests that the observed quadrupole splitting is not caused by the absorbers, but is a property of the source. In order to obtain a single line source we made a ^{83}RbHF$_2$ and a ^{83}RbI source, which were measured with the ^{83}Kr<u>Al</u> absorber. Whereas the spectra taken with ^{83}RbHF$_2$ show very broad lines, the ^{83}RbI spectra show a narrowing of the line. As a next step we dehydrated the ^{83}RbI source at 100 °C for half an hour under H$_2$ flow, which narrows the line even more. These spectra are shown in Fig. 3(c) and (d).

The quadrupole component in ionic ^{83}Rb Mössbauer sources is probably caused by electronic after-effects in the decay of ^{83}Rb to ^{83}Kr and/or a disturbance of the cubic symmetry by hydration. We can minimize this effect by choosing less electro-negative ions and by dehydration of the source.

Another conclusion must be that the ^{83}Kr<u>Al</u> absorber yields nearly a single line spectrum after annealing at 700 K, indicating that the Kr bubbles have grown to such a size that the fraction of interface atoms is negligible with respect to the bulk fraction. Re-analysis of the as-implanted spectrum then yields about equal intensities of the bulk and interface component, which corresponds to an average size of 3.5 nm for the Kr precipitates. The original analysis yielded 1.5 nm, in better agreement with the TEM estimates. The present result compares well with the X-ray result of Andersen *et al.*[9] on an 8-10% Kr sample. We plan to repeat these measurements with a better source in the near future. Also, we will use implanta-

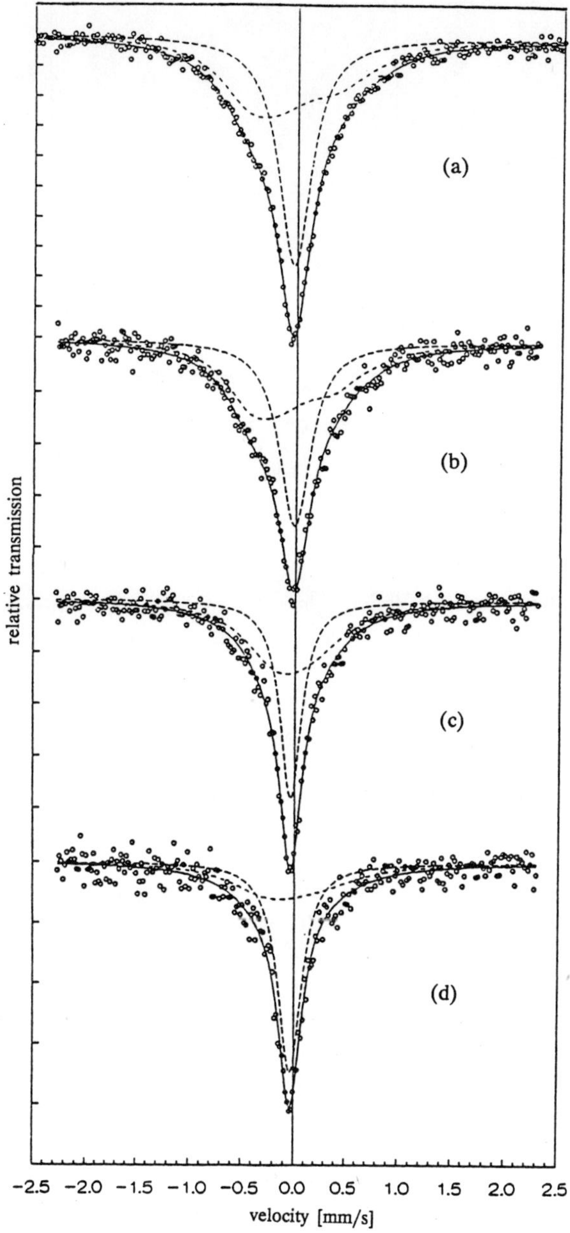

Fig. 3 Mössbauer spectra taken at 4.2 K; (a) and (b) were taken with an ^{83}RbCl source with a solid Kr and ^{83}Kr$\underline{Al}$ absorber respectively; (c) and (d) were taken with an ^{83}RbI source and an ^{83}Kr$\underline{Al}$ absorber. The ^{83}RbI source of spectrum (d) was dried at 100 °C for 30 min under H_2-flow.

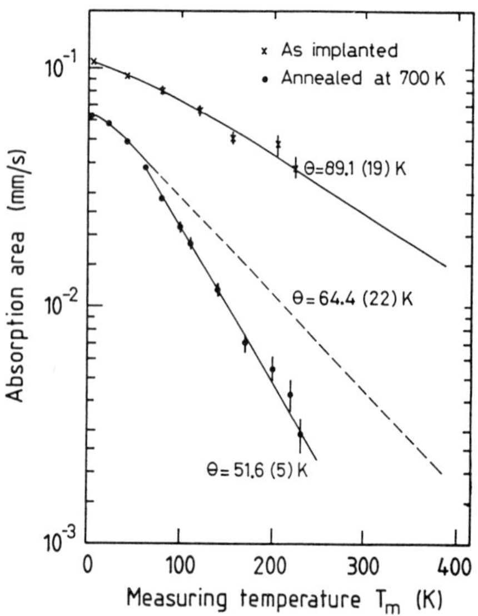

Fig. 4   Total Mössbauer absorption area vs. temperature of measurement. Fitted simple Debye model curves are also shown with the corresponding characteristic temperature (Θ) values.

tion of ^{83}Rb in such a sample, in order to investigate the interface directly.

We note that the spectra do not show evidence for a single-line component with $S = 0.68(3)$ mm/s, which was ascribed to substitutional Rb(Kr) atoms in Al [10].

Using the measured quadrupole interaction we derive $V_{zz} = +2.4 \ 10^{17}$ V cm^{-2} for the interface EFG. The sign is in agreement with that for a metal-vacuum interface [11]. The value for V_{zz} should be interpreted as an average over terrace sites and various edge sites. Although the spectra have insufficient resolution to observe these sites directly, their existence is consistent with the relatively large line width of the quadrupole component.

Spectra were taken as a function of temperature after implantation and after each annealing step. No systematic changes in the intensity ratio of the single line to quadrupole component were found, indicating that the recoilless fraction of both components was nearly the same. The total absorption area as a function of T is shown in Fig. 4 for various annealing temperatures. Except for the measurements at the highest annealing temperature all data could be fitted very well assuming a simple Debye model. The resulting characteristic temperatures Θ are displayed as function of T_A in Fig. 5. A large decrease is observed from 89(2) K after implantation to 63.7(1.0) K after $T_A = 648$ K. The last value is close to the value 61(2) K we measured from the absorption area of a solid Kr layer at 4.2 K and also in agreement with an earlier evaluation of low-pressure recoilless fraction data on solid Kr [12]. Essentially the same value, Θ = 64.4(2.2) K, is obtained from the three lowest points (T < 60 K) after annealing at 700 K. In contrast, in this case the region 60 < T < 230 K could be fitted quite well with a much lower value: Θ = 51.6(5) K; see Fig. 4. This is practically the value of Θ at the melting point at ambient pressure derived from a thermodynamic analysis [13].

In all cases the area showed a smooth dependence on T up to the highest temperatures. In particular, no drop is observed at 115.4 K, the melting temperature of Kr at 1 bar. We derive from the data that at least 90% of the Kr atoms remain in the solid phase up to 230 K, even after annealing for 20 min at 700 K.

In Fig. 5 also the isomer shift of the single line at 4.2 K is plotted as a function of T_A. A gradual decrease of 0.061(6) mm s^{-1} is observed towards the measured value of the isomer shift for solid Kr at 1 bar: $S = -0.032(2)$ mm s^{-1}. Using the known value of $\Delta \langle r^2 \rangle$ for this

236

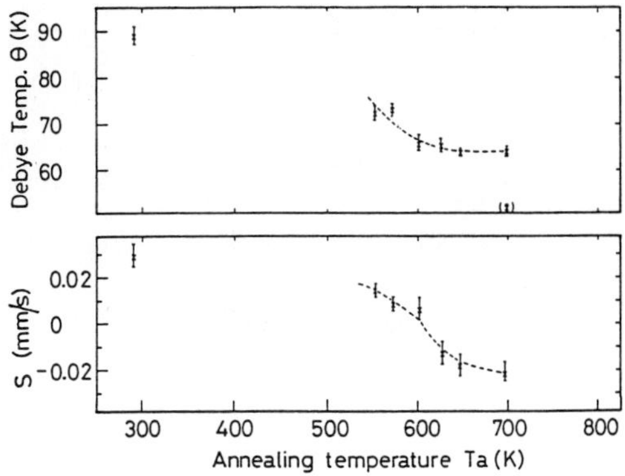

Fig. 5 Behaviour as a function of annealing temperature of characteristic Debye temperature and isomer shift vs. ^{83}RbCl.

transition [10], this corresponds to a decrease in the electron density at the nucleus of 0.51(9) a_0^{-3}. The isomer shift of the quadrupole component showed the same trend, although with larger errors.

The behaviour of the various Mössbauer parameters on annealing can be understood on the basis of two phenomena: the size of the precipitates increases and the density in the precipitates decreases. Unlike the TEM experiments, distinct irreversible changes are observed already after annealing at 553 K, which is 59% of the melting temperature of Al. At a comparable Kr dose the TEM experiments observe melting of the precipitates around 620 K, i.e. 66% of the melting temperature T_m. The situation is different from that in the fcc metals Cu and Ni where melting [14] and a decrease of the Kr density [15] is observed around 40% of the melting point of the host metal. In our case the first stage of bubble growth is probably governed by trapping of thermally created vacancies. This mechanism should become operative around 0.5 T_m, i.e. 470 K for Al. In this way the internal pressure in the bubble can reach equilibrium with the surface tension in the Al interface. For a spherical bubble this would mean: $p = 2\gamma/r$, where γ is the surface energy density (≈ 1 Nm^{-1} for Al) and r the bubble radius. Apart from the numerical constant this equation also holds for other shapes. Combination with the equation of state of Kr [6] leads to the conclusion that in this mechanical equilibrium regime an increase in the temperature leads to an expansion of the bubble and a corresponding decrease of the bubble pressure. This is a reversible process as long as the total number of Kr atoms in the bubble remains constant. However, at the same time bubbles will grow by agglomeration of Kr atoms. The corresponding increase in size leads to a lower equilibrium pressure and density in the bubble.

Cooling after annealing then leads first to increase in pressure, down to the temperature where individual vacancies can no longer escape from the precipitate. Below this vacancy freezing temperature T_v the density in the bubble is constant, apart from a small effect from the host lattice. Our recoilless fraction measurements are performed in this regime. This explains why the Debye model yields such a good description of these measurements: at constant molar volume the phonon spectrum will change only slightly with increasing temperature and the moments of the phonon spectrum will stay practically constant.

For the case of Kr the Lindemann criterion for melting can be written as $<x^2>^{1/2}_M = 0.115$ R_{NN} where $<x^2>^{1/2}_M$ is the RMS amplitude of the Kr atoms and $R_{NN} = 0.412$ nm is the nearest neighbour distance, both taken at the melting point T_m [13]. For the as-implanted sample we obtain $T_m \approx 680$ K by extrapolation of our recoilless fraction data. In practice the start of the melting of bubbles is observed by TEM at a somewhat lower temperature (620 K), because some bubble growth and decrease of Kr density takes place already before the melting point is

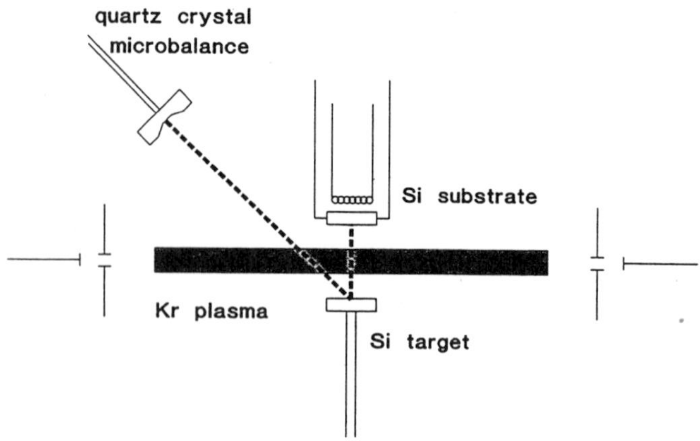

Fig. 6 Schematic drawing of plasma sputter deposition apparatus.

reached. The Lindemann criterion combined with our data predicts that after annealing at 700 K melting starts at ≈ 230 K, i.e. at the highest measurement temperature. This is still twice the melting temperature at ambient pressure. We conclude that this criterion, which works well in the case of bulk rare gases, yields estimates consistent with the experimental data.

4.   Krypton in Silicon

It is interesting to find out if the incorporation of noble gases in silicon shows the same features as in metals, in particular the formation of highly pressurized bubbles within the bulk of the crystallites. Apart from the purely scientific interest, such investigations have technological relevance, because noble gases will be incorporated in Si during plasma etching and plasma deposition procedures. Compared with metals, Si shows the complication of amorphization during sputtering/implantation and subsequent recrystallization at temperatures around 600°C. This may have a large effect on the behaviour of noble gases in this material.

Although our original aim was to study Kr precipitates by Mössbauer spectroscopy, we have as yet no results. Because we need a ^{83}Kr dose of $\geq 10^{18}$ cm^{-2}, we have to produce very thick layers (≈ 50 μm) or layers enriched in ^{83}Kr. We started the production of such layers and characterized them with different techniques.

Samples were made with a High Dose Desorption Spectrometer at the IRI in Delft. In this apparatus it is possible to ionize and implant gases at energies up to 3 keV and to deposit layers of the material on a substrate (Fig. 6). Electrons are emitted from filaments and accelerated by plates while focused by a magnetic field. The gas to be implanted can be introduced at a certain pressure. The electrons have an energy of 100 - 300 eV and ionise the gas. Substrate and target, both single crystal silicon, are at negative potential with respect to the plasma. Noble gas ions from the plasma and neutrals from the target are deposited on the substrate. The thickness of the sputtered layer can be monitored with a quartz crystal microbalance.

First a thick layer of 8 μm Si and Kr was deposited on a Si<100> wafer at a temperature of 363°C and an implantation energy of 200 eV. Several techniques were used to characterize this sample. Rutherford backscattering (RBS) using 1 MeV He$^+$ was applied to the as-implanted sample. Fig. 7 shows that the Kr concentration is uniform through the layer at 3.5 at%. Also a minor pollution of Ta (0.3%) can be seen, caused by sputtering of the sampleholder. In Fig. 8 an X-ray diffraction spectrum is shown, using the Cu-K$_\alpha$ line. Except from a strong Si <400> reflection from the substrate (outside the region shown) the spectrum shows only an amorphous hump from the deposited layer. In addition a forbidden <200> reflection is observed, probably related to strain in the substrate.

238

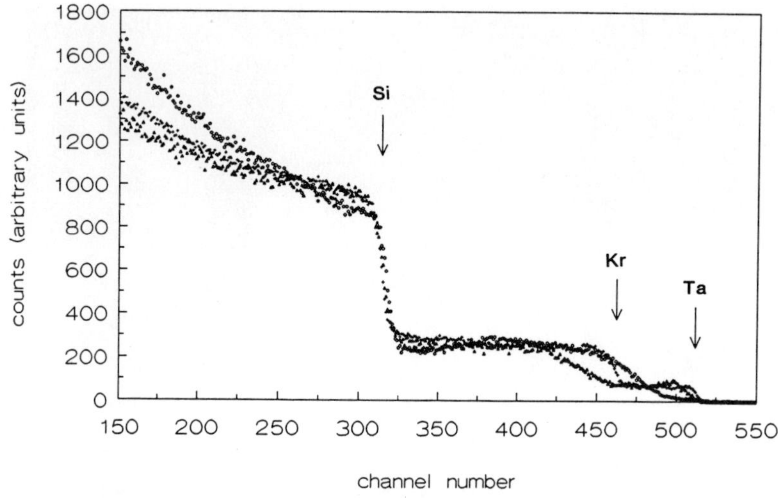

Fig. 7 Rutherford backscattering spectrum using 1 MeV He+ on an 8 μm Kr_Si_ layer deposited on a Si <100> wafer.
+: as implanted layer made at a temperature of 363 °C
Δ: layer annealed for 7.5 hours at 575 °C and 30 min at 900 °C
O: layer annealed for 30 min at 1200 °C.

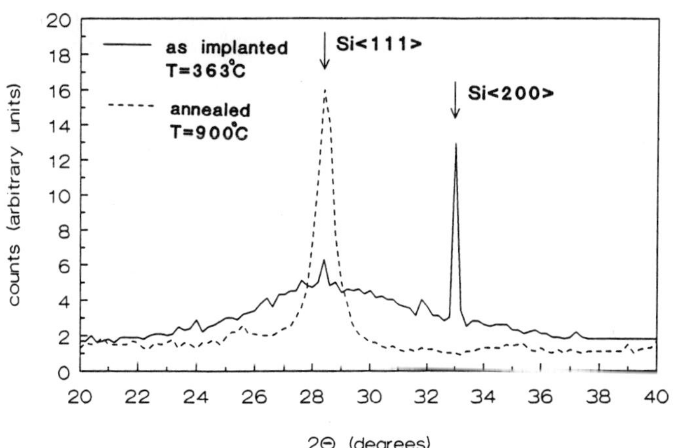

Fig. 8 X-ray diffraction spectrum using the Cu-K$_\alpha$ line taken on an 8μm thick layer annealed at 900 °C.

We annealed this sample for $7\frac{1}{2}$ hours at 575°C and for 30 minutes at 900°C. The RBS spectrum shows that the Kr concentration dropped to 2.9% and that the Kr diffused out of a thin surface layer (approximately 150 nm). The X-ray spectrum now shows all the peaks of Si, so recrystallisation of the amorphous layer has taken place. The forbidden <200> reflection has disappeared, implying that the annealing decreased the strain in the substrate. The grain size can be deduced to be ≈ 12 nm. We also used TEM (40 keV) on this sample (Fig. 9). Very large bubbles with a diameter 600 nm can be observed, still containing Kr. Also bubbles with approximate diameter 100 nm are detectable in the layer. The layer consists of microcrystals with an approximate size of 10 nm, in agreement with the X-ray estimate. The diffraction pattern shows the polycrystalline structure of the layer. We also used SEM to investigate the layer and the surface. Large blisters of 50 μm were observed, some of which extend to the interface between layer and substrate.

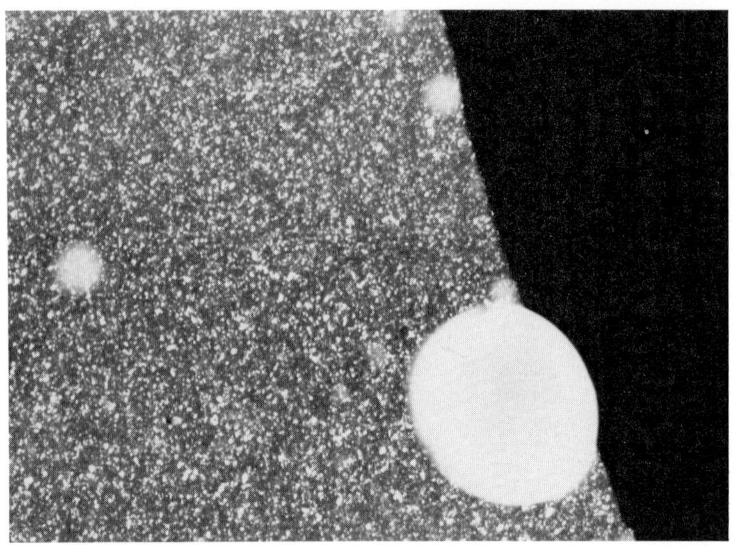

Fig. 9   Dark field TEM picture of the Kr$\underline{Si}$ sample annealed at 900 °C. The diameter of the large bubble is $\approx$ 600 nm.

After this we annealed the sample at 1200°C for 30 minutes. The SEM pictures show that the layer consists of large cavities and has completely deteriorated. However, the RBS spectrum shows that the Kr concentration did not drop significantly. We conclude that most of the Kr is captured inside Si grains with diameters of about 10 nm. The small grain size is probably caused by the fact that epitaxial regrowth from the substrate is inhibited by the presence of too many krypton atoms at the interface between the regrown layer and the sputtered layer [16]. Consequently recrystallization starts inside the layer at many different seeds, leading to fine-grained material.

In an attempt to produce monocrystalline or at least large grain polycrystalline material, a relatively thin (100 nm) layer was grown on a Si<100> wafer at an elevated temperature of 640°C and an implantation energy of 200 eV. The Kr-part of the RBS spectrum of this sample is shown in Fig. 10. The Kr content varies from a maximum of 3.5 at% at a depth of 250 nm to a minimum of 1.3 at% at a depth of 50 nm. This variation is caused by instabilities of the system at the start of the deposition process. The He$^+$ beam was aligned along a <100> and <110> axis of the substrate and also along a 'random' direction. No channelling effect is observed, which means that the layer is not monocrystalline. X-ray diffraction on this layer was performed with the incident beam 2° out of the plane of the sample, varying the detector angle. The X-ray spectrum shows, apart from a Si<311> reflection at $2\Theta = 56.1°$, an additional peak at $2\Theta = 54.5°$ (Fig. 11). No other reflections were observed. If we assign the second peak to a Kr <311> reflection, it corresponds to a lattice constant of 5.57 Å and a molar volume of 26.0 cm^3/mol. According to the equation of state [6] this corresponds to a pressure of 1.1 GPa at 300 K. The grain size is estimated to be 10 nm from the width of the Si reflection. The second reflection has a comparable width, which is difficult to reconcile with the idea that it is caused by Kr precipitates within the bulk of the Si grains. By rotating the sample in the axial plane we found that the two reflections are visible only in an azimuthal range of $\approx$ 10°. These results demonstrate that there is texture in the layer with some grains accidentally oriented in the right way to produce a reflection. If our assignment of the second peak is correct, it implies that the Kr precipitates are epitaxial with respect to the Si grains.

In an attempt to obtain a higher Kr concentration we made a layer by pulsation of the substrate potential (duty cycle 10% and period 1 ms) at 600 eV and at a temperature of 580°C.

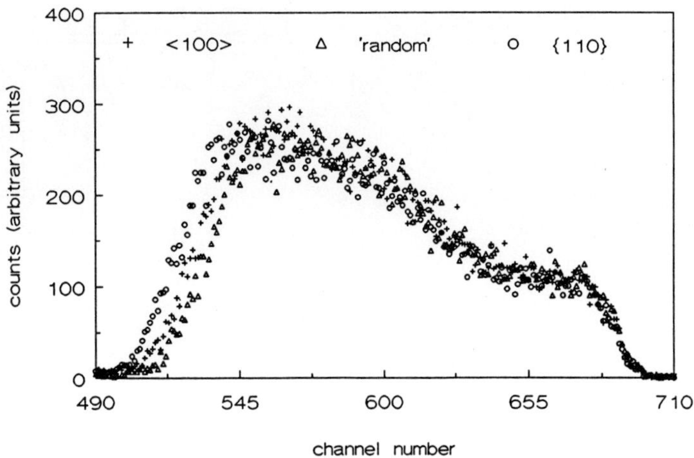

Fig. 10 The Kr-part of the Rutherford backscattering spectrum of a 100 nm Kr<u>Si</u> layer made at 640 °C on a Si <100> wafer, for various angles of incidence of the 1 MeV He$^+$ beam.x

This means that at only 10% of the time Kr atoms are implanted in the layer, at a relatively high energy. The thickness of the layer was 7.3 μm. The variation of the Kr concentration perpendicular to the layer surface was investigated using EDS on a cross-section of the layer. The EDS spots were not bigger than 300 nm. The results are shown in Fig. 11. The maximum concentration is 9% in the middle of the layer. The concentration in the first 0.5 μm agreed within 10% with the RBS result.

5.   Conclusions

^{83}Kr Mössbauer spectroscopy is a suitable tool for the study of inert gas inclusions in solids. It provides information on the local environment of the noble gas atoms. Consequently the method can see both Kr atoms in the bulk and on the interface. Moreover, the vibrational properties of the noble gas atoms can be investigated. The incorporation of Kr in sputtered Si layers was investigated using various techniques. The results show that Kr concentrations up

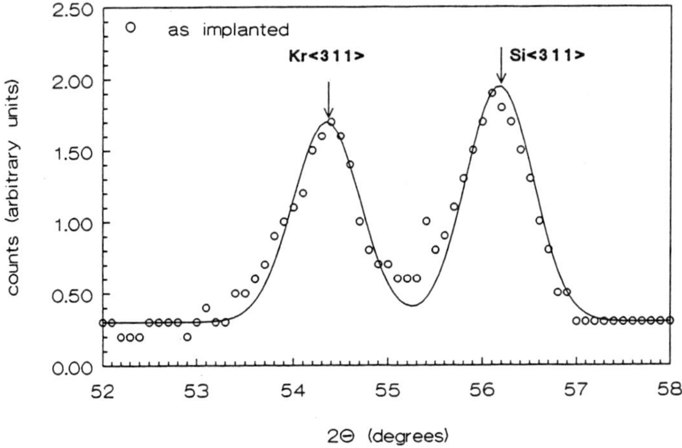

Fig. 11 Part of an X-ray diffraction spectrum on the 100 nm Kr<u>Si</u> layer using the Cu-K$_\alpha$ line. A Si <311> reflection at 2Θ = 56.1° and probably a Kr <311> reflection at 2Θ = 54.5° are shown.

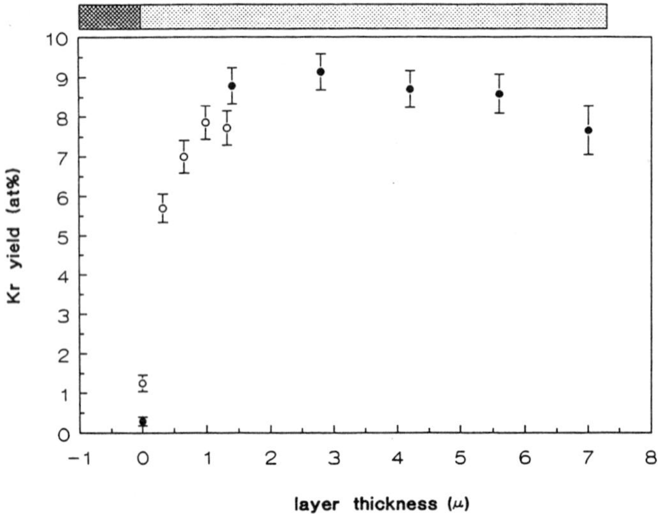

Fig. 12 Kr-concentration perpendicular to the surface of a 7.3 μm thick Kr<u>Si</u> layer deposited at 580 °C by pulsation of the substrate potential (see text). The data were taken with EDS on a cross-section of the layer.

to 9% are possible. The resulting Si grains are only $\approx$ 10 nm in size; still there are clear indications that most Kr atoms are incorporated in the grains, in the form of small bubbles.

ACKNOWLEDGEMENTS

We thank the technical staff of the KVI and F.Th. ten Broek for their help in producing the [83]Rb sources, and J. de Roode for construction and operation of the sputter deposition set up. Dr. P. Bronsveld, Dr. G. Boom, F. van der Horst, U. Nieborg, R. Torfs and Dr. P.J.M. Smulders contributed to various aspects of this work. We also acknowledge useful discussions with Prof.dr. J.Th.M. de Hosson and Dr. E. Johnson.

REFERENCES

1. J.G. Stevens and V.E. Stevens, Mössb. Data Index 1976 (IFI Plenum, New York) 105 (1978).

2. J.H. Holloway, G.J. Schrobilgen, S. Bukshpan, W. Hilbrants and H. de Waard, J. Chem. Phys. **66**, 2627 (1977).

3. R.C. Birtcher and W. Jäger, Nucl. Instr. Meth. **B15**, 435 (1986).

4. R.C. Birtcher and W. Jäger, Ultramicroscopy, **22**, 267 (1987).

5. I. Hashimoto, H. Yorikawa, H. Mitsuya, H. Yamaguchi, K. Takaishi, T. Kikuchi, K. Furuya, E. Yagi and M. Iwaki, J. Nucl. Mater. **149**, 69 (1987).

6. C. Ronchi, J. Nucl. Mater. **96**, 314 (1981).

7. R.K. Crawford and W.B. Daniels, J. Chem. Phys. **55**, 5651 (1971).

8. G.L. Zhang and L. Niesen, J. Phys. Cond. Matter **1**, 1145 (1989).

9. H.H. Andersen, J. Bohr, A. Johansen, E. Johnson, L. Sarholt-Kristensen and V. Surganow, Phys. Rev. Lett. **59**, 1589 (1987).

10. W.J.J. Spijkervet, F. Pleiter and H. de Waard, Hyp. Int. **8**, 173 (1980).

11. B. Lindgren, Hyp. Int. **34**, 217 (1987).

12. B. Kolk, Phys. Lett. **A35**, 85 (1971).

13. R.K. Crawford, Rare Gas Solids vol II, ed. M.L. Klein and J.A. Venables (London: Academic) 771 (1977).

14. J.H. Evans and D.J. Mazey, J. Phys. F.: Met. Phys. **15**, L1 (1985).

15. K.O. Jensen, M. Eldrup, N.J. Pedersen and J.H. Evans, J. Phys. F.: Met. Phys. **18**, 1703 (1988).

16. P. Revesz, M. Wittmer, J. Roth and J.W. Mayer, J. Appl. Phys. **49**, 5199 (1978).

^{133}Xe MÖSSBAUER STUDY OF NEON INCLUSIONS IN MOLYBDENUM

H. Pattyn,[1] P. Hendrickx[1] and S. Bukshpan[2]

[1] Instituut voor Kern-en-Stralingsfysika, University of Leuven
B-3001 Leuven, Belgium

[2] Soreq Nuclear Research Center, Israel

ABSTRACT

Mössbauer studies using microscopic local probes like ^{133}Xe and ^{83}Kr have been undertaken to elucidate the nature and growth of very small inclusions. Besides a small fraction of the probe atoms that reside on substitutional sites—we try to reduce this fraction as much as possible—probe atoms are found inside the bubbles and at interface positions. The inside probes, through the measurement of the ~ Debye temperature, allow us to draw conclusions on the specific volume and, with the EOS on the local pressure. The interface probes give us an idea on the mismatch and the local order. A combination of both fractions allows us to deduce the mean size of the inclusions. XRD measurements have been started to complement the Mössbauer based information. Results on Ne, Kr, Xe and Cs will be presented.

1. Introduction

The study of the precipitation of noble gases (mainly He) in metals has been going on for decades primarily because of the material behaviour problems encountered in the nuclear reactor design. Examples of this interest can be found in the work of Anderman *et al* [1] and of many others. The rapidly growing interest in this field during the recent years however is based on the observation in 1984 by Templier *et al* [2] and by vom Felde *et al* [3] of solid, epitaxial precipitates of heavier noble gases implanted in metals. These pioneering observations had triggered many research teams to study many different aspects of these inclusions. Many authors [4–6], [and others] have deduced the internal pressure of the bubbles from lattice spacing determination, making use of the known equation of state. Cox *et al* [7, 8] followed the bubble size as a function of dose and as a function of annealing temperature. Further insight in the creation and properties of the inclusions was gained from a number of recent publications. Marochov *et al* [9] showed that the density of inclusions is not determined by the damage but by the implanted dose. The bubble geometry was determined in [10]. The dislocation punching mechanism was studied in an experimental [11, 12] as well as a theoretical way [13]. Even studies on defects in bubbles have been reported [14]. All the rare gases have been employed in these investigations. The Ne in Mo system was already studied in 1985 by Luukkainen et al [15].

The first application of the Mössbauer Effect to the present subject of solid noble gas precipitates was published by Zhang and Niesen [16], who measured the size of Kr inclusions in Al and observed the melting of these inclusions. They rightly stressed the advantages of combining Mössbauer measurements with the more classic techniques in elucidating the formation and properties of the bubbles. The location of the Mössbauer probes inside the inclusion or at the interface with the metal shows up in two clearly different components in the spectra. Obtaining spectra at different temperatures allows one to extract the Debye Temperature, Θ_D, of the two sites. Knowledge of the Debye Temperature of the bubble interior is of paramount importance since it gives access - if the Grüneisen constant is known - to valuable information like the molar volume and, through the E.O.S., the pressure. The melting temperature can then be evaluated through the Lindemann criterion.

At the same time we have used the Mössbauer resonance of ^{57}Fe, formed in the decay of ^{57}Co to gain insight in the pressure that exists inside the metal matrix. We therefore implanted the ^{57}Co probes into Al to the same depth where the Kr bubbles were to be formed and used the ^{57}Fe isomer shift to monitor the local pressure inside the aluminium matrix caused by the highly compressed Kr inclusions [17].

In the present work we report on the study of Ne inclusions in Mo by utilising the Mössbauer effect in ^{133}Cs formed in the radioactive decay of ^{133m}Xe. The Mo matrix was selected for several reasons. It has a very high shear modulus, 116 GPa, which should lead to large pressures in the bubble, furthermore the Mössbauer isomer shift scale for ^{133}Xe implanted in Mo is quite large, enabling good resolution, and finally the system of ^{133}Xe implanted in Mo has been thoroughly studied by our research team using the Mössbauer Effect [18,19].

2.   Experimental

Radioactive ^{133}Xe (decaying to the 81 keV Mössbauer level in ^{133}Cs), is an attractive isotope for bubble research using the Mössbauer Effect because first, it has a set of nuclear parameters that make it reasonably suited, secondly, it is commercially available on a regular basis in activities that are large enough to be processed by an isotope separator and to be implanted and thirdly, a single line ^{133}Cs absorber (CsCl) can be prepared easily. It is then straightforward to perform 'source' Mössbauer measurements using the noble gas ^{133m}Xe atoms as local probes for the bubble formation. We introduced the probe atoms in molybdenum by implanting the radioactive ^{133m}Xe ions to a dose of 5 10^{13} atoms/cm^2 with an ion energy of 75 keV. Subsequently the sample was postimplanted with stable Ne at an ion energy of 16 keV, chosen for optimal overlap with the Xe probe distribution, to a dose of 2 10^{16} at/cm^2. One of the problems encountered in our experiments is that after the implantation is finished the Mössbauer Effect measurements showed that an appreciable fraction of ^{133m}Xe resides on substitutional sites in the Mo matrix. This corresponds to a substitional component in the Mössbauer spectra of over 10%. To overcome this problem in future experiments the Mo foil will be heated during the implantation to 200°C which is above the stage III temperature in molybdenum, where vacancies are migrating and, as has been shown in previous studies, at this temperature the substitional fraction diminishes considerably and most of the ^{133m}Xe migrates towards the Ne inclusions.

The absorber, 300 mg ^{133}Cs/cm^2 CsCl, was mounted at the bottom of the liquid He container where its temperature was around 4.2 K. The source was attached to a constant acceleration drive and was mounted in the exchange gas tube, thus allowing its temperature to be set independently from the absorber temperature. The series of spectra collected in this study has been fitted with the fitting program 'Mosaut' [20] that allows the processing of several spectra with a consistent set of parameters.

3.   Results and Discussion

Mössbauer spectra were obtained at source temperatures of 4.2, 20, 30, 40, 50, 60 and 70 K. We fitted the spectra with the lowest number of components which proved to be three: 2 single lines and 1 quadrupole split component. The fitted spectra are plotted in Fig. 1. In the fitting

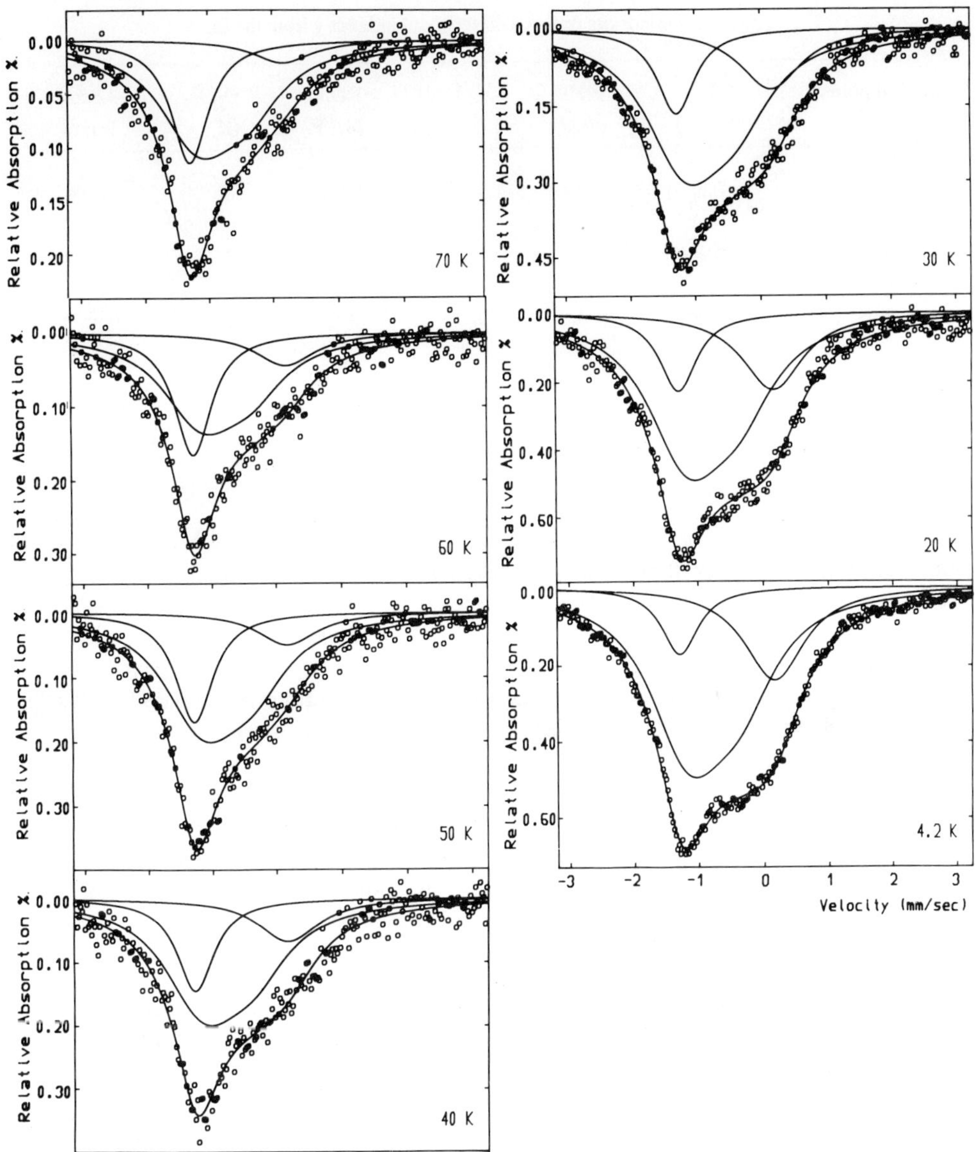

Fig. 1   Mössbauer spectra of ^{20}Ne^{133m}XeMo taken at 7 different source temperatures together with the fitted curves.

procedure the values of the isomer shifts, the quadrupole interaction and the linewidths for each component were kept consistently at the same value throughout all temperatures. We have chosen the asymmetry parameter η of the quadrupole split component to be 0. We will discuss the nature of this component further on and conclude that it relates to the interface, where a distribution of η values around 0 is not unreasonable; moreover, the spectral resolution does not permit to fit it independently. The characteristics of the components are presented in Table 1. The sign of I.S. and Q.I. is that which resulted from the fit, without any change.

In forthcoming work we will present results of measurements on bubbles of other noble gases and show that their spectra look similar to the present ones. The component with an I.S. of -1.27(5) mms^{-1} is well known from the work on low dose ^{133}Xe implantations in Mo

Table 1. Parameters deduced directly and indirectly from the fit.

Component	I.S. (mm/s)	Width (mm/s)	Q.I. (mm/s)	Rel.Int. (4.2 K)	Θ_D (K)	Absol. fract.
Bubble	+0.10(5)	1.21(10)	—	22%	165	18.7%
Interface	−0.90(8)	1.49(12)	−1.91(9)	68%	196	36.8%
Substit	−1.27(5)	0.71(7)	—	10%	262*	2.1%
Invisible						42.4%

*cfr reference [18]

[21, 22]; within the error bars, it has the same I.S. as the substitional component. Its identification as substitutional probe atoms is corroborated by its narrow linewidth, 0.71(7) mms^{-1}, that is close to the natural linewidth - as can be expected for a cubic surrounding where the nearby vacancies are probably cleaned up by the presence of the bubbles. Of prime importance is the component that has an I.S. of +0.10(5) mms^{-1}. This large positive value has never been observed in Mössbauer measurements on ^{133}Cs where the largest till now is +0.05 for Cs_2CO_3. Under the assumption that the I.S. of the latter compound corresponded to the charge state Cs^{+1}, the I.S. calibration for ^{133}Cs has been deduced by Boyle and Perlov [23]. This argument indicates that our component with the most positive I.S. measured up to now has to correspond to a ^{133}Cs atom that has a charge of +1. Since the Mössbauer transition occurs in ^{133}Cs after the parent isotope ^{133}Xe has decayed, the newly formed Cs atom has obviously not succeeded in finding an electron and is left ionised. Similar situations have been observed in many Mössbauer Matrix Isolation measurements. All this leads to the conclusion that our component is the signal of Xe probe atoms trapped inside the neon bubbles, where they are 'matrix isolated'. The considerably large linewidth is due to the small dimensions of the bubble and the subsequent distribution of electric field gradients caused by the neighbourhood of the metal interface to the second and third layer of the Ne inclusion, thereby leading to inhomogeneous broadening of the line.

The third component has an I.S. of −0.90 mm/s and a quadrupole splitting of −1.91 mm/s. This splitting is caused by the interaction of the quadrupole moment of the excited state, −0.22 barn, and an electric field gradient with a V_{zz} value of 2.7 10^{18} V/cm^2. The spectral resolution does not allow a unique sign determination. We assign this component to the Cs probe atoms residing on the molybdenum surface, where a local electric field gradient arises because the cubic symmetry is broken. Such an assignment finds support in the fact that similar values of V_{zz}(latt) were measured at the sites of Na impurities residing on a W surface [24] and Cd on Cu [25]. If we divide the V_{zz} value that was obtained in the present case of Cs on a Mo surface by (1-γ_∞) we obtain a V_{zz}(latt) value of 2.2 10^{16} V/cm^2, which is rather close - taking into account that these are three different metals - to the ones extracted from the above-mentioned references, 0.8 10^{16} for the W surface and 3 10^{16} for the Cu surface. This similarity lends strong support to the view that the quadrupole split component is to be identified as probe atoms sitting at the interface. A similar component was observed in the Mössbauer study of Zhang and Niesen [16] and in the recent P.A.C. study of ^{111}In implanted in Fe, preloaded with Kr inclusions by Decoster *et al* [26]. The large linewidth of this component is plausible since different 'surface' sites are populated giving rise to a different lattice mismatch. The large linewidth may also be indicative of a distribution of electric field gradients. The rather large

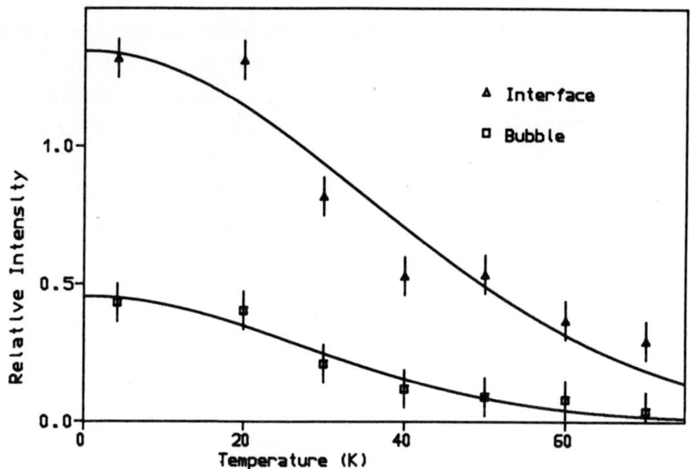

Fig. 2 Plot of the temperature dependence of the relative intensity of the interface and of the bubble Mössbauer component. The fitted curves are represented by the solid lines.

negative value of the I.S. of this component may be explained qualitatively as follows. In ref. [22] the authors have shown that the I.S. of substitutional ^{133}Xe in metals scales with a power of the bulk moduli. The probe atoms of the quadrupole split component are located at an interior wall of the Mo crystal where they experience a compressed inert gas at the other side. The 'mean' bulk modulus at this site, while still proportional to the undisturbed one, will have decreased, leading to an I.S. that is still characteristic for a compressed metallic site but has diminished. This view is supported by our preliminary results on bubbles in W where the I.S. of the Q.S. component scales approximately to that ofsubstitutional.

The temperature dependence of the resonance absorption enables the deduction of the Mössbauer characteristic temperature. The relative intensities of the bubble associated components are displayed against the source temperature in Fig. 2. The usual expression (the low temperature approximation) giving the recoilless fraction as a function of T and Θ_D is fitted to the experimental points, which are proportional to the recoilless fraction of the source at the measuring temperature. The resulting Θ_D of the bubble line and the interface line are given in Table 1. The Θ_D for the substitutional site is taken from reference [21] because the temperature range in the present study was not large enough to extract a reliable value.

Mössbauer studies with a probe like ^{133}Xe that has a small recoilless fraction in source and absorber require that for completeness of the analysis the absolute fractions of the components can be calculated from their relative fractions and from Θ_D, the recoilless fraction of the absorber, the non-resonant background and the thickness of the absorber. The absolute fractions, deduced in this way, are also shown in Table 1. It is striking that the total fraction of the three components which we discussed up to now totals to far less than 1. At this point it is important to realise that the high characteristic Mössbauer temperature for the Cs impurities, indicative for the strengthening of the neon lattice, holds only for a fraction of the Ne bubbles in Mo, the so-called visible fraction. The missing fraction, or the so-called 'invisible' component, amounts to 42.4%. This large component that escapes detection in our spectra, due to its small Θ_D - lower than about 120 K - and correspondingly vanishing recoilless fraction, corresponds most probably to probe atoms inside large bubbles, exhibiting low pressure, or partly filled bubbles. Sites where a probe atom finds itself associated with a small vacancy cluster are far less probable because the total substitutional site population is only 2.1% and we expect that the former and the latter sites should be more or less equally populated. That there be a large fraction of large bubbles at low pressure finds confirmation in other Ne work that has been performed in metals with smaller elastic moduli where liquid bubbles were suggested, e.g. Ne in Al by vom Felde *et al* [3]. Also Luukkainen *et al* [15] found the result of their work on Ne in Mo compatible with about half of the Ne atoms residing outside the small inclusions. The subsequent discussion will stress the small size of the 'visible' inclusions and their high pressure in strong contrast to the 'invisible' bubbles which experience a low pressure and seem to be gaseous or liquid at room temperature. All bubble work up to now associates the latter properties to bubbles that have grown to large sizes. These observations seem to indicate that the as-implanted size distribution has a bimodal size distributions discussed recently by Rest and Birtcher [27], the latter seem to appear only after annealing.

One of the important assets of Mössbauer spectroscopy on a noble gas probe is that information can be deduced on the size of bubbles too small for TEM detection. The absolute fraction of probe atoms inside the inclusions on the one hand and at the interface on the other hand -forgetting the 'invisible' fraction for a moment - gives us the chance to calculate the mean radius of these visible inclusions. On the basis of the simplifying assumption that the visible bubbles are spherical, we obtain a radius R normalised to the lattice constant a:

$R/a = 3.2$ for bcc inclusions

and $R/a = 2.6$ for fcc inclusions

These radii are very small if one thinks of the large fraction of large 'invisible' bubbles. The R/a values of the visible inclusions are not modified if the probes both inside and at the surface have a vanishing small Debye temperature. In the case however that the Q_D of the probes in the surface layer is independent of the radius, the upper limit for the R/a values of the visible inclusions become respectively 10.0 and 8.0. A second asset of Mössbauer measurements is the possibility to deduce the local pressure in the inclusions from Q_D, if the E.O.S. and the Grüneisen coefficient of the noble gas are known. We deduce the Debye (or characteristic Mössbauer) Temperature from the dependence of the relative intensities of the components on the source temperature. The Q_D of the bubble site was given before as 165 K. This value pertains to the ^{133}Cs impurity in the solid Ne and in order to obtain the specific volume of the compressed gas in the bubble it should be translated to the Q_D of Ne. In general this translation requires the knowledge of the ratio of the impurity host-force constant to the host-host force constant and the changes in the phonon spectrum. Since neither are known, we shall try to use the result to obtain some estimate on the force constants ratio by using an independent estimate of the pressure in the bubbles. In the paper by Luukkainen *et al* [15] the density and size of Ne bubbles in Mo was studied for different local concentrations of Ne in Mo. For the local concentration equivalent to our experimental conditions, $C_{max} \approx 7$ at. %, the Ne density was established to be 0.101 at./Å^3. This corresponds to a specific volume of 5.94 cm^3/mole and by way of the E.O.S. of Hama [28] a pressure of about 9 GPa is obtained. If we apply now the Grüneisen relation $Q_D = Q_{D0} \cdot \exp(g_0 DV/V_0)$ [29] where Q_{D0}, V_0 and g_0 are the Debye temperature, the specific volume and the Grüneisen coefficient of solid Ne at normal pressure ($Q_{D0} = 75.1$ K, $V_0 = 13.394$ cm^3/mole, $g_0 = 2.51$) [30] we obtain for the Q_D of the compressed Ne: $Q_D = 303$ K. The formulae which relates the Debye temperature of the impurity to that of the host in the Debye model is given by:

$$\Theta_{D\ host} = \Theta_{D\ impurity} \sqrt{\frac{M_{impurity}}{M_{host}}} \sqrt{\frac{A_{host-host}}{A_{impur-host}}}$$

where $M_{impurity}$ and M_{host} are the respective impurity and host masses and $A_{host-host}$ and $A_{impur-host}$ are the host-host and the impurity-host force constant. By substituting the values of the Debye Temperature and the atomic masses we obtain a force constant ratio of $A_{impur-host}/A_{host-host} = 1.95$.

The result that the $A_{impur-host}$ is larger than $A_{host-host}$ is reasonable. The impurity Cs$^+$ which has the Xe electron configuration and is positively charged will experience Van der Waal's forces with the surrounding Ne atoms which should be larger than the Ne-Ne forces. It is known that the attraction between alkali metal ions and rare gas atoms produces molecular species in the gas phase.

The result obtained here shows that the unique possibilities of the Mössbauer effect for force constant determination when combined with a good estimate of the pressure in the bubble.

If we put specific volume and the Debye temperature in the Lindemann criterion we can estimate the melting temperature to be 233 K. It is unclear how this estimated value relates to the melting of the inclusions if we remember that a large fraction of the Ne atoms are residing at the Ne-Mo interface. From the specific volume inside the inclusions we can now estimate the lattice constant: it has decreased by some 24%. This is in turn gives us the value of the mean radius of the inclusion in A; the most probable (the one that is deduced from the 'visible' populations) radius as discussed earlier is now R = 9(2) Å for bcc or fcc ordering. This very small size of the inclusions, accessible to Mössbauer measurements with ^{133}Xe, suggests—as mentioned before—that a bimodal size distribution may exist, where we have on the one hand the very small 'visible' inclusions and on the other hand the certainly much larger 'invisible' ones. Using the figures given above, we can calculate that there are about 309 atoms in a visible inclusion and that the mean density of inclusions in the depth region of $R_p - \Delta R_p$ to $R_p + \Delta R_p$ is about $6 \ 10^{18}/cm^3$. Similar densities have been measured by TEM measurements.

4.   Conclusions

We have discussed an early application of the Mössbauer Effect to the field ofprecipitates of unsoluble atoms in metals. We observed very small inclusions of Ne in Mo that are at or below the detection limit of the TEM technique which has the widest application in this field. The presence of larger bubbles in our sample could only be deduced through a careful analysis that showed that a component was absent from the Mössbauer spectra. The knowledge of the Debye Temperature let us deduce the specific volume of the small inclusions and, with the help of the E.O.S., the local pressure and the size. The atomic density in the inclusions has been estimated as well as the density of inclusions in the implanted layer. The present study should make clear that microscopic hyperfine interaction technique can be employed in the study of inclusions in metals as interesting and powerful complement to the more classic techniques.

REFERENCES

1. A. Anderman and W.G. German, Phys. Stat. Sol. **30**, 283 (1968).

2. C. Templier, C. Jouen, J-P. Rivière, J. Delafond and J. Grilhé, R C Acad. Sci. Paris **299**, 613 (1984).

3. A. vom Felde, J. Fink, T. Muller-Heinzerling, J. Pfluger, B. Scheerer, G. Linker and D. Kaletta, Phys. Rev. Lett. **53**, 922 (1984).

4. J.H. Evans and D.J. Mazey, J. Phys. F: Met. Phys. **15**, L1 (1985).

5. R. Khanna, A.K. Tyagi, R.V. Nandedkar and G.V.R. Rao, Scr. Metall. **20**, 181 (1986).

6. J.C. Desoyer, C.Templier, J. Delafond and J. Garem, Nucl. Inst. Meth. Phys. Res. **B19/20**, 450 (1987).

7. R.J. Cox, P.J. Goodhew and J.H. Evans, Acta. Metall. **35**, 2497 (1987).

8. R.J. Cox and P.J. Goodhew, Journ. Nucl. Mat. **154**, 233 (1988).

9. N. Marochov and P.J. Goodhew, Journ. Nucl. Mat. **158**, 81 (1988).

10. H.G. Fritsche, D. Bonchev and O. Mekenyan, Phys. Stat. Sol. (b) **148**, K101 (1988).

11. W.G. Wolfer, Philos. Mag. **A58**, 285 (1988).

12. K. Kamada, A. Sagara, H. Kinoshita and H. Takahashi, Rad. Eff. **106**, 219 (1988).

13. G. Knuyt, M. D'Olieslager, L. De Schepper and L.M. Stals, Philos. Mag. **A58**, 243 (1988).

14. S.E. Donnelly, C.J. Rossouw and I.J. Wilson, Rad. Eff. **97**, 265 (1986).

15. A. Luukkainen, J. Keinonen and M. Erola, Phys. Rev. **B32**, 4814 (1985).

16. L. Zhang and L. Niesen, J. Phys.: Condens. Matter. **1**, 1145 (1989).

17. H. Pattyn, P. Hendrickx, J. Odeurs and S. Bukshpan, Mater. Sc. Eng. **A 115**, 103 (1989).

18. E. Verbiest, H. Pattyn and J. Odeurs, Nucl. Instrum. Meth. **182 & 183**, 815 (1981).

19. H. Pattyn and E. Verbiest, to be published.

20. E. Verbiest, Comp. Phys. Commun. **29**, 131 (1983).

21. E. Verbiest, PhD thesis K.U. Leuven (1983).

22. I. Dézsi, H. Pattyn, E. Verbiest and M. Van Rossum. Phys. Rev. **B 39**, 6321 (1989).

23. A.J.F. Boyle and G.J. Perlow, Phys. Rev. **149**, 165 (1966).

24. R.F. Haglund, D. Fick, B. Horn adn E. Koch, Hyp. Int. **30**, 73 (1986).

25. G. Schatz, R. Fink, T. Klas, G. Krausch, R. Platzer, J. Voigt and R. Wesche, Hyp. Int. **49**, 395 (1988).

26. P. Decoster, G. De Doncker and M. Rots, Phys. Rev. **B 41**, 6165 (1990).

27. J. Rest and R.C. Birtcher, Journ. Nucl. Mater. **168**, 312 (1989).

28. J. Hama, Phys. Lett. **105A**, 303 (1984).

29. H. Schlosser, P. Vinet and J. Ferrante, Phys. Rev. **B 40**, 5929 (1989).

30. M.S. Anderson, R.Q. Fugate and C.A. Swenson, Journ. Low Temp. Phys. **10**, 345 (1973).

OVERPRESSURIZED INERT GAS CLUSTERS IN Al AND Si OBSERVED BY EXAFS SPECTROSCOPY

G. Faraci

Dipartimento di Fisica
Universita' di Catania
Corso Italia 57
95129 Catania
Italy

ABSTRACT

I present some recent experimental investigation on inert gas clusters, obtained by ion implantation of Al and Si. The use of the Extended X-ray Absorption Fine Structure (EXAFS) spectroscopy has provided information on the structural and thermodynamical behaviour of the clusters: at room temperature, overpressurized crystalline Ar clusters are detected in the as-implanted samples, while Xe agglomerates are in the solid phase only after annealing. In both samples the evaluation of the Debye-Waller factor has provided an independent determination of the Debye temperature which has been found to be in excellent agreement with the value deduced from the overpressure using the nearest-neighbour distance.

1.   Introduction

In recent years the properties of inert gases in solids have generated considerable theoretical and experimental interest [1-13]. Small agglomerates of rare gases introduced into a metallic or semiconducting solid matrix by ion implantation, have been studied by several experimental techniques such as transmission electron microscopy [4-7], electron energy loss spectroscopy [1], Doppler shift attenuation [10], small angle x-ray scattering [12] and EXAFS spectroscopy [13]. The main interest of these researches is clearly related to the large phenomenology associated with these systems and to the investigation, for instance, of the conditions under which clusters are formed, their phase and the relative lattice parameter, the overpressure acting on the clusters as a function of both the matrix and the size of the bubbles. A general review of inert gas bubbles in metals by Templier is presented elsewhere in this volume and can be consulted for more details. In the present paper I want to summarize the results obtained on overpressurized Ar and Xe clusters formed in an Al or Si matrix by implanting the corresponding ions at high dose. I adopted the EXAFS technique because it is complementary to other spectroscopies in that it provides structural information on nearest-neighbour distance and such parameters as Debye-Waller factor and Debye temperature. For the sake of clarity this technique is briefly summarised but many additional useful details can be found in the review article by Lee et al [14].

2.   The Technique of EXAFS

The EXAFS technique refers to oscillations of the X-ray absorption coefficient which extends up to 1000 eV above the edge of the central atom, according to the expression:

$$\chi(k) = \frac{\mu - \mu_0}{\mu_0} = -\sum_j \frac{N_j}{k r_j^2} \left| f_j(k, \pi) \right| \sin(2k\, r_j + \varphi_j)\, e^{-2\sigma_j^2 k^2} e^{-2 r_j / \lambda_j(k)} \tag{1}$$

This equation is due to the interference of the outgoing photoelectron wave function with the waves backscattered by N_j neighbours located at a radial distance r_j from the excited atom. The back-scattered amplitude f_j is modulated by a phase function which is broadened by the Debye-Waller-like factor $\exp(-2\sigma_j^2 k^2)$ and damped by an exponential term $\exp(-2r_j / \lambda_j(k))$ accounting for the inelastic losses. From the experimental data, which are proportional to $\mu(E)$ if the fluorescence detection is used one should first remove the background and compute $\mu(k)$ as a function of the wave vector k obtained from the experimental threshold E_o as:

$$k = (2m(\hbar\omega - E_o))^{1/2}/\hbar \tag{2}$$

A Fourier transform of the oscillatory part of $\chi(k)$ allows the determination of the different shells of neighbours after the phase correction which can be done by comparison with a model compound. The amplitude A(k) for a single shell of N identical atoms compared to the analogous amplitude of the reference compound by means of the logarithm of their ratio should give a straight line as a function of k^2.

This last plot enables the relative evaluation of the coordination number and of the Debye-Waller factor for the shell under analysis. The technique, here roughly sketched, is then very powerful because it can be applied to disordered materials as well as to liquid or amorphous systems. It allows the easy discrimination between different phases giving valuable information on the bond length, on the number of neighbours and on the atomic vibrations.

3.   Results and Discussion

I compare now some experimental results obtained at room temperature for Ar clusters in Al and Si and at 300 K and 70 K for Xe clusters in Si. The samples have been implanted at high dose (10^{16}-10^{17} at/cm^2) and at an energy of the order of 100 keV [13,15]. The Xe X-ray absorption spectra were recorded in the energy range 4780-5100 eV by monitoring the Xe L_3 fluorescence yield at the Frascati Synchrotron X-ray PULS line, while the Ar spectra were obtained on the K-edge in the range 3150-3500 eV at the LURE in France.

As the implanted ions penetrate into the matrix a few hundred angstroms under the surface, with the eventual formation of clusters, the system under investigation behaves as a very dilute ensemble of rare gas atoms in a thick layer beneath the matrix surface[†]. For this reason it was necessary to maximise the fluorescence yield using a suitable experimental apparatus [13, 15]. Fig. 1 are illustrates typical spectra obtained for our samples. As can be seen in the figure, at all doses and implantation energies, the Ar absorption coefficient shows the same features even though the corresponding jump is lower at higher implanting energy as the penetration of the Ar projectiles is deeper into the matrix while the escape depth of the detected electrons is

---

[†] I point out that after the implantation at 100 keV, the substrate is amorphised by the rare gas ions and can be regrown by an annealing process.

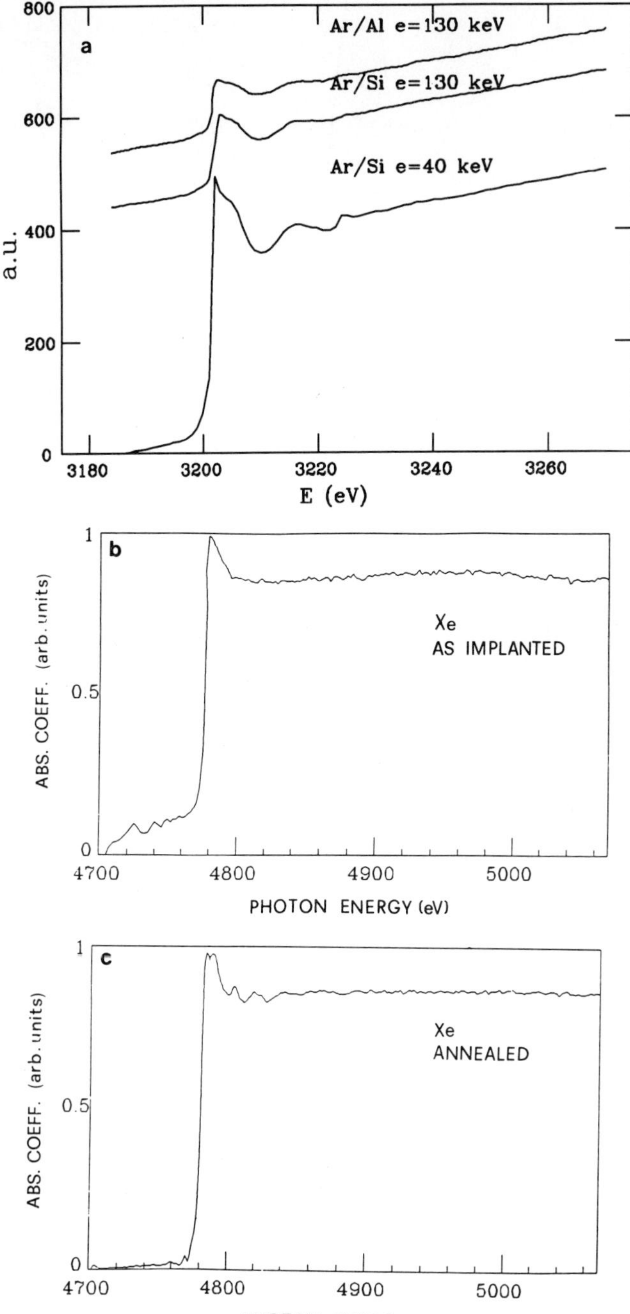

Fig. 1 a) Absorption spectra at room temperature of Ar implanted in Al and Si at various implanting energies. b)Absorption spectrum at room temperature of Xe as implanted in Si. c)Absorption spectrum at 70 K of Xe after annealing.

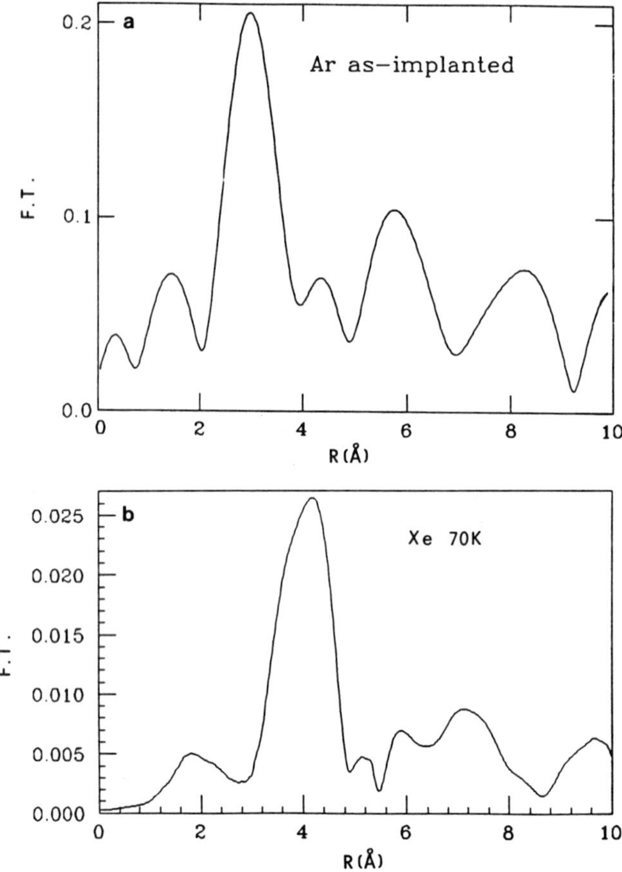

Fig. 2   a) Magnitude of the Fourier transform of the Ar/Al EXAFS spectrum.
b) Magnitude of the Fourier transform of the Xe/Si EXAFS spectrum.

of the order of 300 angstroms. On the contrary, at all doses, the Xe absorption spectra exhibit no EXAFS oscillations on the high energy side of the L3 edge, revealing that the Xe atoms are randomly distributed in the host matrix either in gas-phase bubbles or in atomic form. At doses higher than $5\ 10^{16}$ ions/cm^2 the Xe-implanted samples have shown EXAFS oscillations after annealing (600 °C, 30 min). For this last Xe sample, low temperature (70 K) spectra have been obtained (Fig 1). In Fig 2 are shown the Fourier transforms for two typical Ar and Xe samples, for which we have detected significant EXAFS oscillations. The first peak in the plots provides the rare gas nearest-neighbours (nn) distance in real space after the phase shift correction. This correction can be obtained by comparison with the reference spectra published for Ar and Xe at 5 K by Malzfeldt et al [16]. In the figures there is also clear evidence of the second and third shells of an fcc crystal and this implies that both Ar and Xe clusters are in the crystalline phase. Quantitative determination of the first shell position, coordination numbers and Debye-Waller factors are reported in tables I and II for all the examined samples.

The lattice parameter is always found to be smaller than in the (bulk) solid rare gas, indicating that the clusters are overpressurized because of the host matrix pressure. For Xe the pressure is larger at 300 K than at 70 K, since the mean pressure acting on the condensed bubbles decreases at lower temperature. The values of the pressure however, are to be considered as average values since the spread of the cluster size is of course quite large as confirmed by transmission electron microscope measurements [3]. Further, I note that the

254

Table 1 First coordination shell distance for Ar samples together with the value of the reference solid Ar. The relative coordination numbers, Debye-Waller factors and the corresponding average overpressure acting on the clusters are also reported.

SAMPLES	$R_{F.T.}$ (A)	N/N_0	$\Delta\sigma^2$ (A^2)	P (kbar)
solid Ar	3.26±0.01	1	0	--
Ar/Si	2.84±0.05	.058	−.0010±.0005	44
Ar/Al	2.95±0.05	.208	+.0136±.0005	25

Table 2 First coordination shell distance for Xe samples together with the value of the reference solid Xe. The relative coordination numbers, Debye-Waller factors and the corresponding average overpressure acting on the clusters are also reported.

SAMPLES	$R_{F.T.}$ (A)	N/N_0	$\Delta\sigma^2$ (A^2)	P (kbar)
Xe (cryst)	4.35±0.01	1	0	--
Xe (300 K)	3.94±0.05	.08	....	35
Xe (70 K)	4.23±0.05	.28	+.008 ± .001	5.5

large surface to volume ratio makes it difficult to obtain a good EXAFS signal as the surface atom coordination is about half the bulk coordination. In addition, we should consider the large amount of rare gas in a gas-like phase, and very small aggregates not in crystalline phase.

This is confirmed by the low ratio N/No between the coordination number for the clusters and for the crystal, which would imply (see tables) that only a small percentage of inert gas atoms have bulk coordination.

In the tables are also shown the relative Debye-Waller factors between the Debye-Waller factor of the implanted sample, σ^2, and that of the solid rare gas at 5 K, σ_c^2. To find the absolute value σ^2 (T, D) we have calculated, in the Debye approximation, following the Beni and Platzman equation [17], σ_c^2, using the Debye temperature for the corresponding crystal. The obtained values are:

$$\sigma^2(Ar/Al) = 30.6\ 10^{-3}\ \text{Å}^2$$
$$\sigma^2(Ar/Si) = 16.0\ 10^{-3}\ \text{Å}^2$$
$$\sigma^2(Xe/Si) = 15.5\ 10^{-3}\ \text{Å}^2 \quad (T = 70\ \text{K}).$$

If we use again the same Beni and Platzman equation we find that these last figures correspond to an increased value of the Debye temperature for solid Ar of D=212 K (Ar/Al), D=300 K (Ar/Si), D=80 K (Xe/Si, T=70 K). This implies that the clusters are actually very hard as they should be at high pressure. It should be emphasized that although a very low coordination number, in principle, does not affect the Debye-Waller factor it could render its evaluation quite difficult.

An independent check of the reliability of this result can be obtained by studying the thermodynamic behaviour of the rare gas at high pressure, and comparing the EXAFS Debye temperature for overpressurized clusters with the Debye temperature obtained for the rare gas crystal under the same pressure.

Using the results of Holt and Ross [18] and the equation of ref 19 we obtain the following thermodynamical expression for the Debye temperature of pressurized Ar:

$$D(V) = D_o \, (V_o/V)^{1/2} \, (1-V/V_o) \exp \, (2.2) \qquad\qquad (3)$$

The corresponding values using the contracted lattice parameter are: i) $D = 207$ K (Ar/Al), ii) $D = 300$ K (Ar/Si).

Analogous results are obtained for Xe at low temperature, using the parameters reported by Zisman et al, [20] and by Asaumi [21]: $D = 78$ K (Xe/Si, T=70 K).

The reported values can be compared with the independent results deduced above and in fact there is very good agreement between them and the data published in the literature [2], confirming also the mean pressure acting on the clusters evaluated in the present work and reported in the tables.

4.   Conclusion

In conclusion it has been shown that EXAFS spectroscopy can give valuable and complementary information to that obtainable from transmission electron microscopy.

REFERENCES

1. A. vom Felde, J. Fink, Th Mueller-Heinzerling, J. Pflueger, B. Scheerer, G. Linker and D. Kaletta, Phys. Rev. Lett. **53**, 922 (1984).

2. C.J. Rossouw and S.E. Donnelly, Phys. Rev. Lett. **55**, 2960 (1985).

3. S.E. Donnelly and C.J. Rossouw, Nucl. Instrum. Methods Phys. Res. **B13**, 485 (1986).

4. C. Templier, B. Boubeker, H. Garem, E.L. Math and J.C. Desoyer, Phys. Stat. Sol. **A92**, 511 (1985).

5. C. Templier, H. Garem and J.P. Riviere, Phil. Mag. **A53**, 667 (1986).

6. R.C. Birtcher and W. Jaeger, J. Nucl. Mater. **135**, 274 (1985).

7. M. Wittmer, J. Roth and J.W. Mayer, J. Appl. Phys. **49**, 5207 (1978).

8. R. Revesz, M. Wittmer, J. Roth and J.W. Mayer, J. App. Phys. **49**, 5199 (1978).

9. A.G. Cullis, T. E. Seidel and R.L. Meek, J. Appl. Phys. **49**, 5188 (1978).

10. A. Luukkainen, J. Keinonen and M. Erola, Phys. Rev. **B3**, 4814 (1985).

11. J.C. Desoyer, J. Delafond, C. Templier and H. Garem, Nucl. Instrum. Methods Phys. Res. **B19/20**, 450 (1987).

12. H.G. Haubold, Radiat. Eff. **78**, 385 (1983).

13. G Faraci, A R Pennisi, A Terrasi and S Mobilio, Phys. Rev. **B38**, 13468 (1988).

14. See for instance P.A. Lee, P.H. Citrin, P. Eisenberger and B. M. Kincaid, Rev. Mod. Phys. **53**, 769 (1981).

15. G. Faraci, S. La Rosa, A.R. Pennisi, S. Mobilio and G Tourillon, in press. Phys Rev B.

16. W. Malzfeldt, W. Niemann, P. Rabe and R. Haensel in *Proceedings of the Third International Conference on EXAFS and Near Edge Structure*, eds K.O. Hodgson, B. Heaman and J.E. Penner-Hahn, Springer Proc. in Physics, Vol II (Springer, Berlin, 1984), p445.

17. G. Beni and P.M. Platzman, Phys. Rev. **B14**, 1514 (1976).

18. A.C. Holt and M. Ross, Appl. Phys. Lett. **39**, 894 (1981).

19. L W Finger, R.M. Hazen, G. Zou, H.K. Mao and P.M. Bell, Appl. Phys. Lett. **39**, 892 (1981).

20. A.N. Zisman, I.V. Aleksandrov and S.M. Stishov, Pis'ma Zh. Eksp. Teor. Fiz. **40**, 253 (1984) [JETP Lett. **40**, 6 (1984)].

21. K. Asaumi, Phys. Rev. **B29**, 7026 (1984).

BEHAVIOUR OF KRYPTON ATOMS IMPLANTED INTO ALUMINIUM AS INVESTIGATED BY A CHANNELLING METHOD

E. Yagi

The Institute of Physical and Chemical Research
Wako-shi, Saitama 351-01, Japan

ABSTRACT

The behaviour of Kr atoms in aluminium has been investigated by an ion channelling method on Al specimens implanted at room temperature at 50 keV to various doses from 10^{14} to 10^{16} Kr/cm^2. For the $1\,10^{14}$ Kr/cm^2 -implantation the Kr atoms are distributed over random (R), substitutional (S), tetrahedral (T) and octahedral (O) sites. With increasing dose the fractions of the S-, T-, and O-site occupancies decrease, whereas the fraction of the R-site occupancy increases. For the $1\,10^{16}$ Kr/cm^2 -implantation solid Kr precipitates are formed. The T- and O-site occupancies are interpreted to be a result of the formation of KrV$_4$ and KrV$_6$ complexes due to the trapping of 4 and 6 vacancies by a Kr atom, respectively. The R-site occupancy is ascribed to the Kr atoms associated with larger clusters consisting of more than 6 vacancies (cavities). On annealing of the $1\,10^{15}$ Kr/cm^2- implanted specimen, the T- and O-configurations become unstable between 380 and 470K, and some fraction of Kr atoms become located at different sites (O'-sites). At 593 K the T-, O- and O'-site occupancies disappear, and the S- and R-site occupancies increase. On annealing at 683 K, formation of solid krypton is observed. From these results it is proposed that for the Kr implantation at room temperature, Kr-vacancy complexes such as KrV$_4$, KrV$_6$ and larger ones are formed in the early stage of implantation and they act as nucleation centres for the precipitation of Kr atoms.

1. Introduction

It has been demonstrated that heavier inert gas atoms (Ar, Kr and Xe) implanted into metals at ambient temperatures precipitate into a solid phase (often referred as solid inert gas bubbles) when the implantation dose is high [1,2], and that they have an epitaxial fcc structure in fcc matrices (reviewed in ref. [3]). The growth of such bubbles has been extensively studied mainly by transmission electron microscopy (TEM) and an x-ray diffraction method [3]. These studies have provided important information about the behaviour and structure of the inert gas bubbles. Electron microscopy, however, is sensitive to bubbles greater than a certain minimum size, usually after the interesting initial growth of bubbles has occurred, because of the limit in resolution of a conventional electron microscope. Thus, the TEM observation is more effective to study the bubble growth rather than the initial stage of precipitation, i.e. nucleation. On the other hand, a channelling method can provide direct information on the lattice location of implanted atoms, and is therefore a useful method to study phenomena such as bubble nucleation. Several experiments have been made on the behaviour of Kr atoms implanted into aluminium [4-8]. In the present paper, the results of an investigation by the channelling method on the implantation-dose dependence of the lattice location of Kr atoms in aluminium and their behaviour on annealing are reported, including the previously reported results.

2. Experimental

The specimens used for the channelling experiments were Al single crystal slices about 1 mm thick of 99.999% purity. The largest face of the specimen was nearly parallel to a (110) plane. It was chemically and electrolytically polished, and then annealed in a vacuum of 10^{-6} Torr at 823K for 7h. The presence of a thin oxide layer slightly increases the fraction of dechannelled ions, but it has no serious effect on the determination of lattice locations of Kr atoms. Kr^+- implantation was carried out at room temperature at 50 keV at a dose rate of about $3.2\ 10^{12}$/cm^2.s with various doses from $10^{14}Kr$/cm^2 to $10^{16}Kr$/cm^2. The crystallographic direction for implantation was chosen so as to avoid the implantation under channelling conditions. For the study of implantation-dose dependence of the lattice location of Kr atoms, specimens cut from the same single crystal rod were used. The annealing experiments were made on a specimen which was cut from a different crystal rod and implanted with a dose of $1\ 10^{15}Kr$/cm^2.

The range and the FWHM of the distribution of implanted Kr atoms were estimated to be about 28 nm and 25 nm, respectively. Assuming that the Kr atoms are distributed homogeneously over the thickness of FWHM, the average concentration of Kr atoms would amount to 0.66 at% for the $1\ 10^{15}$ Kr/cm^2 implantation. In order to obtain information on the lattice location of Kr atoms, a channelling angular scan was made at room temperature with respect to the <100>, <110> and <111> axes by means of backscattering with a 1.0 MeV He$^+$ beam accelerated by a tandem type accelerator. The beam was collimated to give a divergence of less than 0.076°. The irradiated area of the specimen was 0.78 mm^2 and the beam current was 1-2 nA. In the channelling angular scan, to minimize the irradiation effect by the analysis-beam, the measurement was started at an incident angle Ψ for the parallel incidence to the channel ($\Psi=0°$).

3. Experimental Results

3.1. Dose dependence

Channelling angular profiles for the specimens implanted with doses of $1\ 10^{14}$, $4\ 10^{14}$, $1\ 10^{15}$ and $1\ 10^{16}$ Kr/cm^2 are shown in Fig. 1. For the former three doses the angular profiles of He ions backscattered by Kr atoms (Kr-angular profiles) exhibited a shallow dip with approximately the same half-width as that of the corresponding Al-dip for the <111> and <100> channels, and a central peak superposed on a shallow dip for the <110> channel. The relative depth of the Kr-dip with respect to the depth of the corresponding Al-dip is smaller for the <100> channel than for the <111> channel. In the <100> Kr-profile for the $4\ 10^{14}Kr$/cm^2 and $1\ 10^{15}Kr$/cm^2 implantations a very small central peak seems to be superposed on a dip.

These results suggest that Kr atoms are distributed over random (R), substitutional (S), tetrahedral (T) and octahedral (O) sites. The T- and O–sites are shadowed by the <111> atomic rows. For the <110> channel they are located at off-centre positions and at the centre of the channel, respectively, and therefore are responsible for the observed central peak in the Kr- angular profile. For the <100> channel, the T-sites are located at the centre of the channel, while the O–sites are shadowed by the <100> atomic rows. Therefore, the T-site occupancy contributes to a central peak and makes the <100> Kr-dip shallower than the <111> Kr-dip. For the $1\ 10^{16}Kr$/cm^2- implantation the backscattering yields from Kr atoms (Kr-yields) were the same independent of the incident angle. This result indicates that almost all Kr atoms are located at R-sites, i.e., in the form of precipitates. TEM observation revealed the presence of solid Kr crystals [9].

The fractions of Kr atoms at various kinds of site were estimated by fitting the calculated angular profiles to the observed ones. The calculation was made by the multi-string model as in a previous paper [7]. The dose dependence of the distribution of Kr atoms is shown in Fig. 2. The number densities of Kr atoms located at various kinds of site were estimated by

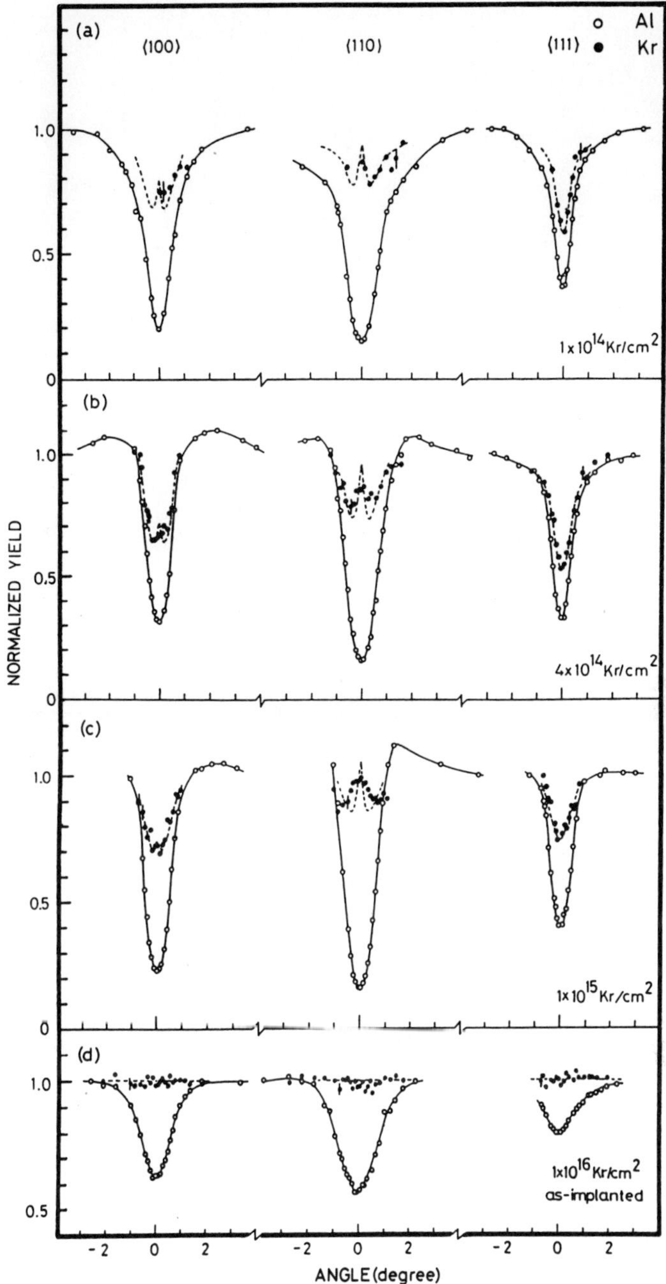

Fig. 1 Channelling angular profiles of backscattering yields of He ions from Al and Kr atoms on the as-implanted Al specimens for various implantation doses. The dotted curves are examples of calculated ones for the distribution with (a)36% at R-, 34% at S-, 15% at T- and 15% at O–sites, (b) 30% at R-, 40% at S-, 10% at T- and 20% at O-sites, (c) 57% at R-, 23% S-, 3% at T- and 17% at O-sites, and (d) 100% at R-sites.

multiplying the implantation dose by the fraction of the corresponding occupancies. Examples of calculated profiles are indicated by dotted curves in Fig 1.

For the 4 10^{15}Kr/cm^2-implantation (the result is not shown in Figs. 1 and 2), the Kr yields exhibited angular profiles similar to those observed for the 10^{14}– 10^{15}Kr/cm^2 implantation,

259

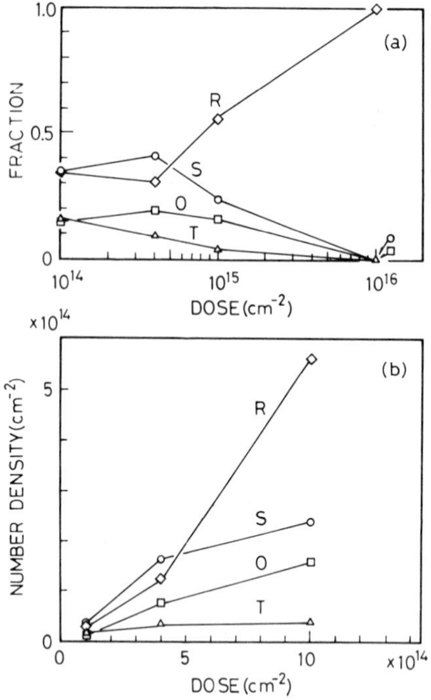

Fig. 2 Dose dependence of the distribution of Kr atoms. (a) The fraction of Kr atoms located at various kinds of site, and (b) their number densities.

i.e. a shallow dip for the <l00> and <111> channels and a central peak for the <110> channel. However, the <100> and <111> dips are slightly narrower (by about 20%) and they have approximately the same relative depth. In addition, the shallow dip observed in the <110> channel disappeared [10]. These results suggest that the fractions of T-, O- and S-site occupancies decreased considerably, and some fraction of Kr atoms came to be located at new sites slightly displaced from O–sites (O*-sites). The magnitude of this displacement is roughly estimated to be 0.03 nm. This configuration change is considered to be due to the clustering of Kr atoms which are located at O–sites.

3.2. Annealing

The changes of angular profiles on annealing at 371, 473, 593, 683 K for 30 min are shown in Fig. 3. These profiles were obtained on another 1 10^{15} Kr/cm^2-implanted specimen at different spots. Fig. 4 shows the values

$$f^{<hkl>} = \frac{1 - \chi_0^{Kr<hkl>}}{1 - \chi_{min}^{Al<hkl>}} \tag{1}$$

ie., the relative depth of the Kr-dip with respect to the Al-dip for an <hkl> channel, as a function of the annealing temperature, where χ_{min}^{Al} is the normalized minimum yield in the Al-dip and χ_0^{Kr} is the normalized Kr-yield for a beam incident parallel to the channel, i.e., at $\Psi=0°$. This fraction has been often used as the measure of the fraction of impurity atoms shadowed by the <hkl> atomic rows of host atoms. If the Kr atoms are distributed over only R-, S-, T and O–sites, it should be that $f^{<111>} > f^{<100>}$.

260

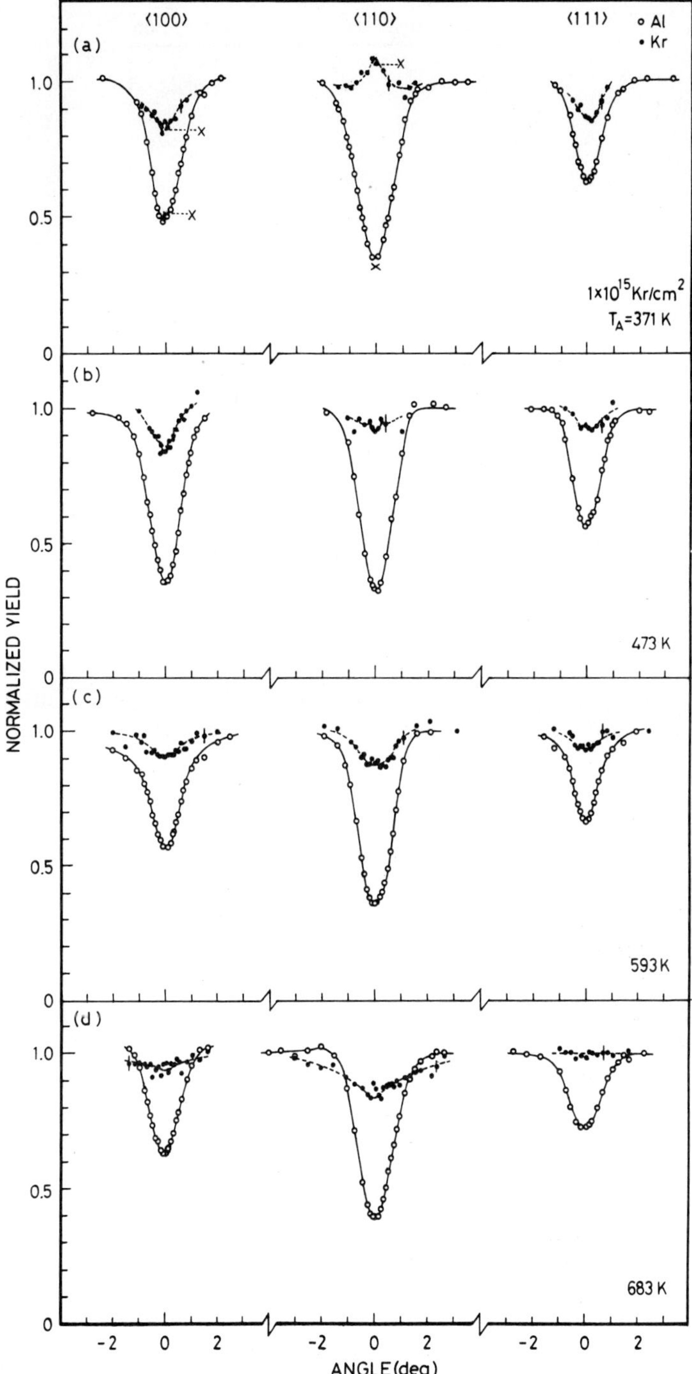

Fig. 3 Channelling angular profiles of backscattering yields of He ions from Al and Kr atoms obtained after annealing at various temperatures (T_A) for 30 min on the 10^{15}Kr/cm^2 -implanted Al specimen. The full curves and dotted curves were drawn only to guide the eye.

Angular profiles were not measured for as-implanted state, but only χ_{min}^{Al} and χ_0^{Kr} were measured for the <100> and <110> channels. They are indicated by cross symbols in Fig.3(a). After annealing at 371 K the angular profiles still exhibited the same features as those for the

261

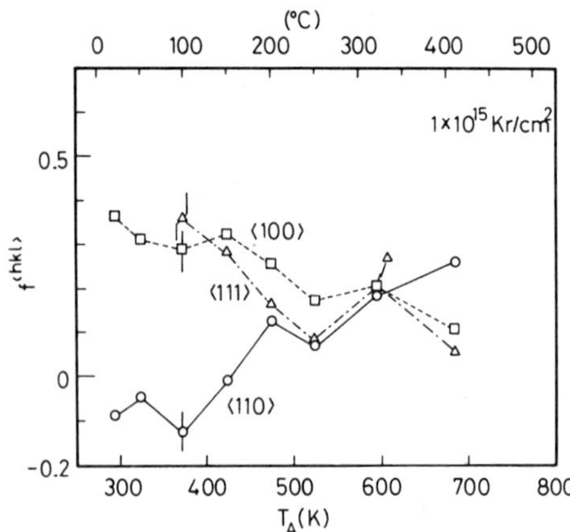

Fig. 4   The change of the fraction $f^{<hkl>}$ (eq. (1)) with the annealing temperature.

as-implanted state described above, indicating that Kr atoms are located at R-, S-, T- and O-sites. The fractions of site occupancies were estimated to be 62-63% at R-, 13-14% at S-, 3-4% at T- and 20–22% at O–sites.

The annealing at 473 K changed the <110> Kr-angular profile markedly; the central peak disappeared and a double peak appeared at $\approx \pm$ (0.2-0.3)°, being superposed on a shallow dip. On the <111> Kr-dip a very small central peak seems to be superposed. This result indicates that a large fraction of the T- and O–site occupancies disappeared while some fraction of Kr atoms came to be located at new sites (O'-sites). These sites are different from the O*-sites. The relative depth $f^{<111>}$ decreased much more than $f^{<100>}$, and it followed that $f^{<111>} < f^{<100>}$ (Fig. 4). The result that the fraction of Kr atoms shadowed by <100> atomic rows, $f^{<100>}$, is larger than those by <111> rows, $f^{<111>}$, seems to suggest the O'-sites to be those shadowed by <100> atomic rows. However, as the observed fine structures in Kr profiles are small, further discussion on the lattice location of the O'-site and the distribution of Kr atoms by profile fitting will not be made here.

Upon annealing at 593 K each of the <100>, <110> and <111> Kr-profiles changed to a simple dip with the same half-width as that of the corresponding Al-dip and the same relative depth of 20%, i.e., $f^{<100>} = f^{<110>} = f^{<111>} = 0.2$. These results indicate that 80% of Kr atoms are located at R-sites and 20% at S-sites.

After annealing at 683 K, the <110> Kr-profile exhibited a shallow dip about 1.5 times broader than the Al-dip. The <100>Kr-dip was also about 1.5 times broader than the Al-dip but much shallower. The <111> Kr-yields are approximately independent of the incident angle. As described in a previous paper [8] these results can be interpreted as the precipitation of Kr atoms into an epitaxial solid phase. The broader Kr-dip was observed most conspicuously for the <110> channel which has the strongest lattice steering effect on channelling of He ions.

4.   Discussion

According to a TRIM code calculation, about 500 Frenkel defects are produced per one Kr atom on average by 50 keV Kr-implantation into aluminium. At room temperature, interstitials are highly mobile and a large fraction of them disappear by recombination with vacancies. The remaining interstitials cluster to form dislocation loops as observed in electron irradiation experiments [11]. Since vacancies can move slowly at room temperature as the irradiation proceeds, their concentration becomes high enough for clustering. Assisted by trapping at the implanted Kr atoms, a large number of cavities are produced [12].

262

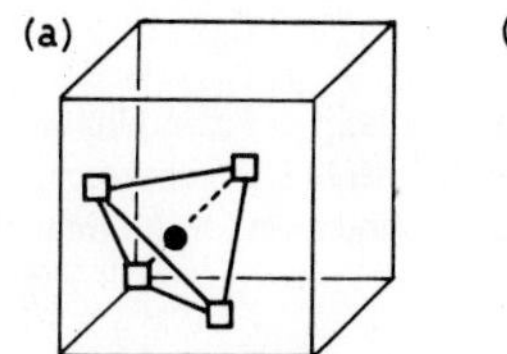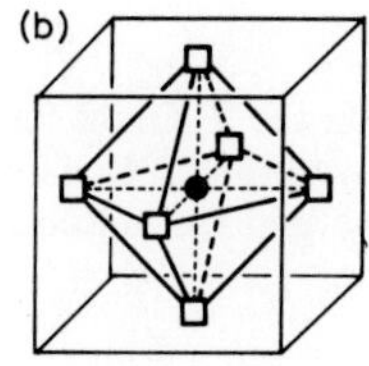

Fig. 5 Proposed Kr-vacancy complexes for (a) T- and (b) O-site occupancies. Open squares and filled circles represent vacancies and Kr atoms, respectively.

The T- and O–site occupancies are considered to result from the trapping of multiple vacancies by Kr atoms. Channelling experiments on the impurity-vacancy interaction in copper and aluminium indicated that large-sized impurities tend to interact strongly with vacancies [13]. Since the atomic radius of a Kr atom is 0.2 nm, which is about 40% larger than that of an Al atom, the T- and O-site occupancies of Kr atoms are interpreted to be a result of the Kr atom displacement to T- and O-sites by trapping 4 and 6 vacancies as shown in Figs. 5(a) and (b), respectively. With increasing dose from 10^{14} to $4\,10^{14}\,Kr/cm^2$ the fraction of the O–site occupancy increases, whereas that of the T-site occupancy decreases. This behaviour can be explained as the growth of the T-site configuration (KrV_4) to the O–site configuration (KrV_6) by trapping still more vacancies introduced during implantation.

The R-site occupancy can be ascribed to Kr atoms associated with larger vacancy clusters (cavities) consisting of more than 6 vacancies. As reported in a previous paper, with increasing dose from $4\,10^{14}$ to $1\,10^{15}Kr/cm^2$ the number density of R-site occupancy markedly increased (Fig. 2) and, correspondingly, cavities become observed [8].

On annealing at 593 K the T- and O–site occupancies disappeared, and the fractions of the S- and R-site occupancies became larger than those in the 371 K-annealed specimen. This result indicates the decomposition of small Kr-vacancy complexes such as T-, O- and O'-configurations to isolated Kr atoms located at Al lattice sites and vacancies. As free vacancies are highly mobile at 593 K, most of them disappear through migration to surfaces or are trapped by cavities.

After annealing at 473 K the fractions of the T- and O-site occupancies decreased and a different site occupancy (O'- site occupancies) appeared. This suggests that some fraction of T- or O–configurations are changed to the O'-configurations. As it is thought that after annealing at 371 K there remain no isolated vacancies, and the T- and O-configurations cannot move retaining their configurations, this configuration change is neither due to trapping of isolated vacancies by the T- or O-configurations nor their clustering. Therefore, the annealing behaviour might be explained as follows.

On annealing at 473 K some fraction of the T- and O-site occupancies are dissociated into KrV pairs and vacancies, and a fraction of mobile KrV pairs and vacancies encounter remaining undissociated T- or O–configurations to form the O'- configurations. Another fraction of KrV pairs and vacancies migrate to be trapped by the cavities, and, accordingly, the fraction of R-site occupancy increases. At 593 K the small complexes such as T-, O- and O'-configurations are decomposed to isolated Kr atoms (S-sites) and vacancies. Thus, the T- and O-configurations become unstable at temperatures between 380 and 470 K.

Regarding the Kr-implantation at room temperature, it can be said that at the initial stage of implantation Kr-vacancy complexes such as KrV_4 KrV_6 and larger ones (cavities) consisting of more than 6 vacancies are formed. When the implantation proceeds these complexes grow to larger ones by trapping vacancies, and, being assisted by vacancy motion, Kr atoms migrate to be trapped by them [7]. Such Kr-vacancy complexes act as nucleation centres for the precipitation of Kr atoms. With increasing dose, the clustering of Kr atoms located at O-sites also takes places around $4\,10^{15}Kr/cm^2$.

ACKNOWLEDGEMENTS

The author is grateful to M. Iwaki and Y. Okabe for their help in Kr-implantation, T. Aruga for his help in the TRIM-code calculation, T. Kobayashi and T. Urai for their technical assistance, I. Hashimoto and H. Yamaguchi for their help in TEM observation, and The Mitsubishi Foundation for the financial support.

REFERENCES

1. A. vom Felde, J. Fink, Th. Müller-Heinzerling, J. Pflüger, B, Scheerer, G. Linker, and D. Kaletta, Phys. Rev. Lett. **53**, 922 (1984).

2. C. Templier, C. Jaouen, J.-P. Rivière, J. Delafond, and J. Grilhé, C. R. Acad. Sci. Paris 299 Ser.II 613 (1984).

3. C. Templier, this conference.

4. H.H. Andersen , J. Bohr, A. Johansen, E. Johnson, L. Sarholt-Kristensen, and V. Surganov, Phys. Rev. Lett. **59**, 1589 (1987).

5. E. Yagi, Phys. Stat. Sol. **A104**, K13(1987).

6. E. Yagi, M. Iwaki, K. Tanaka, I. Hashimoto, and H. Yamaguchi, Nucl. Instr. and Meth. **B33**, 724 (1988).

7. E. Yagi, Nucl. Instr. and Meth. **B39**, 68 (1989).

8. E. Yagi, I. Hashimoto, and H, Yamaguchi, J. Nucl. Mater.**169**, 158 (1989).

9. I. Hashimoto, H. Yorikawa, H. Mitsuya, H. Yamaguchi, K. Takaishi, T. Kikuchi, K. Furuya, E. Yagi, and M. Iwaki, J. Nucl. Mater. **149**, 69 (1987).

10. E. Yagi, in preparation.

11. M. Kiritani, N. Yoshida, and H. Takata, J. Phys. Soc. Jpn. **36**, 720 (1974).

12. J. Takamura, Y. Shirai, K. Furukawa, and F. Nakamura, Mater. Sci. Forum **15-18**, 809 (1987).

13. M.L. Swanson, L.M. Howe, and A.F. Quenneville, J. Nucl. Mater. **69-70**, 372 (1978).

X-RAY DIFFRACTION STUDIES OF Kr AND Pb INCLUSIONS IN ALUMINIUM

J. Bohr[1], L. Gråbæk[1], H.H. Andersen[2], A. Johansen[2], E. Johnson[2], L. Sarholt-Kristensen[2], V. Surganov[3], I.K. Robinson[4], D. Broddin[5] and G. Van Tendeloo[5]

[1] Physics Department
Risø National Laboratory
DK-4000 Roskilde
Denmark

[2] Physics Laboratory
H.C. Ørsted Institute
DK-2100 Copenhagen Ø
Denmark

[3] Minsk Radioengineering Institute
6 Brovka St.
220069 Minsk
USSR

[4] AT&T Bell Laboratories
Murray Hill
New Jersey 07974
USA

[5] University of Antwerp (RUCA)
Groenenborgerlaan 171
B-2020 Antwerp
Belgium

ABSTRACT

Krypton and lead inclusions in aluminium prepared by ion implantation have been studied by x-ray diffraction. The krypton inclusions undergo a morphology change with temperature as the interfacial facets undergo a roughening transition. This in itself reflects a gradual misalignment of the inclusions. Lead inclusions on the contrary stay perfectly aligned until the disappearance of crystallinity. The melting transition for the lead inclusions shows a large hysteresis with both superheating and supercooling.

1. Introduction

Oil floating on water is an example of the phase separation of an immiscible system which has been known since ancient time. Micro emulsions are an example which have achieved much interest in modern times. Likewise, inclusions in hosts are a result of the phase separation which occurs after attempts to mix immiscible elements. Such phase separation can progress either as a spinodal decomposition or by nucleation. Inclusions are a result of the latter. The

Fundamental Aspects of Inert Gases in Solids
Edited by S.E. Donnelly and J.H. Evans, Plenum Press, New York, 1991

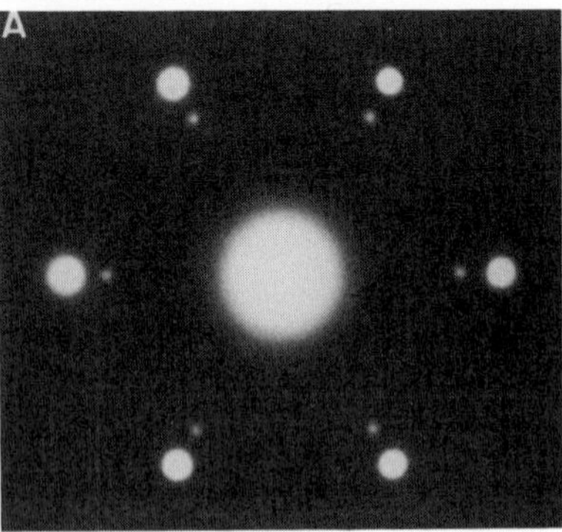

Fig. 1   Electron diffraction pattern from an aluminium single crystal implanted with $2 \cdot 10^{19}$ Pb^+/m^2 at an energy of 60 keV. In the diffraction picture the two set of bright spots clearly have the same symmetry with respect to the central spot. The outermost set originates from the aluminium host whereas the innermost set originates from the lead inclusions. The incident electron beam is along the [111] direction.

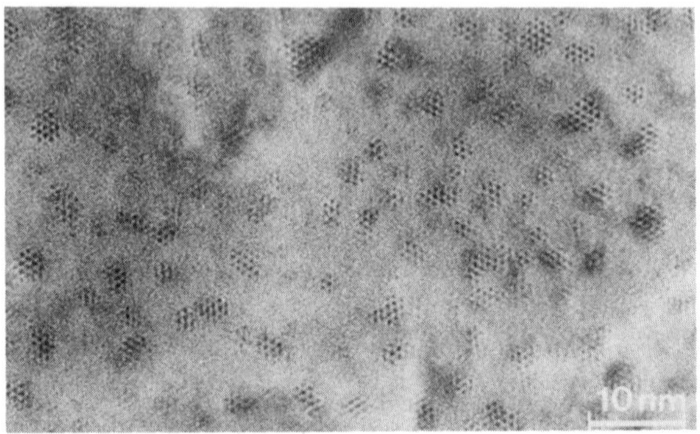

Fig. 2   Electron micrograph from an aluminium single crystal implanted with $2 \cdot 10^{19}$ Pb^+/m^2 at an energy of 60 keV. The small pointed areas are moire patterns formed from the overlap of the aluminium lattice and the lead lattice of an inclusion. The incident electron beam is along the [111] direction.

unstable mixtures of immiscible or almost immiscible elements can be obtained in many ways. Typical sample preparation methods are ion beam mixing (e.g. Ossi [1]), ion assisted deposition (e.g. Hubler [2]), ion implantation (e.g. Birtcher and Liu [3]) and quenching from the melt (e.g. Saka *et al.* [4]). Such systems allow for the study of a wide range of physical phenomena, such as diffusion and growth processes (e.g. Rothaut *et al.*[5], Marochov *et al.* [6]), melting and solidification of inclusions (e.g. Gråbæk *et al.* [7]) or thin layers fully embedded in a second type of material (e.g. Willens *et al.* [8]). Also phenomena at the interface between the inclusions and the host matrix can be studied. Examples are roughening transitions and the initial stages of melting (Andersen *et al.* [9], Gråbæk [10]). In this paper we review results for aluminium single crystals ion implanted with krypton or with lead. The main emphasis of these studies are fundamental physical problems such as the melting, solidification and growth of the small gas or lead inclusions.

Most earlier inclusion studies have been performed using transmission electron microscopy and electron diffraction. Electron microscopy allows the direct observation of single inclusions and is able to provide a detailed analysis of the shape and size of these inclusions. Using this technique it is also possible to determine the size distribution for the inclusions. Electron diffraction patterns, particularly along high symmetry zone axes, can nicely reveal the epitaxial relationship between the inclusions and the host material. In the particular case where both the host material and the inclusions have the same (cubic) crystal structure, two sets of diffraction spots will appear. Figure 1 shows a [111] electron diffraction pattern from an aluminium single crystal; the intense outer reflections arise from the Al-matrix while the weaker inner satellites originate from the Pb-inclusions. The direct image at relatively low magnification (Fig. 2) clearly reveals a moiré pattern due to the overlap between Al-host material and Pb-inclusion. As the ratio between the lattice parameter for lead and aluminium is between 6:5 and 5:4, the distance between the rows in the moiré pattern is approximately 8 Å. High resolution images showing atomic scale details on the Al-lattice hardly revealed any more information on the detailed shape of the Pb-particles.

2. Technique

The two samples discussed are disks of single crystal aluminium cut with a (111) surface, 9 mm in diameter and ≈ 1 mm thick. After being mechanically and electrolytically polished one was implanted with Kr and one with Pb in a 4 mm wide strip across the sample. The Kr sample was implanted to a fluence of 2×10^{20} m^{-2} and subsequently to 1.5×10^{20} m^{-2}; the energies were 200 and 100 keV, respectively. The Pb sample was implanted to a fluence of 2×10^{20} m^{-2} at an energy of 150 keV. The implantations were performed at a beam current of 20-30 μA/cm^2 which is sufficiently low to avoid significant heating of the sample during the implantation.

We have studied the samples using x-ray diffraction, most of the work being carried out at the 12 kW RIGAKU rotating anode at Risø. The anode is equipped with a double-axis diffractometer with a focusing graphite monochromator, as sketched in Figure 3. In the study of facets at the lead inclusions we used the AT&T four circle diffractometer on beamline X16B at the National Synchrotron Light Source at Brookhaven National Laboratory in USA. The advantage of the four circle instrument is the large flexibility allowing for an easy change of scattering plane, making it possible to access large regions of reciprocal space, the basic principle, however, is still as sketched in Figure 3.

X-rays were used to get a detailed knowledge of the position and shape of the diffraction peaks originating from the inclusions. From the peak position we are able to determine the lattice parameter which in turn allows us to determine the pressure of the inclusions, for the noble gases this was done using the Ronchi equation of state (Ronchi [11]). Analysis of the peak shape gave information about the size distribution of the inclusions. From our measurements it was also possible to determine the integrated intensity in the diffraction peaks, allowing us to calculate the Debye-Waller factor. Further the disappearance/appearance of intensity made it possible for us to determine when the inclusions melted/solidified.

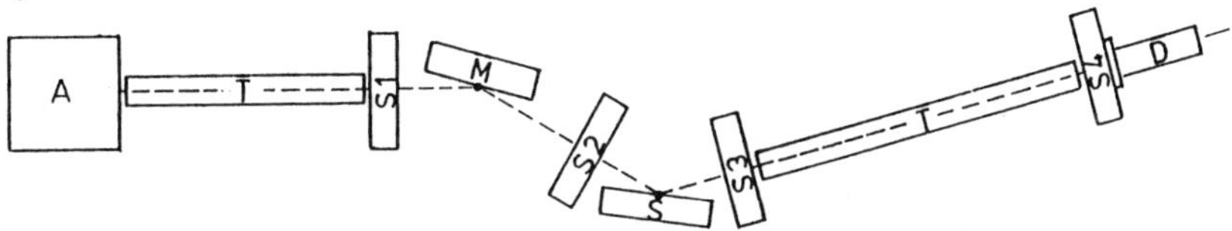

Fig. 3   Schematic drawing of the double axis diffractometer at Risø. A) is the rotating anode, M) is the monochromator, S) is the sample and D) is the detector. S1 ... S4 are slits used to define the beam and reduce the background and T) are evacuated flight tubes.

3. Krypton Inclusions

In the krypton sample a dual size distribution of the inclusions was observed. At low temperatures both the smaller (≈ 35 Å in diameter) and the larger (≈ 90 Å in diameter) inclusions were crystalline and epitaxially aligned with the surrounding aluminium crystal (Andersen *et al.* [9]). Such epitaxial alignment of gas inclusions had previously been observed (Templier *et al.* [12], vom Felde *et al.* [13], Evans and Mazey [14], Birtcher and Jäger [15]). The bimodal size distribution was revealed by a detailed study of the Kr diffraction peak which could be satisfactorily fitted by a sum of two Gaussians with different positions (lattice parameters), widths (inclusion size) and heights. The proof that this is the case comes about when the fitting is done systematically for the complete set of data obtained at temperatures in the interval 12-620 K, the results of which are shown in the Figures 4-7.

Figure 4 shows the peak intensity for the two Gaussian components of the (111) Kr peak. As can be seen the larger (≈ 90 Å) inclusions melt abruptly at a temperature close to the triple point (116 K). The intensity rapidly vanishes over a finite temperature interval about 4 K wide. This originates from the pressure broadening of the transition due to the restricted volume of the inclusions. The smaller inclusions (≈ 35 Å) remain solid up to at least 620 K, the highest temperature of the experiment. The slow decrease of the peak intensity is due to a gradual loss of the epitaxial alignment. This can be seen in figure 5 where the transverse widths are depicted. The large increase in transverse width above 250 K is due to misalignment. As the radial width (Fig. 6) is nearly constant so is the size of the inclusions. Melting can also be excluded as the integrated intensity remains constant at these temperatures. Figure 6 shows the radial widths of the two Gaussians; they are roughly constant with temperature confirming the dual size distribution of the inclusions in this sample.

Figure 7 shows the lattice parameter of the krypton inclusions as determined from the fit to two Gaussians. From the lattice parameter the inclusion pressure is determined using the Ronchi equation of state [11]. For the larger Kr inclusions the pressure is found to be close to ambient (consistent with a melting temperature close to the triple point). In contrast the smaller Kr inclusions are subject to a considerable pressure, increasing to 1.94 GPa at 508 K. These small pressurized inclusions were seen gradually to lose the alignment with the aluminium as the temperature was raised, see Figure 5. The gradual loss of alignment is thought to be due to a roughening of the krypton aluminium interface, assuming that it is the faceted nature of the interface which is responsible for the epitaxial alignment. Alternatively, the loss of epitaxial alignment may be due to a fluid layer at the interface.

4. Lead inclusions

The lead inclusions studied are somewhat larger than the krypton inclusions, typically between 100 and 300 Å in diameter and they are not pressurized (< 0.18 GPa, Gråbæk *et al.* [7]). The lead inclusions also grow epitaxially with the surrounding aluminium, as demonstrated by Thackery and Nelson [16]. Epitaxial alignment of metal inclusions has also been reported for other systems, Silcock[17] and Saka *et al.*[4].

During melting and solidification of the lead inclusions we observed a large superheating (up to 67 K) and supercooling (at least 18 K). This was observed in six successive heating cycles, showing that it is an intrinsic physical phenomenon (see Fig. 8). As can be seen in the figure, the temperature interval from melting to solidification decreases from one temperature cycle to the next; this is due to inclusion growth. The truncated octahedron morphology of the inclusions was confirmed by the observation of rods (streaks) of intensity from the {111} and {100} facets. Figure 9 shows the intensity distribution in the (111) rod obtained from a series of scans across the rod.

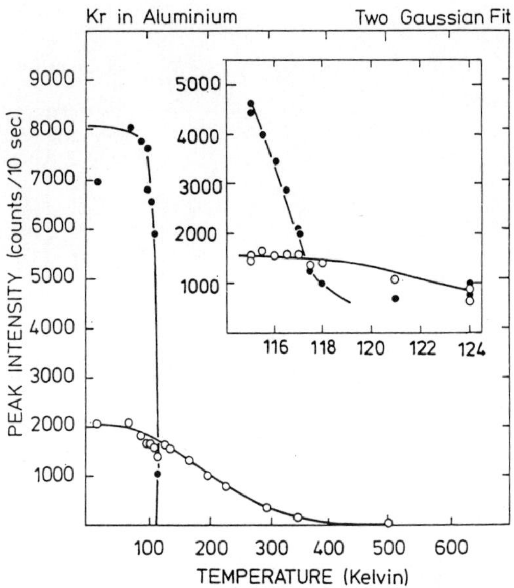

Fig. 4   The peak intensity of the (111) krypton peak. (•) data for the peak component originating from the larger inclusions, (o) data for the peak component originating from the smaller inclusions. The inset is a detail figure for the temperature region where the larger inclusions melt. The lines are guides to the eye.

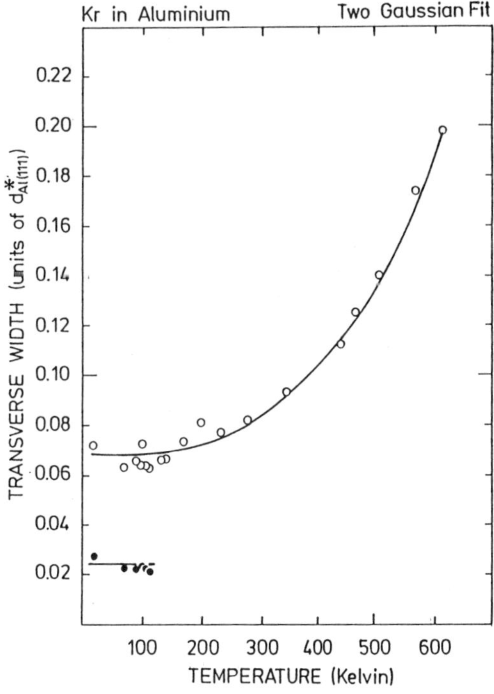

Fig. 5   The transverse width of the (111) krypton peak, in units of $d^*_{Al(111)}$ (distance to (111) Al in reciprocal space). (•) data for the peak component originating from the larger inclusions, (o) data for the peak component originating from the smaller inclusions. The lines are guides to the eye.

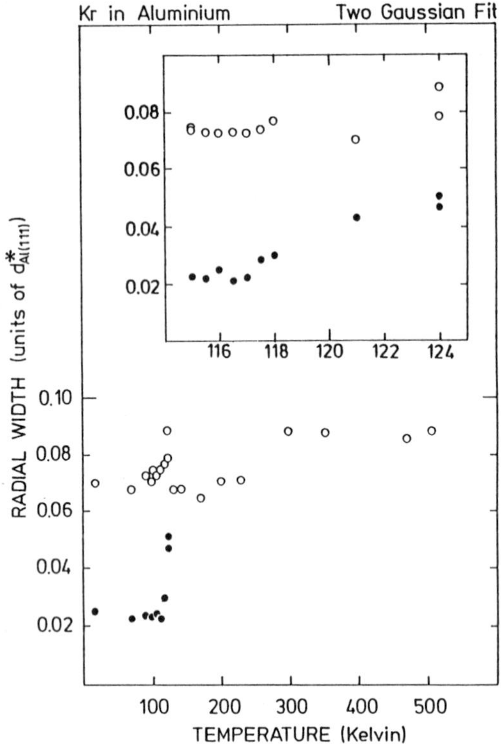

Fig. 6 The radial width of the (111) krypton peak in units of $d^*_{Al(111)}$ (distance to (111) Al in reciprocal space). (•) data for the peak component originating from the larger inclusions, (o) data for the peak component originating from the smaller inclusions. The inset is a detail figure for the temperature region where the larger inclusions melt.

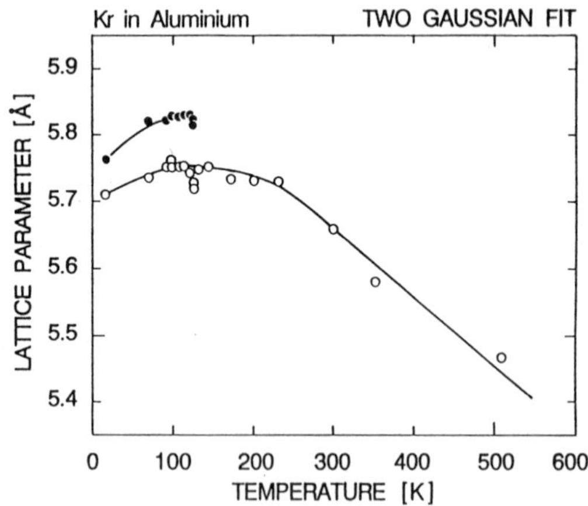

Fig. 7 Lattice parameter for the two inclusion populations. (•) data for the larger inclusions, (o) data for the smaller inclusions. The lines are guides to the eye.

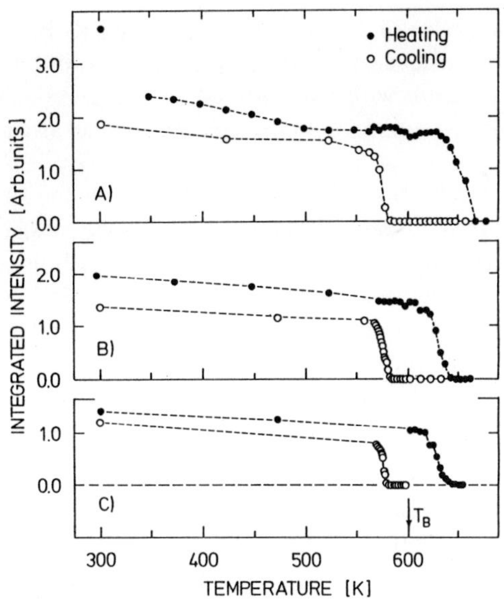

Fig. 8   Integrated intensity for the (111) lead peak. (•) data obtained on heating and (o) data obtained on cooling. A) data for the first heating cycle, B) data for the third heating cycle and C) data for the sixth heating cycle. Note the gradual decrease of the width of the hysteresis loop around the bulk melting point T_B, 88 K in A) decreasing to 68 K in C). The lines are guides to the eye.

In the first heating cycle (Figure 8A) the temperature is 40 K above the bulk melting point of lead (T_B = 601 K) before the integrated intensity drops below that expected from a simple Debye-Waller behaviour. At 67 K above T_B the integrated intensity goes to zero indicating that all the lead inclusions are melted. We expect the lack of free surface to be the most important effect causing the increase in the melting temperature. To melt a crystal it is necessary to create some defect/disorder—for the confined lead crystals this is difficult as one or more aluminium atoms have to move into the host matrix in order to obtain the space needed for the defect/disorder to appear. Another mechanism which may contribute to the increased melting temperature is the energy gained by the epitaxial alignment and the faceting. This stabilization is lost in the liquid phase.

In all heating cycles (Fig. 8) the integrated intensity decreases over a finite temperature interval, signifying gradual melting. The main reasons for this gradual decrease in intensity are: 1) The size distribution among the inclusions; the inclusions melt according to their size, with the largest inclusions melting at the lowest temperature. 2) Pressure broadening; using the Clausius-Clapeyron equation (Landau and Lifshitz [18]) we find the pressure gradient obtained on the melting of lead, dp/dt = 0.0129 GPa/K. From this we see that a pressure of 0.18 GPa (worst case) contributes 14 K to the width of the melting transition.

For the lead inclusions we also observe supercooling of at least 18 K, most likely due to the inclusion size. For a liquid to solidify, a nucleation grain of a certain critical radius r_c is needed where:

$$r_c = 2\gamma_{sl}T_B/L_0\Delta T \tag{1}$$

γ_{sl} is the interface energy between solid and liquid lead and has a value of 33.3 mJ/m^2, L_0 is the latent heat of fusion with a value of 2.71×10^8 J/m^3, T_B is the bulk melting temperature and ΔT is the supercooling. For inclusions of size ≈ 250 Å to solidify, the supercooling should be at least 11 K. This is in agreement with the observed 18 K supercooling. Note that the above argument implies that the lead-aluminium interface does not act as a nucleation center for the solid.

We have not experimentally been able to determine if it is the superheated or the super-cooled, or even both states, which are metastable, as both states are stable within the time scale of our experiment. In one temperature cycle we tested the stability by keeping the sample superheated at 618 K for 21 hours and supercooled at 588 K for 35 hours. In neither case were changes in the scattering observed during the observation period. However, the above inter-pretation suggests that it is the supercooled phase that is metastable.

5.   Metal and Gas Inclusions, Parallels and Differences

The inclusions we have studied were in the size range 20-100 Å for krypton and 100-300 Å for lead. For both systems the inclusions turn out to be single crystals epitaxially aligned with the host matrix and the inclusions form $\{111\}$ and $\{100\}$ facets as demonstrated by Birtcher and Liu [3] and Moore $et\ al.$ [19]. The resulting shape is that of a truncated octahedron.

The aluminium single-crystal hosts have a lattice parameter of 4.05 Å whereas the inclusions have a significantly larger lattice parameter; 4.95 Å for lead and— depending on the actual pressure—5.46-5.82 Å for krypton. Therefore there must be a large mismatch between the two structures resulting in an incoherent interface. This is consistent with the assumption that the epitaxy is due to the morphology coincidence of the inclusion and the host cavity. Further, the incoherence of the interface is seen in Rutherford backscattering data [9].

The lead inclusions do not show any detectable sign of an excess pressure. At room temperature the lattice parameter determined from our measurements, 4.954 ± 0.004 Å, agrees with the tabulated bulk value of 4.9505 Å [20]. As the average coefficient of thermal expansion is 2.47×10^{-5} and 3.03×10^{-5} [21] for aluminium and lead, respectively, we would expect a pressure build up in the lead inclusions during heating. At T_B we determine the lattice parameter for the lead inclusions to be 4.993 ± 0.004 Å. Calculating the lattice parameter at this temperature we find 4.9957 Å, so only a minor pressure is expected. Taking now the worst case within our experimental uncertainty we find, using the compressibility of lead (2.2×10^{-11} Pa^{-1} [21]), that the pressure at T_B is at most 0.18 GPa.

Both the lead and the small krypton inclusions were observed to melt at temperatures well above the bulk melting point (116 K for Kr and 601 K for Pb). The melting behaviour for the inclusions is determined by the interactions at the interface, the pressure and the size. In an equilibrium system the pressure, P, and radius, r, are connected via the following expression:

$$P = 2\gamma/r \qquad\qquad (2)$$

where γ is the energy of the aluminium/inclusion interface ($\gamma = 1.2$ J/m^2 for the noble gas in-clusions [22]). Both smaller and larger pressures than this equilibrium value are possible. Due to a large difference in the coefficient of thermal expansion for krypton (average 28.9×10^{-5} [23]) and aluminium (2.47×10^{-5} [21]), the inclusion pressure is expected to increase with temperature. For the smaller inclusions such a pressure increase is actually observed, see Figure 7. However, we note also that the inclusion morphology changes with the roughening of the inclusion or the cavity facets and that an eventual liquid layer at the interface also would increase the pressure. The superheating of approximately 480 K for argon inclusions in aluminium reported by Rossouw and Donnelly [24] may in turn also be ascribed to these effects.

A small faceted crystal will undergo a series of roughening transitions as the temperature is increased. The rounding of a given facet proceeds gradually over a temperature interval from

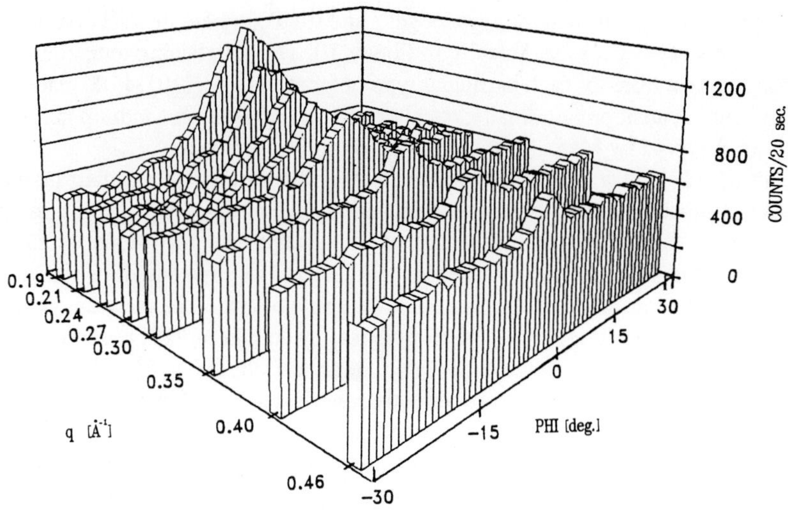

Fig. 9 Three-dimensional plot of the rod originating from the {111} facets at the lead inclusions. Data obtained in eight scans across the [$\bar{1}\bar{1}$1] direction away from the (111) lead peak. q gives the distance to the lead peak PHI is the number of degrees the sample is turned away from the [$\bar{1}\bar{1}$1] direction, the data are obtained at 303 K.

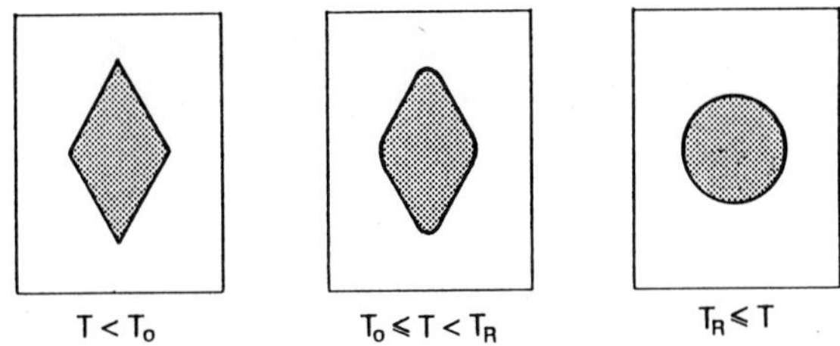

Fig. 10 Schematic diagram showing the roughening transition for a faceted inclusion. The transition is gradual, it initiates at T_0 where rounding starts at the corners, and continues until the inclusion has lost all its facets at T_R.

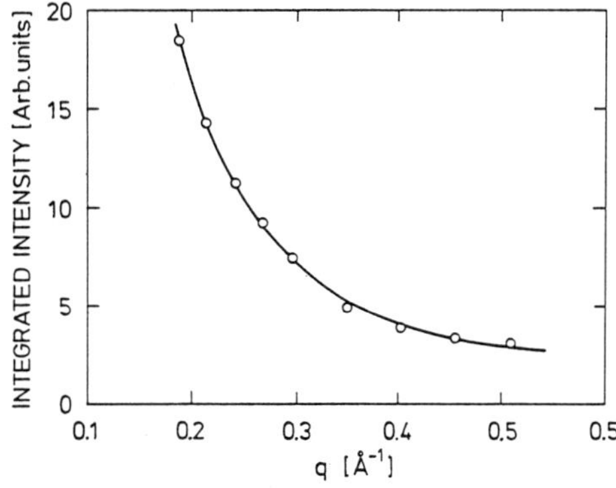

Fig. 11 Background subtracted integrated intensity along the rod originating from the {111} facets at the lead inclusions, room temperature data. The rod is measured along the [$\bar{1}\bar{1}$1] direction. The fully drawn curve is the best fit to the data using the expression by Robinson [28] (see text).

T_0, where the rounding sets in at the facet edges, and till T_R, where the facet has completely disappeared. This is shown schematically in Figure 10. The roughening temperature T_R is a characteristic temperature for each particular type of facet {111}, {210}, {100} etc. [25]. The existence of such transitions at small crystals (< 5 μm) of lead and indium on a graphite substrate was observed by Métois and Heyraud [26] and Heyraud *et al.* [27].

Moore *et al.* [19], in a series of electron micrographs, demonstrate that a roughening transition takes place at the facets of a lead inclusion. Their $Al_{0.945}Pb_{0.05}Si_{0.005}$ sample was produced by rapid solidification and contained quite large lead inclusions, around 500 Å across. These inclusions were clearly faceted, with large {111} and smaller {100} facets. At temperatures from room temperature to 550 °C, a gradual shape change was observed, the inclusions changing from sharply faceted to spherical. Moore *et al.* [19] observed the rounding to start at the {100} facets in the temperature interval from 300 to 350 °C, and continues on the {111} facets until the inclusion is perfectly spherical at 550 °C which is well above the melting point for the inclusion. The roughening transition at the {111} facets therefore takes place at a temperature at which the lead has already melted. However, Moore *et al.* [19], propose that the rounding of the {100} facets is due to melting of the inclusions.

We did a series of experiments where we studied the rods of intensity originating from {111} and {100} facets at the lead inclusions. These streaks/rods of intensity extend from the Bragg peaks originating from the inclusions in directions perpendicular to the facet. The intensity in the rods I_{RODS} decreases with distance from the Bragg peak according to the following relation [28]:

$$I_{RODS} = N_1^2 N_2^2 \frac{(1-\beta)^2}{4[1 + \beta^2 - 2\beta \cos(qa)] \sin^2(\frac{1}{2}qa)} \tag{3}$$

where N_1 and N_2 represent the scattering components parallel to the facet. The remaining part of the expression represents the scattering component perpendicular to the facet, where q is the distance to the inclusion Bragg peak, a is the lattice parameter and β is a roughness parameter for the facet giving the fractional occupancy of the first incomplete layer. At room temperature we measured the integrated intensity along the (111) rod. The above expression can be fitted very nicely to these data (Figure 11), and from this fit we obtain $\beta = 0.13 \pm 0.02$, indicating that the facet is smooth.

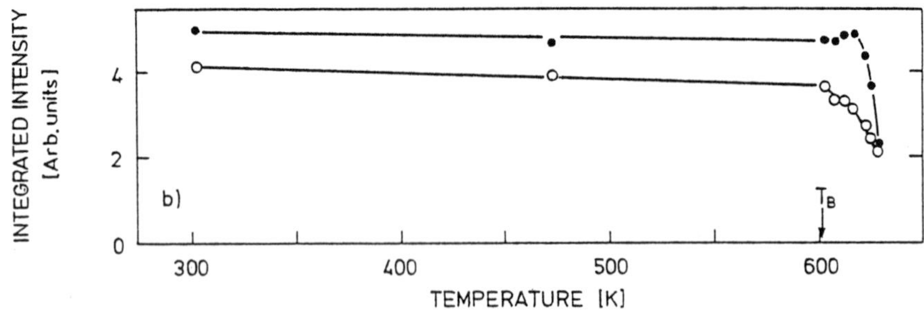

Fig. 12   Integrated intensity in a scan across the (111) rod (•) and across the (100) rod (o). The distances to the lead (111) peak were 0.242 and 0.171 Å^{-1}, respectively. The lines are guides to the eye and the arrow marks the bulk melting point T_B.

In one heating sequence we followed the intensity across the rods at distances of 0.242 and 0.171 Å^{-1} from the Pb (111) Bragg peak, for the (111) and (100) rods respectively. Doing this we actually found small differences in behaviour indicating some kind of transition at the {100} facets at a temperature 15 K below the temperature where the lead inclusions melt and the intensity from the {111} facets begins to decrease as illustrated in Fig. 12. We are not able to deduce from the above experiment whether the transition taking place at the {100} facets was a roughening transition or premelting.

6.   Conclusions

In summary we have seen how inclusions (rare gas or metal) in a host matrix can serve as a model system for the study of fundamental physical phase transformations such as roughening transitions and the melting transition. More specifically, the observed misorientation of krypton inclusions was seen to be a consequence of a roughening transition of the interfacial facets. For lead this roughening transition only takes place at a temperature above the melting temperature of the inclusions. The epitaxial alignment of the lead inclusions is consequently perfectly preserved until melting. The lead inclusions therefore are ideal systems for studying melting and solidification transitions. A large thermal hysteresis is seen at the melting transition with both superheating and supercooling. The superheating is perceived as having its cause in the lack of a free surface and in the additional stability gained by the epitaxial alignment. The fact that the inclusions do not solidify as long as they are smaller than the critical grain size for nucleation gives strong evidence that the solidification takes place inside the bulk of the inclusions rather than on the host interface.

REFERENCES

1. P.M. Ossi, Mat. Sci. Eng. A **115**, 107 (1989).

2. G.K. Hubler, Mat. Sci. Eng. A **115**, 181 (1989).

3. R.C. Birtcher and A.S. Liu, J. Nucl. Mat. **165**, 101 (1989).

4. H. Saka, Y. Nishikawa and T. Imura, Phil. Mag. A**57**, 895 (1988).

5. J. Rothaut, H. Schroeder and H. Ullmaier, Phil. Mag. A**47**, 781 (1983).

6. N. Marochov, L.J. Perryman and P.J. Goodhew, J. Nucl. Mat. **149**, 296 (1987).

7. L. Gråbæk, J. Bohr, E. Johnson, A. Johansen, L. Sarholt-Kristensen and H.H. Andersen, Phys. Rev. Lett. **64**, 934 (1990).

8. R.H. Willens, A. Kornblit, L.R. Testardi and S. Nakahara, Phys. Rev. B**25**, 290 (1982).

9. H.H. Andersen, J. Bohr, A. Johansen, E. Johnson, L. Sarholt-Kristensen and V. Surganov, Phys. Rev. Lett. **59**, 1589 (1987).

10. L. Gråbæk, *X-ray Diffraction Studies of Kr, Xe and Pb Inclusions in Aluminium*, PhD Thesis, Risø-M-2868, Roskilde, DK-4000 Denmark, (1990).

11. C. Ronchi, J. Nucl. Mat. **96**, 314 (1981).

12. C. Templier, C. Jaouen, J.P. Riviére, J. Delafond and J. Grilhé, C.R. Acad. Sc. Paris **299** Serie II 613 (1984).

13. A. vom Felde, J. Fink, Th. Müller-Hinzerling, J. Pflüger, B. Scherer, G. Linker and D. Kaletta, Phys. Rev. Lett. **53**, 922 (1984).

14. J.H. Evans and D.J. Mazey, J. Phys. F**15**, L1 (1985).

15. R.C. Birtcher and W. Jäger, Ultramicroscopy **22**, 267 (1987).

16. P.A. Thackery and R.S. Nelson, Phil. Mag. **19**, 169 (1969).

17. J.M. Silcock J. Inst. Met. **84**, 19 (1955-56).

18. L.D. Landau and E.M. Lifshitz, *Statistical Physics* part 1, pp 255-6, Pergamon Press Oxford (1980).

19. K.I. Moore, K. Chattopadhyay, and B. Cantor, Proc. R. Soc. Lond. A **414**, 499 (1987).

20. R.W.G. Wyckoff, *Crystal Structures*, vol. 1, Wiley & Sons, New York (1963).

21. *Smithells Metals Reference Book* ed. by E.A. Brandes, Butterworths, London (1983).

22. A.R. Miedema, Philips Techn. Rev. **38**, 257 (1978-79).

23. P. Korpiun and E. Lüscher in *Rare Gas Solids*, eds. M.L. Klein and J.A. Venables Academic Press, London (1977).

24. C.J. Rossouw and S.E. Donnelly, Phys. Rev. Lett. **55**, 2960 (1985).

25. M. Wortis in *Fundamental Problems in Statistical Mechanics VI*, Proceedings of the Sixth International Summer School, Trondheim, Norway, 1984, edited by E.G.D. Cohen (Elsevier, New York, 1985), p. 87.

26. J.J. Métois and J.C. Heyraud, Ultramicroscopy **31**, 73 (1989).

27. J.C. Heyraud, J.J. Métois and J.M. Bermond, J. Cryst. Growth **98**, 355 (1989).

28. I.K. Robinson, Phys. Rev. **B33**, 3830 (1986).

GAS DENSITIES IN HELIUM BUBBLES DETERMINED BY SMALL ANGLE NEUTRON SCATTERING

H. Ullmaier

Institut für Festkörperforschung des Forschungszentrums Jülich
Postfach 1913, D-5170 Jülich and Association KFA-EURATOM
Germany

ABSTRACT

Neutron small angle scattering (SANS) combined with transmission electron microscopy (TEM) can yield detailed information on the structure of helium bubbles in metals. Besides conventional information such as size distributions, the density of helium in the bubbles can directly be determined by contrast variation of SANS. This is exemplified by measurements in nickel, implanted with 1200 appm He^3 and He^4, respectively, at room temperature and subsequently annealed at temperatures up to 1173 K. The most important result is the finding that even at the highest annealing temperature (corresponding to about 70% of the melting temperature), the bubbles contain an overpressure which lies about 3 GPa above the value for thermodynamic equilibrium. This has far-reaching consequences on the coarsening behaviour of the bubbles.

1. Introduction

Inert gases in solids are of increasing interest in several fields of modern science and technology: thin films (sputtering and plasma deposition, ion beam mixing), fission reactor fuels and waste disposal, fusion materials development (embrittlement by (n, α)-produced He, sputtering and surface modification of plasma-facing components, tritium storage) and sintering techniques (hot isostatic pressing). In most cases the incorporation of inert gases in solids leads to detrimental changes of their properties. They are caused by the nucleation and growth of bubbles which inevitably form at elevated temperatures because of the very small solubility of inert gases in solids.

A key parameter for the formation of bubbles is the pressure p of the gas contained in them—and closely related to it via the equation of state—the gas density ρ. An experimental determination of these quantities is thus of great interest. Techniques applied to this problem include TEM (equating the measured swelling to the amount of introduced gas), electron energy loss spectroscopy [1], VUV-absorption [2] and positron annihilation [3]. All these methods are either afflicted with large uncertainties and/or are restricted to very limited parameter ranges. We therefore introduced small angle neutron scattering (SANS) [4] as a method providing a rather straightforward determination of helium densities in bubbles in metals. Since the principle of the experimental method and its evaluation procedure has been given in detail in Ref. [4], I give here (Section 2) only a brief description of the most important features. In Section 3 some recent results [5] on the coarsening behaviour of helium bubbles in nickel during post-implantation annealing are presented. A short discussion of the results and their connection to related investigations [6-8] is given in Section 4.

2. Small Angle Neutron Scattering

2.1. General

In SANS, scattered neutrons are recorded as a function of a small momentum change Q which is determined by the scattering angle θ and the neutron wavelength (typically around 1 nm)

$$Q = \frac{2\pi}{\lambda} 2\sin\frac{\theta}{2} \approx \frac{2\pi}{\lambda}\theta \tag{1}$$

At small Q-values, i.e. $Q < 2\pi/a$ (with a being the atomic distance), the atomic structure is smeared out. In the Born approximation the differential scattering cross section per unit volume, $d\Sigma/d\Omega$ (Q) for randomly distributed helium bubbles is given by:

$$\frac{d\Sigma}{d\Omega}(Q) = nV^2 F(Q)\,(\Delta\rho)^2 \tag{2}$$

with n the number density, V the volume and $F(Q)$ the form factor of the helium bubbles (more precisely, Eq. (2) is an integral over the size distribution of the helium bubbles). The contrast $\Delta\rho$ is determined by the coherent scattering lengths b and the atomic volumes Ω of helium (He) and metal (M), respectively:

$$\Delta\rho = \frac{b_M}{\Omega_M} - \frac{b_{He}}{\Omega_{He}} \tag{3}$$

From the scattering cross sections, the size distribution of the helium bubbles can be extracted by standard evaluation procedures (e.g. by fitting log-normal distributions [4] or by using a spline function method [9]). Besides this conventional information, SANS results can also be used to determine the helium density $\rho_{He} = \Omega_{He}^{-1}$ in the bubbles by applying the contrast variation method.

2.2. Contrast variation

A determination of the quantity of interest, Ω_{He} from the He scattering law in Eq. (2) is virtually impossible because the quantities n, V and $F(Q)$ are not known accurately. In addition, a precise absolute calibration of the scattered intensity would be necessary. However, by utilizing contrast variation, Ω_{He} can be determined in a simple and accurate way. The method is based on the requirement that two specimens with identical bubble structure but different scattering isotopic composition, I_1 and I_2, are available. Then, the intensity ratio, α given by:

$$\alpha = \frac{\dfrac{d\Sigma}{d\Omega}(Q, I_1)}{\dfrac{d\Sigma}{d\Omega}(Q, I_2)} = \frac{|\Delta\rho(I_1)|^2}{|\Delta\rho(I_2)|^2} \tag{4}$$

contains only Ω_{He} as an unknown quantity since the coherent He scattering lengths and Ω_M in Eq. 3 are accurately known while n, V and $F(Q)$ cancel out.

Two different scattering contrasts can be achieved in a number of ways: (1) two specimens with identical structure but implanted with different gas isotopes (He^3/He^4) or (2) consisting

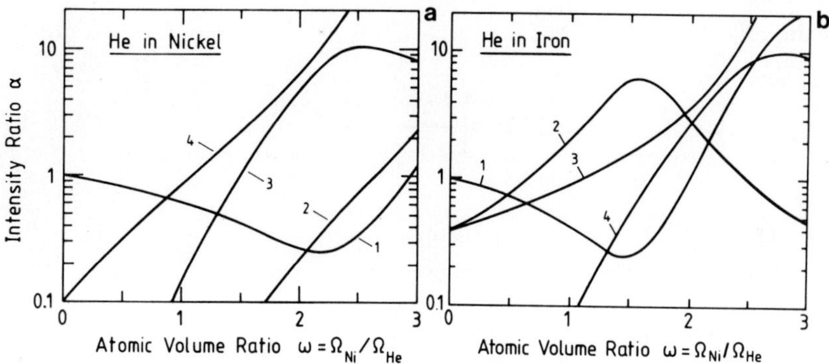

Fig. 1 The ratio α of the neutron scattering intensities from two specimens with identical bubble structure but different isotopic composition is a unique (and known) function of the ratio ω of the atomic volumes of metal and helium, respectively. Two examples are shown for various contrast combinations: (a) He^3/He^4 in Ni^{58} (1), He^4 in Ni^{60}/Ni^{58} (2), He^3 in Ni^{60}/Ni^{58} (3) and He^4 in Ni/Ni^{62} (4); (b) He^3/He^4 in Fe (1), magnetic and nuclear contrast of He^3 in Fe (2) and of He^4 in Fe (3) and He^3 in Fe and polarized neutrons (4) [10].

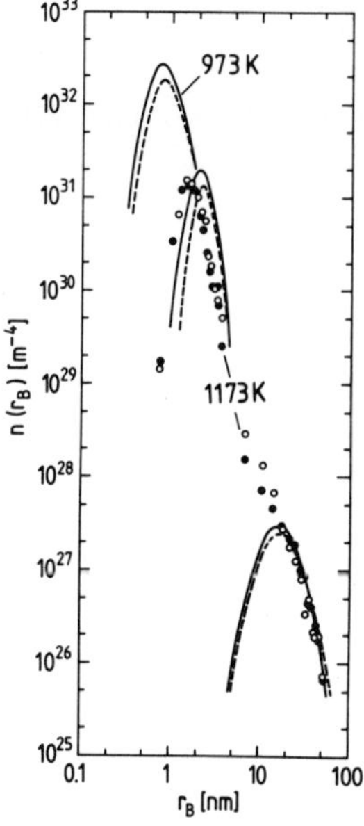

Fig. 2 Size distributions $n(r_B)$ in specimens implanted with He^3 (dotted lines) and He^4 (solid lines), respectively, and annealed up to 973 K and 1173 K, respectively, as evaluated from the SANS data. Also shown are the results obtained from TEM for 1173 K (He^3, He^4, for details see text).

of different matrix isotopes (e.g. Ni^{58}/Ni^{60}), (3) one ferromagnetic specimen using nuclear and magnetic contrast and (4) different orientations of polarized neutrons with respect to the applied magnetic saturation field. Fig. 1 gives examples of a as a function of $\omega = \Omega_M / \Omega_{He}$ for nickel and iron, respectively. Details of these possibilities are discussed in Ref. [10]. Here we concentrate on the combination He^3/He^4 in Ni^{58} for which experimental results are presented in the next section.

In this case (curve 1 in Fig. 1a):

$$\alpha = \left| \frac{1 - \omega b_{He^3}/b_{Ni^{58}}}{1 - \omega b_{He^4}/b_{Ni^{58}}} \right|^2 \tag{5}$$

For $\omega = \Omega_{Ni}/\Omega_{He} = 0$ (void), $\alpha = 1$. With increasing helium density, i.e. increasing ω, α decreases continuously to about 0.25 at $\omega = 2$, the maximum value of physical significance.

3. Results for Nickel

The experimental results described were obtained from two Ni^{58} specimens into which 1200 appm He^3 and He^4, respectively, had been implanted at room temperature. After a first SANS measurement the specimens (together with a helium-free reference sample) were annealed in high vacuum at $T_a = 673$ K for 2 hours before the next SANS measurement was performed. This procedure was repeated for 10 more isochronal annealing steps up to a temperature of 1173 K. After the last SANS measurements TEM specimens were prepared.

Since the experimental details, the raw data from SANS and TEM and their evaluation procedures are described in Ref. [5], only the most important results of these evaluations are briefly presented here. Fig. 2 shows examples of size distributions as determined by SANS ($T_a = 973$ K and 1173 K) and TEM ($T_a = 1173$ K), respectively. For $T_a \leq 973$ K, the measured $d\Sigma/d\Omega(Q)$ relations suggest monomodal size distributions, whereas at higher annealing temperatures they point to bimodal distributions. TEM revealed [5,7,11] that the latter are due to spatially separated populations of "small" bubbles in the bulk (Fig. 3a) and "large" bubbles near grain boundaries and free surfaces (Fig. 3b). A comparison of the measured size distributions shows that they are almost identical for the He^3 and He^4 specimens, i.e. the basic requirements for the use of Eq. 4 are well fulfilled.

From the measured size distributions, the total bubble densities C_B and the mean radii $\bar{r}_B$ can be calculated. Their development after each annealing step is shown in the Arrhenius-type plots of Figs. 4 and 5. The data can be approximated by straight lines whose slopes correspond to energies of about 0.35 eV for $\bar{r}_B$ of the "small" bubbles, 1.1 eV for $\bar{r}_B$ of the "large" bubbles and -1.0 eV for C_B of the "small" bubbles.

Fig. 6 shows examples of the measured intensity ratios, α, as a function of the scattering vector Q after different annealing steps. At low temperatures no helium bubbles are formed and the scattering is due to irradiation-induced defects. Since no significant differences in the structure and the annealing of the radiation damage induced by He^3 and He^4 bombardment are expected, their scattering should be identical. This is indeed confirmed by finding $\alpha \approx 1$ for $T_a \leq 673$ K (Fig. 6a). Bubbles of size visible in TEM begin to appear at annealing temperatures around 773 K [7]. They are expected to be filled with helium of very high density. If we assume a density ρ_{He} of around $2 \cdot 10^{29}$ m^{-3} (corresponding to a pressure of 5 GPa [12], we expect an α-value of about 0.26 (Fig. 1). The fact that this value is close to the observed value for $Q > 0.15$ Å^{-1} (Fig. 6c) indicates that the scattering from specimens annealed at $T_a = 773$ K is already dominated by the helium bubbles. This conclusion should hold even better for higher annealing temperatures where the increasing bubble size (Fig. 4) leads to a strong increase in their scattering cross-section ($d\Sigma/d\Omega \propto \bar{r}_B^6$!). Moreover, the radiation induced defects (mainly dislocation loops) coarsen, i.e their (weak) scattering is shifted to smaller Q-values. It is therefore fairly safe to use the measured-values for $T_a \leq 823$ K for the determination of helium densities via Eq. (5) and Fig. 1. As evident from Fig. 6, α decreases slightly with increasing Q for $Q \geq 0.05$ Å^{-1}. In this Q-range the scattering is dominated by the "small" bubbles in the matrix. This weak Q-dependence is a consequence of the non-zero width of the bubble size distributions (Fig. 2); smaller bubbles (scattering at large Q-values) contain helium with higher density (i.e. smaller α-values, see Fig. 1), larger bubbles (scattering at small Q-values)

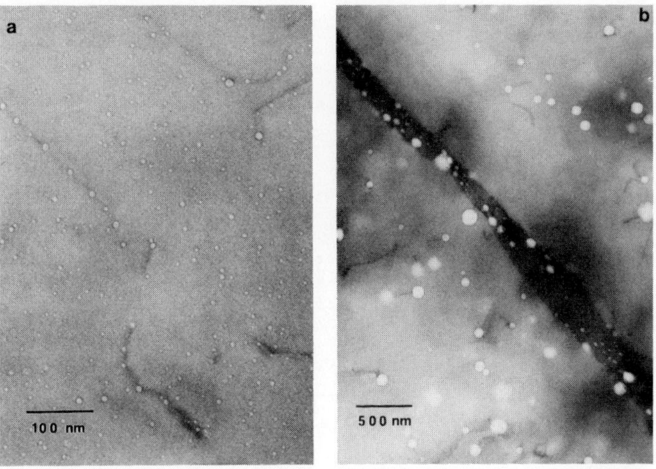

Fig. 3 Helium bubble structures typical for specimens after the last annealing step at 1173 K: (a) bubbles in the matrix and at dislocations and (b) at and in the vicinity of a grain boundary. The grain boundaries are surrounded by bubble denuded zones which are frequently asymmetric (as exemplified here by the bubble-free strip to the right of the grain boundary). Beyond the denuded zones the grain boundaries are surrounded by regions of several µm width containing bubbles which are much larger than those present in the matrix [5].

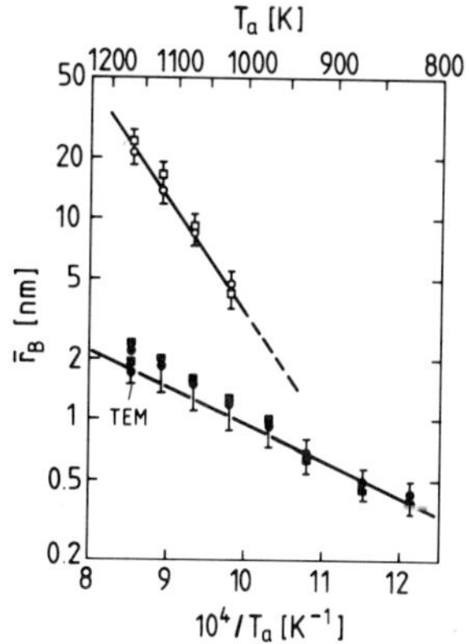

Fig. 4 Mean radii $\bar{r}_B$ of "small" bubbles in the bulk (filled symbols) and "large" bubbles near grain boundaries (open symbols) in specimens implanted with He^3 (■ □) and He^4 (● ○), as a function of the reciprocal annealing temperature T_a. The TEM values determined for T_a=1173 K are given by TEM (⊠ ⊗) [5].

contain helium with lower density (i.e. higher α values). For $0 \leq Q \leq 0.05$ Å^{-1}, there is a stronger decrease of α with increasing Q. For 723 K $\leq T_a \leq 923$ K, this low-Q tail is probably due to scattering by remainders of radiation-induced defects. After being absent at $T_a \geq 973$ K (Fig. 6g), the tail starts reappearing for $T_a = 1023$ K which is attributed to the formation of a population of "large" bubbles around the grain-boundaries (see Fig. 3b) at high annealing temperatures. Inserting the measured α-values (averaged over Q-ranges which correspond to the size distributions of the "small" bubbles) into Eq. (5) yields the ω-values which are plotted in Fig. 7a as a function of the mean bubble radii corresponding to the different annealing temperatures (Fig. 4). Combining these experimental ω (or ρ_{He})-values with the measured size distributions, the total concentration c_{He} of the helium contained in the bubbles can be

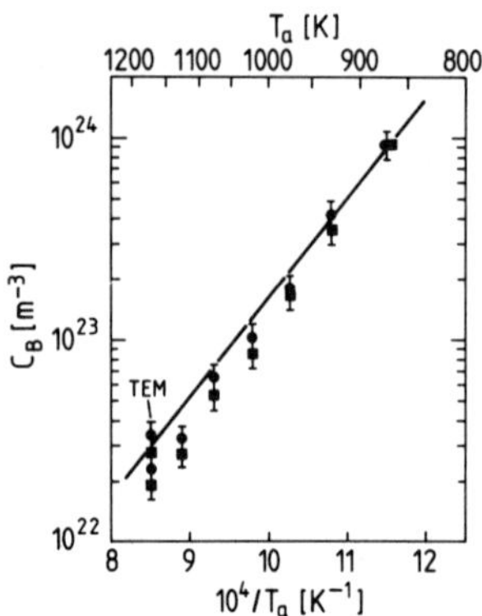

Fig. 5   Total density C_B of "small" bubbles in the bulk of specimen implanted with He^3 ($\blacksquare$) and He^4 ($\bullet$), respectively, as a function of the reciprocal annealing temperatures T_a. For T_a=1173 K the densities have also been determined by TEM ($\boxtimes \otimes$) [5].

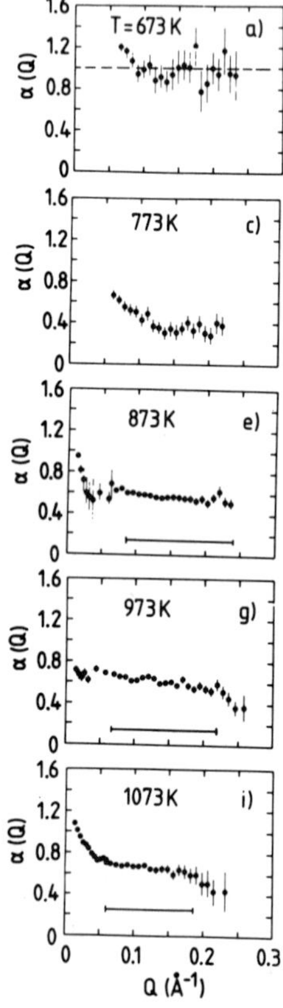

Fig. 6   Scattering intensity ratios α as a function of the scattering vector Q for pairs of Ni^{58} specimens implanted with He^3 and He^4, respectively, and annealed up to different temperatures T_a. The horizontal bars indicate the Q-range over which the α-values are averaged to yield the ω-values given in Fig. 7a [5].

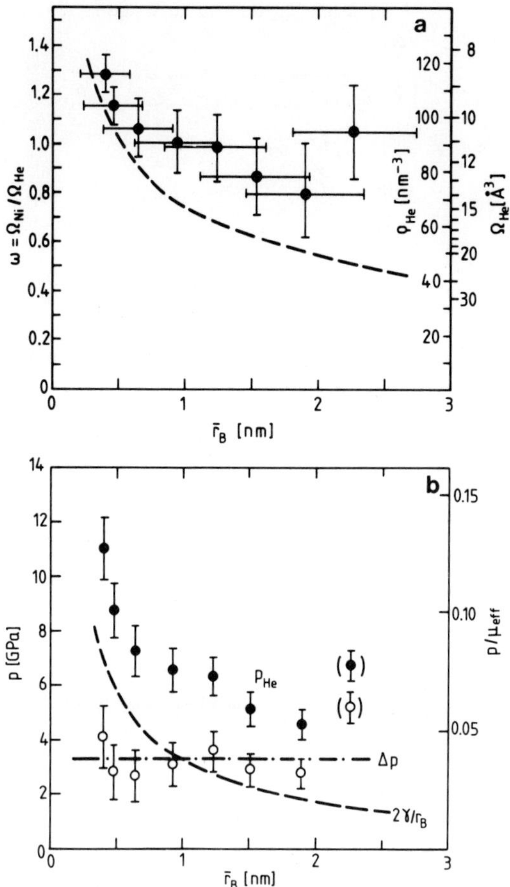

Fig. 7 (a) Atomic volume ratio ω, helium density ρ_{He} and helium atomic volume Ω_{He} as functions of the mean bubble radius $\bar{r}_B$. The filled dots are the values obtained from the measured intensity ratios (Fig. 6) via Eq. (5). The horizontal bars indicate the FWHM of the size distributions, the vertical bars correspond to the α-variation over these widths (see Fig. 6). The experimental uncertainties are smaller (about $\pm 10\%$ for ω and $+0/-10\%$ for $\bar{r}_B$). The dashed curve shows the density in bubbles in thermodynamic equilibrium assuming a surface energy $\gamma = 1.8$ Nm^{-1} (b) Helium pressure p_{He} as a function of the mean radius $\bar{r}_B$, obtained via the Trinkaus-EOS [12] from the experimental helium densities of Fig. 7a. Also shown are the equilibrium pressure $2\gamma/\bar{r}_B$ and the "overpressure" $\Delta p = p_{He} - 2\gamma/\bar{r}_B$.

calculated by:

$$c_{He} = \frac{4\pi}{3} \int_0^{\infty} \frac{\Omega_{Ni}}{\Omega_{He}} \, n(r_B) \, r_B^3 \, dr_B \tag{6}$$

The result is shown in Fig. 8. Using the experimentally determined helium densities of Fig. 7a, Eq. (6) yields a constant value of $c_{He} = 1220 \pm 100$ appm (A in Fig. 8), in agreement with the value deduced from measuring the accumulated charge of the implanted helium ions. This close coincidence of the absolute values is probably fortuitous because the experimental uncertainties in both the charge measurement and the measurement of $n(r_B)$ from SANS are considerable. It is rather the constancy of the values which supports the correctness of the determined ρ_{He} values since outgassing experiments show only negligible losses of helium up

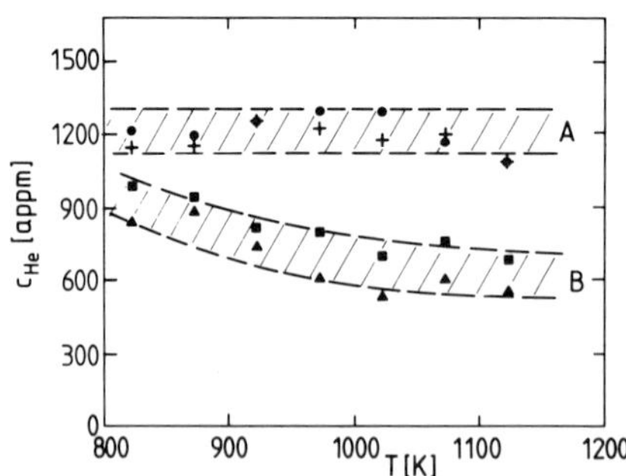

Fig. 8 Helium concentration c_{He} as a function of the annealing temperature T_a ($\bullet$ $\blacksquare$ He^3, Δ + He^4). The data in band A are calculated using the measured size distributions and the experimentally determined helium densities of Fig. 7. For the data in band B equilibrium densities (dashed curve in Fig. 7a) were used. The nominal (implanted) concentration is 1200 ± 100 appm.

to temperatures close to the melting point T_M [13]. This contradicts the decrease in helium inventory which would be obtained if equilibrium densities were used in Eq. (6) (B in Fig. 8).

Having available an equation of state valid for the pressure and temperature range considered here (Trinkaus 1986 [12]), the helium densities given in Fig. 7a can easily be converted into mean gas pressures p_{He} in the bubbles. The result is given in Fig. 7b. In the smallest bubble population investigated, with a mean radius of about 0.4 nm at $T_a = 823$ K, the mean pressure is around 11 GPa which is roughly 0.13 of the effective shear modulus μ of Ni (for comparison, the mechanical stability limit of bubbles, determined by the threshold pressure for loop-punching, is $\approx 0.5\,\mu$ [14,15]). For bubbles with $\bar{r}_B \approx 2$ nm at $T_a = 1123$ K, p_{He} has dropped to about 5 GPa. The "overpressure", i.e. the difference $\Delta p = p_{He} - 2\gamma/\bar{r}_B$ is found to be around 3 GPa, rather independent of the bubble radius.

4. Discussion

In the past it was tacitly assumed that annealing of helium containing metals at temperatures $T_a \geq 0.6\,T_M$ results in bubbles which attain a state close to thermodynamic equilibrium, i.e. their internal gas pressure p_{He} is balanced by the surface tension $2\gamma/r_B$ (γ is the surface free energy). However, a recent compilation of data on bubble coarsening in Ni [7] and their theoretical interpretation [6] led to the conclusion that the bubbles in the bulk retain a considerable overpressure up to annealing temperatures as high as 1200 K or more. Qualitatively, this is attributed to the inability of the dislocations to supply sufficient vacancies for a substantial relaxation of the pressure which in turn impedes the triggering of fast Ostwald ripening as it occurs near abundant vacancy sources such as free surfaces or grain boundaries. Coarsening in the bulk therefore proceeds by relatively slow migration and coalescence processes—the bubbles remain small and maintain their overpressure. Although the arguments presented in Refs. [6] and [7] are rather convincing, there was no direct evidence for the existence of overpressurized bubbles.

This deficiency is removed by the results of the present work: The helium densities determined via Eq. (5) from the measured intensity ratios—even if a maximum uncertainty of about 10% is assumed—are clearly above the values expected from bubbles in thermodynamic equilibrium (Fig. 7a). In addition to confirmation from comparison of the helium inventories (c.f. Eq. (6) and the following discussion), there is a further strong indication of the existence of overpressurized bubbles: the slope of the $\ln C_B$ versus $1/T_a$ plot (Fig. 5) is about a factor of 3 higher than that of the $\ln \bar{r}_B$ versus $1/T_a$ plot (Fig. 4), pointing to a constant helium density rather than equilibrium densities which would yield a factor of 2. Finally, a recent evaluation of positron annihilation data [3] yields helium densities which are compatible with the values given in Fig. 7a.

Combining all the evidence presented, the existence of an "excess pressure" in helium bubbles formed in Ni during annealing can be considered as warranted. This unexpected finding together with a theoretical analysis of the effect of pressure on the coarsening behaviour [6] could resolve several discrepancies in the data obtained from different investigations, such as, for example, large differences in the bubble structure observed (i) by low energy (keV) and high energy (MeV) helium implantation, respectively, and (ii) for low (≤ 100 appm) and high (≥ 500 appm) helium concentrations, respectively. (For a detailed study of the concentration dependence of bubble coarsening see Refs. [8,16,17]). Furthermore, in many preceding TEM investigations apparent discrepancies between the quantity of helium introduced in the specimen and the (smaller) quantity contained in the bubbles were found and usually explained by: "… part of the helium must reside in invisible clusters"—a statement which seems to be obsolete now.

Several other questions remain to be answered. They include an explanation of the very small slopes 1/m of the increase of $\bar{r}_B$ with time in isothermal annealing experiments (in some cases m as large as 10 is observed [18] whereas the highest theoretical value is 6 for surface diffusion controlled migration and coalescence). Furthermore, it is not fully understood why the dislocations present in the bulk (even at $T_a = 1173$ K, the dislocation density in this study was about $7\ 10^{13}$ m^{-2} [5]) are unable to supply sufficient vacancies for relaxing the overpressure of the "small" bubbles in the bulk. This effect seems to depend on the impurity content of the material: whereas high pressure-impeded coarsening in the bulk and a consequent split-up into "small" and "large" bubbles are observed in pure Ni and a "pure" FeNiCr alloy, preliminary results indicate that these features are absent in commercial AISI 316 L stainless steel [17].

ACKNOWLEDGEMENTS

I am grateful to F. Carsughi, P. Fichtner, W. Kesternich, H. Schroeder, D. Schwahn, H. Trinkaus and B. Viswanathan for valuable discussions.

REFERENCES

1. W. Jäger, R. Manzke, H. Trinkaus, R. Zeller, J. Fink and G. Grecelius, Rad. Effects **78**, 315 (1983).

2. S.E. Donnelly, A.A. Lucas and J.P. Vigneron, Rad. Effects **78**, 337 (1983).

3. B. Viswanathan and K. Jensen, this volume.

4. D. Schwahn, H. Ullmaier, J. Schelten and W. Kesternich, Acta Metall. **31**, 2003 (1983).

5. Qiang-Li, W. Kesternich, H. Schroeder, D. Schwahn and H. Ullmaier, Acta Metall. et Mater. **38**, 2383 (1990).

6. H. Trinkaus, Scripta Metall. **23**, 1773 (1989).

7. V.N. Chernikov, H. Trinkaus, P. Jung and H. Ullmaier, J. Nucl. Mat. **170**, 31 (1990).

8. H. Schroeder and V.N. Chernikov, this volume.

9. M. Magnani, P. Puliti and M. Stefanon, Nucl. Instr. and Methods **A271**, 611 (1988).

10. D. Schwahn, Mat. Res. Soc. Symp. Proc. **166**, 443 (1990).

11. W. Kesternich, D. Schwahn and H. Ullmaier, Scripta metall. **18**, 1011 (1984).

12. H. Trinkaus, Radiat. Effects **78**, 189 (1983).

13. H.J. von den Driesch and P. Jung, High Temp.-High Pressures **12**, 635 (180).

14. H. Trinkaus and W.G. Wolfer, J. Nucl. Mat. **122&123**, 552 (1984).

15. W.G. Wolfer, Phil. Mag. **A58**, 285 (1988).

16. H. Schroeder and P.F.P. Fichtner, Proc. Intern. Conf. on Fusion Reactor Materials, Kyoto, Dec. 1989; J. Nucl. Mat. **179–181** , 118 (1991).

17. P.F.P. Fichtner, H. Schroeder and H. Trinkaus, to be published.

18. F. Carsughi, D. Schwahn and H. Ullmaier, to be published.

INERT GAS BUBBLES IN METALS:
HIGH-TEMPERATURE BUBBLE EVOLUTION

INERT GAS BUBBLE COARSENING MECHANISMS

H. Schroeder, P.F.P. Fichtner* and H. Trinkaus

Institut für Festkörperforschung
KFA Jülich
Postfach 1913
D-5170 Jülich
Germany

ABSTRACT

In order to discriminate the different coarsening mechanisms of inert gas bubbles in metals, i.e. Ostwald ripening (OR) and migration and coalescence (MC), we have studied the growth of helium bubbles in two different metal systems choosing the experimental parameters such that specific invariances of the different mechanisms could be tested. Specimens of AISI 316 stainless steel and of a ternary model alloy, Fe-15Cr-15Ni, were helium implanted at room temperature to helium concentrations from 2-6600 appm and then annealed at 1023 K and 1073 K, respectively, over a wide range of times, 1.2 ks-12.1Ms (0.30h-3360h). In both materials we found regimes in which one of the mechanisms was dominant and regimes in which the mechanism changed from one to the other. Observed deviations from the predicted invariance behaviour may be attributed to an explicit dependence of the extremely high pressures inside small inert gas bubbles on helium concentration and time.

1.   Introduction

In metals and metal alloys inert gases have an extremely low solubility. As a consequence the inert gas which usually is deposited into materials as an undesired side effect of nuclear and ion-beam technologies segregates into bubbles. This very often leads to undesired and/or harmful changes of the materials properties. Therefore it is important to know the mechanisms of the evolution of the bubbles—nucleation and growth—because of the technological importance of effects such as helium embrittlement [1] or because of the interesting basic physical aspects (very high gas pressures and densities) of inert gases in solids [2].

Despite of the large number of papers in the literature resulting in a large database for different inert gases in many different materials there is still a standing controversy about the question: which bubble coarsening mechanism—migration and coalescence (MC) or resolution and reabsorption of gas atoms (and vacancies), i.e. Ostwald ripening—is the controlling one under which experimental conditions? The most common experimental approach to answer this question is the study of bubble size distributions (characterized by mean radius, $\bar{r}$, and standard deviation, Ω) and their evolution with annealing time t_a. Very often the working mechanism is determined by a comparison of the logarithmic derivative $\partial \ln(\bar{r})/\partial \ln(t_a)$ with theoretical predictions for the different coarsening mechanisms. As this determination is not unambiguous (as shown in section 2) we have applied a different strategy in a recent paper [3] for the first time: specific invariance conditions for both of the different coarsening mechanisms have been studied as suggested by Trinkaus [4]. In section 2, a brief description

---

*On leave from: Departamento de Engenharia Electrica, Escola de Engenharia, UFRGS, Porto Alegre, Brazil

of the coarsening mechanisms and their invariance conditions is given. Experimental details are described in section 3, while the results (including those of the previous work [3] for completeness) are presented and discussed in section 4. It should be mentioned that the data in this paper represent bulk properties in contrast to many other studies which were performed close to the surface, e.g. in thin films or by using low energy implantation ($E \leq 100$ keV).

2.   Coarsening Mechanisms

To clarify our approach we present in this section a brief description of the coarsening models and the respective invariances. If the coarsening is controlled by MC the change of the average number of helium atoms in a bubble with average size $\bar{r}$ and average pressure $\bar{p}$, $\bar{N}(\bar{r},\bar{p})$, with time yields [4,5]:

$$\frac{d\bar{N}(\bar{r},\bar{p})}{dt_a} \propto \bar{r}\, D_B(\bar{r},\bar{p})\, c_{He} \qquad \text{[MC]} \qquad (1)$$

$D_B(\bar{r},\bar{p})$ is the diffusion coefficient of the bubble as a whole, which depends on the basic diffusion mechanism such as surface diffusion (SD), volume diffusion (VD) or vapour transport (VT) through the bubble, c_{He} is the atomic concentration of the totally implanted helium, which is assumed to be completely collected in the bubbles. The pressure $\bar{p} = \bar{p}\,(\bar{r}, c_{He}, t_a, T)$ may be an explicit function of several parameters such as mean radius, $\bar{r}$, implanted helium concentration, c_{He}, annealing time, t_a, and temperature, T. As we have performed annealing experiments at constant temperature we can neglect the T-dependence. If $\bar{p}$ is not explicitly dependant on the annealing time t_a we can separate the dependencies on $\bar{r}$ and t_a. Simple integration with respect to t_a yields:

$$\int_0^{\bar{r}} dr' \, \frac{dN(r',\bar{p})}{dr'} \, \frac{1}{r' D_B(r',\bar{p})} \propto c_{He}\, t_a \qquad \text{[MC]} \qquad (2)$$

In the case that $\bar{p}$ is not explicitly a function of c_{He}, the integral in equation (2) and therefore $\bar{r}$ itself only depend on the product $c_{He}\, t_a$ but not on c_{He} and t_a separately, i.e. $\bar{r}$ is *invariant* with respect to changes of c_{He} and t_a as long as the product ($c_{He}\, t_a$) is kept constant. Only for very few cases can the integral be solved, e.g. for ideal gas in equilibrium bubbles (but this is a very bad approximation in the size range of the experimental data, i.e. 1 nm $\leq \bar{r} \leq 8$ nm) or for constant pressure. For these cases the integration results in a power law:

$$\bar{r}^{\,n} \propto D_X\, c_{He}\, t_a \qquad \text{[MC]} \qquad (3)$$

D_X is the diffusion coefficient of the underlying diffusion mechanism (x=SD, VD or VT). A similar expression can be given for the case of OR [4,6]:

$$\frac{d\bar{N}(\bar{r},\bar{p})}{dt_a} \propto \bar{r}\, D_{He}(\bar{r},\bar{p})\, \hat{c}_{He}(\bar{r},\bar{p}) \qquad \text{[OR]} \qquad (4)$$

$N(\bar{r})$ and $\bar{r}$ have the same meanings as above, D_{He} is the diffusion coefficient of *dissolved* helium atoms in the matrix, $\hat{c}_{He}(\bar{r})$ is the average concentration of dissolved helium atoms in the matrix in the presence of a bubble distribution with average radius $\bar{r}$. If one neglects an explicit dependence of $\bar{p}$ on t_a and T for the same reasons as above, separation and integration with respect to time yields:

290

Table 1 Coarsening Mechanisms and Exponents (slope = 1/n)

COARSENING MECHANISM	ATOMIC PROCESS OR EQ. OF STATE (e. o. s.)	EXPONENT PARAMETER n	
		equilibrium bubbles	constant volume
migration and coalescence (MC) $r = (D_b \cdot c_{He} \cdot t_a)^{1/n}$	surface diffusion	5	6
	volume diffusion	4	5
	vapour transport	3	5
ostwald rippening (OR) $r = (D_{He} \cdot \hat{c}_{He} \cdot t_a)^{1/n}$	ideal gas	2	3
	realistic e. o. s.	2 – 6	3

r = mean bubble radius $\hat{c}_{He}$ = dissolved He conc.
t_a = annealing time D_b = diffusivity controling bubble migration
c_{He} = implant. He conc. D_{He} = diffusivity of He in the matrix

$$\int_0^{\bar{r}} dr' \frac{d\bar{N}(r',\bar{p})}{dr'} \; \frac{1}{r' D_{He}(r',\bar{p})\hat{c}_{He}(r',\bar{p})} \propto t_a \qquad [OR] \qquad (5)$$

If $\bar{p}$ is not an explicit function of the implanted helium concentration, c_{He}, the integral and therefore $\bar{r}$ depend only on t_a. Hence at constant t_a, $\bar{r}$ is invariant with respect to c_{He}. Once more, the integration is possible analytically only for very simple cases. Assuming ideal gas behaviour and equilibrium bubbles (i.e. the pressure p is balanced by $2\gamma/r$ where γ is the specific surface energy), integration yields:

$$\bar{r}^n \propto D_{He} \exp(-H^s/kt)\, t_a$$

$$\propto \exp(-H_{Diss}/kT)\, t_a \qquad [OR] \qquad (6)$$

H^s is the enthalpy of solution for helium, $H_{Diss} = H^s + H^m$ is the dissociation enthalpy, and H^m is the migration enthalpy of the helium atoms in the matrix.

Table 1 shows the exponent n for some simple conditions for both mechanisms. It can be easily realized that an unambiguous identification of the mechanism cannot be derived from the often used exponent $n = (\partial \ln(\bar{r})/\partial \ln(t_a))^{-1}$.

Therefore we have used the invariances [4] outlined above as a guideline for our experiments. The expected behaviour is schematically shown in Figs. 1a and 1b. In the case of OR, $\bar{r}$ is independent of the totally implanted helium concentration, c_{He}. Hence, if $\bar{r}$ is plotted vs. c_{He}, $\bar{r}$ is a constant for constant annealing time, while in the case of MC $\bar{r}$ monotonically increases with c_{He} under the same conditions (see dashed lines in Fig. 1a). On the other hand, if the product ($c_{He}\, t_a$) is kept constant $\bar{r}$ is also a constant in the case of MC, while for OR the mean radius is monotonically decreasing with increasing c_{He} because ($c_{He}\, t_a$) = constant implies that $t_a \propto 1/c_{He}$ (see Fig. 1b). The expected behaviour of $\bar{r}$ vs. c_{He} for a transition from MC to OR for both of the described invariances is represented by full lines in Figs. 1a and 1b, respectively.

3. Experimental

The specimens were prepared from 100 µm thick foils of two different materials: a) commercial AISI 316 austenitic steel, solution annealed at 1323 K in high vacuum for 2 hours, and aged at 1073 K for 24 hours to avoid drastic microstructural changes such as precipitation during post-implantation annealing treatments, b) a ternary model alloy Fe-15%Ni-15%Cr, which was solution annealed at 1323 K. Samples from these specimens were homogeneously implanted with ^{4}He using a 28 MeV ^{4}He^{++} beam, energy-degraded by a rotating wheel with

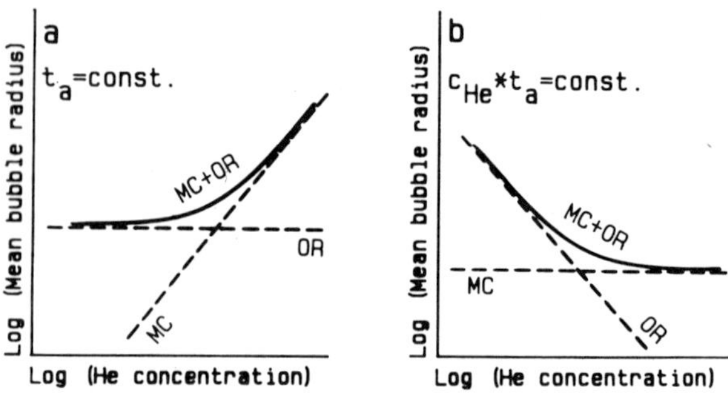

Fig. 1   Schematic presentation of the invariances of the mean radius $\bar{r}$ with respect to the implanted helium concentration, c_{He}, for the different coarsening mechanisms.
a) Ostwald ripening (OR): t_a = const., b) Migration and Coalescence (MC): $c_{He} \cdot t$ = const.

Table 2 Experimental results (annealing temp. 1023 K): AISI 316 SS. The upper values correspond to mean radius (nm) and the lower ones to the standard deviation of the bubble size distribution.

He conc. (appm)	annealing time (hours)							
	1.05	3.33	10.5	33.3	105	334	1022	3359
2							(3.75)	
6.3						(3.27)		
20					3.00 0.45		5.78 2.27	7.46 2.54
63				1.87 0.42				7.76 3.21
200			1.82 0.55		2.96 0.94		4.03 1.10	8.17 3.22
630		1.90 0.85		2.43				5.97 1.43
2000	1.89 0.68		2.94 1.22		4.73 1.35			7.05 2.08

Table 3 Experimental results (annealing temp. 1073 k): Fe–15Ni–15Cr. The upper values correspond to mean radius (nm) and the lower ones to the standard deviation of the bubble size distribution (nm).

He conc. (appm)	annealing time (hours)							
	0.33	1.0	3.33	10	33.3	100	333	1000
2							3.16 0.82	
6.7						3.54 0.99		4.08 1.27
20					2.04 0.63		3.16 1.06	
67				1.55 0.43		2.09 0.60		3.42 1.16
200			1.47 0.39		1.78 0.45		2.32 0.56	
667		1.16 0.39		1.81 0.46		2.24 0.65		2.69 0.59
2000	1.28 0.36		1.77 0.48		2.21 0.61		2.61 0.59	
6667		1.50 0.44	2.14 0.58			2.52 0.59	2.63 0.55	2.85 0.58

aluminium foils of different thicknesses to allow homogeneous helium deposition as a function of depth. During implantation the samples were maintained at room temperature in vacuum [7]. The implanted helium concentrations were varied over a wide range: a) 2-2000 appm for 316 SS, b) 2-6660 appm for the ternary alloy. 3 mm disks were punched out from the implanted samples and annealed at 1023 K (316 SS) and 1073 K (Fe-Ni-Cr), respectively, during different time intervals from 1 to 3360 hours in high vacuum ($p \leq 10^{-3}$ Pa).

Specimens for transmission electron microscopy (TEM) investigations were prepared using the double jet electropolishing technique. The TEM observations were performed at 120 keV by imaging the bubbles in underfocused kinematical conditions, such that they appear as white disks surrounded by dark fringes. Bubble size distributions were made from enlarged micrographs using a particle size analyser.

4.   Results and Discussion

Table 2 (316 SS) and Table 3 (Fe-Ni-Cr) present our experimental results of mean radius $\bar{r}$ and standard deviation of the size distributions obtained for each implanted He concentration and annealing time investigated. In general the distributions were determined from populations of about 300 to 1000 bubbles. No distinction between bubbles trapped at dislocations and those free in the matrix have been made. The observations were done in regions far from grain boundaries or free surfaces. In the case of 2 and 6.3 appm samples, the mean radii are quoted between parenthesis because less than 30 bubbles were counted. The overall estimated error in bubble size is about 15%.

4.1.   AISI 316 SS

In Fig. 2 the mean radius, $\bar{r}$, is plotted versus the implanted helium concentration, c_{He}, for different, but constant annealing times, t_a, which correspond to the columns in Table 2. This plot is a test of the invariance of $\bar{r}$ with respect to c_{He} for OR.

While for short annealing times (10.5 h and 33.5 h) one can conclude from the limited data that most probably MC is the dominant mechanism, because $\bar{r}$ is increasing with increasing c_{He}, OR seems to be the working mechanism at very long annealing times (t_a=3360 h) because $\bar{r}$ is independent of c_{He} (see Fig. 1a for comparison). For these long anneal times additional intragranular precipitation was observed which may have influenced the growth of the bubbles. (This was one reason for using the simpler model alloy for additional experiments at higher temperature). The plot for t_a=105 h in Fig. 2 looks very similar to the combined curve in Fig. 1a, indicating a transition from OR at low helium concentrations to MC at high ones. Although the database to show this transition curve is not large, Fig. 2 shows a convincing sequence from MC at low t_a to OR at high t_a. In addition, previous data also on a 316 steel from

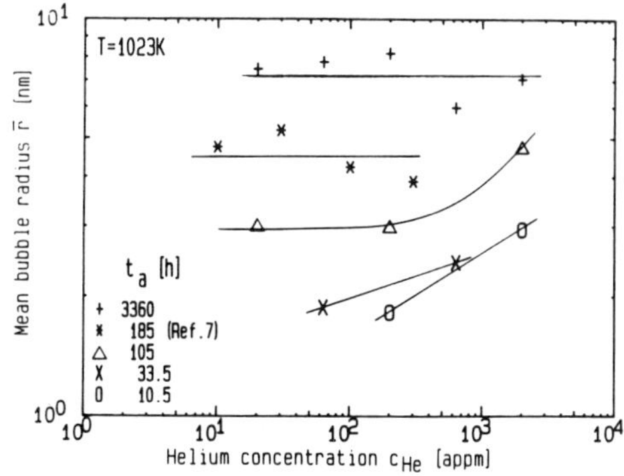

Fig. 2   AISI 316 SS: Mean bubble radius, $\bar{r}$, vs. implanted helium concentration, c_{He}, for different annealing times, t_a.

Rothaut et al. [7], who determined the $\bar{r}$ dependence in a narrower concentration range, fit very well to the data presented in the present work . From Fig. 2 one has to conclude that the transition from OR to MC shifts to larger concentrations with increasing annealing time.

In Fig. 3 the mean radius $\bar{r}$ is plotted once more against the implanted helium concentration c_{He} for constant, but different values of the product ($c_{He}\, t_a$) in order to test the invariance for MC. These data correspond to the diagonals from bottom left to top right in Table 2. The curves guiding all three data sets look very similar to the combined curve in Fig. 1b indicating a transition from OR at low concentrations to MC at high concentrations. For ($c_{He}\, t_a$) = 2.1 10^3 appm·h the shape of the curve is most convincing for a transition although the data for very small concentrations have possibly large errors because of the poor statistics. For concentrations higher than 20 appm, the mean radius is invariant indicating clearly that MC is the underlying coarsening mechanism. The transition from OR to MC moves to higher concentrations with increasing $c_{He}\, t_a$ parameter (or increasing time for constant c_{He}). This is the same conclusion as for the plot in Fig. 2.

Fig. 4 presents the 'usual' plot, mean radius, $\bar{r}$, vs. annealing time, t_a. The data correspond to the rows in Table 2 for the helium concentrations 20-2000 appm. If one fits these data by a power law, $\bar{r} \propto t_a^{1/n}$, there is a steady decrease of n from 6 to about 3 with decreasing concentration. Nevertheless it is impossible to identify a coarsening mechanism from Fig. 4 for two reasons: (i) For both coarsening mechanisms the observed values fall within the predicted ranges (see Table 1). (ii) As we have shown in Fig. 2 and 3 for constant helium concentration, the coarsening mechanism is changing within the applied range of annealing time for nearly all the concentrations investigated. This means, either that the time dependence of $\bar{r}$ cannot be described by one power law only, if the exponents for the different coarsening mechanisms are different (a two exponent fit would have very large uncertainties due to the small number of data points), or it can be described by one power law, but nevertheless two different coarsening mechanisms may be working having the same n.

Using the strategy of different invariances for the different coarsening mechanisms, migration and coalescence (MC) and Ostwald ripening (OR), we were able to identify unambigously the working coarsening mechanisms in bubble growth experiments at 1023 K after room temperature helium implantation in the AISI 316 stainless steel. We also could show that for many of the helium concentrations investigated there is a transition from MC to OR with increasing annealing time. In general, the bubble size at this transition is concentration dependent and increases with increasing helium concentration. In contrast, the usually applied method to identify the coarsening mechanism only by the time dependence of the mean radius is not sufficient.

4.2. Fe-15Ni-15Cr

Similar plots have been made for the ternary model alloy. Fig. 5 shows the test of the invariance for OR, i.e. mean radius, $\bar{r}$, vs. the implanted helium concentration, c_{He}, for different, but constant annealing times, t_a, at 1073 K (corresponding to the data in the columns of Table 3). A comparison of Fig. 5 with Fig. 1a yields deviations from the predicted behaviour. At low helium concentrations we find decreasing bubble size with increasing c_{He} instead of the invariant size characteristic of OR. At large c_{He} the bubble radii increase with increasing helium concentrations, i.e. these branches can be identified as controlled by MC. With increasing annealing time, this corresponds also to larger bubble sizes, the turnover to MC controlled behaviour is shifted to higher concentrations, as expected from the 316 results, and the slope of the MC branches decreases.

Fig. 6 shows the test for the MC invariance condition, i.e. mean radius, $\bar{r}$, vs. implanted helium concentration, c_{He}, for different, but constant values of the product ($c_{He}\, t_a$) (these data correspond to the diagonals in Table 3). The comparison with Fig. 1b identifies OR at low and MC at high c_{He}, but once more with increasing deviations from the expected invariant behaviour in the MC branches with decreasing bubble size or with decreasing annealing time at

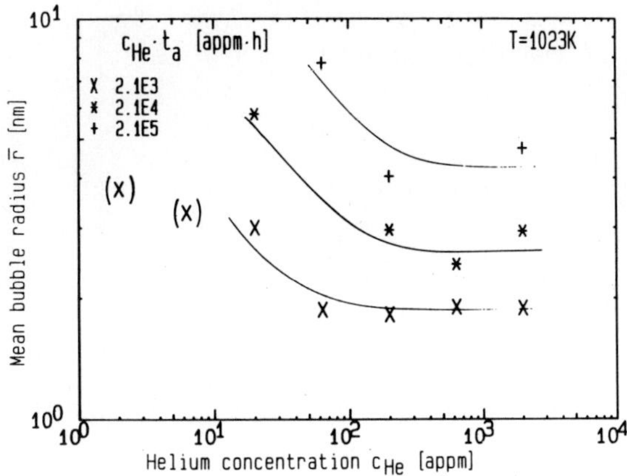

Fig. 3 AISI 316 SS: Mean bubble radius, $\bar{r}$, vs. implanted helium concentration, c_{He}, for different values of the product $c_{He} \cdot t_a$.

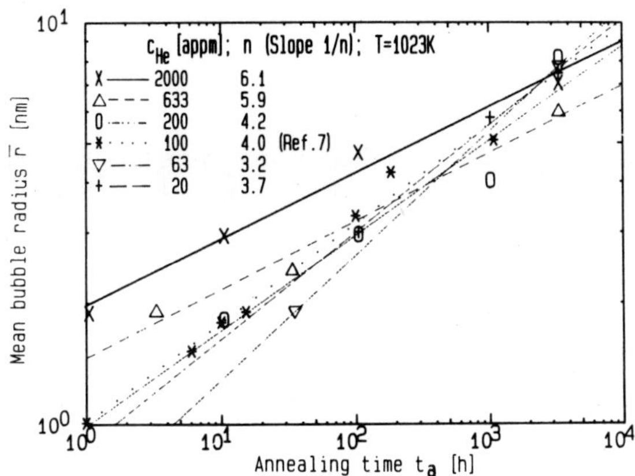

Fig. 4 AISI 316 SS: Mean bubble radius, $\bar{r}$, vs. annealing time, t_a, for different helium concentrations, c_{He}; $1/n$ is the slope of a least square fit.

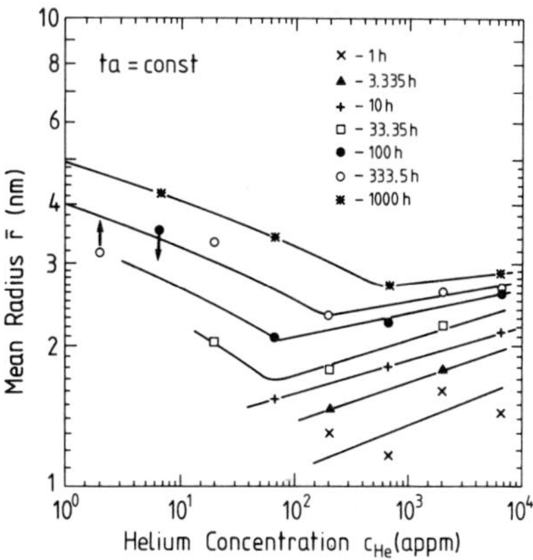

Fig. 5 Fe-15Ni-15Cr: Mean bubble radius $\bar{r}$, vs. implanted helium concentration, c_{He}, for different annealing times, t_a.

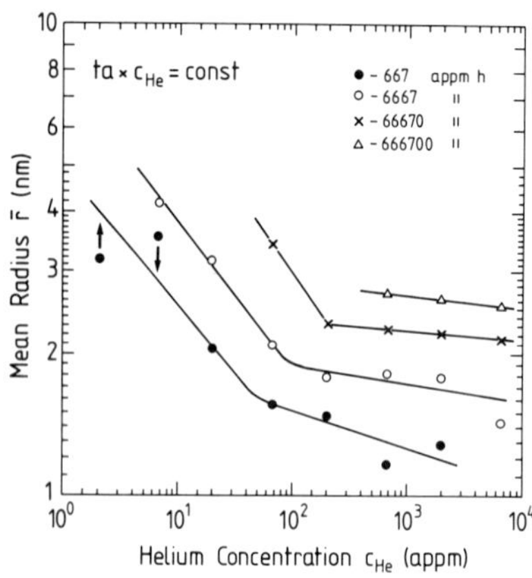

Fig. 6   Fe-15Ni-15Cr: Mean bubble radius, $\bar{r}$, vs. implanted helium concentration, c_{He}, for different values of the product $c_{He} \cdot t_a$.

constant concentration. As in Fig. 5 the transition from OR to MC controlled behaviour is shifted to higher c_{He} with increasing t_a. The data points in Figs. 5 and 6 are identical. All those lying in the OR controlled regime in Fig. 5 also belong to this regime in Fig. 6. The same can be stated for those points in the MC controlled branches in both plots. As deviations from the predicted invariant behaviour have been found for each regime in the corresponding tests (OR in Fig. 5 and MC in Fig. 6) all the data points deviate from the predicted behaviour although some of them seem to follow the predictions qualitatively (OR regime in Fig. 6 and MC regime in Fig. 5). What is the reason for these deviations?

Deviations from the invariance conditions mean that $\bar{r}$ depends on c_{He} even though t_a or $(c_{He}\, t_a)$ are kept constant, i.e.

$$\left| \frac{\partial \ln(\bar{r})}{\partial \ln(c_{He})} \right|_{t_a \text{ or } (c_{He} \cdot t_a)} \neq 0 \tag{7}$$

As discussed in connection with the invariance conditions (see eqs. (2) and (5)) an explicit dependence of the gas pressure $\bar{p}$ on c_{He} will destroy the stated invariances. What are the conditions for such a dependence? It can be imagined that $\bar{p}$ depends on c_{He} if the bubbles are not in equilibrium with the thermal vacancy concentration, in particular if they are overpressurized: $p \geq p_{equil} = 2\gamma/r$. There are some recent experimental results indicating overpressure in He bubbles in bulk Ni [8,9]. In addition Trinkaus [10] has developed models of the influence of (over-)pressure on coarsening. As a result, high pressures can substantially suppress the coarsening rates, but to a different extent for OR and MC, respectively. This may lead to shifts in the (size, concentration, time, temperature) regimes controlled by the different mechanisms. As the overpressure should increase with increasing concentration, the suppression of the coarsening should be more severe at higher concentrations. This is consistent with the trends in Figs. 5 and 6.

The 'usual' plot $\ln(\bar{r})$ vs. $\ln(t_a)$ at different helium concentrations, c_{He}, is shown in Fig. 7. For short annealing times, the differences between the different concentrations are much larger than for large annealing times. The main reason for this is the continuous decrease of the slope with increasing t_a, especially at the high concentrations. While at low t_a the slope corresponds to n of about 7, at times larger than 50h n ≥ 10. This also demonstrates the large retardation of the coarsening in the ternary alloy which is consistent with the predictions of the Trinkaus model [10] on the effect of very high pressures which are close to the GPa range even

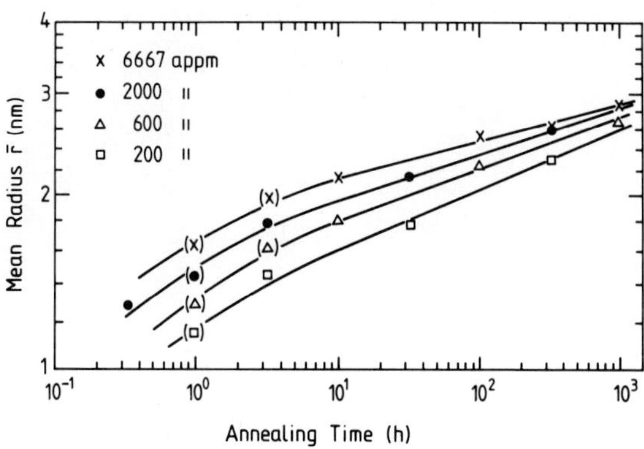

Fig. 7   Fe-15Ni-15Cr: Mean bubble radius, r̄, vs. annealing time, t_a, for different helium concentrations, c_{He}.

for equilibrium bubbles, if r = 2 nm. The curves in Fig. 7 also would be in agreement with the assumption that the nucleation of new ledges on the facets of the bubbles is controlling the growth rate because then r̄ $\propto \ln(t_a)$. Indeed, even the small bubbles of the materials investigated are facetted, but because they have many facets of different orientations they appear nearly spherical. The high number of different facets in the small bubbles suggests an argument against ledge nucleation because each facet consists only of a few atoms. It would then seem very unlikely that there is a high barrier against nucleation of new facets.

5.   Conclusions

In the stainless AISI 316, migration and coalescence as well as Ostwald ripening could be identified as mechanisms controlling coarsening of helium bubbles using the new strategy of invariant behaviour. OR is dominant at low helium concentrations and/or long annealing times, while MC dominates coarsening at short times and/or high concentrations. The results in the ternary model alloy Fe-15Ni-15Cr deviate from the predicted invariant behaviour. As a possible explanation, a concentration dependence of the helium gas pressure has been suggested leading to overpressurized bubbles in this system. This overpressure may retard the coarsening. As this argumentation is reasonable, but speculative to some extent, future work should concentrate on showing a) the existence of overpressurized bubbles in this system and b) the reasons for the overpressurization.

ACKNOWLEDGMENTS

One of the authors, PFPF, would like to thank the support from the Alexander von Humboldt Foundation and from the Brazilian National Research Council (CNPq).

REFERENCES

1. H. Schroeder, W. Kesternich and H. Ullmaier, Nucl. Eng. and Design/Fusion 2 (1985) 65.

2. S. E. Donnelly, Rad. Effects **90**, 1 (1985).

3. H. Schroeder and P.F.P. Fichtner, J. Nucl. Mat. **179–181**, 1007 (1991).

4. H. Trinkaus, Rad. Effects **78**, 189 (1983).

5. P.J. Goodhew and S.K. Tyler, Proc. R. Soc. Lond. **A377**, 151 (1981).

6. A.J. Markworth, Met. Trans. **4**, 2651 (1973).

7. J. Rothaut, H. Schroeder and H. Ullmaier, Phil. Mag. **A47**, 781 (1983).

8. V.N. Chernikov, H. Trinkaus, P. Jung and H. Ullmaier, J. Nucl. Mat. **170**, 31 (1990).

9. Quiang-Li, H. Schroeder, W. Kesternich, D. Schwahn and H. Ullmaier, Acta Met. Mat. **38**, 2383 (1990).

10. H. Trinkaus, Scripta Met. **23**, 1773 (1989).

INFLUENCE OF REAL GAS BEHAVIOUR ON THE OSTWALD RIPENING OF INERT GAS BUBBLES IN BULK MATERIALS

P.F.P. Fichtner*, H. Schroeder and H. Trinkaus

Institut für Festkorperforschung
Forschungszentrum Jülich
P.O. Box 1913, D-5170 Julich
Germany

ABSTRACT

We have studied the influence of real gas behaviour on the Ostwald ripening kinetics of inert gas bubbles in metals using a computer simulation approach. The time evolution of He, Ar and Xe bubble populations and particularly of their mean radii (r_m) were analysed and compared with the corresponding ideal gas behaviour. For all cases, significant qualitative and quantitative differences in the coarsening under real and ideal gas conditions were found. These differences increase with increasing atomic number of the gas and occurs in a bubble size range of $r_m < \approx 70$ nm. The results are generalized in an attempt to provide an universal description of the Ostwald ripening kinetics of gas bubbles for real gas behaviour.

1. Introduction

Microstructural modifications such as precipitation and coarsening of gas bubbles are typical thermal evolutions in materials containing inert gases, even for concentrations as low as fractions of atomic parts per milion (appm). The coarsening of gas bubbles has been discussed and experimental results have been interpreted in terms of two main mechanisms: bubble migration and coalescence (MC) [1,2] on one side, and resolution and reabsorption of gas atoms and vacancies (also named Ostwald ripening, OR) [3-5] on the other side. A procedure to discriminate between these mechanisms has been proposed [6] and applied recently for the cases of He in AISI 316 stainless steel [7] and in a $Fe_{70}Ni_{15}Cr_{15}$ model alloy [8]. These experiments covered a set of parameters comprising He concentrations, c_{He}, in the range $2 \le c_{He} < 7000$ appm and annealing times, t_a, in the range of $0.3 \le t_a < 3500$ hours at temperatures $T_a \approx 1000$ K, which resulted in mean bubble radii in the range of $1 \le r_m \le 8$ nm depending on the material and on the c_{He} and/or t_a values. As a general result, it was found that both mechanisms (MC and OR) may govern the coarsening in certain regimes of the experimental parameters: MC rather dominates the coarsening in the regime of high c_{He} and/or short t_a, while OR is the dominant mechanism in the regime of low c_{He} and/or large t_a.

For bubbles in the size range of $r<8$ nm (as mentioned above), the condition of thermodynamic equilibrium (implying that the gas pressure p within a bubble of radius r is balanced by the surface tension of the medium, $p=2\gamma/r$, where γ is the specific surface energy) yields values of the pressure in the GPa range in which the ideal gas law is a poor approximation and a more sophisticated equation of state (EOS) is needed [9]. In contrast, previous analysis of the OR kinetics of gas bubbles have only considered ideal gas behaviour [3,4].

---

*On leave from: Departamento de Engenharia Electrica, Escola de Engenharia, UFRGS, Porto Alegre, Brazil

Fundamental Aspects of Inert Gases in Solids
Edited by S.E. Donnelly and J.H. Evans, Plenum Press, New York, 1991

On the other hand, an analytical theory incorporating real gas properties has been proposed [10] and preliminary results have been given previously [6]. In addition, a more systematic analysis of the OR kinetics for such conditions, using a computer simulation approach, has been performed recently [11] for the specific case of helium in metals. In the present contribution we extend this work to the other inert gases. In this sense we attempt to provide a universal description of the OR kinetics of equilibrium gas bubbles for real gas behaviour.

2.   Basic Assumptions and Equations

For completeness, we summarize here the basic assumptions and equations adopted in the simulations [11]: (1) the system is assumed to be infinite and homogeneous, consisting of a parent phase and a dispersed second phase formed by gas bubbles occupying a small volume fraction so that a diffusional mean field approach may be used; (2) the dissolved gas within the matrix forms an ideal solution; (3) equilibrium thermodynamic concepts may be applied even to small bubbles containing only a few gas atoms; (4) the bubbles are spherical (althought a generalization to non-spherical bubbles is straightforward); (5) the bubbles are in local thermodynamic equilibrium $(p=2\gamma/r)$.

Under these assumptions the rate of change with time t of the number of gas atoms N in a given bubble of radius r may be writen as:

$$dN^*/dt^* = r^*\Delta c_g^* , \tag{1}$$

which is expressed in terms of reduced variables (denoted by *) as suggested in [6]. The reduced time t^* comprises all properties of the dissolved gas in the matrix,

$$t^* = (3/\Omega)\ (2\gamma/kT)^2\ D_g\ \exp(-G_g^s/kT)\ t . \tag{2}$$

D_g is the gas diffusivity, G_g^s is the Gibbs free enthalpy of solution, Ω is the matrix atomic volume, and kT has the usual meaning. The properties of the gas phase are accounted for in N^*, r^* and Δc_g^* defined as:

$$N^* = n^{-2}z^{-3} = (3/4\pi)\ (2\gamma/kT)^{-3}\ N, \tag{3}$$

$$r^* = (nz)^{-1} = (2\gamma/kT)^{-1}\ r, \tag{4}$$

and

$$\Delta c_g^* = [<r\ \exp(\mu_g(r)/kT)>/<r>] - [\exp(\mu_g(r)/kT)] \tag{5}$$

where n is the atomic density of the gas phase, $z=p/(nkT)$ is the compressibility factor defining the deviation from the ideal gas behaviour $(z=1)$, and μ_g is the chemical potential of the gas phase. Δc_g^* is proportional to the difference between the dissolved gas concentration in the mean field (first term on the right-hand side of Eqn. 5) and the amount of gas dissolved close to the surface of a bubble with radius r (second term on the right-hand side of Eqn. 5). The brackets in Eq. 5 mean ensemble averaging and render Eqn. 1 into a rather complicated integro-differential equation when real gas behaviour is accounted for [6].

In order to describe the gas properties in terms of n(p,T) and μ_g(p,T) we considered the semi-empirical high density EOS for ^{4}He given in ref. 6 (Trinkaus EOS) and a more simplified EOS derived on the basis of a hard sphere interatomic potential approach by Carnaham-Starling (C-S EOS) [12]. Such a hard sphere EOS provides a generalization to the other inert gases in terms of a characteristic dimension of the system (volume of the hard sphere) which

300

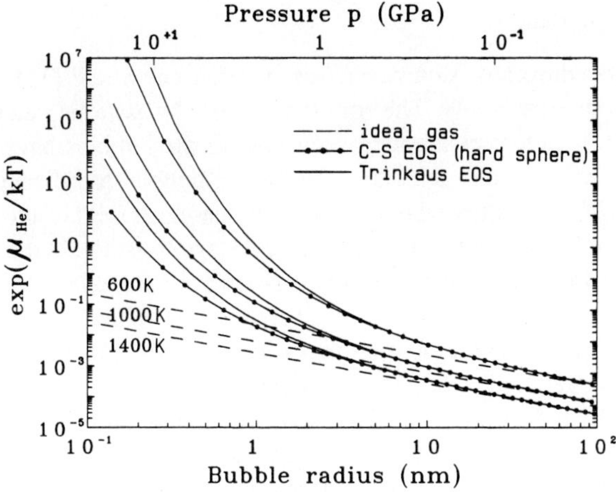

Fig. 1   Dependence of the dissolved He concentration $c_g \propto \exp(\mu_{He}/kT)$ in the vicinity of an equilibrium bubble on the bubble radius r or internal pressure p for a typical metallic matrix (γ=2 N/m) at temperatures T in the range $600 \leq T \leq 1400$ K.

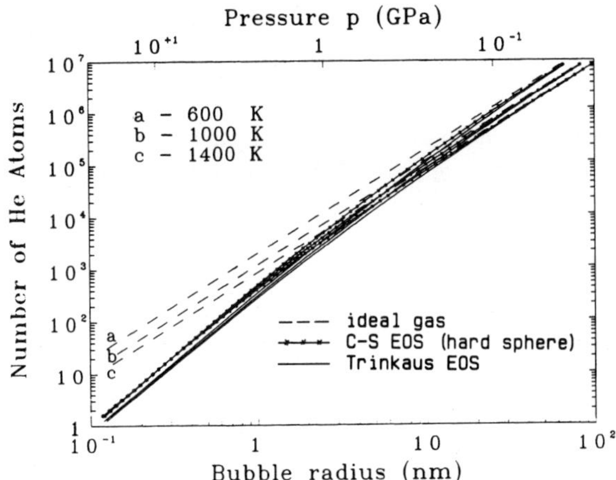

Fig. 2   Number density of He atoms in equilibrium bubbles at different temperatures ($600 \leq T \leq 1400$ K) as a function of the bubble radius r or internal pressure p = $2\gamma/r$ for a metallic matrix with γ=2 N/m.

may be obtained from the empirical values of the atomic volumes of the liquid phase at melting [13].

In Figs. 1 and 2 the dependences of $\exp(\mu_g/kT)$ and $N=(4\pi/3)r^3 n(p,T)$ on the bubble radius r are illustrated for equilibrium He bubbles in a metallic matrix (γ=2 N/m) at various temperatures assuming ideal gas, Trinkaus EOS and C-S EOS, respectively. There are significant differences between the real and ideal gas approximations (especially in Fig. 1), but less pronounced differences between the two real gas cases. These discrepancies are expected to manifest themselves in the coarsening kinetics.

We point out that in a hard sphere approach the compressibility factor z becomes formally infinite at closest packing. Such divergent behaviour is expected for small bubble radii which are out of the range of parameters considered in Figs. 1 and 2. We also note that the differences between a hard sphere approach and an EOS based on more realistic interatomic potentials are expected to be most significant for the case of He but to decrease with increasing atomic number for the other gases.

3. Numerical Analysis

The simulation procedure has been described in detail elsewhere [11] and therefore in this paper we present a summary only. The simulation starts by assuming an arbitrary bubble size distribution (BSD), which is discretized such that all the bubbles have different radii. In a small time interval Δt^*, chosen such that, at most, one bubble may disappear during this time interval, all the bubbles are allowed to interact with the mean field of the dissolved gas atoms as described by Eqn. 1 assuming that the mean field stays constant. In the next step, new mean values (r_m, c_m, etc.) are calculated and the cycle is repeated. Results consisting of mean values and BSD histograms are printed out at given preset times. When the number of bubbles has decreased to a given limit, a repopulation procedure restores this number to the initial value, maintaining the characteristics of the ensemble at that time. As a consequence, a good definition of the BSD is preserved at all times of the simulation.

The computer code was tested by applying it to the ideal gas case for which the stable asymptotic quasi-steady-state solution has been analysed previously [4]. An excellent agreement between simulated and analytically predicted behaviour, comprising the temporal evolution of the mean radius as well as the BSD, shows that the code is working properly.

4. Results and Discussion

In order to illustrate the influence of the real gas behaviour on the OR kinetics we first consider the He case using the Trinkaus EOS and compare the results with those for ideal gas assuming otherwise the same simulation conditions. We have started the simulations taking the analytical quasi-steady-state BSD for the ideal gas [4] with $r_m=1$ nm as the initial BSD. The time evolution of the mean radius $r_m^*(t^*)$ is shown in Fig. 3 for both, ideal gas (IG) and real gas (RG), and for various temperatures in the range $923K \leq T \leq 1223K$, assuming a typical metallic matrix ($\gamma = 2$ N/m). The mean radius for the RG case grow much faster and approaches the IG results at $r_m \approx 50$ nm (see scale on the right). We remark that the deviations from the constant slope observed on the initial stages of the IG curves are artefacts due to the finite initial $r_m=1$ nm value of the BSD, but that the r_m growth follows strictly the analytical behaviour.

Fig. 3 also illustrates the scaling obtained with the use of the reduced coordinates t^* and r^* (Eqs.2 and 4), which provides some kind of master curve depending only weakly on the temperature.

To test whether $r_m^*(t^*)$ for real gas may depend on the particular shape of the initial BSD we have performed simulations considering different initial BS, such as a gaussian distribution (with standard deviation $\sigma = 0.25$ nm) and an empirical distribution with a long tail on the large bubble side. The results obtained show that within small fluctuations the curves are very similar to those in Fig. 3. Hence, the $r_m^*(t^*)$ curves as presented in Fig. 3 for the real gas case actually represent a more general behaviour. This is an important point to be emphasized, because it shows that the time dependence of the mean radius cannot be described in terms of a power law of the type $r_m \propto t^{1/n}$. For example, an evaluation of such a power law in terms of the logarithmic derivative $[d\log(r_m)/d\log(t)]^{-1}$ yields values for the exponent parameter continuously changing from $n \approx 7$ at short times to the ideal gas limit $n=2$ at long times.

Similar to the growth rate, the BSD is also changing continuously with time. This is illustrated in Fig. 4 for the the data corresponding to the higher temperature case in Fig. 3. Starting from the analytical quasi-steady-state BSD for IG (Fig. 4a), the RG-BSD quickly develops a shape characterized by a short tail towards smaller r values and a relative sharp cut at large r values (Fig. 4b). With increasing time (Figs. 4c-d) the RG-BSD becomes wider but is still much narrower than the corresponding IG-BSD for the same r_m and same number of bubbles (solid curves). Even for $r_m \approx 60$ nm the real gas behaviour still influences the shape of the BSD (Fig. 4f).

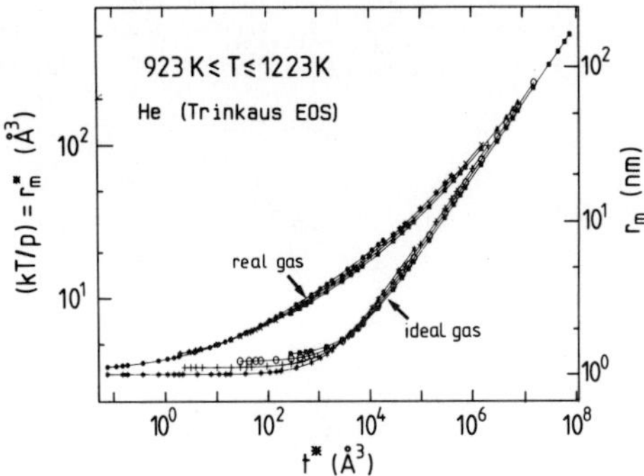

Fig. 3 Time evolution of the mean radius r_m in terms of the reduced variables t^* (Eq. 2) and r^* (Eq. 4) for ideal and real gas behaviour (Trinkaus EOS) at various temperatures, assuming the same matrix properties. The real gas system shows a much faster initial growth but converges to the ideal gas behaviour at large r_m.

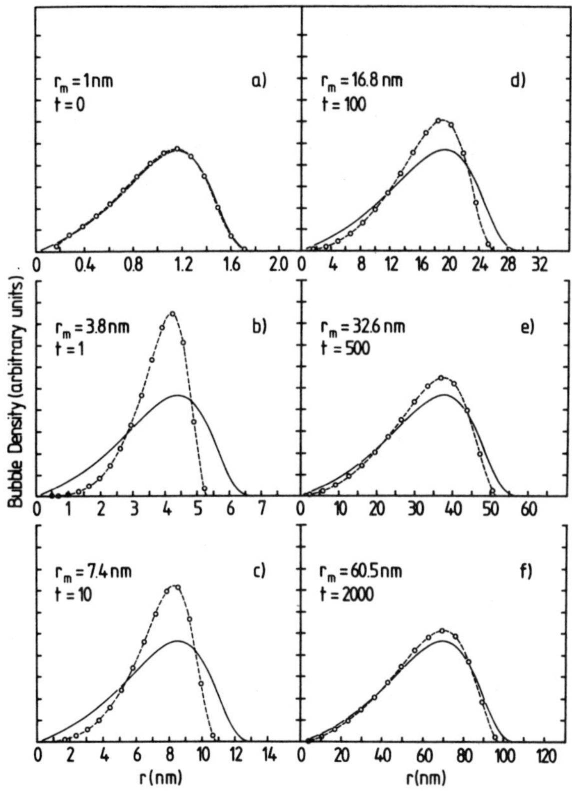

Fig. 4 Time evolution of the bubble size distribution (BSD) assuming real gas behaviour and an initial BSD corresponding to the ideal gas behaviour. These plots correspond to the data for the T=1223 K in Fig.3. The continuous line represents a quasi-steady-state ideal gas BSD with the same r_m and number of bubbles. The real gas BSD are much sharper than the corresponding ideal gas BSD and are characterized by a short tail towards small bubbles and a relativelly sharp cut at the large bubble side.

303

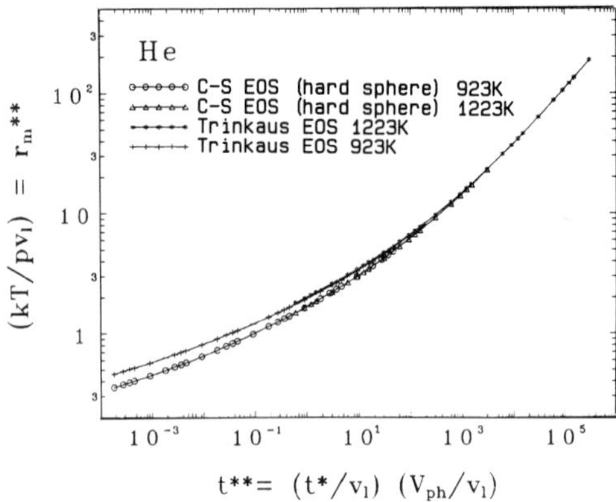

Fig. 5   Time evolution of the mean radius r_m in terms of the reduced variables t_m^{**} (Eq. 6) and r^{**} (Eq. 7) for an initial BSD with a gaussian shape (r_m=0.25 nm and ***s=0.025 nm) using Trinkaus EOS and C-S EOS. This figure illustrates the scaling provided by such reduced variables and shows that a hard sphere EOS provides a good approximation in describing the coarsening kinetics.

The temporal evolution of the mean radius of helium bubbles under real gas behaviour, either assuming the semi-empirical high density EOS of Trinkaus [6] or the less sophisticated hard sphere EOS of Carnaham-Starling (C-S) [12], is illustrated in Fig. 5. In order to cover smaller r_m values we started these simulations assuming an initial BSD of a gaussian shape with r_m=0.25 nm and $\sigma = 0.025$. The small differences between the curves show that, in describing the OR kinetics, such a hard sphere EOS already provides a good approximation even for the case of He for which the largest discrepancies between a hard sphere and a more realistic approach is expected.

In addition, Fig. 5 also shows a better scaling for the different temperatures (compared with Fig. 3). This scaling is obtained by means of a new normalization into dimensionless reduced variables

$$t^{**} = (t^*/v_1)\ (V_{ph}/v_1) \tag{6}$$

and

$$r^{**} = r^*/v_1, \tag{7}$$

which accounts for the temperature (and mass) dependence of the dissolved gas in terms of the phase space volume, V_{ph}, and the dimensional parameter of the hard sphere EOS in terms of the empirical volume per gas atom in the liquid phase at melting, v_1. In this way, a hard sphere EOS which depends only on one dimensional parameter and not on specific details of the interatomic potential may provide an universal description of the OR kinetics. In effect, Fig. 6 shows how $r_m^{**}(t^{**})$ curves for different temperatures and gas elements condense into a single master curve, which is also independent of the specific values of the diffusivities and solubilities characteristic of each element. We point out that such scaling behaviour also applies to the BSD, and therefore the results for He are also representative for the other inert gases.

On the other hand, under such conditions, a similar master curve for the ideal gas behaviour cannot be obtained because the IG EOS does not depend on a dimensional parameter such as v_1. Therefore IG and RG behaviour are compared best in terms of the ratio $r_m^{RG}(t)/r_m^{IG}(t)$ as a function of $r_m^{RG}(t)$, by which the time dependence is scaled out. This is illustrated in Fig. 7 for

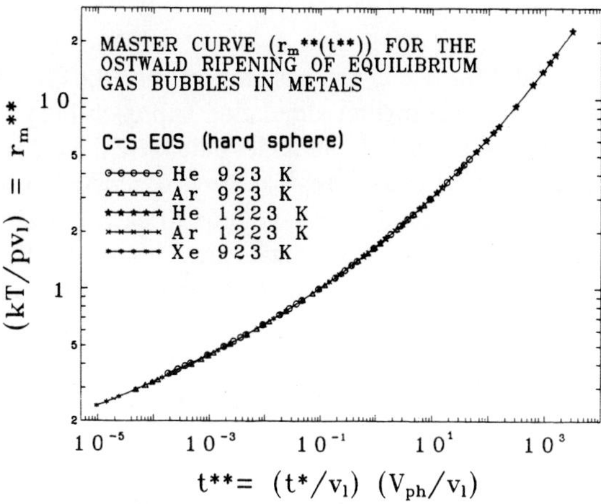

Fig. 6   Time evolution of the mean radii r_m for He, Ar and Xe bubbles for various temperatures, in terms the reduced variables r_m^{**} and t^{**} using the same initial conditions as in Fig. 5 and C-S EOS. This figure illustrates the scaling into a single master curve resulting from the normalization to the hard sphere melting volume v_l and the phase space volume V_{ph}.

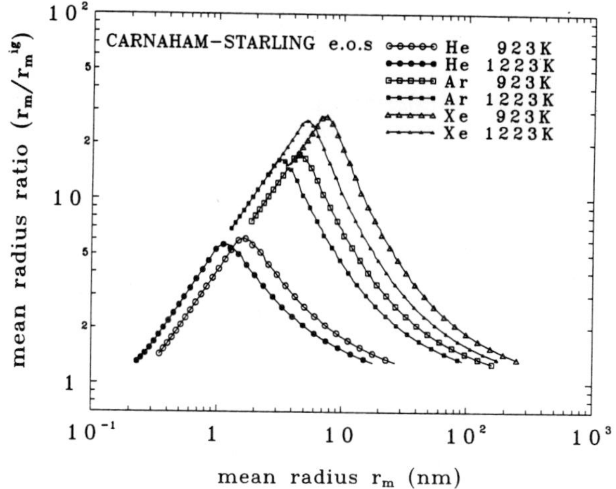

Fig. 7   Mean radii ratio r_m^{RG}/r_m^{IG} as a function of r_m^{RG} for He, Ar and Xe at various temperatures.

He, Ar and Xe. Because the simulations start with the same BSD for both IG and RG cases, the ratios r_m^{RG}/r_m^{IG} are equal to 1 at t=0 (not shown). During a transient the r_m^{RG}/r_m^{IG} ratios increase up to maximum values characteristic of each element and then start to converge asymptotically to 1 as the growth rate for RG approaches the IG behaviour at large r_m values. We remark that the specific r_m^{RG}/r_m^{IG} values for small r_m may depend on the choice of the initial BSD, in particular because the growth rate for the IG case is rather sensitive to the initial BSD. However, it is important to emphasize that, regardless of initial transients, the discrepancies between RG and IG behaviour are substantial and increase with increasing atomic number of the inert gas. In addition, the r_m regions of the maximum discrepancies are shifted towards larger r_m values. We also note that the maximum ratios and therefore the maximum discrepancies occur within a range of $1 \leq r_m \leq 20$nm, which is the range typically considered in experimental observations by transmission electron microscopy.

6. Conclusions

In summary, to analyse the influence of real gas properties on the Ostwald ripening kinetics of gas bubbles we have applied a computer simulation approach [11] and have compared the results with those for the ideal gas behaviour under otherwise the same simulation conditions.

The real gas behaviour of all inert gases has been described in terms of a hard sphere equation by attributing a proper hard sphere volume to each gas atom. In considering the coarsening kinetics, a hard sphere equation of state provides reasonable approximations when compared with those based on a more accurate semi-empirical high density equation of state as available for He. In addition, the hard sphere volumes (defined by the empirical values of the atomic volume of the liquid phase at melting) as well as the other gas properties (mass, diffusivity and solubility) can be scaled into reduced coordinates so that an universal description of the Ostwald ripening of gas bubbles under real gas behaviour is achieved in one master curve.

In comparison with ideal gas, the results for real gas considering He, Ar and Xe show large differences in the size range characterized by $r_m < 70$ nm. Within this range, the time dependence of the mean radii cannot be described by a single power law as for the ideal gas case. Under real gas behaviour the mean radii reached within a given annealing time are substantially larger and the bubble size distributions are significantly narrower than in the corresponding ideal gas case. These differences become increasingly pronounced with increasing atomic number of the gas.

ACKNOWLEDGEMENTS

One of us, P.F.P.F., would like to acknowledge the support from the Alexander von Humbold Foundation (Bonn, F.R. Germany) and from Conselho Nacional de Pesquisa (CNPq - Brazil).

REFERENCES

1. P. J. Goodhew and S.K. Tyler, Proc. Royal Soc. **A377**, 151(1981).

2. Z. H. Luklinskov, G. von Bradsky and P.J. Goodhew, J. Nucl. Mater.**135**, 206 (1985).

3. G. W. Greenwood and A. Boltax, J. Nucl. Mater., **5**, 234 (1972).

4. A. J. Markworth, Met. Trans. **4**, 2651 (1973)

5. J. Rothaut, H. Schroeder and H. Ullmaier, Phil. Mag. **A47**,781(1983).

6. H. Trinkaus, Rad. Effects **78**, 189 (1983).

7. H. Schroeder and P.F.P. Fichtner, J. Nucl. Mater. **179-181**, 1007 (1991).

8. H. Schroeder and P.F.P. Fichtner, this volume.

9. S. E. Donnelly, Rad. Eff. **90**, 1-47 (1985).

10. H. Trinkaus, unpublished work.

11. P.F.P. Fichtner, H. Schroeder and H. Trinkaus, submited to Acta Met.

12. N.F. Carnaham and K.E. Starling, J. of Chem. Phys. **51**, 635 (1969).

13. In *Rare Gas Solids*, ed. by M.L.Klein and J.A.Venables, Acad. Press (1977).

RECENT EXPERIMENTAL STUDIES ON THERMAL AND IRRADIATION-INDUCED RESOLUTION OF GAS ATOMS FROM BUBBLES IN SOLIDS

J.H. Evans

RS Materials and Chemistry Division
Harwell Laboratory
Didcot, Oxon., OX11 0RA, UK

ABSTRACT

Studies of inert gas behaviour in solids have generally been concerned with the removal of gas atoms from solution, their precipitation into bubbles and the subsequent response to annealing. Much less effort has been paid to routes through which gas atoms can *leave* bubbles. This paper briefly describes two of these resolution routes, irradiation-induced and thermal, with emphasis on recent experiments. Historically, discussion on irradiation induced resolution has been mainly confined to fission gas behaviour in uranium dioxide reactor fuel but there are close connections with precipitate dissolution and ion beam mixing. Here we present preliminary experimental results to demonstrate ion beam resolution effects from inert gas bubbles during implantation. Examples for experiments on Zr and UO_2 will be given. The second resolution route, thermal resolution, has been discussed for both UO_2 and for metals in recent years. In the former case, there are reasons for believing that thermal resolution plays an important part in fission gas release in temperature transients, while in the latter case it is one of the possible components (together with migration and coalescence) involved in bubble coarsening. However, separating the two components is not always easy. The paper will summarise recent experiments in which the thermal resolution of helium from bubbles in gold, silicon and molybdenum have been demonstrated.

1.   Introduction

It is well known that the high heat of solution of inert gas atoms in metals and most solids gives a strong driving force for precipitation. The resulting gas-atom clustering and bubble formation, and the subsequent response of bubbles to annealing, has led to a large number of interesting investigations in the last three decades. Work up to 1982 is reviewed in the Jülich symposium [1] and is updated in the present workshop proceedings. In contrast, generally less attention has been paid to the routes—specifically irradiation-induced resolution and thermal resolution—through which gas atoms can leave bubbles. However, in particular areas, both these routes have been of interest. For example, since the early '60s there has been considerable discussion [e.g. 2-4] on the role of irradiation-induced resolution in understanding the steady state release of fission gases from irradiated UO_2 nuclear fuel while in the same field, thermal resolution has been applied in recent years [5-8] to the modelling of fission gas release during temperature transients. In metals, thermal resolution has been used as the first step in the Ostwald ripening process [9,10] to describe the coarsening of bubble populations during thermal annealing. In the present volume, Schroeder *et al.* [11] cover this aspect in more detail while Matzke [12] discusses inert gas behaviour in UO_2.

Fundamental Aspects of Inert Gases in Solids
Edited by S.E. Donnelly and J.H. Evans, Plenum Press, New York, 1991

In the two main sections of this paper, both modes of resolution will be described briefly. In the section on irradiation-induced resolution the main aim will then be to present preliminary experimental results demonstrating the plausible role of the phenomena for inert gas atoms in bubbles in Zr and UO_2. In the section on thermal resolution, a summary will be given of recent experiments in gold [13] and silicon [14], in which the resolution of helium from bubbles has been demonstrated directly. In addition, new evidence for the same phenomenon in molybdenum will be given.

2. Irradiation Induced Resolution

2.1. Background

Interest in this topic has almost been entirely directed at bubble behaviour in UO_2 and stemmed from the need to understand the processes which led to fission gas release from the fuel. Although some release is due to the knock-out of near-surface fission gas, the main route involves first the transport of fission gas from the matrix to grain boundaries [15] and then the coalescence and interlinkage of grain boundary bubbles to give pathways to the surface [16]. A large literature exists covering the many different aspects of the gas release process; the few references given in the present paper can only be representative. Early studies assumed simple diffusion of fission gas atoms to boundaries but this picture had to be revised when electron microscopy demonstrated that a significant fraction of gas was precipitated as a high density of small bubbles [17]. It was recognised that the amount of fission gas arriving at boundaries through the high density of intragranular bubble traps would be very small, far below the steady state experimental gas release measurements. Resolution of the gas from bubbles during the irradiation was therefore postulated as a method of significantly increasing the gas population in the matrix and enhancing the gas fraction reaching the deeper grain boundary bubble traps.

Experimental observations on resolution were initiated in 1966 by the TEM results of Whapham [17] who subjected a previously irradiated UO_2 sample containing 10nm diam. matrix bubbles and 10-30nm diam. bubbles on dislocations to a re-irradiation with neutrons at 100°C. After this treatment no bubbles were seen, directly demonstrating that bubbles could be destroyed by the irradiation environment. Clearly the results can be interpreted in terms of the re-irradiation resoluting the gas and giving a new distribution of bubbles (presumably submicroscopic), characteristic of the new irradiation temperature.

Further experimental work [e.g. 18-21] corroborated Whapham's results and led to estimates of a re-solution parameter to describe the rate at which gas atoms in bubbles were removed back to the matrix by the irradiation environment. This parameter, usually denoted by b, was defined as the probability per unit time of a gas atom returning to solution, with the half-life of a gas atom in a bubble being given by (ln 0.5)/b.

As might be expected, considerable discussion ensued on the plausible mechanisms of resolution that could be induced by the 50-100 MeV fission fragments. As reviewed by Turnbull [4], the mechanisms ranged from complete bubble destruction in the fission track (4 to 8 mm long) down to simple knock out of gas atoms from bubbles by primary knock-ons. Discussion included the influence of bubble radius; the observed bubble substructures of uniform small bubbles unexpectedly suggested that large bubbles were more easily resoluted - otherwise large bubbles would survive more easily to give a situation analogous to Ostwald ripening. However, large bubbles did survive on grain boundaries while there were many other complexities involved such as the role of resolution on bubble nucleation. Ronchi and Elton [22] have recently re-examined resolution mechanisms and have questioned their efficiency.

308

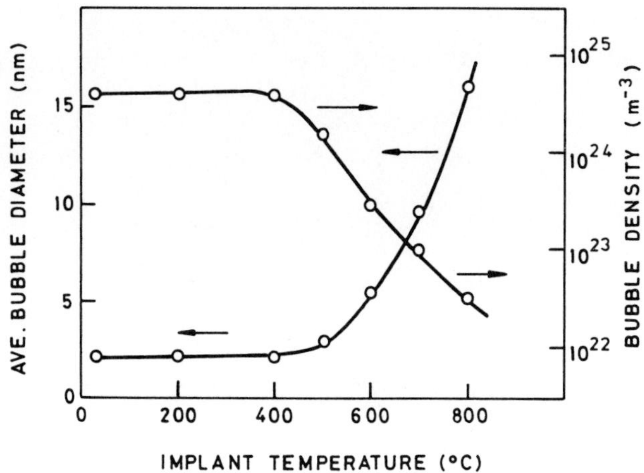

Fig. 1   The variation of Kr bubble parameters in Zr as a function of implant temperature.

2.2. *Present approach*

The purpose of the present work was to investigate whether, in principle, ion beam techniques could be used to demonstrate resolution processes. Naturally it was not supposed that fission track effects would be simulated but there was some attraction in obtaining evidence of direct knock-on effects. The technique used was to set up a population of bubbles by ion implantation and then reimplant with the same or different ion to see if changes in bubble substructure parameters could be induced. It is self-evident that if irradiation-induced resolution is taking place to any extent, then it will already have been operating in the initial implant to set up the bubble structures. Therefore, changes due to the second implant would only be seen if the two implant temperatures were in regions where different bubble nucleation characteristics were expected.

During investigations into krypton precipitation in the hexagonal-close-packed metals, titanium and zirconium [23-25], it was found that the bubble concentration dropped significantly toward the upper end of the 20 to 800°C available range of implant temperatures. Results for zirconium are given in Fig. 1 and provided a template to check effects of temperature changes on bubble parameters and the speed by which they might change to the new steady state values.

2.3. *Experimental details*

High purity zirconium specimens, 3mm diameter, were pre-dimpled by electropolishing and implanted using a 200 keV Danfysik accelerator. For a survey of bubble substructures as a function of temperature for a dose of 10^{16} 200 keV krypton ions/cm^2, samples were subsequently electropolished from the rear surface to give electron transparent regions in the regions of high krypton content. The krypton mean range was 70nm according to TRIM calculations. Results of bubble parameters as a function of variable implant temperature are given in Fig. 1. As in results for molybdenum [26], the inert gas bubble concentration was independent of temperature up to moderate temperatures before dropping fairly sharply. For implantation at temperatures up to 500°C, the Kr within bubbles was in the solid phase.

For the resolution experiments, various techniques were tried. Initially, the second resolution implant was made on thin foils which had already been examined. Although these

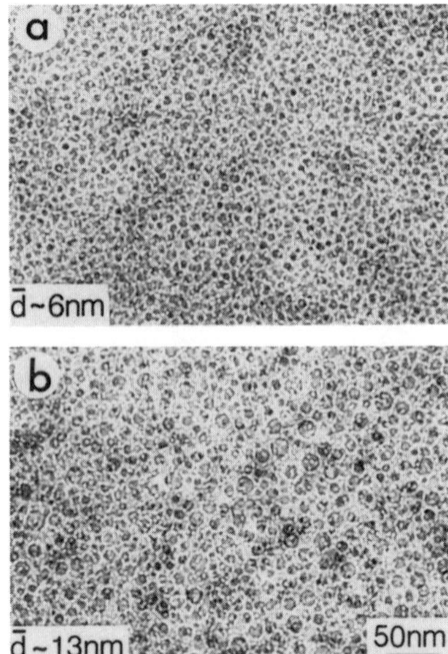

Fig. 2   Irradiation induced resolution: a sample with a high density of small Kr bubbles in Zr formed by implantation at 200°C has been reimplanted at 800°C with 2 10^{15} 200 keV Xe ions/cm^2 through a grid. Comparison of areas shadowed (a) and open (b), directly illustrate the effect of resolution.

showed positive effects, foil bending was a problem in getting good information from thin areas. The optimum technique for our purposes turned out to be one in which 100 keV Kr (range 37 $\pm$ 20nm) was implanted to give the initial bubble substructures, followed by a second implant of either 200 keV Kr or Xe ions (ranges 70 and 49 nm respectively) through a grid. Thinning for electron microscopy examination took place after the second implant. Direct comparison of bubble substructures was then possible across the grid shadow while other effects, for example, annealing during the second hot implant, acted equally on both areas and therefore could be ignored.

An example is given in Fig. 2 of an experiment on a sample with a high density of small bubbles formed in a 200°C implant of 10^{16} 100 keV krypton ions/cm^2 and then reimplanted through a grid at 800°C. In (a) we show the shadowed area where the bubble diameter was 6nm; clearly the starting substructure of 2 to 3 nm diameter bubbles has been influenced by the 800°C anneal in the second implant. In (b), however, the extra influence of the xenon implant is very apparent with the bubble diameters at 13nm, exactly that expected for a 800°C implant, Fig. 1, for a similar total implant dose. One useful advantage of using xenon as the resoluting ion was that its presence could be detected analytically with EDX in the electron microscope, confirming the areas which were open and shadowed. In addition, we had found in earlier experiments that xenon gave more pronounced resolution effects than krypton.

Several experiments were carried out for the reverse of the above case, where large bubbles were produced by 800°C implants, followed by implants at low temperatures, again through a grid. There was no doubt that the large bubbles could be destroyed, at least partially, in the second implant and were renucleated as a high density of small bubbles characteristic of the lower temperature. The resoluting ion also formed part of the new population. One difficulty with this experiment was that imaging the population of large bubbles remaining among the high density of smaller bubbles was far harder than expected, thus making the changes in the population and size of the large bubbles less obvious. However, careful measurements on

310

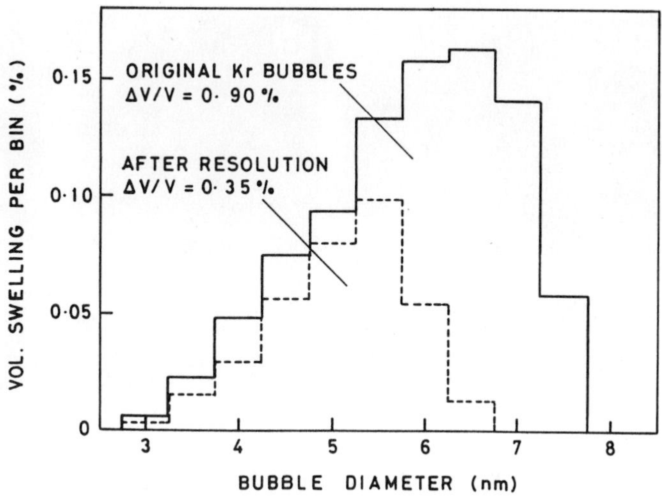

Fig. 3   Histogram of bubble sizes in open and shadowed areas in Zr foil containing large bubbles (d = 6.5nm) and subjected to a re-soluting dose of 1 10^{15} 200keV Xe ions/cm^2 at 200°C.

regions on either side of the mask shadow covering the same specimen thickness allowed size spectrum changes to be obtained as shown in Fig. 3. The initial Kr bubbles were formed at 800°C with a dose of 2 10^{15} 100keV Kr ions and resoluted with an identical dose of 200 keV Xe ions at 200°C. It was quite clear that the size spectrum has been altered with a marked loss in the larger size bubbles. Translated into bubble volume, the swelling of 0.9% in the original large bubbles had been sharply reduced to 0.35%, with the gas being transferred to a high density of small bubbles, d ≈ 2nm diameter, not shown in the histogram.

A result qualitatively similar to the above was reported by Dauben *et al.* [27] in 1986. They produced relatively large bubbles by helium implantation of Fe-12Cr at 973 K and then irradiated with Fe ions at 573 K. A new population of small bubbles was observed to form, particularly in the vicinity of the original bubbles, and attributed to the displacement cascade induced release of helium from these sources.

2.4. *Discussion*

The experiments carried out show that marked resolution effects occur when implant conditions are changed to new temperatures (up or down) at which different nucleation parameters are operating. It would seem that the incoming ions must displace the gas atoms within bubbles and return them to the lattice to be reprecipitated on a new scale, characteristic of the new temperature. Although the results presented here are limited, we believe that they demonstrate that irradiation resolution could be studied systematically by ion implantation methods. However, some drawbacks are apparent: for example with TEM, depth information is being integrated so that if processes are apparently uncompleted—as for the results in Fig. 3—some uncertainties could remain on the influence of depth.

One aspect, easily overlooked, is that bubble behaviour under resolution conditions cannot ignore the vacancies from which the bubble is formed. It is not impossible that all the gas could be resoluted from a bubble to be captured elsewhere while the vacancies are unaffected; in other words the cavity might be transformed from a bubble to a void. For situations where

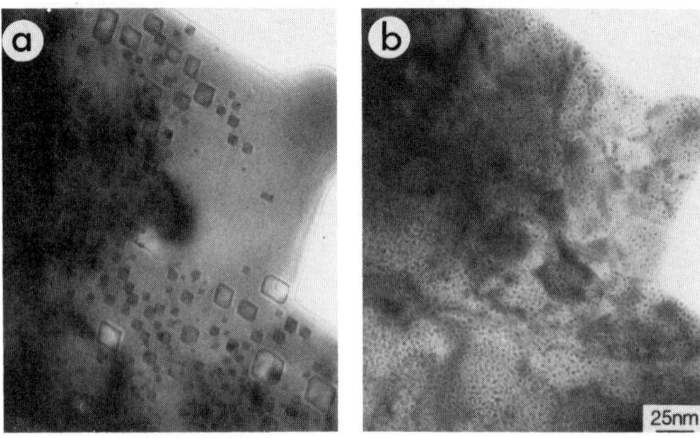

Fig. 4   Ion beam induced resolution in UO_2: the large bubbles in (a) have been resoluted, (b), by a dose of $2\ 10^{15}$ 200 keV krypton ions/cm^2 at ambient temperature.

the resoluting implant is at a high temperature in the thermal vacancy range, there is no problem; after gas atom ejection, equilibrium will quickly be re-established by the net removal of vacancies.   This is probably the situation for the first experiment described (Fig. 2). However, the question remains for the case where the second implant temperature is relatively low.   The original (larger) cavities can only shrink if they capture more interstitials than vacancies but it is not immediately evident how this is done.   At an early stage it is probable that resoluted gas atoms effectively tie up some of the vacancies in clusters to leave the equivalent interstitials to diffuse to the cavities but how long this process continues after the new bubble population is nucleated is unclear.

It ought to be noted that the resoluting doses in the present experiments were chosen to be relatively high in order to emphasise the effects.   A TRIM calculation on zirconium putting a 5nm krypton layer at a depth of 40 nm, and simulating a 200 keV xenon ion implant, suggested that a total of $2\ 10^{15}$ ions/cm^2 would have given the krypton atoms a displacement dose of 25 dpa.

Clearly the phenomenon, and the processes involved, are closely allied with those of ion beam mixing and precipitate dissolution. Results relevant to the present work on heavy inert gas implants are those of Templier *et al.* [28] where they produced solid bubble 'alloys' of implanted xenon and krypton in aluminium and measured the lattice parameter of the resulting alloy. No difference in the results was found when the order of implantation was reversed. This is consistent with complete atomic mixing of the species, via resolution, having taken place. The conclusion in [28] that the result demonstrates all the gas to be in the solid bubbles does not follow since complete mixing would also apply to any population of submicroscopic clusters.

Another aspect of interest is that nucleation of bubbles cannot be regarded, at least for the heavier inert gases, as a process occuring at an early stage and then being fixed with only growth processes taking place. It seems instead that the final observed bubble density must represent an equilibrium between the nucleation and growth processes and the bubble shrink-age due to gas atom resolution.

Finally, since the topic of radiation-induced resolution began with UO_2, we end this section with an example of ion beam induced resolution of bubbles in this material, Fig. 4. The same thin foil area, originally with large xenon bubbles, is seen after a resolution dose of $2\ 10^{15}$ 200 keV Kr ions / cm^2 at ambient temperatures. Again the potential of ion beam techniques to study resolution is clear.

3. Thermal Resolution

3.1. Background

The thermal resolution of a second phase, whether a precipitate or a bubble, within a parent matrix is clearly intimately linked up with the solubility of the second phase constituent. Historically, the inert gases were considered in most materials to have very high heats of solution [e.g. 29,30] so that thermal resolution has generally been considered not to be too important. While this may be true for many aspects—as shown by the strong tendency for inert gas precipitation into bubbles—there are situations in which this may not be so. In metals, the thermal resolution of helium from bubbles has been used to explain the coarsening of bubble distributions during annealing [9,10]. For smaller helium clusters, resolution is well known in the guise of thermal detrapping and is the well-established basis of the technique of thermal desorption spectroscopy to study the early stages of helium-vacancy cluster growth in metals (e.g. [31]).

Thermal resolution of inert gases has also been considered for UO_2 as a mechanism of fission gas release in post-irradiation annealing (or in temperature transients in reactor fuel). It appears that simple gas atom diffusion in the presence of a high bubble concentration, or the migration of small bubbles with or without temperature gradients, cannot satisfactorily explain the amount of gas reaching grain boundaries to eventually escape to the fuel surface. Clearly there is a direct analogy here with the problem of gas release from UO_2 under steady state conditions which was solved, as described earlier, by realising that irradiation induced resolution was important. In the case of temperature transients, it was suggested in early discussions of this problem [e.g. 32] and more recently by MacInnes and co-workers [5-8], that thermal resolution of the gas from bubbles could be a vital mechanism to provide an effective source term for gas atom migration to the grain boundaries.

In metals, discussion has concentrated on the roles of the two mechanisms in controlling bubble growth kinetics during annealing (i.e. thermal resolution, migration and coalescence) and the methods of distinguishing these mechanisms via the evolution of bubble size distributions, e.g. modal radii, etc. Separating the two components of bubble coarsening has not always been an easy task. However, the article in this volume by Schroeder *et al.* [11] shows a significant advance has now been made. To complement this, the aim in the present section is to briefly review two investigations, in gold and in silicon, in which thermal resolution of helium has been demonstrated in a fairly direct way. In addition, recent results for molybenum are introduced.

3.2. Thermal resolution of helium from bubbles in silicon

This work [14] was interesting in that it usefully combined helium desorption spectroscopy (HDS) with transmission electron microscopy (TEM). Silicon specimens were implanted with 10keV helium to doses of $2 \times 10^{17}/cm^2$ at ambient temperatures to give high densities of small helium bubbles. The bubble structures were thus very similiar to those after helium implants into metals—though in this and subsequent work [33,34], no evidence of bubble lattices have been seen in silicon. To study the response of bubbles to thermal annealing, specimens were heated in an electron microscope hot stage. Up to 1000 K the bubbles were very stable but above this temperature there were clear signs of migration with visible coalescence and loss at the nearby thin foil surfaces. In fact, to anticipate the HDS results below, by this time the bubbles had lost their helium and are better described as cavities. Figure 5(a) shows the cavity evolution in one sample observed after step-wise anneals at an average heating rate of 0.05 K/s.

The thermal desorption spectra on similar samples is illustrated by the example in figure 5b obtained with a heating rate of 2 K/s. The large release between 600 and 900 K was due to

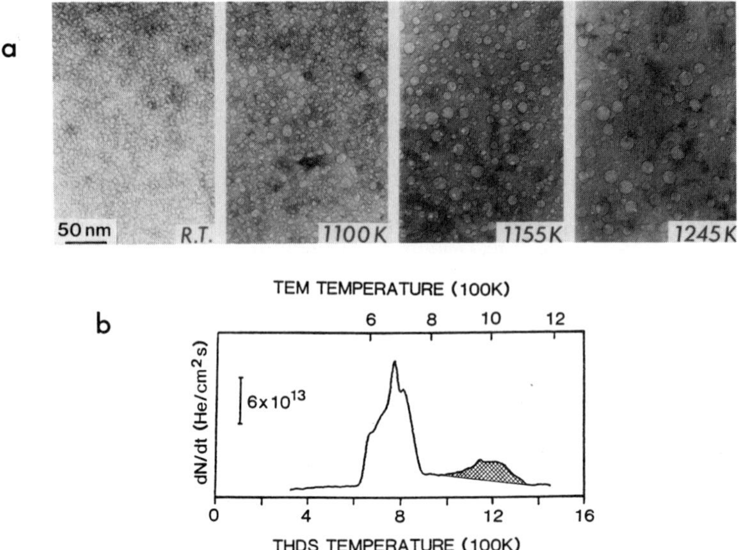

Fig. 5 (a) Sequence of electron micrographs showing the apparent coarsening of helium bubbles in Si during annealing. However, helium desorption results (b) indicated that in the second peak (shaded), helium permeation was taking place to give a bubble-to-void transition. On the TEM temperature scale, this transition was completed by 1100 K.

blistering effects but it became clear that after correction for the 40-fold difference in heating rate, the helium desorption peak centred at 1200 K corresponded to about 1000 K in the TEM anneal. Thus there was clear helium release at a temperature where bubble migration had hardly started, strongly suggesting that helium was permeating out of bubbles via thermal resolution and that the subsequent observations were on empty cavities, or voids, rather than bubbles. The result was confirmed by the technique of using small 'probe' doses of low-energy helium which then became trapped at the voids. Repeating the desorption gave the same helium release peak as seen before. This technique was eventually used to follow the annealing (at higher temperatures) of the voids themselves.

Additional corroboration of these thermal resolution results came from the discovery that van Wieringen and Warmoltz [35] had published permeation results on helium in silicon in 1956 where they showed that the activation enthalpy of permeation was a relatively low 1.73 eV. In the helium probe dose work [14], calculated helium desorption rates due to helium permeation from the bubbles were fitted almost exactly using this enthalpy value.

The permeation of helium out of bubbles in silicon at 1000 K ($0.6T_m$) occurs at a relatively low temperature, at least compared to metals. Van Wieringen and Warmoltz [35] showed that the 1.73 eV permeation energy was partitioned into 0.47 eV for the heat of solution and 1.26 eV for the helium activation energy for diffusion. The latter result is consistent with the results of van Veen *et al.* [36] showing intrinsic helium mobility at 400 K. One final point of interest regarding the transition of the helium bubbles to voids was the absence of any cavity shrinkage. This absence is associated with the anomalously high vacancy formation energy and low vacancy concentation found in silicon at high temperatures.

3.3. Thermal resolution of helium from bubbles in gold

In metals, as already stated, the separation of the thermal resolution and migration and coalescence components in bubble annealing is not easy. However, if bubble migration can be suppressed then it is self-evident that any thermal resolution effects will become more

314

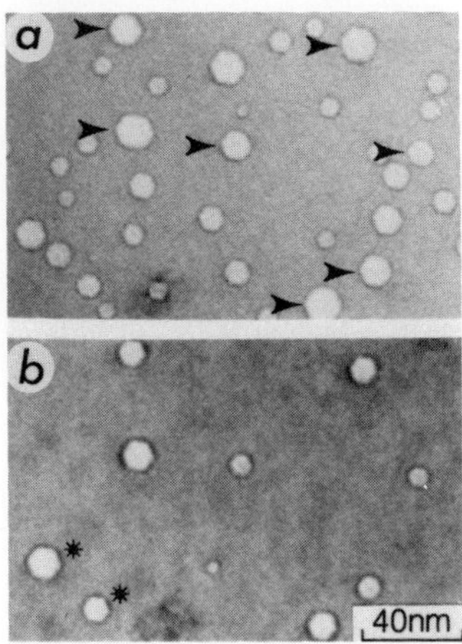

Fig. 6   Evidence for bubble shrinkage in gold at 1200 K: the bubbles arrowed in (a) have all reduced their diameter in (b) after 5 min. annealing. (The size increase in the starred bubbles in (b) is due to near-neighbour coalescence.)

clearly transparent. In a short investigation of helium bubble annealing in gold, Evans *et al.* [37,13] fortuitously found such a situation. Thin gold foils had been helium implanted with 200 eV helium ions at ambient temperature to give a layer of small bubbles near the entrance surface. In two foils, produced by evaporation, helium bubble migration was studied during in situ hot stage electron microscopy. Conventional behaviour was seen with bubble diffusion being sufficient at 600K (0.45 T_m) to allow a significant fraction of bubbles to migrate the short distance to the surface. The work allowed an estimate of surface diffusion in gold to be made [37].

The results of interest here were found in a second set of thin foils, produced by gel precipitation. During annealing, although bubble coalescence was seen, changes in bubble concentration were now taking place at temperatures well above those found previously. It emerged that the coalescent events were between close neighbours and that surprisingly no bubble migration was observed. Step-wise anneals to successively higher temperatures at 50° intervals eventually showed at 1200 K that distinct bubble shrinkage was occurring [13]. This is seen clearly in Fig. 6.

These results, made on two foils, showed directly that thermal resolution of helium from bubbles was taking place in gold at 1200 K ($0.9T_m$). The shrinkage of the bubbles is of course in direct contrast to the behaviour in silicon but is easily explained by the virtually instantaneous evaporation of thermal vacancies to maintain thermal equilibrium. However, no Ostwald ripening of bubbles was seen since the foil surfaces acted as a major sinks for desorbed helium. In a bulk sample, the faster shrinking small bubbles would have fed the larger to give a classic ripening situation.

Attempts were made to model the observed shrinkage results, for both substitutional and interstitial diffusing helium. In the former case, the early bubble shrinkage could be satisfied with a value of 3.42 eV for the sum of migration and formation enthalpies of a substitutional helium (HeV). However, this seemed too high when compared with a combination of an experimental value of 1.7 eV for H_{HeV}^m [38] and calculations of 1.0 eV for H_{HeV}. This implied value of 2.7 eV for the helium permeation was a better fit to the observed later stages of bubble shrinkage. For helium diffusing interstitially, the early stages of shrinkage were fitted to an activation enthalpy for permeation of 2.95 eV. When the relative metal melting points are considered, this compares well with the value of 3.5 eV derived for stainless steel by Rothaut *et al.* [9] on the basis of bubble size evolution with annealing.

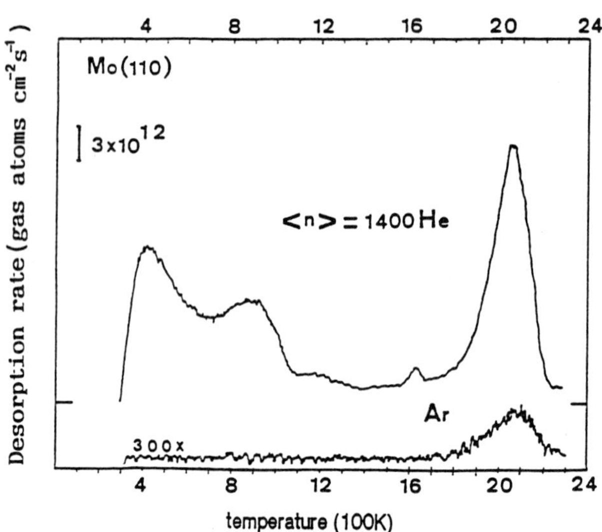

Fig. 7 Desorption curves for argon and helium in molybdenum from near-surface bubbles each containing one argon atom and an average of 1400 helium atoms. The shape of the helium desorption curve and the higher temperature for the argon peak is consistent with thermal resolution of the helium from the bubbles.

It is worth noting that in recent work on aluminium, in which hydrogen bubbles had been formed by ion implantation, Furuno *et al.* [39] found analogous results to those described above. In their case, the bubble shrinkage took place as hydrogen resolution occurred.

3.4. Thermal resolution of helium from bubbles in molybdenum

Recent results by van Veen and Evans [40] using thermal desorption spectroscopy (TDS) appear to give very good evidence for the thermal resolution of helium from bubbles in molybdenum. This evidence comes from the shape of the desorption curve associated with the disappearance of near-surface bubbles, but in addition the conclusion has been tested in a rather novel way.

The helium bubbles of interest were formed by decorating near-surface traps with low energy non-damaging helium implants. As shown for tungsten by Kornelsen [41] and for molybdenum by the Delft group [e.g. 42], the initial traps, often vacancies created by keV energy helium implants, or occasionally substitutional inert gas atoms such as argon, krypton or xenon, act as unsaturable traps for the subsequent incoming helium. Systematic work has illustrated the growth of these helium-vacancy clusters via first the trap mutation process and later loop punching [42–43]. Although the morphology of the larger clusters starts as a platelet in both molybdenum and nickel (see summary in [44]), conversion to a single bubble or a cluster of bubbles can occur either during implantation or during a medium temperature anneal. In molybdenum, it has been shown that platelets containing less than 2000 helium atoms will convert to a single bubble by about 1000 K.

The systematic changes in helium desorption spectra from molybenum as n, the average number of helium atoms per trap, is increased, is illustrated in Fig. 1 of ref. 42. What is very apparent at values of n above 200 is the growth of a desorption peak at between 1800 and 2000 K which was attributed to the degassing of bubbles. The exact mechanism was not discussed at the time but it was generally assumed that the peak was associated with the migration of bubbles to the nearby surface.

More recently, partly prompted by the interest in thermal resolution as outlined already in this paper, the bubble degasing peak was examined more critically. It quickly became clear

316

that the peak shape did not have the characteristics of a diffusion controlled mechanism where a desorption curve having a relatively sharp rise and long tail would be expected, with the loss of less than half the diffusing defect by the peak maximum. In fact the peak asymmetry was in the opposite direction, with a sharp drop on the trailing edge. Nevertheless, as a secondary check, some modelling was carried out assuming surface controlled bubble diffusion. It was quickly found that unless E_s and D_o, the surface diffusion energy and pre-exponential, were both prohibitively high, the model curves had half-widths far in excess of the measured curves.

As a result of this work, it was concluded that bubble migration could not play any important role in the bubble desorption peak. In examining the alternative of thermal resolution of helium from bubbles, it was found that this indeed would give the observed peak shape since the resolution should accelerate as bubbles shrink and increase their internal pressure in maintaining equilibrium. An interesting test for thermal resolution was then performed in which single substitutional argon atoms were used as the trapping centres for helium atoms, implanted at low energies (150 eV) as already described. In this way, bubbles containing a known average number of helium atoms could be formed, each containing one argon atom. By monitoring helium and argon simultaneously during the desorption peak, it was easy to see qualitatively that if helium escapes bubbles by thermal desorption, then it must peak at an earlier temperature than the argon, which, by remaining in the bubbles, is an exact signature for the final bubble disappearance.

The results in Fig. 7 show exactly this effect, with the He and Ar peaks being separated by about 50 K. It is difficult to explain this separation in other ways such as surface evaporation or by any process associated with bubble migration. In the latter case, both argon and helium should be desorbed simultaneously unless the bubble size spectrum is wide; in that case, the assumption that small bubbles have a higher diffusivity would lead to an argon peak below that of helium. All aspects of the results are therefore consistent with the thermal resolution from bubbles playing a dominant role in the helium release during desorption.

3.5. Discussion

The experimental examples of thermal resolution of helium from bubbles presented here appear to be fairly direct demonstrations of the phenomenon. This is particularly true for the silicon and the gold results. Although it should have been possible to have predicted the silicon results had the permeation data already been at hand, in gold the results depended on the unexpected absence of bubble migration seen in other samples. Reasons for this absence are uncertain, but it is worth mentioning that recent ion implant experiments in silicon [33] and aluminium [45] have shown that added impurities can suppress bubble migration. Clearly the controlled suppression of bubble migration could be useful if further direct evidence of thermal resolution is required for particular metals.

For molybdenum, it is interesting that the permeation of helium from the bubbles is taking place as low as $0.71\ T_m$, well below the $0.9\ T_m$ needed in gold. It may be significant that the desorption peak (the H peak) corresponding to the release of helium from a vacancy, is also appreciably lower on an homologous temperature scale for the bcc metals relative to the fcc metals [31].

4. Summary

This paper has discussed the two main routes for gas atom resolution from inert gas bubbles in solids. For irradiation-induced resolution, results have been presented demonstrating that ion implantation techniques have some potential in studying this area systematically. It is possible that the part played by such resolution during inert gas implantation, particularly for the heavier inert gases, has yet to be fully explored in the context of solid bubble pressures and bubble growth processes that still are subjects of topical discussion [46,47].

In the field of thermal resolution, previous results on silicon, gold and molybdenum have been summarised to highlight the potential of this mode of release for helium trapped in bubbles.

ACKNOWLEDGEMENTS

The significant contribution of Tom Van Veen to the thermal resolution work discussed in this paper is gratefully acknowledged.

REFERENCES

1. Proc. Symp. on Fundamental Aspects of Helium in Metals, Jülich, 1982 (Ed. H. Ullmaier) Rad. Effects **78** (1983).

2. M.V. Speight, Nucl. Sci. and Eng. **37**, 180 (1969).

3. R.S Nelson, J. Nucl. Mater. **31**, 153 (1969).

4. J.A. Turnbull, Rad. Effects **57**, 243 (1980).

5. D.A. MacInnes and I.R. Brearley, J. Nucl. Mater. **107**, 123 (1982).

6. I.R. Brearley and D.A. MacInnes, J. Nucl. Mater. **118**, 68 (1983).

7. P.T. Elton, P.E. Coleman and D.A. MacInnes, J. Nucl. Mater. **135**, 63 (1985).

8. I.R. Brearley, P.E. Coleman, P.T. Elton and D.A. MacInnes, Proc. Workshop on Fission Gas Behaviour in Safety Experiments, Caderache 1983; Eur. Appl. Research Reports **5**, 1159 (1984).

9. J. Rothaut, H. Schroeder and H. Ullmaier, Phil. Mag. **A47**, 781 (1983).

10. H. Trinkaus, Rad. Effects **78**, 189 (1983).

11. H. Schroeder, P.F.P. Fichtner and H. Trinkaus, this volume.

12. Hj. Matzke, this volume.

13. J.H. Evans, A. van Veen and M.W. Finnis, J. Nucl. Mater. **168**, 19 (1989).

14. C.C. Griffioen, J.H. Evans, P.C. de Jong and A. van Veen, Nucl. Instr. Methods **B27**, 417 (1987).

15. Hj. Matzke, Rad. Effects **53**, 219 (1980).

16. M.O. Tucker, Rad. Effects **53**, 251 (1980).

17. A.D. Whapham, Nucl. Applications **2**, 123 (1966).

18. A.M. Ross, J. Nucl. Mater. **30**, 134 (1969).

19. J.A. Turnbull and R.M. Cornell, J. Nucl. Mater. **37**, 355 (1970).

20. J.A. Turnbull and R.M. Cornell, J. Nucl. Mater. **41**, 156 (1971).

21. M.O. Marlowe and A.I. Kaznoff, Proc. Symp. on Nuclear Fuel Performance, British Nucl. Energy Soc., London 1973, pg. 79.1.

22. C. Ronchi and P.T. Elton, J. Nucl. Mater. **140**, 228 (1986).

23. J.H. Evans and D.J. Mazey, J. Nucl. Mater. **138**, 176 (1986).

24. D.J. Mazey and J.H. Evans, J. Nucl. Mater. **138**, 16 (1986).

25. J.H. Evans, A.J.E. Foreman and R.J. McElroy, J. Nucl. Mater. **168**, 340 (1989).

26. D.J. Mazey, B.L. Eyre, J.H. Evans, S.K. Erents and G.M. McCracken, J. Nucl. Mater. **64**, 145 (1977).

27. P. Dauben, R.P. Wahi and H. Wollenberger, J. Nucl. Mater. **141-143**, 723 (1986).

28. C. Templier, H. Garem, J.C. Desoyer and J. Delafond, Scripta Met. **20**, 1705 (1986).

29. D.E. Rimmer and A.H. Cottrell, Phil. Mag. **2**, 1345 (1957).

30. C.F. Melius, W.D. Wilson and C.L. Bisson, Rad. Effects **53**, 111 (1980).

31. A. van Veen, Mater. Sci. Forum **15-18**, 3 (1987); also this volume.

32. J.R. MacEwan and P.A. Morrel, Nucl. Appl. **2**, 158 (1966).

33. J.H. Evans, A. van Veen and C.C. Griffioen, Nucl. Instr. Methods **B28**, 360 (1987).

34. S. Romani and J.H. Evans, unpublished results.

35. A. van Wieringen and N. Warmoltz, Physica **22**, 849 (1956).

36. A. van Veen, P.C. de Jong, K.R. Bijkerk, H.A. Filius and J.H. Evans, Proc. Symp. on Fundamentals of Beam-Solid Interactions, Boston 1987 (Eds. M.J. Aziz, L.E. Rehn and B. Stritzker), MRS Symp. Proc. **100**, 231 (1987).

37. J.H. Evans and A. van Veen, J. Nucl. Mater. **168**, 12 (1989).

38. V. Sciani and P. Jung, Rad. Effects **78**, 87 (1983).

39. S. Furuno, K. Hojou, H. Otsu, K. Izui, N. Kamigaki and T. Kino, J. Nucl. Mater. **179–181**, 1011 (1991).

40. A. van Veen and J.H. Evans, in preparation.

41. E.V. Kornelsen, Rad. Effects, **13**, 227 (1972).

42. A. van Veen, J.H. Evans, W.Th.M. Buters and L.M. Caspers, Rad. Effects, **78**, 53 (1983).

43. J.H. Evans, A. van Veen and L.M. Caspers, Rad. Effects, **78**, 105 (1983).

44. M. D'Olieslaeger, G. Knudt, L. de Schepper and L.M. Stals, this volume.

45. R.J. Cox, P.J. Goodhew and J.H. Evans, Nucl. Instr. Methods **B42** (1989) 224.

46. S.E. Donnelly, this volume.

47. H. Trinkaus, this volume.

HELIUM BUBBLE NUCLEATION IN ALUMINIUM IRRADIATED WITH 600 MeV PROTONS

F. Paschoud,[1] M. Victoria[1] and R. Gotthardt[2]

[1] *Paul Scherrer Institut, CH–5232 Villigen, Switzerland*

[2] *Ecole Polytechnique Fédérale de Lausanne, IGA, CH-1015 Lausanne, Switzerland*

ABSTRACT

During 600 MeV proton irradiation, helium, hydrogen and other impurities are produced simultaneously with the displacement damage. Transmission electron microscopy was used to characterize the bubble structure in aluminum after different irradiation temperatures (390 K–750 K) and doses (0.9 dpa–6.0 dpa). For low irradiation temperature, between 390 K and about 500 K, the observed homogeneous bubble structure was interpreted with a diatomic nucleation model. In the temperature regime between 500 K and 600 K, a bimodal bubble size distribution was observed in one specimen. For higher irradiation temperature, up to 750 K, two different behaviours appear: in the first case, specimens show the same behaviour as was observed at lower irradiation temperatures; in the other case, a lower bubble density (up to 50 times lower) of bigger bubbles was observed. This bifurcation at high irradiation temperatures is not yet well understood.

1.   Introduction

The production of helium in materials by nuclear reactions, for instance (n, α), is an important problem to be studied for the future fusion reactor. Indeed helium is mostly insoluble in metals and then participates in the nucleation of bubbles, voids and grain-boundaries cavities. These effects are mainly studied either by neutron irradiation [1] or by helium implantation in metals [2] sometimes simultaneously with ion irradiation (dual-beam irradiation) in order to produce displacement damage as a rate close to the fusion conditions [3].

In our experiment, 600 MeV protons were used to produce simultaneously displacement damage, helium and other impurities in aluminium. A detailed description of the interaction between a proton and a target's atom is given in [4]. The main points can be summarized as follows:

The first step of the interaction is an intranuclear cascade where particles of nucleus are ejected. The nucleus is in a highly excited state, and further evaporates protons, neutrons or alpha particles. The complete reaction is called a spallation reaction. The product of this reaction is an atom with a recoil energy (about 2 MeV in aluminium) which will be dissipated inside the matrix by a sequence of collisions, called a displacement cascade. This atom has a lower atomic number than the original one and is called transmutation atom. Therefore, the primary defects produced by 600 MeV protons are: helium, hydrogen, transmutation elements (impurities) and vacancy and interstitial defects.

The aim of this work was to characterize the bubble structure produced under 600 MeV proton irradiation and to try to understand the bubble nucleation process at the different irradiation temperatures.

2.   Experimental Procedure

2.1. Proton irradiation

The specimens used in this investigation were high purity aluminium foils (99.9999%) rolled to a thickness between 0.07 to 0.125 mm and subsequently fully recrystallized by annealing for 1 hour at 820 K in a vacuum of 10^{-3} Pa. The specimens were irradiated in the PIREX I (Proton IRradiation EXperiment I) facility built in the 600 MeV proton beam of the PSI (Paul Scherrer Institut, Villigen Switzerland). The gaussian beam profile created an irradiated zone with an elliptical cross section. The beam density was between 3.4 and 11 $\mu A/mm^2$ which corresponds to a displacement rate between 2.6 10^{-6} and 8.4 10^{-6} dpa/s (in aluminium). The production rate of transmutation elements in aluminium were calculated by the HETC code [5]. Amongst other impurities a quantity of 215 ± 20 appm/dpa of helium was retained in the specimens [6]. A detailed description of the installation is given elsewhere [4]. The aluminium specimens were irradiated at doses of 0.9 dpa to 6.0 dpa and at temperatures ranging from 390 K to 750 K.

2.2. TEM observations

Discs of 3 mm diameter were punched out of the irradiated zone whose centre was determined by autoradiography. The discs were thinned in a 20% perchloric acid–ethanol solution at -20 °C. An electron microscope operating at 100 kV was used to observe the specimens. For the determination of the bubble density C_b, the thickness of the specimen was measured with stereoscopic micrographs. The error in the bubble density was estimated to be 20%.

3.   Results

3.1. Helium bubble structure

In order to characterize the bubble structure formed in aluminium during 600 MeV proton irradiation, about 30 specimens were observed. Some results (mechanical properties, defect structure, thermal defect stability) on these specimens have already been reported [7,8].

The proton irradiation produced a homogenous bubble distribution inside the grains and bubbles of different sizes on the grain boundaries, depending on the misorientation. In the case of a high irradiation temperature, above 600 K, bubbles were also observed at the surface of precipitates formed during irradiation and probably composed by transmutation atoms. Preliminary NMR experiments have shown that no hydrogen, produced during irradiation, was retained in the specimens [9]. The bubbles are mainly filled with helium.

In this paper, we concentrate on the helium bubble nucleation process inside the grains by studying the bubble density C_b as a function of the irradiation temperature.

3.2. Temperature dependence

In Fig. 1, we plot the logarithm of the bubble density as a function of the inverse of the irradiation temperature. In this Arrhenius plot, two domains can be distinguished.

In domain I—low irradiation temperature (390 K to 520 K)—the bubble density follows the same behaviour for all the specimens. Assuming a linear behaviour, the slope of the line gives the apparent activation energy for the nucleation $E_{nucl}^a = 0.36 \pm 0.12$ eV.

At higher irradiation temperature in domain II, above 600 K, the behaviour of the bubble density splits in two branches. On one hand, a low density of big bubbles were observed (lower branch). On the other hand, some other specimens exhibit a behaviour with a high density of small bubbles, as a continuation of the domain I behaviour (higher branch). For instance, Fig. 2 shows the different bubbles structures of two specimens irradiated at similar temperatures and doses (respectively 650 K, 1.2 dpa and 660 K, 1.1 dpa). Specimen 60 has a low density, 2.2 10^{20} m^{-3} of big bubbles, 27.7 nm diameter, whereas specimen 92 has a higher

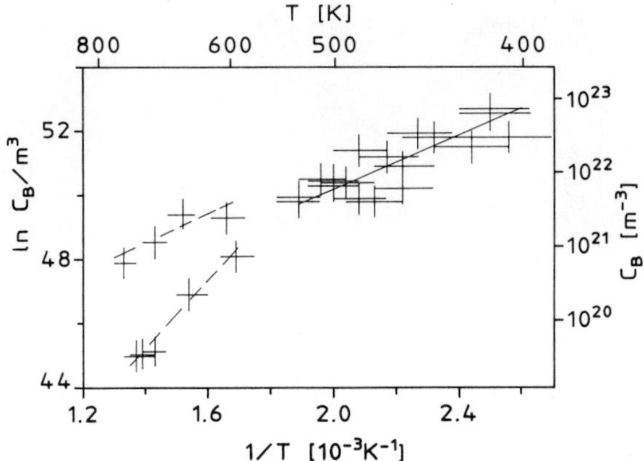

Fig. 1 Logarithm of the bubble density as a function of the inverse of the irradiation temperature (observations made by B.N. Singh at Risø Nat. Lab. Denmark, D. Gavillet [6] and F. Paschoud [15])

density, $2.9\ 10^{21}\ m^{-3}$, of smaller bubbles, 4.4 nm diameter. These differences increase with increasing irradiation temperature.

In the transient regime, for an irradiation at 590 K and a dose of 1.6 dpa, a bimodal size bubble distribution was observed (Fig. 3). If we suppose a cut–off diameter of 12 nm, the density of the small and big bubbles are respectively $2.1\ 10^{21}\ m^{-3}$ and $4.2\ 10^{20}\ m^{-3}$.

4.   Discussion

These observations of the bubble structure can now be used to test the nucleation models. For this we will separate the discussion in two main parts, like the behaviour of the He bubble structure: domain I where we test the diatomic Trinkaus model and domain II where only suggestions can be given.

4.1.   Helium bubble nucleation in domain I

The diatomic nucleation model as developed by Trinkaus [10] has two main hypotheses: two helium atoms form a stable and immobile nucleus and the nucleation is controlled by the diffusion of helium under irradiation. In these conditions the asymptotic bubble density (which can be compared with our observations) is given by the following expression:

$$C_b \approx 2C_b^*\left(1 + \left(\frac{4\pi r_t^2 \gamma}{3kT}\ \ln\left(\frac{t}{2t^*}\right)\right)^{\frac{1}{2}}\right)$$

(1)

where t is the irradiation time, γ the surface energy, r_t the trapping radius between two He atoms, k the Boltzmann constant, T the temperature, t^* the time when the nucleation rate is maximum. C_b^* is the bubble density at the time t^* and is given by:

$$C_b^* \approx \left(\frac{P_{He}}{4\pi r_t D_{He}^{eff}}\right)^{\frac{1}{2}}$$

(2)

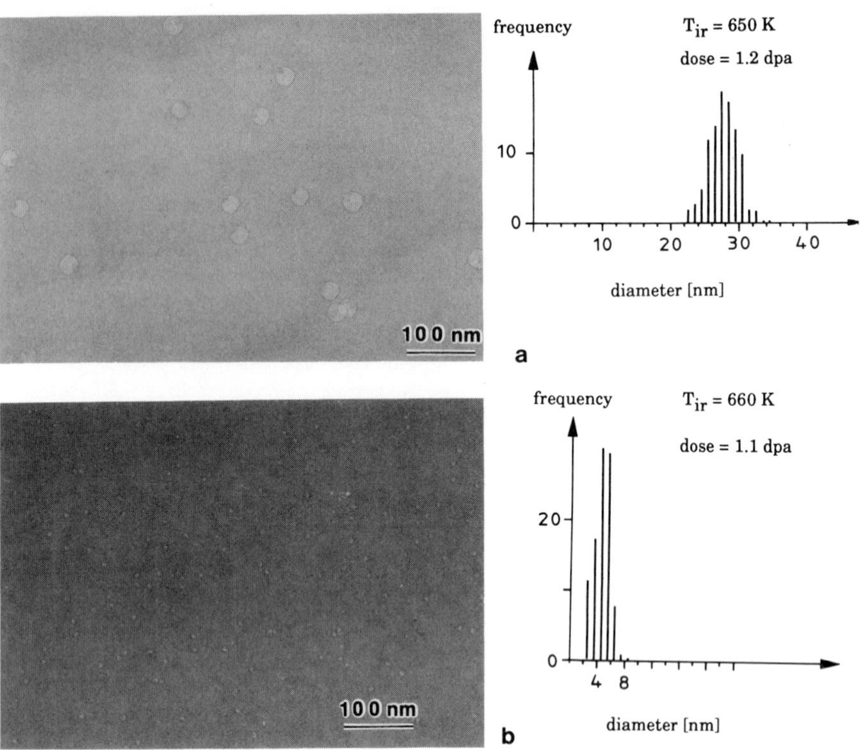

Fig. 2 The two bubble structures in specimens irradiated at high temperature: a) specimen 60, low density of big bubbles (mean diameter = 27.7 nm, density = 2.2 10^{20} m^{-3}) and b) specimen 92, high density of small bubbles (mean diameter = 4.4 nm, density = 2.9 10^{21} m^{-3}).

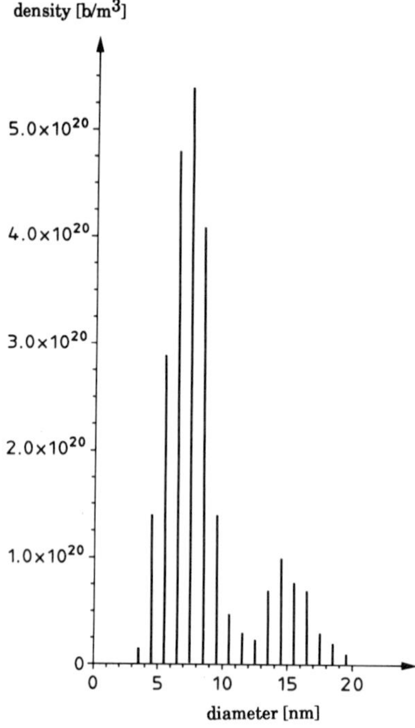

Fig. 3 Bimodal bubble size distribution observed in a specimen irradiated at 590K and 1.6 dpa (from B.N. Singh, Risø Nat. Lab).

324

where P_{He} is the He production rate and D_{He}^{eff} the effective He diffusion coefficient.

The dependence of the bubble density on the irradiation temperature is mainly given by the term C_b^{*}, and therefore the apparent activation energy given by this model E_n^a is one half of the effective He diffusion energy:

$$E_n^a = 1/2 \; E_{He}^{m,ef} \tag{3}$$

In this framework for domain I, the experimental behaviour of the bubble density as a function of the irradiation temperature gives an apparent activation energy for the nucleation of 0.36 ± 0.12 eV which gives an effective helium diffusion energy of 0.72 ± 0.24 eV. This value has to be compared with the different possibilities of helium diffusion energy under irradiation. In addition to the vacancy mechanism [11], helium which is assumed to be in substitutional position, can be detrapped from its vacancy and migrate rapidly as an interstitial until it is trapped again by a defect. The helium atoms can be detrapped either by thermal energy [12], or by a self–interstitial produced during irradiation (replacement reaction) [12] or by a collision process induced by a cascade (displacement reaction) [13]. Comparing the experimental and theoretical values, only the replacement reaction mechanism gives a value within the experimental error. In fact, for this mechanism $E_{He}^{m,ef} = E_v^m = 0.62$ eV (in the case of aluminium).

In conclusion, for irradiation temperature in the range of 400 K to 500 K (domain I), the helium bubble nucleation can be explained by a diatomic nucleation model where helium diffuses by a replacement reaction mechanism.

4.2. *Helium bubble nucleation in domain II*

Since the experimental error in the measurement of the bubble density is about 20%, it is important to show that the two branches observed at high irradiation temperatures correspond to observations beyond the scattering of the measurements. As can be seen in Fig. 1, the separation of the two branches in domain II increases with increasing temperature, the difference reaching a factor 20 at the highest irradiation temperature. Moreover, for a dose of about 2 dpa, the difference in bubble diameter is more than a factor of 5. The observed differences are therefore well above the the experimental scattering.

The helium inventory confirms these differences in terms of the gas pressure in the bubbles. Indeed, a comparison can be made of the amount of helium produced during irradiation He_{ir} with the amount of helium inside the bubbles He_{ap}. This latter amount is calculated from the bubble size distribution using the Trinkaus equation of state [14], assuming the bubbles to be in equilibrium ($P=2\gamma/r$) at the irradiation temperature. He_{ir} is the mean helium content as analysed after irradiation (215 $\pm$20 apmm/dpa [6]). Assuming that all the helium is in the bubbles (this hypothesis has been discussed extensively elsewhere [15]), the ratio He_{ap}/He_{ir} gives information about the pressure state of the bubbles. For the specimens in the high branch the ratios are between 0.4 and 1.3 showing overpressurized or close to equilibrium bubbles (as in the low irradiation temperature domain), whereas this ratio is from 1.2 to 2.1 for the lower branch indicating bubbles that are at equilibrium or underpressurized. Bubbles in each of the two branches are therefore in different states of pressurization.

Two kinds of problems can appear during irradiation: i) the beam stability (specimens irradiated at 5 dpa stay about 5 weeks in the beam) and ii) the temperature control at the begining of the irradiation.

    i) The spatial stability of the beam has been checked on a specimen irradiated at 505 K and 4.7 dpa by measuring the local activity on the specimen and by comparing this value with the average beam profile recorded during irradiation [6]. This comparison shows that the spatial beam stability is better than 0.2 mm.

There are also beam interruptions during the experiments but no correlation was found with the two different bubble structures.

ii) The evolution of the cooling temperature at the begining of irradiation (important period for the nucleation) is similar for the specimens of the two branches, no difference could be systematically detected.

Up to now, it is not very clear what is the cause of this effect. Nevertheless, some observations may help the interpretation. Indeed, the observation of a bimodal bubble size distribution is probably a crucial point. Such distributions are usually observed after dual beam irradiation [16] and are interpreted in terms of a critical bubble radius r_c [17]. This critical radius corresponds to the size where the bubbles grow by bias–driven growth. The lower size is associated with bubbles whereas the bigger size is a void component.

In our case we can regard the distributions of the lower branch as coming from an initial bimodal distribution where the smaller bubbles are not stable and so disappear leaving only the void component of the distribution, confirmed by the helium inventory. The point which is not clear, is how the material chooses its behaviour between the two branches.

5.   Conclusions

In this work we tried to understand helium bubble formation under 600 MeV proton irradiation. Depending on the irradiation temperature, two domains can be distinguished: domain I between 390 K and about 520 K and domain II above 600K.

In domain I, the bubble formation can be explained with a diatomic nucleation model where the kinetics are controlled by helium diffusion. Our experimental results reveal that helium diffuses under irradiation via a replacement mechanism where helium is detrapped from its vacancy by an interstitial and migrates rapidly as interstitial until it is trapped again by another vacancy.

In domain II, the results of the transmission electron microscopy observations show clearly the existence of two different bubble size distributions in different specimens. This point is confirmed by the observation of a bimodal distribution in a specimen irradiated at 590 K (intermediate regime). All this information and the helium inventory suggest an interpretation with a mechanism where the system has to choose between two types of behaviour.

ACKNOWLEDGEMENTS

The authors would like to thank B N Singh and D Gavillet for providing all the necessary information about their TEM observations.

REFERENCES

1. K. Farrell and J.T. Houston, proc. conf 750989 on *Radiation Effects and Tritium Technology for Fusion reactors* **2**, 209 (1976).

2. T. Muroga, H Watanabe, K Araki and N Yoshida, J. Nucl. Mater. **155–157**, 1290 (1988).

3. P. L. Lane and P. J. Goodhew, Phil. Mag. **A48**, 965 (1983).

4. K. Ono, M. Inoue, T. Kino, S. Furuno and K. Izui, J. Nucl. Mat. **133 & 134**, 477 (1985).

5. J. W. Muncie and D. J. Mazey, J. Nucl. Mater. **154**, 204 (1988).

6. K.Farrell and E.H. Lee, Radiation–Induced Changes in Microstructure: 13th Symp. ASTM STP 955, 1 498 (1987).

7. W.V. Green, M. Victoria and S.L. Green, J. Nucl. Mat. **133 & 134**, 58 (1985).

8. S.L. Green, *Damage and impurity calculations relating to PIREX II*, EIR technical internal report, TM– 22–85–69 (1985).

9. D. Gavillet, R. Gotthardt, J–L. Martin, S.L. Green, W.V. Green and M. Victoria, *Effects of Radiation on Materials* 12th Symp. ASTM STP 870, **1**, 394 (1985).

10. D. Gavillet, M. Victoria, W.V. Green, R. Gotthardt and J–L. Martin, J. Nucl. Mat. **155–157**, 992 (1988).

11. F. Paschoud, R. Gotthardt and S.L. Green, *Radiation–Induced Changes in Microstructure*: 13th Symp. ASTM STP 955, **1**, 478 (1987).

12. J–L. Vergey–Gaugry, D. Gavillet, W.V. Green, M. Victoria, J–L. Martin and R. Gotthardt, EIR technical internal report, TM–22–85–56 (1985).

13. H. Trinkaus, Rad. Effects **101**, 91 (1986).

14. V. Sciani and P. Jung, Rad. Effects **78**, 87 (1983).

15. H. Trinkaus, J. Nucl. Mat. **118**, 39 (1983).

16. N.M. Ghoniem, S. Sharafat, J.M. Williams and L.K. Mansur, J. Nucl. Mat. **117** 96 (1983).

17. H. Trinkaus, Rad. Effects **78**, 189 (1983).

18. F. Paschoud, *Contribution à l'étude de la microstructure de l'aluminium irradié par des protons de 590 MeV*, Thesis No. 834, E.P.F.L, (1990).

19. J.A. Spitznagel, W.J. Choyke, N.J. Doyle, R.B. Irwin, J.R. Townsend and J.N. McGruer, J. Nucl. Mat. **108 & 109**, 537 (1982).

20. L.K. Mansur and W.A. Coghlan, J. Nucl. Mat. **119**, 1 (1983).

NEW ASPECTS OF GAS-INDUCED SWELLING IN HELIUM-IMPLANTED NICKEL DURING ANNEALING

V.N. Chernikov,[1] P.R. Kazansky,[1] H. Trinkaus,[2] P. Jung,[2] and H. Ullmaier[2]

[1] *Institute of Physical Chemistry*
of the Academy of Sciences of the USSR
Leninsky Pr. 31, 117915, Moscow, USSR

[2] *Institut für Festkörperforschung des Forschungszentrums*
Jülich, Postfach 1913, D-5170 Jülich
BRD and Association KFA-EURATOM

ABSTRACT
We review results of an extensive study on bubble coarsening in nickel and provide some useful additional data which round off our picture deduced from the combined experimental evidence and from theoretical considerations: During annealing the bubble structure near the surface and near grain boundaries coarsens faster than in the bulk, leading to separated populations of large and small bubbles, respectively. The coarsening mechanism of the large bubbles near boundaries is identified as Ostwald ripening, whereas in the bulk coarsening proceeds by slow migration and coalescence, i.e. the bulk bubbles remain small and keep their overpressure.

1.   Introduction

Although investigations of metals loaded with helium at low temperatures and subsequently annealed ("post-irradiation experiments") do not reproduce the conditions of materials in a fission or fusion reactor environment, they yield valuable information on the coarsening mechanisms of helium bubbles. An analysis of these processes yields basic data of helium in metals which are in turn needed for modelling [1] the macroscopic property changes [2] caused by helium in nuclear environments.

We therefore started an extensive study of the formation of helium bubbles in nickel (as a model material for austenitic stainless steels) during post-implantation annealing, employing different experimental methods (TEM, SANS, THDS) in close interaction with theoretical analysis. The conclusions drawn from this earlier work [3–5] are summarized in section 2. In section 3 new results on the time-dependence (isothermal annealing) of the bubble sizes and the widths of the high-swelling zones near boundaries are presented. Furthermore, an estimate of the critical helium content is given, above which a splitting into two populations of large near-surface bubbles and small bulk bubbles takes place (section 4). Finally, a brief summary of the progress achieved and an outlook for the future work, necessary to answer some remaining questions is given in section 5.

2.   Summary of Previous Results

Recently it has been recognized [6–8] that two spatially separated populations of bubbles appear during annealing of nickel specimens which had been implanted with helium at room

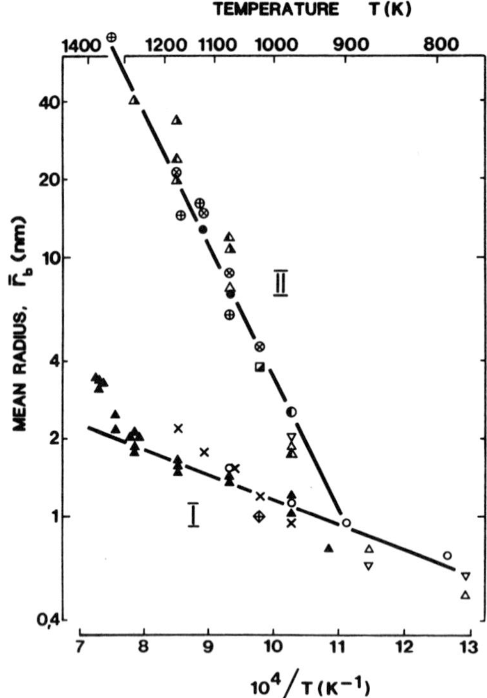

Fig. 1   Mean radius $\bar{r}_B$ of helium bubbles as a function of the annealing temperature T_a in the bulk of helium loaded thick foils ($\blacktriangle$, $\bigcirc$, $\times$, $\oplus$), near the surface ($\blacktriangle$, $\triangle$, ∇, $\bullet$, $\otimes$), in the volume of thin foils ($\mathbb{O}$) and near certain grain boundaries in thick foils ($\blacktriangle$):

$\blacktriangle$, $\blacktriangle$ ref. [3,8] supplemented with latest data of authors;

$\blacktriangle$ ref. [9] (in part);

$\triangle$, ∇ ref. [8];

$\times$, $\otimes$ ref. [5] (SANS data);

$\bigcirc$ ref. [10];

$\bullet$ ref. [6];

$\mathbb{O}$ ref. [11];

$\oplus$ ref. [12] ($c_{He} \approx 100$ appm).

Points denoted by $\oplus$ and $\blacksquare$ relate to the layers implanted with: 500 keV He$^+$ up to $c_{He}^{max} \approx 4.75$ at.% [7], and 24 MeV He$^+$ up to $c_{He}^{max} \approx 0.1$ at.% (SS 316) [13], respectively. Solid lines are results of calculation on the basis of: eqn.(1) for $v_{He} = 8$ Å^3 ($c_{He} = 10^{-3}$) —curve I and eqn.(4) for $p = 3$ GPa — curve II; $t_a = 1$h.

temperature. A subsequent compilation of bubble radii and densities vs. reciprocal temperature [3] revealed that all published data on nickel containing between 500 and 5000 appm He indeed lie within two clearly separated narrow bands (Figs. 1 and 2): a high density of small bubbles in the bulk (branch I) and a low density of large bubbles near free surfaces and most grain boundaries (branch II). Independent measurements of the helium densities employing a contrast variation method of small angle neutron scattering (SANS) [5] showed that the small bubbles in the bulk are not "thermal equilibrium bubbles" but maintain a considerable "overpressure" even at annealing temperatures beyond 1173 K (corresponding to 0.7 of the melting temperature). This unexpected finding was very recently backed up by positron annihilation (PAT) results [14]. The analysis of the experimental data and their interpretation by a theoretical model [4] led to the following picture of the evolution of the bubble structure during high temperature annealing of helium containing nickel:

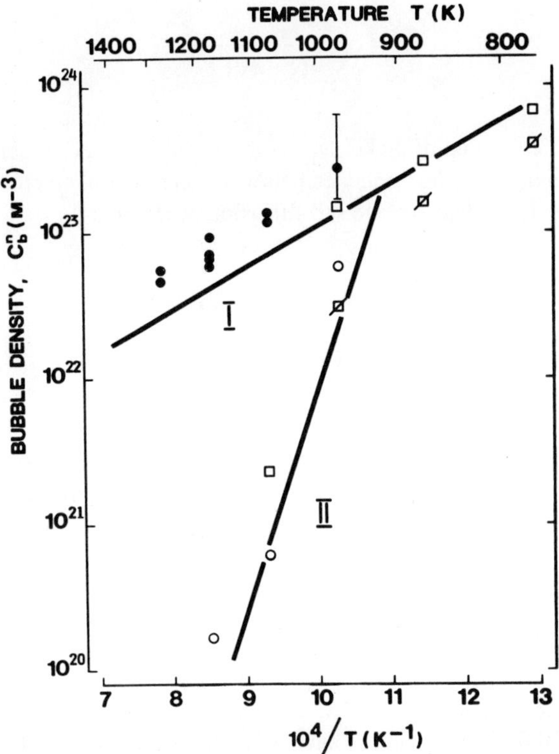

Fig. 2 Volume density of bubbles $\bar{C}_B^n$ normalized to $c_{He} = 0.1$ at.% in the bulk (●) and near the surface (○,□,⊠) as a function of the annealing temperature T_a: ○ , ● ref. [3]; □,⊠ ref. [8]. The curves correspond to the ones in fig.4 with additional assumptions: S_o = const—curve I, c_{He} = const.–curve II.

(1) For high implanted helium concentrations, annealing starts from bubble nuclei containing helium under high pressure. High pressures reduce the coarsening rate: Weakly in the case of migration and coalescence, strongly in the case of Ostwald ripening.

(2) In regions of sufficient vacancy supply, i.e. near free surfaces and most grain boundaries, the high pressures can relax whereby Ostwald ripening speeds up. Although the activation energy for this process is high (permeation energy controlled), it soon dominates and leads to large bubbles (branch II in Figs. 1 and 2). The width L of this high swelling region should increase with increasing annealing temperature T_a and time t_a (see next section).

(3) In the bulk, with helium concentrations above around 100 appm, the dislocations are not able to supply sufficient vacancies for a substantial relaxation of the pressure, i.e. triggering of fast Ostwald ripening cannot occur. Coarsening proceeds by moderately reduced bubble migration and coalescence with a low activation energy (surface diffusion controlled). The bubbles remain small and keep their overpressure (branch I in Figs. 1 and 2).

The above statements are corroborated by the following quantitative comparisions between experimental data and theory:

(a) for the small bulk bubbles, the temperature dependences of the mean radii and the bubble densities (branches I in Figs. 1 and 2) are compatible with those for coarsening by surface diffusion controlled bubble migration and coalescence:

$$\bar{r}_B \approx [\, 6\Omega^{1/3} v_{He} c_{He} D_s(p)\, t_a\,]^{1/6} \tag{1}$$

$$\bar{C}_B = c_{He} \frac{v_{He}}{\Omega} \frac{3}{4\pi \bar{r}_B^3} \tag{1a}$$

where Ω is the atomic volume of nickel, v_{He} is the atomic volume of helium in the bubbles, assumed to be constant, c_{He} is the implanted helium concentration, t_a is the annealing time and $D_s(p)$ is the pressure-dependent surface self-diffusion coefficient

$$D_s (p) = D_{s0} \exp (-H_s(p)/kT) \tag{2}$$

$$H_s(p) \approx H_s(0) + p\Omega \tag{3}$$

Inserting literature values for Ω, D_{s0}, $H_s(0)$ and using $v_{He} = 0.008$ nm^3 as the only fit-parameter (for details see Refs. [3] and [4]), good agreement between experimental data and theory is obtained (solid lines I in Figs. 1 and 2). This concerns both the slopes (corresponding to an apparent activation energy $H_I = H_s(p)/6 \approx 0.25$ eV for $r_B(T)$)) and the absolute values of $\bar{r}_B$ and C_B.

(b) In the above analysis a constant atomic volume v of He helium in the bubbles was assumed as a first approximation. The value of $v_{He} = 0.008$ nm^3 which gave the best fit to the experimental data is considerably below the values for bubbles in thermodynamic equilibrium (for which, e.g., v_{He} would be 0.020 nm^3 for $\bar{r}_B = 2$ nm). The existence of such overpressurized bubbles has been directly confirmed by SANS measurements [5] where values between 0.009 (for $\bar{r}_B = 0.5$ nm) and 0.014 (for $\bar{r}_B = 2$nm) have been found, i.e. somewhat higher than assumed above. Such an apparent discrepancy is not serious considering the crudeness of the theoretical model and the experimental uncertainties.

(c) For the large bubbles near surfaces and grain boundaries, the temperature dependencies of the mean radii and the bubble densities (branches II in Figs. 1 and 2) are compatible with those for coarsening by vacancy dissociation controlled Ostwald ripening

$$\bar{r}_B \approx \left(\frac{8}{9} \frac{\gamma}{k} \frac{\Omega}{T_a} D_{SD}(p) t_a \right)^{1/3} \tag{4}$$

where γ is the surface free energy and $D_{SD}(p)$ is the pressure dependent volume self-diffusion coefficient

$$D_{SD}(p) = D_{v0} \exp(-H_{Diss}(p)/kT) \tag{5}$$

$$H_{Diss}(p) \approx H_{SD} + p\Omega$$

where D_{v0} is the pre-exponential factor and H_{SD} the activation energy for self diffusion in Ni. Inserting again literature values for Ω, γ, D_{v0} and H_{SD} and using p=3 GPa as the only fit parameter, good agreement between experiment and theory is obtained (solid lines II in Figs. 1 and 2), concerning both the slopes (corresponding to an apparent activation energy $H_{II} = H_{Diss}(p)/3 \approx 1.1$ eV) and the absolute values of $\bar{r}_B$ and C_B.

Further evidence for the correctness of the interpretation of the experimental data can be obtained from the *time* dependence of $\bar{r}_B$ and of the width L of the regions containing large bubbles, respectively. The results of such isothermal annealing experiments are given below.

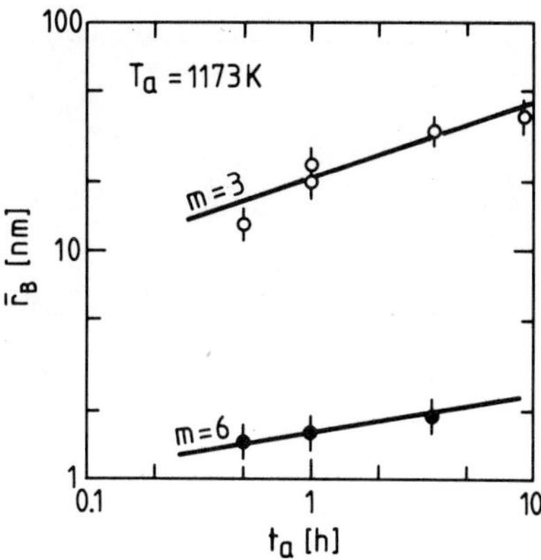

Fig. 3   Mean radius $\bar{r}_B$ as a function of annealing time t_a at 1173 K for bubbles near the surface (○) and in the bulk (●). The solid lines indicate slopes of m =3 (eqn. 4) and 6 (eqn. 1), respectively.

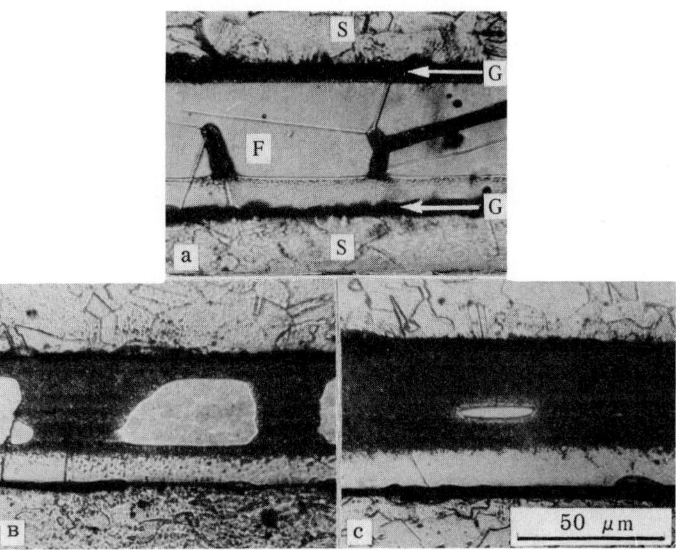

Fig. 4   Cross-sections of nickel foils loaded with helium after annealing: at 1173 K for 1h (a), for 3.5h (b) and at 1373 K for 1h (c). Foils (F) are secured between nickel spacers (S) with epoxy glue (G). Bright field image. Gray contrast against bright background indicates zones of the coarse gas porosity. Irradiated (entrance) surface is at top of figure.

3.   Time Dependence of Bubble Structure

Fig. 3 gives the experimental values of the mean radii $\bar{r}_B$ as a function of annealing time t_a for bubbles in the bulk and near boundaries, respectively, at 1173 K. Theory predicts a time dependence of $\bar{r}_B \propto t^{1/m}$, with m= 6 for bulk bubbles (eqn. 1) and m=3 for near-surface bubbles (eqn. 4), respectively. Comparison of the experimental data and the theory (solid lines in Fig. 3) shows reasonable agreement.

Fig. 4 gives three typical examples of cross-sections of nickel foils which were used for a metallographic determination of the widths L of the regions containing large bubbles near

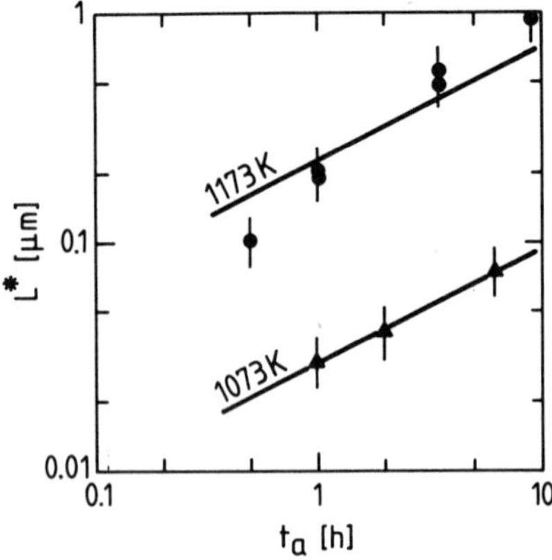

Fig. 5   Normalized width $L^* = L(S-S_0)^{1/2}$ as a function of annealing time t_a at $T_a = 1073$ and 1173 K respectively. The solid lines indicate the predictions of eqn. (7a).

surfaces (high swelling zones). A rough theoretical estimate of L can be obtained by assuming that vacancy absorption is restricted to a sharply defined moving boundary separating the region of high swelling S to that of low swelling S_0 [3, 9, 15]:

$$L = \left(2\, D_v\, t_a\, (c_v^e - c_v^0)/(S-S_0) \right)^{1/2} \tag{7}$$

$$L^* = L\,(S-S_0)^{1/2} = \left(2\, D_v\, c_v^e\, t_a\, (1 - c_v^0/c_v^e) \right)^{1/2} \tag{7a}$$

where D_v is the vacancy diffusion coefficient for Ni, c_v^e is the vacancy concentration in thermal equilibrium and $c_v^0 < c_v^e$ is the vacancy concentration at distances $x > L$ from the vacancy source (i.e. in the "bulk"). In Fig. 5 experimental values for L^* extracted from cross-section micrographs (see Fig. 4) and swelling determined by TEM are plotted. They can be described by eqn. (7a) using the literature value for $D_v\, c_v^e = D_{SD} = 9\ 10^{-5}\exp(-3.36\ 10^4/T_a)\ [\mathrm{m^2 s^{-1}}]$ and taking $c_v^0/c_v^e = 0.75$ for $T_a = 1173$ K and 0.90 for 1073 K as fit parameters to obtain the solid lines in Fig. 5. Considering the crudeness of the model, the above absolute values of the "vacancy deficit" $(1 - c_v^0/c_v^e)$ should not be taken literally. Nevertheless, their order of magnitude and their tendency to increase with increasing temperature are reasonable. Another question of interest concerns a possible redistribution of the homogeneously implanted helium concentration during the evolution of the different bubble populations in the regions $x<L$ (near surface) and $x > L$ (bulk), respectively. A redistribution could be suspected from the large differences in swelling (e.g. for $T_a = 1173$ K and $t_a = 1$h, $S \approx 1\ 10^{-2}$ and $S_0 \approx 2\ 10^{-3}$) and had indeed to be inferred if the bubbles in both regions were assumed to be under equilibrium pressure, i.e. $p = 2\gamma/r_B$. However, if one takes into account that the bubbles in the bulk are overpressurized (cf. section 2b) and uses the density values reported in [5], the actual helium concentrations in the near surface regions and the bulk are equal within the experimental uncertainty. In other words, no migration of helium over macroscopic distances has to be assumed in order to comply with the different swelling in the different regions.

334

4.   Critical Helium Concentration

Whereas all available results displayed in Figs. 1 and 2 show the splitting up into two spatially separated populations during annealing, there is one set of data [12] where only "large" bubbles are observed both near boundaries and in the bulk. In Ref. [3] this result was attributed to the low implanted helium content (100 appm) of the specimens in Ref. [12] as compared to c_{He}=500–5000 appm employed in the other investigations.

It was suggested that there exists a critical helium concentration c^*_{He} below which sufficient vacancies are available from the implantation damage or from dislocations to relax the initial high pressure in the bubbles to the value of maximum Ostwald ripening. For $c_{He} > c^*_{He}$, the "internal" vacancy supply is insufficient for triggering fast Ostwald ripening and the population of large bubbles develops only in regions near "external" vacancy sources such as free surfaces or grain boundaries.

A rough estimate of a lower bound for c^*_{He} can be obtained by comparing the concentration of implanted helium atoms

$$c_{He} = P_{He}\,\tau \tag{8}$$

with the concentration of implantation-induced vacancies, c^r_v, which survive recombination by interstitials and are thus available for bubble formation (cf. e.g. [16])

$$c^r_v \approx (K\,\tau\,\bar{\rho}_d\,\Omega/2\pi R)^{1/2} \tag{9}$$

with P_{He} helium implantation rate, K displacement rate, τ implantation time, $\bar{\rho}_d$ dislocation density averaged over τ, R radius of spontaneous recombination of Frenkel pairs. Employing an equation of state for He [17], it can be shown that a ratio c_{He}/c^r_v, of around 1 is necessary to relax the bubble pressure to about 7 GPa, the estimated value for maximum Ostwald ripening in Ni [4]. Combining this criterion with Eqs. (8) and (9) yields

$$c^*_{He} \approx (K/P_{He})^{1/2}\,(\bar{\rho}_d\,\Omega/2\pi R)^{1/2} \tag{10}$$

Inserting $\bar{\rho}_d = 5\ 10^{14}\mathrm{m}^{-2}$, $\Omega = 1.1\ 10^{-29}\mathrm{m}^3$, R = 0.8 nm and $K/P_{He} = 10^2$ as reasonable estimates, a critical helium concentration c^*_{He} of about 100 appm is obtained. Considering the large uncertainties in estimating $\bar{\rho}_d$ and the drastic simplifications in the derivation of eqn. (9), the agreement with the experimental value of somewhat above 100 appm is satisfactory. There are two recent experimental results which confirm the existence of a critical gas concentration:

(1) In a systematic investigation of the influence of the helium concentration on the coarsening mechanism of bubbles in the bulk of AISI 316 stainless steel [18] and a FeNiCr alloy [19], a distinct transition from Ostwald ripening at low c_{He} to migration and coalescence at high c_{He} was found for $c^*_{He} \approx 100$ appm.

(2) Implantation of (100–500) keV Ne$^+$ ions to c_{Ne} = 47500 appm into nickel and subsequent annealing at 1023 K [20] resulted in the formation of a high density of tiny bubbles, whereas large bubbles with a very low density were obtained for c_{Ne} =1750 appm. We may interpret this different behaviour by the different concentrations, being above c^*_{Ne} in the first case and below c^*_{Ne} in the second case: assuming that $(K/P)_{Ne} \approx 30\ (K/P)_{He}$ [20] yields $c^*_{Ne} \approx 3000$ appm, i.e. a value between the high and low concentrations used in the experiments of ref. [20].

5. Summary and Outlook

Our concerted experimental and theoretical investigations on the coarsening of helium bubbles during annealing has led to a rather detailed understanding of the underlying processes and could explain some apparent discrepancies between data obtained under different conditions such as:

(a) Large differences in the bubble structure observed after low energy (keV, i.e. near surface deposition of He) vs. high energy (MeV, i.e. bulk deposition) implantation,

(b) large differences in the bubble structure observed for high (≥ 500 appm) and low (≤ 100 appm) helium concentrations, and

(c) an apparent surplus of the amount of He introduced over the amount in bubbles calculated by assuming equilibrium bubbles and usually attributed to helium in "invisible clusters"—which are now no longer necessary.

However, several other questions remain to be answered. They include an explanation of an apparent temperature dependence of the slope m in a double log plot of $\bar{r}_B$ vs. time in isothermal annealing experiments. Whereas at high temperatures the observed slopes m are compatible with the theoretical predictions (Fig. 3), m as large as 10 is observed at medium temperatures [21]. Furthermore, it is not fully understood why the dislocations in the bulk (even at $T_a = 1173$, the dislocation density is around $10^{14} \mathrm{m}^{-2}$ in He-containing Ni) are unable to supply sufficient vacancies for relaxing the overpressure in the "small" bubbles in the bulk. This effect seems to depend on the impurity content of the material: whereas high pressure-impeded coarsening in the bulk and a consequent split-up into "small" and "large" bubbles are observed in pure Ni and a "pure" FeNiCr alloy, preliminary results indicate that these features are absent in commercial stainless steel [18].

REFERENCES

1. H. Trinkaus, J. Nucl. Mat. **133 & 134**, 105 (1985) and Radiat. Effects **101**, 91 (1987).
2. H. Ullmaier, Nucl. Fusion, **24**, 1039 (1984).
3. V.N. Chernikov, H. Trinkaus, P. Jung and H. Ullmaier, J. Nucl. Mater. **170**, 31 (1990).
4. H. Trinkaus, Scripta Metall. **23**, 1773 (1989).
5. Qiang-Li, W. Kesternich, H. Schroeder, D. Schwahn and H. Ullmaier, Acta Metall. et Mater. **38**, 2383 (1990) and H. Ullmaier, this volume.
6. W. Kesternich, D. Schwahn and H. Ullmaier, Scripta Metall. **18**, 1011 (1984).
7. M. Marochov, L.J. Perryman and P.J. Goodhew, J. Nucl. Mater. **149**, 296 (1987).
8. V.N. Chernikov, A.P. Zahkharov and P.R. Kazansky, Doklady AN SSSR **295**, 1119 (1987) and **304**, 870 (1989).
9. V.N. Chernikov, P.R. Kazansky, A.P. Zakharov and A.V. Markin, Doklady AN SSSR **311**, 108 (1990).
10. P. Ehrhart, A. Gaber and W. Jäger, Acta Metall. **35**, 1943 (1987).
11. E.Ya. Michlin, V.Ph. Chkuaseli, Yu.N. Sokursky and G.A. Arutyunova, Atomnaya Energiya **56**, 144 (1984).
12. J. Laakmann, Löslichkeit und Blasenwachstum von Helium in Metallen unter hohem Druck, Berichte der Kernforschungsanlage Jülich, J L-2007 (1985) ISSN 03660885.
13. K. Shiraishi and K. Fukai, J. Nucl. Mater. **117**, 134 (1983).
14. B. Viswanathan and G. Amarenda, this volume.
15. V.V. Slezov and P.O. Mchedlov-Petrosyan, Phys. Met. Metal. **46/2**, 26 (1979).
16. S.I. Golubov, Fizika Metallov i Metallovedenije **60**, 428 (1985).
17. H. Trinkaus, Rad. Effects **78**, 189 (1983).
18. H. Schroeder and P. Fichtner, Proc. 4th Intern. Conf. on Fusion Reactor Materials, Kyoto (1989), J. Nucl. Mater. **179–181**, 1007 (1991).
19. H. Schroeder and P. Fichtner, this volume.
20. N. Marochov and P.J. Goodhew, J. Nucl. Mater. **158**, 81 (1988).
21. F. Carsughi, D. Schwahn and H. Ullmaier, to be published.

PHASE TRANSFORMATIONS OF ARGON IN BUBBLES FORMED IN NICKEL DURING LOW AND HIGH ENERGY ARGON ION BOMBARDMENT

D.B. Kuzminov and V.N. Chernikov

Institute of Physical Chemistry, Academy of Sciences of the USSR
Leninsky Prospekt 31, 117915, Moscow, USSR

ABSTRACT

Small precipitates/bubbles of argon in a nickel matrix have been formed by 0.6 keV Ar[+] ion bombardment at 870 K. These bubbles contain fluid argon which transforms to the solid state at 840 K. The appearance of bubbles and the lattice parameter of solid inert gas at room temperature were obtained using TEM. Gas desorption experiments revealed a sharp increase in the argon desorption rate at 840 K on cooling the irradiated sample which was related to the argon solidification in the bubbles. A quantitative model is also proposed according to which the melting of heavy inert gas inside overpressurized bubbles results from their abrupt volume increase by the emission of dislocation loops.

1. Introduction

In 1984 Templier et al.[1] and vom Felde et al.[2] showed that rare gas atoms (Ar and Xe) introduced into Al by room-temperature ion bombardment segregated there into bubbles in a crystalline state. Later, the formation of bubbles containing heavy inert gases (HIG) in a crystalline state (Ar[c], Kr[c], Xe[c]) as a result of high-energy ion implantation at 300 K was registered in other metals too [3–7]. The formation of this type of inclusion as a result of low–energy ion bombardment (at ion energies <1 keV) at moderate irradiation temperatures has not been observed and has been considered to be impossible because of rather low ion penetration depth R_p = 1.5–2.0 nm and a relatively high value of sputtering rate. In spite of a number of studies available on crystalline HIG in metals, until now there are no adequate models of melting and crystallization of HIG[c] under temperature variations.

In the first part of this work we present a new concept of melting of HIG[c] (solid precipitates) in over-pressurized gas bubbles, and also a mechanism of HIG transformation within bubbles into a fluid state, which is believed to be dominant. In the second part of the paper we report on formation of solid argon bubbles in nickel as a result of low-energy ion bombardment. A mechanism for fluid argon crystallization in over-pressurized gas bubbles accompanied by gas release on cooling is suggested.

2. Experimental Procedure

Single crystalline Ni specimens were chosen as objects of investigation: massive ones with surface normal <110> and 50 μm thick discs with surface normal of <112> type. The irradiation of the specimens with 0.6 keV Ar[+] ions at different temperatures and heating of room-temperature implanted foils were carried out in a specially constructed experimental cell mounted in a RIBER oil-free UHV system. The system comprised a sample heating unit,

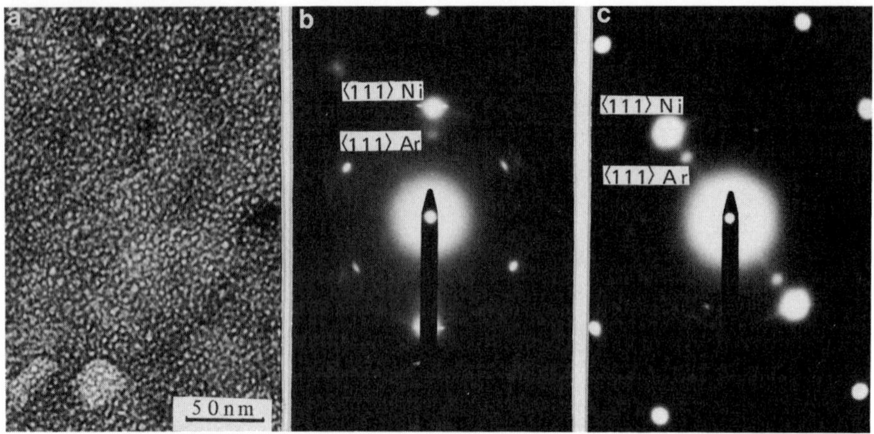

Fig 1. Small cavities in near-surface layers of Ni foils irradiated with 30 keV Ar+ ions to a dose of 2 10^{20} ions m^{-2} at 300 K as revealed by TEM:
(a) bright-field underfocused image (Δf= -0.8 μm);
(b) selected area diffraction pattern, which correspond to (a); the orientation relation is {112}Ni ∥ {112}Ar;
(c) the same as (b), but after single heating.

low-energy ion guns, and a quadrupole mass-spectrometer which allowed both to register inert gas desorption transients and to perform depth profiling of near-surface regions of irradiated bulk specimens.

One group of Ni foils was irradiated with low-energy Ar$^+$ ions at elevated temperatures to a dose of 8 10^{21} ions m^{-2}. The other group of foils was implanted at 300 K with 30 keV Ar+ ions up to a dose of 2 10^{20} ions m^{-2}; some of these foils were then heated up to 820 K at a rate of 10 K/s . From here on, all the specimens were back-thinned and jet electropolished from the side opposite to the irradiated one. In this manner the objects were prepared for studying of Ar-loaded near-surface layers in TEM. This analysis was performed at room temperature in a Philips EM-400T electron microscope operated at 120 kV.

3. Melting of Inert Gases within Bubbles of Crystalline Metal Matrix

3.1. Argon bubbles in nickel after 30 keV ion implantation at 300 K

A dense population of small cavities with a mean diameter d= 3.0 nm was revealed in the near-surface layers of irradiated specimens using phase contrast imaging in TEM (Fig. 1a). Examination of microdiffraction patterns (Fig. 1b) and a dark-field analysis led us to the conclusion that argon in the cavities was present in the crystalline state and had an fcc lattice with an average lattice parameter a= 0.470 ± 0.003 nm. The latter corresponds to the volume density of Ar atoms ρ= 3.9 ± 0.2 10^{28} atoms m^{-3}. According to the extrapolated equation of state (EOS) suggested by Ronchi [8] the pressure of Ar in these bubbles is of the order of 6 GPa. At the same time the surface energy for a cavity of d=3nm in Ni does not produce a pressure which exceeds 2.7 GPa. Hence, Ar-filled cavities were assigned as over-pressurized.

Diffraction pattern analysis of specimens prepared from single heated foils (Fig. 1c) pointed out that the heating caused the Ar lattice parameter to grow to the average value a = 0.495±0.005 nm. The latter corresponds to the atomic density in these bubbles ρ= 3.3 ± 0.1 10^{28} atoms m^{-3} and internal pressure p ≈ 2.4 GPa according to the Ronchi EOS . Taking into account the data obtained by Evans and Mazey [3] on the phase transformation of Kr into fluid within bubbles in the Kr/Ni system at T= 825–875 K, one can assume that the heating of irradiated Ni specimens in our experiment (up to 820 K) also led to the melting of

338

Table 1. Properties of crystalline inert gases in implanted metals.

System HIG/Me	Mean bubble diam., $\bar{d}$, nm	Solid inert gas properties			References
		Lattice unit $\bar{a}$, nm	Pack. dens. $10^{28} m^{-3}$	T_m, K	
Ar/Ni	3.0	0.470±0.003	3.9	–	This work
Ar/Al	2.7±0.5	0.516±0.005	3.2	730±20	Donnelly & Rossouw [5]
Kr/Ni	3.0	0.5	3.2	825–870	Evans & Mazey [3]
Kr/Al	3.2	0.534±0.003	2.63	620	Hashimoto, Yorikawa et al.[6]
Kr/Mo	4.5*	0.51	3.2	920	Evans & Mazey [4]
Xe/Al	3.6+0.5	0.604+0.004	1.9	670	Donnelly & Rossouw [5]

*)The value was calculated from the micrograph given in [4].

Ar in bubbles. Then the increase of the lattice parameter and the corresponding decrease of the hydrostatic pressure of Ar can be attributed to the process of Ar melting on heating to 820 K and its crystallization on subsequent cooling. The quantitative experimental data on different HIG /metal systems including HIG melting points are summarized in Table 1 and will be used below.

3.2. A model of melting and discussion

The calculated plots for Ar^c, Kr^c, Xe^c, showing the change of the average pressure p(T) of HIG^c in bubbles with the matrix temperature T (for Ar, Kr, Xe in Ni, Al and Mo) are shown in Fig. 2. The calculation of these curves has been performed on the basis of Ronchi EOS for experimental data quoted in Table 1, i.e. starting from the values of a_{HIG} at 300 K. The thermal expansion of the metal was taken into account, so that the p(T) curves are, in fact, "quasi-isochores". The accuracy of these quasi-isochores depends on the accuracy of a_{HIG} determination. For studies where the error is not indicated, the p(T) plots are shown by solid lines, whereas for those stating the a_{HIG} accuracy they are represented by bands. The abscissae of the points shown by full circles in the same Fig. 2 correspond to the intervals/values of HIG^c melting temperatures T_m found in the experiments.

The HIG^c melting curves in p–T coordinates are described by the Simon type equation [9]:

$$p = A (T - T_o)^c + B \qquad (1)$$

where A, B, c, T_o are empirical coefficients. Their numerical values for Ar, Kr and Xe are given in Table 2. The HIG^c melting curves $p(T)|_m$ are shown by dashed lines in Fig. 2. One might think that the melting temperature of a particular HIG^c in a bubble can be determined as the abscissa of the point of intersection of a previously calculated p(T) curve with the Simon melting curve (1). However, it is seen that in many cases considered here, the melting temperatures T_m determined in this manner differ appreciably from the experimentally registered values. Below we outline physical reasons for these discrepancies, and suggest a model which enables us to predict accurately the temperature of phase transformations.

When heating, the phase transformation of HIG^c into fluid is possible only under the condition of an increase of the volume it occupies. Following the Clausius-Clapeyron equation this increase is:

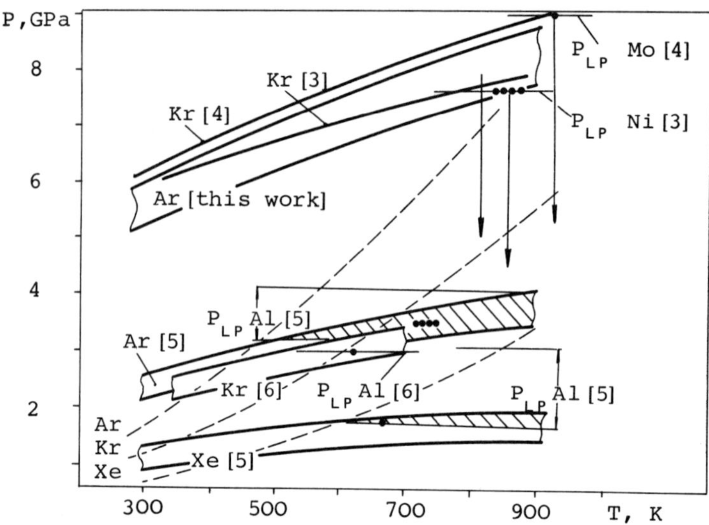

Fig 2. Summary diagram of p(T) plots: Ronchi EOS for various HIG with the use of data in Table 1 are indicated by thick solid lines; melting curves $p = p(T)|_m$ calculated by Eq. (1) are indicated by dashed lines; critical pressure curves/bands for dislocation loop punching $P_{LP} = P_{LP}(T)$ according to Greenwood et al. [16] with the use of data in Table 1 are indicated by thin solid lines. The abscissae of the points denoted by full circles correspond to the experimentally registered values of HIG melting temperatures.

$$\Delta V_{pt} = \frac{H_m}{T_m}\left.\left(\frac{\partial p}{\partial T}\right)\right|_m^{-1}$$

(2)

where H_m is the latent heat of melting (the H_m/T_m ratio being for all elementary substances $10.5 \pm 2.1 J\ mol^{-1}\ K^{-1}$ [10]) and $\partial p/\partial T|_m$ is the derivative of the melting curve at T_m. Hence, solid gas contained in a closed cavity with fixed volume when reaching melting parameters (p_m, T_m) starts overheating and the phase transformation is delayed. One can expect that probably a phenomenon like this was observed by Rossouw and Donnelly [11]. It is appropriate to note that the values of ΔV_{pt} assessed by means of eqn.(2) exceed substantially those of thermal expansion of the bubble volume which occur due to overheating by a few hundred degrees (!)

In general, the volume growth of a gas-filled cavity on heating can proceed in two ways: 1) by trapping of thermal vacancies or 2) by punching out interstitial type dislocation loops which can be realized under conditions of "vacancy deficit". The vacancy deficit is originated either due to insufficient mobility of vacancies or because of a slow advance of vacancy fronts (from vacancy sources) through the HIG enriched zone, as shown by Marochov *et al.* [12] and Chernikov *et al.* [13-15] for He/Ni systems. Which mode of cavity volume growth and HIG melting will be realized, depends in general on the metal, initial gas pressure and the rate of heating. The melting of Kr in bubbles in a Ni matrix observed by Evans and Mazey [3] might have occurred due to accumulation of thermal vacancies as was proposed by the authors. However, with the above in mind a condition of sufficient vacancy mobility can be regarded only as a necessary one for realization of the first mechanism. On the other hand, melting of HIG within bubbles was observed also at temperatures rather low for the first mechanism to occur, e.g. in Kr/ Mo, Ni systems (see Table 1). In these cases dislocation loop punching by

Table 2. Coefficients A, B, c, T_o in Simon-type equation [9]; see eqn. 1.

INERT GAS	A	B, MPa	c	T_o, K
Ar	0.499	148.4	1.43	30.18
Kr	0.336	177.8	1.44	38.10
Xe	0.571	129.6	1.30	95.45

bubbles due to the inner pressure increase on heating is the only mechanism for the volume expansion of bubbles. This affirmation is corroborated by the quantitative estimates given below. According to Greenwood et al. [16] the dislocation loop punching (LP) occurs on reaching, inside a bubble, a pressure:

$$p_{LP} = \frac{2\gamma}{R} + \frac{\mu b}{2\pi (1-v) R_{DL}} \ln\left(\frac{R_{DL}}{r_o}\right) \tag{3}$$

where R is the bubble radius, R_{LP} is the radius of a dislocation loop ($R_{LP} \approx R$), μ is the shear modulus, b is the Burgers vector, r_o is the core radius of a dislocation (of the order of b), γ is the surface specific energy of a metal, v is Poisson's ratio. As a result of punching a dislocation loop the bubble volume increases by $\Delta V_{DL} = \pi R^2 b$, the internal pressure and the atomic density falls and this may result in the melting of a HIG^c. Indeed, for all the HIG^c/metal systems listed in Table 1, the condition $\Delta V_{DL} = (3-4) \Delta V_{pt}$ is satisfied (referring to a single bubble). Thus, the melting point of HIG^c in a given bubble may be attained when reaching the critical gas pressure p_{LP}. For systems quoted in Table 1, T_m values can be defined as the abscissae of the intersection points of p(T) curves based on Ronchi EOS with corresponding relationships $p_{LP} = p_{LP}(T)$ calculated by means of eqn. (3), taking into account thermal expansion. In Fig. 2, p(T) curves are shown by thin solid lines or bands between thin lines (depending on the knowledge of the accuracy of initial bubble diameter determination). The values of metal constants μ, γ, b and v from the review by Donnelly [17] are adopted.

It is clearly seen that all the experimental values of melting temperatures (Table 1) fall within the temperature intervals determined by the above formulated diagram procedure (Fig. 2). According to the same model the pressure drop in a bubble due to dislocation loop punching is sufficient for turning HIG^c to the fluid state. Indeed, the length of each vertical arrow in Fig. 2 corresponds to the difference of HIG pressure before and after dislocation loop ejection (for Ar, Kr/ Ni and Kr/ Mo systems), and as clearly seen, the arrows intersect the corresponding Ar and Kr Simon melting curves. Some discrepancies between calculated and experimental values of T_m do not exceed 50 K, and can be attributed to the errors in bubble parameter determination. Besides, in the case of Al one cannot ignore, *a priori*, the possibility of the bubble volume expansion due to thermal vacancies, i.e. by the first mechanism.

On the basis of the suggested model the experimental results obtained in [3] appear to be clear and logical in the frame of our interpretation. First, according to [3] the temperatures of melting and subsequent crystallization of Kr (the latter is not specified) did not coincide; secondly, the heating and cooling of irradiated Ni resulted in a reduction of Kr^c atomic density in bubbles from $\approx 3.2 \ 10^{28}$ atoms m^{-3} to $\approx 2.7 \ 10^{28}$ atoms m^{-3} . According to our model the melting of Kr^c in [3] at 825-875 K results in the bubble pressure drop down to ≈ 4 GPa (Fig. 2). The average pressure of Kr in bubbles will decrease on cooling in accordance with the shape of p = p(T) curve defined by the Ronchi EOS until the intersection with the melting curve of

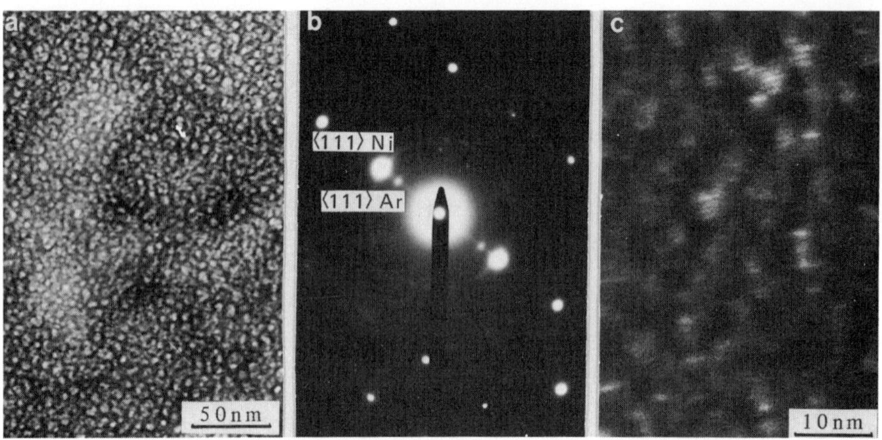

Fig 3. Subsurface layer of single crystalline Ni, irradiated with Ar^+ ions (E=0.6 keV, D= $8 \cdot 10^{21}$ ions m^{-2}) at 870 K as revealed by TEM:

(a) bright-field image in phase contrast, under focus (Δf= -0.8 μm);

(b) dark-field image obtained with a <111>Ar reflection with capturing of a part of matrix reflection of the same type;

(c) diffraction pattern, {112}Ni ∥ {112}Ar.

Kr. The abscissa of this point corresponds to the temperature of crystallization T_c = 730 K. This temperature is in fact lower than the melting temperature T_m determined in [3]. The drop of Kr pressure takes place due to the bubble volume expansion by ΔV_{DL} = 1.8 nm^3 which leads to a reduction of the atomic density by a factor of 1.13. This means that the resultant density is approximately $2.8 \cdot 10^{28}$ atoms m^{-3} which is comparable with the value found in [3].

A similar analysis of the behavior of Ar / Ni system enables us to conclude that after secondary crystallization the atomic density of Ar in bubbles at 300 K becomes $\approx 3.5 \cdot 10^{28}$ atoms m^{-3}, which is also quite close to the value found in our experiment, namely $\approx 3.3 \pm 0.1 \cdot 10^{28}$ atoms m^{-3}.

4. Solid Argon Bubbles in Nickel after Low-Energy Ion Irradiation

4.1. Temperature conditions of solid argon precipitation in 0.6 keV ion-bombarded Ni

TEM analysis showed that the irradiation to a dose of $8 \cdot 10^{21}$ ions m^{-2} at a temperature 870 K (0.5 T_m) led to the formation in the near-surface region of the Ni specimen of a dense population of argon bubbles (c_b= $1.6 \pm 0.4 \cdot 10^{24}$ m^{-3}) in the size range from 2.5 to 10 nm and with the mean diameter $\bar{d} \approx 3.5$ nm (Fig. 3a). In addition, one could clearly discern in diffraction patterns, apart from Ni-matrix spots, the extra reflections (Fig. 3b), which corresponded to a crystalline phase of fcc structure and with the mean value of lattice parameter $\bar{a}$ = 0.495 $\pm$ 0.005 nm. Dark-field analysis with the use of these extra reflections led to the conclusion that the new phase discovered was argon which had crystallized endotaxially in bubbles in the nickel matrix. In particular, the analysis of dark-field images formed using <111>-type diffraction beams of Ni and Ar simultaneously allowed us to see moiré fringes (from 3 to 5) within the images of some bubbles (Fig. 3c). These fringes were oriented normally to the direction of both the operative diffraction vectors, having a spacing of about 0.7 nm which is very close to the calculated value (0.706 nm).

The mean fcc argon packing density and its molar volume estimated from the lattice parameter were found to be $\bar{\rho}$ = $3.3 \cdot 10^{28}$ atoms m^{-3} and $\bar{V}_m$ = $1.83 \cdot 10^{-5}$ m^3 mol^{-1} respectively. The associated argon pressure in the bubbles was obtained from the equation of state of

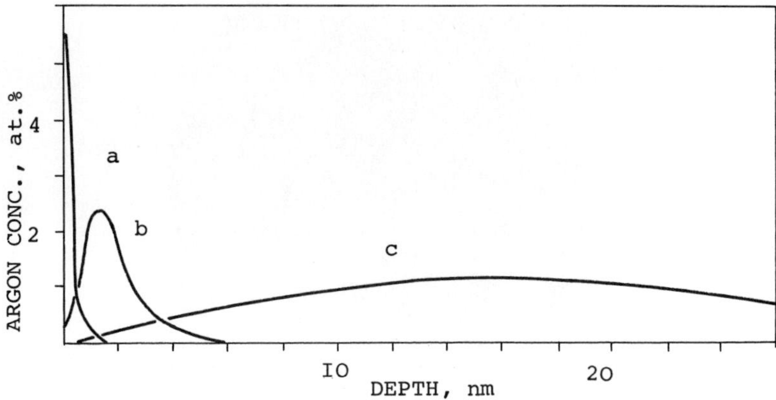

Fig 4. Depth distributions of Ar in Ni after irradiation of Ni(110) with 0.6 keV Ar+ ions to a dose of 8 10^{21} ions m^{-2} at temperatures: (a) 300 K, (b) 700 K, (c) 870 K.

Ronchi and gave a value of ≈ 2.5 GPa for $\overline{V}_m$. This pressure is sufficient for the formation of the solid phase in the bubbles and comparable with the Laplace pressure for bubbles of 3.5 nm in diameter in nickel. Thus, all visible bubbles can be considered to be close to the equilibrium ones. Nevertheless, taking into account scatter in the bubble size, one can easily come to the conclusion that there are not only solid argon bubbles present in the sample, but also those containing argon at lower pressure, i.e. in the fluid state.

Depth profiling of specimens irradiated with 0.6 keV Ar$^+$ ions at 870 K revealed the location of inert gas atoms at a depth of more than 30 nm (Fig. 4c). Penetration of argon atoms to such a distance (it is considerably larger than R_p) probably takes place as a result of Ar atom migration by a divacancy mechanism, i.e. in the form of ArV$_2$ complexes (migration activation energy $E^m = 1.67$ eV [18]). At irradiation temperatures less than 800 K the width of location layer of argon atoms does not differ significantly from the ion penetration depth R_p (Fig. 4a,b) and no bubbles are visible in TEM. The explanation consists of the following. Diffusive penetration of argon atoms diminishes substantially, and ion sputtering of the implanted surface layers does not allow concentrations of inert gas atoms high enough to form bubbles athermally. Thus the competition between ion trapping and ion-induced gas release leads to a rather high temperature threshold of argon bubble formation in nickel under low-energy ion bombardment.

TEM analysis of nickel samples irradiated at 930 K revealed the formation of mostly faceted argon bubbles in the size range from 10 to 20 nm (Fig. 5) containing argon in the fluid state (in this case no extra reflections were observed on the diffraction pattern). The bubble size growth on the increase of irradiation temperature above 0.5 T$_m$ is obviously caused by trapping of thermal vacancies. Thus, solid argon bubble formation can occur in nickel as a result of low-energy ion irradiation carried out in a narrow temperature range between $\approx$860 K and $\approx$900 K. While the lower threshold depends mainly on diffusive penetration of argon atoms, the upper threshold is limited by the increase of concentration and mobility of thermal vacancies.

4.2. Crystallization of fluid argon in bubbles of nickel matrix

Despite the fact that irradiation and analysis in the TEM were carried out at different temperatures, one can hardly expect detectable bubble shrinkage on cooling from 870 K to room temperature. Also, the mean bubble diameters after 30 keV ion implantation at 300 K (see Fig. 1a) and that after low-energy bombardment at 870 K (Fig. 3a) do not differ

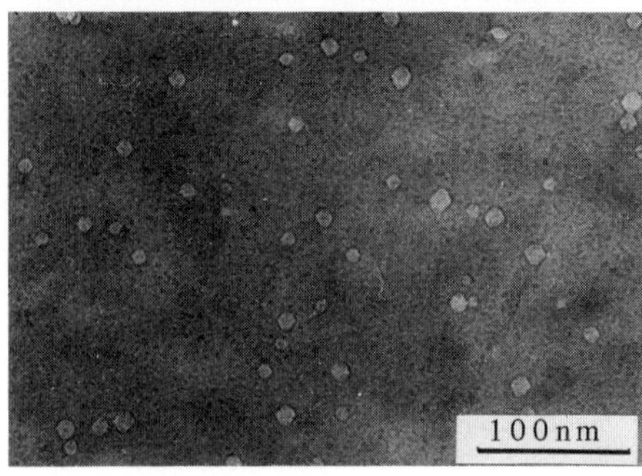

Fig 5. Bright-field image of Ni irradiated with Ar$^+$ ions (E = 0.6 keV, D = 10^{21} ions m^{-2}) at 930 K.

significantly. Hence, one can expect the formation of over-pressurized bubbles in the course of low-energy bombardment at 870 K. In this case, the gas pressure inside bubbles of 3.5 nm in diameter should not exceed 6.8 GPa, as estimated from (3). Comparison of this value with the Simon melting curve (1) allows us to suggest the formation of fluid argon inside bubbles at 870 K and its solidification on subsequent cooling at about 820–840 K.

Gas desorption experiments on cooling down after irradiation at 870 K show that the argon desorption rate increases abruptly, leading to a maximum in the argon desorption transient versus decreasing temperature at about 840 K (Fig. 6c).The intensity of this anomalous desorption peak gradually decreases both on raising and on lowering the irradiation temperature (Fig. 6). Since the appearance of the argon desorption peak is always accompanied by the observation of solid bubbles in TEM and also taking into account the peak maximum temperature, one may conclude that inert gas atom desorption takes place as a result of the fluid to solid argon phase transformation inside these bubbles. A similar anomalous krypton desorption peak (at T≈860 K) was recently observed on cooling of Ni after bombardment with low energy Kr ions at elevated temperature [19,20]. The physics of this phenomenon can be explained in the following way. Argon solidification leads to decrease of its molar volume, which in its turn results in a single bubble volume decrease by vacancy emission into the metal matrix. These vacancies assist some argon atoms to desorb out of a bubble. The estimation of the free-energy change caused by argon crystallization inside a bubble in the Ni lattice containing vacancy-type defects shows that this process becomes energetically favorable, if argon atoms leave a bubble in the form of ArV$_2$ complexes. Since such complexes are highly mobile in nickel lattice at T > 800 K they mostly migrate towards the surface and desorb, thus causing the argon desorption rate to increase versus decreasing sample temperature. The estimated pressure drop inside bubbles makes them practically equilibrium with room-temperature argon packing density of ≈3.3 10^{28} atoms m^{-3}. The last conclusion is in full agreement with TEM data.

5. Summary

In this paper we have presented some experimental results on solid argon precipitation in nickel both after 30 keV ion implantation at room temperature and after 0.6 keV ion bombardment at 870 K, and have suggested new models of argon transformation and accompanying effects at temperature variations.

(i) 30 keV Ar$^+$ ion implantation of Ni at 300 K up to a dose of 2 10^{20} ions m^{-2} leads to the formation in the near-surface layer of over-pressurized bubbles (d ≈ 3.0 nm) containing solid monocrystalline precipitates of argon.

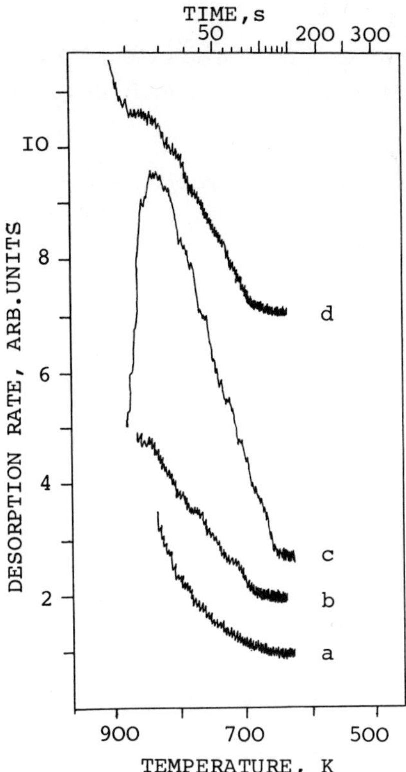

Fig 6. Argon desorption transients at cooling down of a single crystalline Ni specimen irradiated with Ar⁺ ions (E = 0.6 keV, D = 8 10^{21} ions m⁻²) at temperatures: (a) 830 K, (b) 840 K, (c) 870 K, (d) 900 K. The desorption curves have been shifted with respect to each other in order to avoid overlapping.

(ii) 0.6 keV Ar^+ ion bombardment of Ni at 870 K up to a dose of 8 10^{21} ions m⁻² leads to the formation of over-pressurized bubbles (d≈3.5 nm) containing fluid argon which transforms into the solid state on decreasing temperature at T ≥ 840 K . The latter is accompanied by some gas loss from bubbles which makes them nearly equilibrium.

After both kinds of irradiation TEM reveals at room temperature bubbles with single crystalline Ar precipitates having fcc lattice oriented endotaxially with respect to the host Ni matrix.

(iii) Anomalous argon desorption observed during cooling down of Ni, irradiated at 870 K is a result of fluid to solid argon transformation inside bubbles. Argon crystallization is accompanied by a bubble volume decrease due to vacancy emission, which assists the resolution of some argon atoms into the host matrix in the form of ArV_2 complexes, and subsequent argon desorption.

(iv) A new model of heavy rare gas crystalline precipitates melting in metal matrix is suggested. Under conditions of vacancy deficit the melting of a crystalline heavy rare gas in an over-pressurized bubble proceeds during heating due to the abrupt bubble volume expansion by the dislocation loop punching. The knowledge of average bubble radius in a given metal, and a lattice parameter for the rare gas precipitates crystallized therein, permits a reasonably precise calculation of the melting temperature and the resultant inner pressure, which is sufficient to predict the further state of gas in bubbles.

REFERENCES

1. C. Templier, C. Jaouen, J.-P. Rivière, J. Delafond and J. Grilhé, C.R.Acad. Sci., Paris, **299**, 613 (1984).
2. A. vom Felde, J. Fink, Th. Muller-Heinzerling, J. Pfluger, B. Sheerer, J. Linker and D. Kaletta, Phys. Rev. Lett. **53**, 922 (1984).
3. J.H. Evans and D.J. Mazey, J. Phys. **F15**, 21 (1985).
4. J.H. Evans and D.J. Mazey, Scripta Metallurgica **19**, 621 (1985).
5. S.E. Donnelly and C.J. Rossouw, Nucl. Inst. & Meth. **B13**, 485 (1986).
6. I. Hashimoto, H. Yorikawa, H. Mitsuya, H. Yamagushi and K. Takashi, J. Nucl. Mater., **149**, 69 (1987).
7. R.C. Birtcher and W. Jager, Ultramicroscopy **22**, 267 (1987).
8. C. Ronchi, Journ. Nucl. Mater. **96**, 314 (1984).
9. R.K. Crawford, in *Rare Gas Solids*, M.L. Klein and J.A. Venables, ed., Acad. Press, London (1977).
10. *Handbook of Physical and Chemical Units*, A.A. Ravdel and A.M. Ponomoreva, ed., Khimiya, Leningrad (1983).
11. C.J. Rossouw and S.E. Donnelly, Phys. Rev. Lett. **55**, 2960 (1985).
12. N. Marochov, L.J. Perryman and P.J. Goodhew, J. Nucl. Mater. **149**, 296 (1987).
13. V.N. Chernikov, A.P. Zakharov and P.R. Kazansky, J. Nucl. Mater. **155–157**, 1142 (1988).
14. V.N. Chernikov, H. Trinkaus, P. Jung and H. Ullmaier, J. Nucl. Mater. **170**, 31 (1990).
15. V.N. Chernikov and P.R. Kazansky, J. Nucl. Mater. **172**, 155 (1990).
16. G.W. Greenwood, A.J.E. Foreman and D.E. Rimmer, J. Nucl. Mater. **4**, 305 (1959).
17. S.E. Donnelly, Rad. Eff. 90, 1 (1985).
18. C.F. Melius, W.D. Wilson and C.L. Bisson, Rad. Eff. **53**, 111 (1980).
19. D.B. Kuzminov, M.J. Gerchikov, A.M. Panesh and A.P. Simonov, Zhurnal Fiz. Khim. (in Russian) **10**, 1980 (1988).
20. D.B. Kuzimov and V.N. Chernikov, Rad. Eff. Lett. **2**, 159 (1989).

BUBBLE GROWTH MECHANISMS

INERT GAS BUBBLE GROWTH MECHANISM MAPS FOR METALS

P.J. Goodhew

Department of Materials Science and Engineering
University of Liverpool
PO Box 147, Liverpool L69 3BX, UK

ABSTRACT
Bubble growth and shrinkage is of great importance for the use of materials in reactor environments.
Cavity growth mechanism maps are potentially useful because they enable the growth mechanism
and rate to be predicted in regimes where experiments are difficult or impossible. Maps are presented
for helium bubbles in niobium and nickel.

1.   Introduction

Mechanism maps are a means of representing a large amount of data in a way that is readily
assimilated. This makes them useful in numerous areas of science. Deformation mechanism
maps were the first maps of this type to be constructed [1] but since then the mapping process
has been used in many diverse areas.

A mechanism map is useful when there are several distinguishable and independent ways
in which a process can occur. Cavity growth provides a classic example of this since the
population of bubbles or voids can grow by many different processes [2].

All maps must be based on accurate understanding and modelling of the appropriate
mechanisms and on good experimental data. The more complete these are, the more reliable
the maps will be. However, even inexact maps can give some useful general information and,
by identifying the regions where data or theory are poor, they can act as a spur to improvement.

2.   Cavity Growth Mechanism Maps

Mechanism maps were first used to describe particle mobility by Ashby [17] and cavity
growth by Goodhew [3]. The general term 'cavity' is used to include both bubbles and voids.

In order to construct a map, equations describing each mechanism are required. The extent
of each field is then established by evaluating the equations and determining the region within
which each mechanism dominates the growth process. The growth rates given by the
equations must also be added in an appropriate way to give the total growth rate and thus to
permit the plotting of contours of constant rate.

Many cavity growth mechanisms have been modelled in the past and the growth rate for
each mechanism, acting alone, is quite well established [eg 4, 5]. The equations used to plot
the maps presented in this paper are given in Appendix 2. The punching of dislocation loops
and individual interstitials has been omitted since these mechanisms are usually only applicable to very small cavities. Indeed no attempt has been made to include the nucleation of the
bubbles, nor (in this paper) to consider the effect of continuous radiation damage or gas

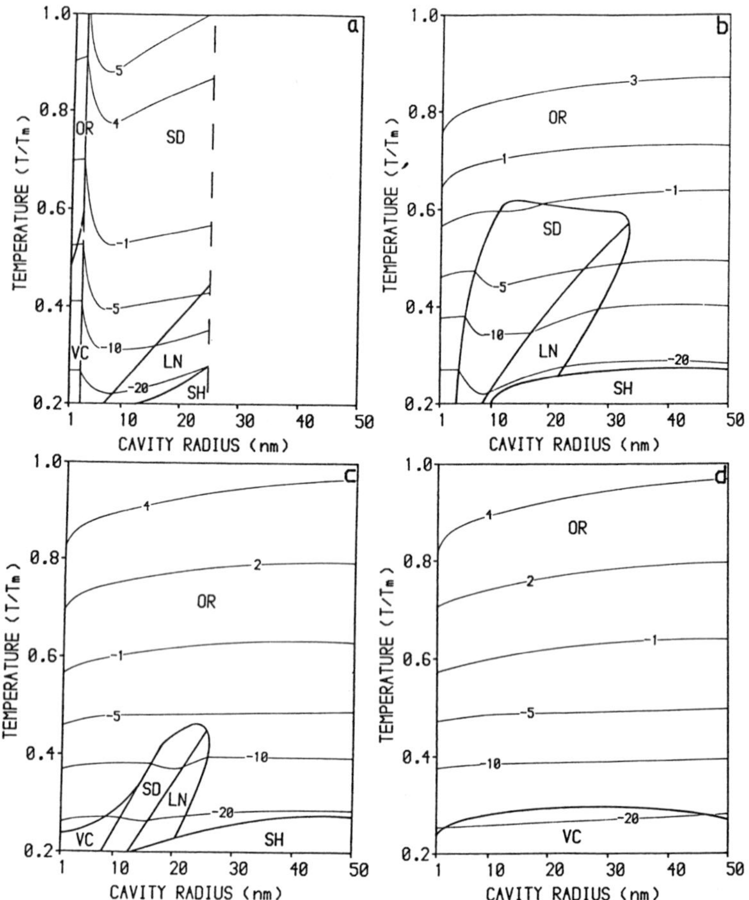

Fig. 1   Maps for the growth of helium cavities in niobium. Gas concentration 1000pm. Cavity density; (a) $10^{22}m^{-3}$, (b) $10^{21}m^{-3}$, (c) $10^{20}m^{-3}$ and (d) $10^{19}m^{-3}$.

supply. A further effect which has not yet been modelled is growth in response to stresses arising, notably in uranium, from irradiation-induced growth. The maps presented here are thus only applicable as yet to a restricted set of conditions, characteristic of "cold implantation and annealing" experiments. It is however easy to see how they could be extended to more "realistic" conditions as and when data of the necessary quality becomes available.

It is necessary to consider the extent to which the various mechanisms which are to be considered can operate in parallel and thus how the growth rates given by each equation should be summed. It is possible to show [2] that in general the individual rates should simply be added, since the mechanisms are not mutually exclusive. For example migration and coalescence can occur while cavities are shrinking by emitting vacancies. This approach has therefore been adopted in computing the total growth rate.

The parameters of most interest in a cavity population are the cavity size distribution, density, total cavity volume and growth rate under the selected conditions. We have chosen to map two of the three main parameters mean cavity size, temperature and cavity density. The method by which the maps are constructed has already been described in detail by Perryman & Goodhew [6, 7].

3.   The Experimental Basis for the Maps

All maps are based on our current understanding of the mechanisms involved and on the experimental data available. The more complete these are, the more accurate and trustworthy the maps. Only a limited amount of experimental data on cavity growth is available in the open literature. In this paper we present maps for cavity growth in two of the best-documented systems; nickel [8, 9] and niobium [4, 10].

Figure 1a is a map at constant cavity density for niobium which shows, as a dashed box, the regime within which experimental data is available. The key to this, and all the other, maps is given in Table 1. Four growth regimes and one shrinkage regime can be distinguished. The map shows fields in which the dominant (ie fastest) rate-controlling mechanism is surface diffusion (SD), vacancy collection (VC), Ostwald ripening (OR) and ledge nucleation (LN). The shrinkage regime (SH) involves essentially the same mechanism as vacancy collection, but there is a net emission of vacancies when the excess pressure in the bubble is negative. The map extends only to cavities of radius about 24 nm since larger cavities than this would touch at the cavity density (10^{22}m^{-3}) selected. It should be emphasised that the map uses the data given in Perryman and Goodhew [7] and could only be fitted to experimental data within the dashed box (where surface diffusion dominates). There is no cavity growth data to support the map in other regions. Therefore, close to the box the map is likely to be accurate but this accuracy will decrease as the physical conditions deviate from those outlined by the box. This is important since all the maps for niobium are based on this same experimental data.

It is notable that in Fig. 1, as in all other maps yet calculated for metals, the vapour transport (VT) and volume diffusion (VD) mechanisms, although considered, do not dominate bubble behaviour at any temperature.

The maps for nickel are also based on the data used by Perryman and Goodhew [7]. The surface diffusion data reported by Maiya and Blakely [11] were used since they provided the best fit with the Marochov results.

It can be seen from the equations given in Appendix 2 that several of the mechanisms are rather sensitive to the choice of materials data. A modest change in some of these data values can have a significant effect. For example, in the worst case, it is possible to completely eliminate Ostwald ripening from the niobium maps (Fig. 1) by changing the pre-exponent for helium diffusion by less than four orders of magnitude, which is approximately the current uncertainty. We have chosen to use data which are not at the extremes of the range of reported values. It is interesting that some evidence for the occurrence of Ostwald ripening in nickel has been reported recently [12].

Table 1 Key to the labels used in the cavity growth mechanism maps. The total growth rates are plotted as contour lines marked in units of log(nm s^{-1}).

	Growth Mechanism
SD	Migration and coalescence limited by surface diffusion
LN	Migration and coalescence limited by ledge nucleation
VD	Migration and coalescence limited by volume diffusion
VT	Migration and coalescence limited by vapour transport
OR	Ostwald ripening
VC	Vacancy collection
SH	Shrinkage (vacancy emission)

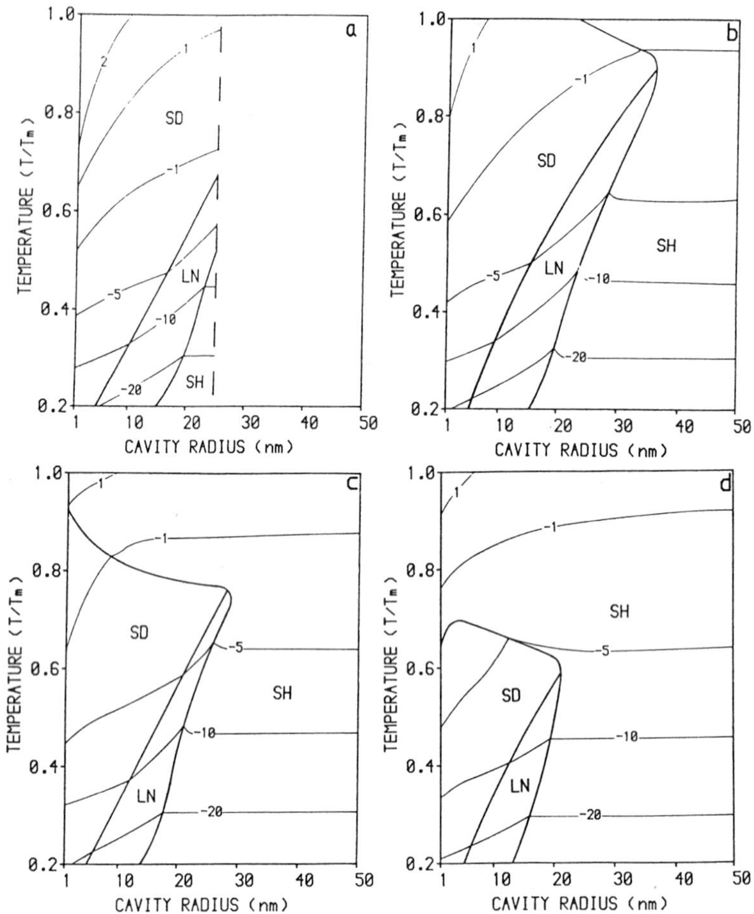

Fig. 2 Maps for the growth of voids in nickel. Cavity density; (a) 10^{22}m^{-3}, (b) 10^{21}m^{-3}, (c) 10^{20}m^{-3} and (d) 10^{19}m^{-3}.

4. The Maps

Several general conclusions emerge from a study of cavity growth maps. Each map requires that all parameters but the two being plotted are kept constant, so an exhaustive study would require a great many maps. Some of the ways in which the maps can be used are illustrated in the figures.

Figure 1 shows four maps for the growth of helium cavities in niobium. These maps form a series of decreasing cavity density. For the highest cavity density, 10^{22}m^{-3}, migration and coalescence is the dominant growth mechanism but as the cavity density decreases the distance between cavities increases and so the rate of coalescence declines. Then growth by the absorption of vacancies or ripening dominates. The contours show that while cavity growth is very rapid above $0.7T_m$ it is insignificant below $0.3T_m$ as even surface diffusion is frozen out.

Figure 2 shows four maps of decreasing cavity density for the growth of voids in nickel. Interestingly, although voids might be expected to shrink by the emission of vacancies (and must eventually do so) the maps show that while they are close together the void population will tend to grow. Although each individual void will generally emit more vacancies than it absorbs, and the total void volume must decrease, void coalescences ensure that the mean size increases.

352

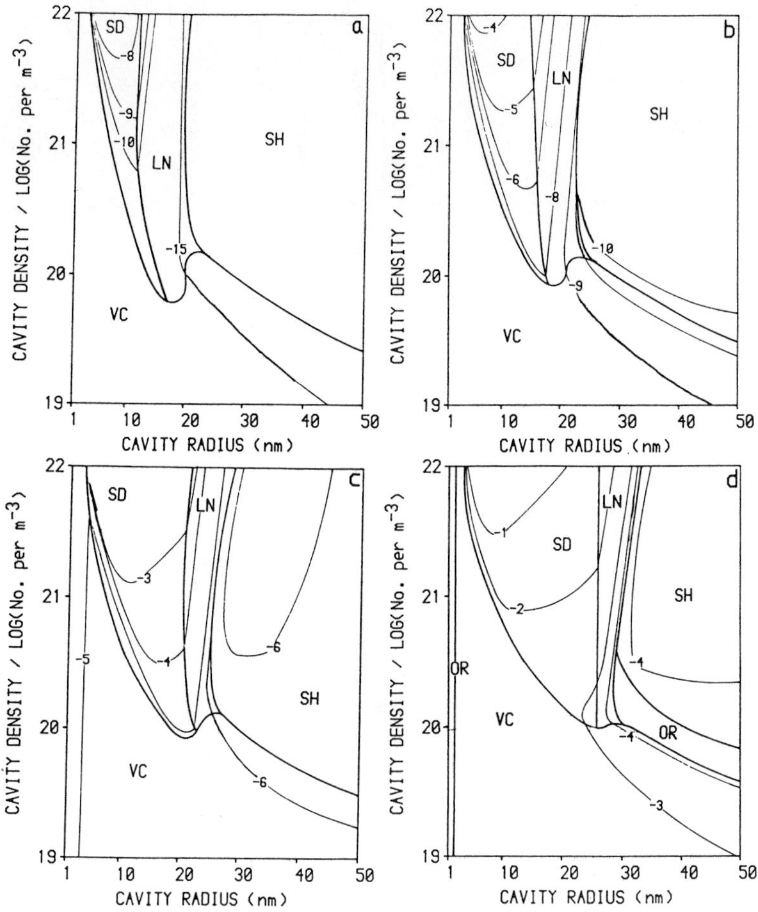

Fig. 3   Helium cavities in nickel. Gas conc. 1000ppm and temperature; (a) 600K, (b) 800K, (c) 1000K and (d) 1200K.

As cavities grow or shrink the cavity density in general changes. It is not easy to trace the evolution of a population of cavities through a series of maps each constructed for a constant cavity density (e.g. Fig. 1 or 2). Figure 3 therefore shows four isothermal maps for the growth of helium-filled cavities in nickel. In these maps the axes are cavity density and mean cavity radius. The total gas content has been maintained at 1000 ppm on the reasonable assumption that inert gas escape is not significant at these temperatures. The major changes to the maps as the temperature is increased are that the growth rates increase (as the contour lines indicate), the surface diffusion limited field expands to larger bubble sizes and (more controversially) Ostwald ripening fields appear at 1200K.

In use, many reactor materials will experience a temperature gradient. This can have a considerable effect on growth by migration and coalescence, since a directional flow is superimposed on the Brownian motion of cavities. In order to calculate the growth rate it is then necessary to know the cavity size distribution, since small cavities, with higher mobilities, will overtake larger slower cavities, thus increasing the coalescence rate. The assumption of a constant cavity size, which is generally acceptable when considering Brownian migration and coalescence [4], would result in the false conclusion that the application of a temperature gradient makes no difference to cavity growth rates. We have therefore assumed a Gaussian size distribution, as discussed in Appendix 2.

353

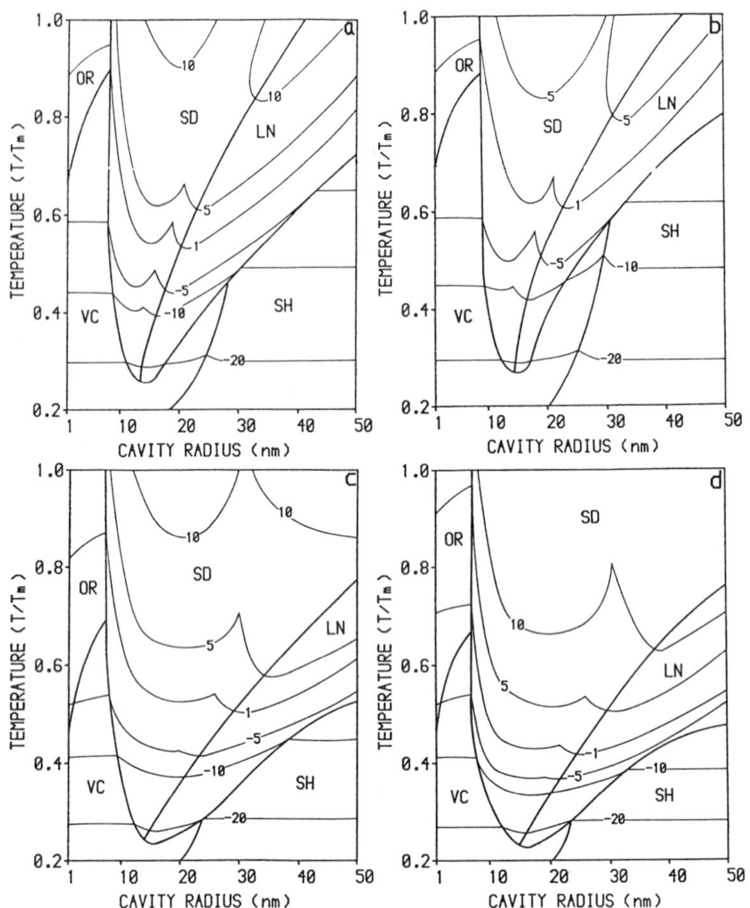

Fig. 4 Helium conc. 1000ppm, cavity density $10^{20}\,m^{-3}$. (a) and (b); Helium cavities in nickel with a temperature gradient of 10^5K/m and 10^3 K/m respectively. (c) and (d); Helium cavities in niobium with a temperature gradient of 10^3 K/m and 10^5 K/m respectively.

Figures 4a and 4b are maps for the growth of helium cavities in nickel when a temperature gradient is present. This has the effect of increasing the growth by migration and coalescence, especially at high temperatures (compare with Fig. 3d). These two maps clearly show that the growth rate is very dependent on the extent of the temperature gradient and indicate that this may be important in reactor shielding. Figures 4c and 4d (compare with Fig. 1c) show that similar conclusions can be drawn for niobium.

In a temperature gradient the majority of coalescences occur when smaller cavities collide with larger cavities. This effect is greatest when the variation in velocity between the smallest and largest bubbles is greatest. The cusps in the rate contours within the SD field in figure 4 arise because under those conditions cavity velocity varies only slightly with cavity size.

Further effects which must be modelled before maps can usefully be employed in the assessment of reactor materials are displacement damage and continuous gas generation. Perryman and Goodhew have given some initial attention to these effects [7].

5. Conclusions

In order to produce realistic maps it is necessary to know all the materials parameters. Since cavity growth experiments are in general difficult and time consuming, data are often not available or are known with little accuracy. It follows that the maps shown in this paper are approximate and serve only to give indications of the growth rate and mechanism.

Figure 4 (although it ignores displacement damage) gives an initial indication of the minimum cavity growth rates to be expected in the first wall of a fusion reactor and clearly shows the temperature range which must be observed to minimise swelling. One encouraging feature of these maps is the upward slope of the rate contours as cavity size increases, indicating that swelling will slow down as the cavities grow.

ACKNOWLEDGEMENTS

The author would like to thank SERC and UKAEA Harwell for financial support and Dr N Marochov, Dr L J Perryman and Dr J H Evans for valuable discussions.

Appendix 1 *Symbols and nomenclature*

a	atomic separation
c^v_e	equilibrium concentration of vacancies
D_b	bubble diffusion coefficient
D_g	diffusion coefficient of a matrix atom through the bubble
D_0	pre-exponent for gas atom diffusion through the matrix
D_s	surface self-diffusion coefficient
D_v	self (vacancy) diffusion coefficient
ε	ledge energy
γ	surface energy
k	Boltzmann constant
L	ledge length
N	number of gas atoms in a bubble
p	vapour pressure
P	gas pressure
q	adatom interaction volume
Q_{He}	activation energy for gas atom diffusion through the matrix
Q_g	heat of transport for vapour transport
Q_s	heat of transport for surface diffusion
Q_v	heat of transport for volume diffusion
r	cavity radius
t	time
T	absolute temperature
v	volume of a gas atom
V	volume of a bubble
y	Nv/V

Appendix 2 *The equations used in the mechanism maps*

In the absence of a temperature gradient Brownian migration occurs by migration and coalescence. The bubble mobilities are [2, 5, 13]:

$$D_b(total) = [D_b(VD) + D_b(SD) + D_b(VT)]L\exp(-Le/kT)/6a$$

$$D_b(VD) = 3D_v a^3/4\pi r^3$$

$$D_b(VT) = 3D_g a^6 p/4\pi kT r^3$$

$$D_b(SD) = \frac{D_s a^4}{2\pi r^4} \frac{\exp[-3qN[1 + 2y + 2y^2\ ]]}{[4\pi r^3[\ (1-y)^2\ (1-y)^3\ ]]}$$

In the presence of a temperature gradient:

$$D_b{}^* = [(Q_v/2)D_b(VD) + (Q_s/2)D_b(SD) + Q_g D_b(VT)]Lexp(-Le/Kt)/6a$$

A Gaussian cavity size distribution was assumed. Measured cavity distributions [4, 8, 9] have standard deviations which lie between 0.1<r> and 0.3<r>, where <r> is the mean cavity radius. Since all values of the standard deviation within this range gave the same calculated rate of growth to one significant figure, the standard deviation was assumed to be 0.2<r>.

Growth rates are derived from bubble mobility via an appropriate equation of the form $dr/dt = const\ D_b/r^n$, where n depends on the rate-controlling mechanism and on whether the cavity can reach equilibrium size between coalescences.

The rate of growth (or shrinkage) of the mean cavity radius by vacancy collection is taken as [14]:

$$dr/dt = (D_v C_v{}^e/r)\{1-\exp[-(P-2/r)a^3/kT]\}.$$

The rate of growth of the mean cavity radius by Ostwald ripening is given as [15]:

$$dr/dt = (D_0{}^{He}/r)\exp(-Q_{He}/kT).$$

The gas pressure was calculated using the equation of state [16]

$$PV/NkT\ =\ (1+y+y^2+y^3)/(1-y)^3.$$

The equilibrium pressure was taken to be $P = 2\gamma/r$.

REFERENCES

1. M.F. Ashby, Acta. Met. **20**, 887 (1972).
2. L.J. Perryman and P.J.Goodhew, Acta. Met. **36**, 2685 (1988).
3. P.J. Goodhew, Scripta. Met. **18**, 1069 (1984).
4. P.J.Goodhew and S.K. Tyler, Proc. Roy. Soc. (Lond) **A377**, 151 (1981).
5. F.A. Nichols, J. Nucl. Mat. **84**, 1 (1979).
6. P.J. Goodhew and L.J. Perryman in *Materials for Nuclear Reactor Core Applications*, British Nuclear Energy Society, London, Vol **1**, 17 (1987).
7. L.J. Perryman and P.J. Goodhew, J. Nucl. Mater **165**, 110 (1989).
8. E.Y. Mikhlin, Y.F. Chkuaseli, Yu.N. Sokurskii and G.A. Arutyunova, Atomnaya Energiya **56**, 144 (1984).
9. N. Marochov, L.J. Perryman and P.J. Goodhew, J. Nucl. Mat. **149**, 296 (1987).
10. S.K. Tyler and P.J. Goodhew, J. Nucl. Mat. **74**, 27 (1978).
11. P.S. Maiya and J.M. Blakely, J. Appl. Phys. **38**, 698 (1967).
12. H. Trinkaus, Scripta. Met. **23**, 1773 (1989).
13. E.E. Gruber, J. Appl. Phys. **38**, 243 (1967).
14. G.W. Greenwood, A.J.E. Foreman and D.E. Rimmer, J. Nucl. Mat. **4**, 305 (1959).
15. A.J. Markworth, Met. Trans. **4**, 2651 (1973).
16. S.E. Donnelly, Rad. Effects **11**, 111 (1985).
17. M.F. Ashby, 1st Risø Int. Symp. on Met. and Mat. Sci., Eds. A.R. Jones and T. Leffers (1980) 325–336.

LOOP-PUNCHING AS A MECHANISM FOR INERT GAS BUBBLE GROWTH IN ION-IMPLANTED METALS

S.E. Donnelly,[1] D.R.G. Mitchell[1] and A. van Veen[2]

[1] *Electronic Materials & Devices Group*
Department of Electronic and Electrical Engineering
University of Salford, Manchester M5 4WT, UK

[2] *Interfaculty Reactor Institute, Delft University of Technology*
2629JB Delft, The Netherlands

ABSTRACT

The punching of dislocation loops by overpressurized inert gas bubbles has long been regarded as the dominant mechanism for bubble growth under implantation at low homologous temperatures. The present work examines the predictions of the loop-punching theory in some detail and presents experimental evidence to show that loop-punching does not appear to operate for Kr and Xe in Al, Cu and Ni. Two alternative models of athermal bubble growth are discussed.

1. Introduction

Small, spherical or facetted bubbles of inert gas are known to occur in a number of metals following inert gas ion implantation. Because of the possible relevance of helium bubbles in metals to first wall problems in the proposed fusion reactor, the origins and consequences of this type of radiation damage have been extensively studied for a number of years.

Before the discovery of room-temperature, solid, inert gas precipitates in ion-implanted metals by Templier et al [1] and vom Felde et al. [2] in 1984 no measurements had been carried out which indicated unequivocally the pressure in small room temperature bubbles. However, the shifts in atomic transitions in the helium in the bubbles as measured by Vacuum Ultra-Violet Absorption Spectroscopy (VUVAS) and Electron Energy Loss Spectroscopy (EELS) [3–4] had indicated that the pressure could be high. Unfortunately, due to the difficulty in modelling the peak shifting processes no consensus was reached by the different researchers involved in the research with pressure estimates ranging from 1–50 GPa for bubbles with radii of the order of 1 nm.

Whilst it was generally assumed that, for implantations at low homologous temperatures, a lower limit to the pressure would be determined by the surface energy of the bubble, factors causing a high pressure limit were much more difficult either to predict or to determine.

In 1950, Seitz discussed the punching of prismatic dislocation loops as a means of stress relief in crystalline materials [5]. Since that time observations have been made of stress relief around inclusions in a variety of materials by the punching of dislocation loops. In particular, in 1959 Lally and Partridge observed loops punched out by what they assumed to be hydrogen bubbles formed during the quenching of magnesium [6]. In later studies Wampler *et al.* observed loops punched by hydrogen bubbles in copper [7] and Evans *et al.* observed loops punched by helium platelets in molybdenum [8]. There is thus ample evidence that under certain conditions stress relief around small inclusions can occur by loop punching and in 1959 Greenwood, Foreman and Rimmer (GFR) had developed this idea as a mechanism of pressure relief for inert gas bubbles in fissile materials [9]. Their analysis of the energetics of loop formation led to the derivation of an expression for the minimum pressure, P, required by a bubble to punch out a dislocation loop with Burgers vector b and core radius r_o ($\approx$b) in a material with shear modulus μ and Poisson's ratio ν. If, as usual, it is assumed that the radius of the dislocation loop is equal to the bubble radius, R, then this expression is:

$$P = \frac{2\gamma}{R} + \frac{\mu b}{2\pi(1-\nu)R} \ln \frac{R}{r_o} \tag{1}$$

where γ is the surface tension or surface free energy of the material in which the bubble occurs and the term $2\gamma/R$ the equilibrium pressure. (it should be pointed out that in their derivation GFR dropped the (1-ν) term)

For a bubble of radius 1 nm this expression yields pressures in the range 5 GPa (Al) to 16 GPa (Mo) if taken at face value. However, in general the loop radius might be expected to be smaller than than the bubble radius and, more importantly, as this expression is derived from

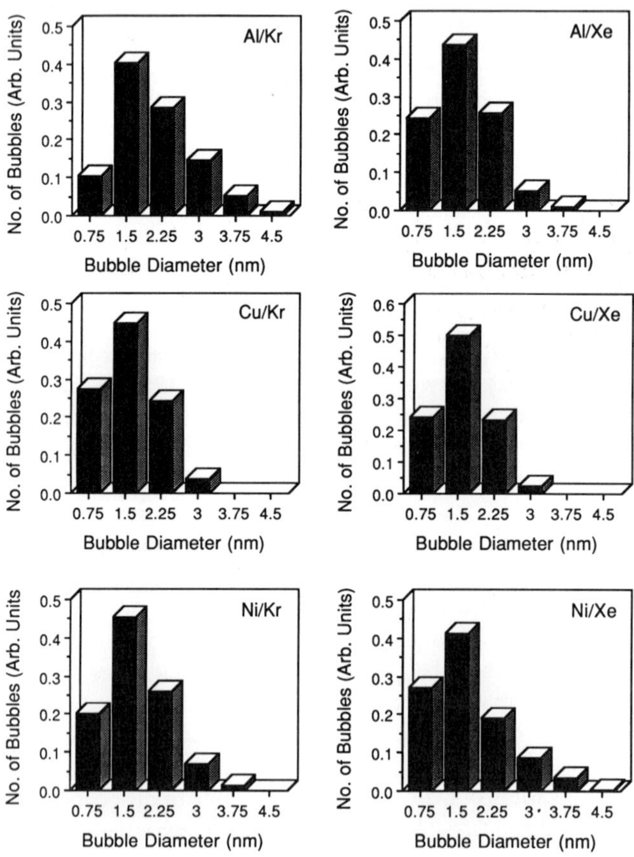

Fig. 1   Some examples of bubble size distributions measured from dark field micrographs using reflections from the solid inert gas contained in the bubbles. Between 200 and 300 bubbles were measured for each distribution. (Specimens implanted at 40 keV to a dose of 2 10^{16} ions/cm^2.)

the difference in energy between bubble and bubble plus loop it takes no account of any energy 'saddle point' resulting from the dynamics of the loop punching process [9]. This expression should thus perhaps be considered as defining a lower limit to the pressure required for loop punching. Trinkaus [10], in fact, suggested that the simpler expression

$$P = \frac{2\gamma}{R} + \frac{\mu b}{R} \tag{2}$$

provided a better estimate of the loop-punching pressure for small bubbles (R < 2 nm). This expression gives pressures considerably higher than the GFR expression yielding, for a bubble of radius 1 nm, values ranging from 9 GPa (Al) to 40 GPa (Mo).

Although other mechanisms of pressure relief have been discussed in the literature, loop-punching was (and still is) widely regarded as the most likely mechanism for the growth of inert gas bubbles in metals at low homologous temperatures. For the last seven years, however, diffraction measurements on metals containing room-temperature solid inert gas precipitates (which are still generally referred to as bubbles) have enabled researchers to quantify the lattice constant (and thus density and—with certain assumptions—pressure) for Xe, Kr and Ar in small bubbles in a number of materials (see contribution by Templier in this volume). Although, for small bubbles in materials with a large shear modulus, equation (2) predicts pressures of 10 GPa or more, in practice lattice constants corresponding to such pressures have never been measured in diffraction experiments.

In the present article we aim to demonstrate quantitatively that a discrete loop-punching mechanism is not consistent with experimental measurements for Kr and Xe in Al, Cu and Ni. Whilst we do not rule out the possibility that the mechanism may operate for isolated bubbles, we will argue that it is not the pressure limiting mechanism for the high bubble concentrations generally observed following ion implantation.

2.   Results and Discussion

In general, with the possible exception of cases where bubble lattices are formed, implantation of a metal with inert gas ions results in the formation of bubbles with a range of diameters. We have used dark field micrographs obtained using reflections from solid inert gas in bubbles to obtain bubble size distributions for Kr and Xe implanted into a number of metals. Fig. 1 shows some sample distributions for Xe and Kr in Al, Cu and Ni and illustrates clearly that a range of bubble sizes occurs in all cases. This point is an important one as it indicates that there is no mechanism which co-ordinates growth of individual bubbles and which results in all bubbles having the same diameter. Growth of an individual bubble by the spontaneous punching of a discrete dislocation loop is assumed to occur when the bubble reaches the threshold pressure given by equation (1) or (2). Thus, even where the bubble size distribution is narrow, this process should result in a range of pressures occurring. For instance if we consider a number of individual bubbles each with the same diameter, some will contain a relatively low pressure, having just punched out a loop, some will contain a pressure just below the loop-punching threshold (i.e. arrival of an additional inert gas atom will trigger the loop-punching) and others will have a pressure between these extremes.

Provided that the equation of state (EOS) of the inert gas is known [11–12], and assuming that it is applicable to bubbles containing small numbers of atoms, it is possible to model the growth of an individual bubble by loop-punching from an initial small equilibrium bubble. We have performed such calculations for bubbles of Kr and Xe in Cu, Ni and Al. The first step of the calculation consists of determining the number of inert gas atoms in a cavity of 0.5 nm radius assuming equilibrium pressure. One atom at a time is then added to this number and on each step the EOS [11–12] is used to recalculate the pressure and thus determine whether the

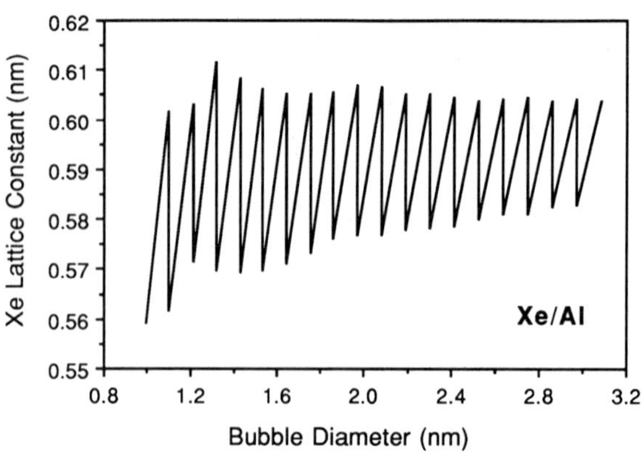

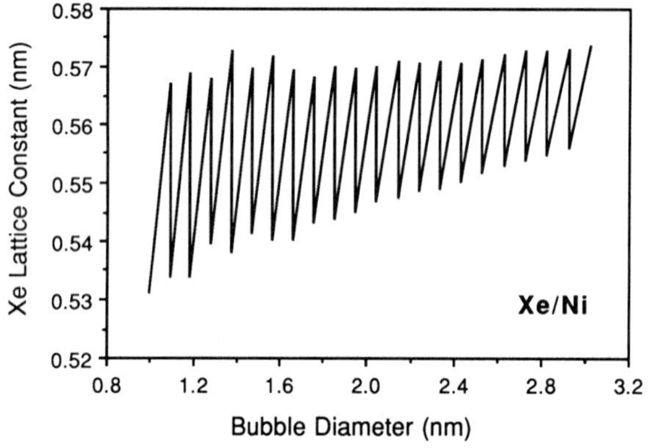

Fig. 2   Theoretical variation of gas lattice constant with bubble size for Xe in Al and Ni.

loop-punching pressure has been exceeded. If the loop-punching pressure has been exceeded then the bubble volume is increased by $\pi R^2 b$. The EOS is again used to determine the resulting density from which the lattice constant is calculated assuming an f.c.c. structure. Finally, the volume is redistributed into a spherical shape and the resulting radius is calculated. This process continues until the bubble radius reaches a predetermined size.

In this way it is possible to plot the (theoretical) variation of inert gas density with bubble radius for a bubble growing by loop-punching (governed by eqn. 1). Fig. 2 shows this information for a Xe-implanted aluminium specimen and a Xe implanted Ni specimen.

The vertical components of the curve result from the addition of Xe atoms to the cavity without volume change (elastic effects are neglected) with the diagonal lines indicating the (discontinuous) change in bubble radius as a loop is punched. Particularly for small bubbles, the significant fractional volume increase which results from loop-punching results in a significant change in lattice parameter (up to 8%).

Fig. 1 indicates that, for instance, implantation of aluminium with 40 keV Xe ions to a dose of $2 10^{16}$ ions/cm^2 results in an (observable) bubble size distribution ranging from 0.75 to 3.75 nm. Inspection of Fig. 3 indicates that if these bubbles have grown by loop-punching as described by eqn. (1) then it is to be expected that bubbles will contain Xe with lattice constants ranging from $\approx.575$ nm to .605 nm. Indeed even for bubbles with identical radii, ≈ 1 nm say, a 5% variation in lattice constant is to be expected. In an electron diffracton experiment this variation in lattice constant would be expected to result in a non-isotropic broadening of the inert gas diffraction spots as illustrated schematically in Fig. 3, due to the fact that the inert gas diffraction pattern can now be considered as the superimposition of

360

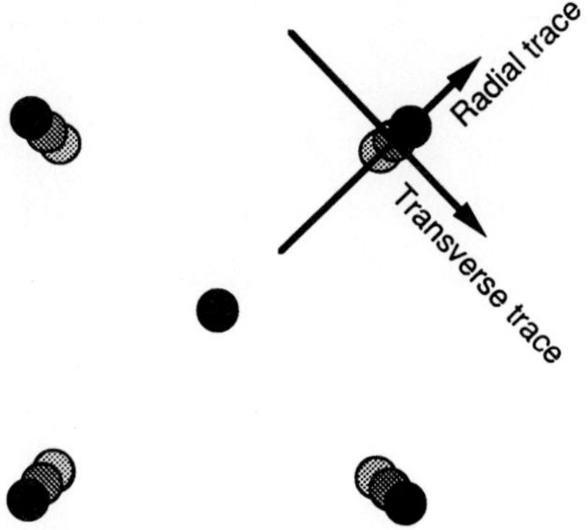

Fig. 3 Illustration of expected form of diffraction pattern when a range of inert gas lattice constants is present in a specimen. (Inert gas diffraction pattern should result from the sum of aligned diffraction patterns from inert gas with different lattice constants).

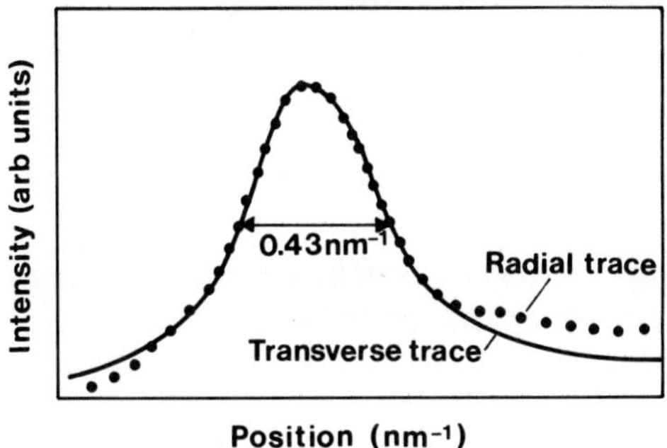

Fig. 4 Experimental densitometer tracings through a Xe {200} reflection taken in the radial and transverse directions as illustrated in Fig. 3.

diffraction patterns from aligned crystallites with different lattice constants. Given the large predicted variation in lattice constant it should be possible to observe such broadening in diffraction patterns from specimens containing solid bubbles.

Fig. 4 compares densitometer tracings through a Xe (200) reflection from a diffraction pattern from the Xe/Al specimen whose bubble size distribution is illustrated in Fig. 2. The densitometer tracings have been taken along two orthogonal vectors as illustrated in Fig. 3. It is clear that there is no measurable difference in width between the transverse and radial tracings either in the case illustrated or in a number of other similar measurements.

To obtain a theoretical densitometer tracing we have to account for the fact that larger bubbles, containing a larger number of gas atoms, make a proportionately larger contribution to the diffracted intensity than smaller bubbles. The bubble size distribution was thus appropriately weighted (by multiplying by bubble volume—which is approximately proportional to the number of atoms in the solid bubble—for each class in the histogram). The mean of the distribution was then determined. This value thus corresponded to the mean bubble size in relation to contribution to the inert gas diffracted intensity. On the assumption that the major factor determining width of diffracted beams from the inert gas bubbles is the shape transform of the bubble, the reciprocal of the mean of this histogram (2.3 nm) has then been used as the full width at half maximum (FWHM) of a Gaussian intensity curve which is illustrated in Fig. 5. The width of the experimental Xe (200) spot was $4.33 \pm .05 \times 10^{-1}$ nm^{-1} yielding a real-space width of 2.31 nm (assuming a spherical bubble) in excellent agreement with the value of 2.3nm quoted above.

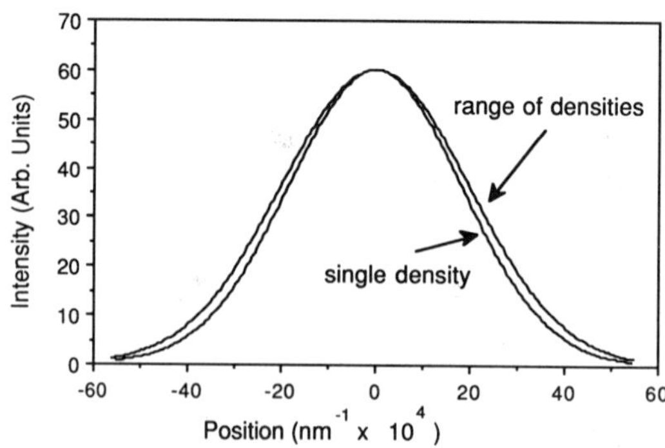

Fig. 5   Calculated densitometer scan through a Xe {200} reflection when a single Xe density and a range of densities is present.

Inspection of Fig. 2 reveals that for Xe bubbles of diameter 2.3 nm in Al one would expect Xe lattice constants ranging from <.55 nm to >.57 nm which (as illustrated schematically in Fig. 3) would be expected to broaden the diffracted beams from the inert gas in the radial direction.

The expected magnitude of this broadening is illustrated in Fig. 5 for the Xe/Al specimen discussed above. Although the expected brodening effect is small it is nonetheless sufficiently large to be be clearly measurable. It is not, however, observed in the densitometer tracings illustrated in Fig. 4 nor indeed in other similar measurements made on a number of specimens.

The radial broadening of the inert gas spots which appears to be an inevitable consequence of the standard loop-punching model is thus not present in the experimental data. Although we illustrate the argument with data from the Xe/Al system, Kr and Xe in Cu, Al, Ni, (and Sn and Co) have recently been studied in our laboratories but no radial broadening has been observed. Also, to the authors' knowledge, no report of such broadening of diffracted beams from solid inert gas in bubbles has ever been published.

In 1986, Evans and Mazey published a summary diagram which compared the pressures predicted by the loop-punching model with experimental data on bubble pressures which were available at that time [13]. The diagram indicated that for most materials, the full loop-punching equation was a reasonable predictor of bubble pressures although some materials (Al, Au Zn and Mo) exhibited pressures which were somewhat lower than the lower theoretical limit. However, as already pointed out by Gerritsen [14], Evans and Mazey omitted the equilibrium pressure (the first term in eqn. (1) above) in their expression for loop-punching pressure and, as a consequence, their theoretical pressures were significantly low.

In Fig. 6 we plot our data for Xe and Kr in Al, Cu and Ni. In each case we also plot the theoretical loop-punching expressions using both eqn. (1) (lower pressure curve) and eqn. (2)—in both cases *including* the equilibrium pressure term. In all cases the pressures which we deduce from accurate measurements of inert gas lattice parameter are significantly lower than either of the loop punching expressions.

As concluded by Evans and Mazey, comparison of the experimental data for the different metals indicates clearly that the experimentally determined pressures scale with shear modulus. This observation does not, however, imply that loop-punching is the pressure limiting mechanism, nor indeed that the shear modulus itself necessarily plays any role in the pressure limiting mechanism. Table 1 gives numerical values for a number of different physical properties of Al, Cu, Ni and Mo. It can be seen that all these quantities scale with shear modulus for these elements. A dependence on any one of these parameters (or one of a number of others) would give the *false* impression of dependence on shear modulus.

The combination of measured pressures considerably below those to be expected from the standard loop-punching picture with an apparent absence of radial broadening of diffracted

362

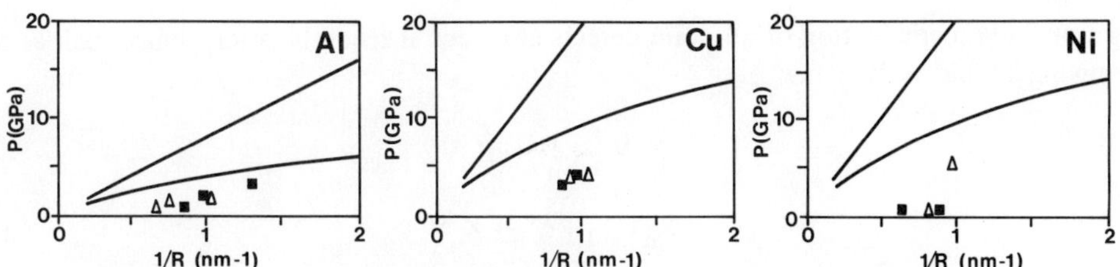

Fig. 6 Solid Kr pressures (triangles) and solid Xe pressures (filled squares) pressures obtained from lattice constant measurements for Al, Cu and Ni plotted against reciprocal bubble radius. The continuous curves have been calculated using equation 1 (lower pressure curve) and equation 2 including the equilibrium pressure curve in each case.

Table 1 A tabulation of some parameters which scale with shear modulus

	Shear Modulus ($\times 10^{11}$ dyne/cm^2)	Surface Tension (dyne/cm)	Melting Temp. (K)	Activation Energy for Self Diffusion (eV)
Mo	12	2390	2890	4.2
Ni	7.7	1890	1726	2.95
Cu	4.8	1410	1356	2.19
Al	2.5	1030	933	1.47

beams from the solid inert gas in small bubbles indicates that the discrete loop-punching mechanism does not, in general, provide an upper limit to pressure in inert gas bubbles in implanted metals. However, there is considerable evidence that this mechanism operates for isolated bubbles or other precipitates. It thus appears that, in more heavily irradiated materials the bubble pressure may not reach the loop-punching threshold due to the availability of other growth mechanisms which result in a lower limiting pressure.

3. Bubble Growth Mechanisms

A number of possible mechanisms for athermal bubble growth (for helium bubbles) were discussed by Evans [15] and later by Donnelly [16]. Mechanisms discussed included the creation of single self-interstitials, interbubble fracture and loop-punching. Consideration of the pressures required for the various mechanisms to operate led to the conclusion that under most circumstances loop punching would occur at the lowest pressure and thus would be the dominant mechanism.

Loop punching models generally assume that the gas agglomerates into clusters which subsequently become in quasi-equilibrium with the surrounding material by the emission of self-interstitial loops. In defect notation:

$$X_n V_m + kX \rightarrow X_{n+k} V_m$$

$$X_{n+k} V_m \rightarrow X_{n+k} V_{m+j} + I_j \tag{3}$$

where X represents an inert gas atom, V a vacancy and I a self-interstitial atom (SIA)—with the subscripts representing the number of vacancies/gas atoms in a cluster/loop.

This approach is valid when a) the gas is introduced without the production of lattice damage i.e. low-energy (subthreshold) He implantation or the 'tritium trick' or b) when there is a perfect recombination of all point defects at a neutral recombination centre such as the growing bubble itself. Thus:

$$X_n V_m + I \rightarrow X_n V_{m-1}$$

$$X_n V_{m-1} + V \rightarrow X_n V_m. \tag{4}$$

In the latter case it is unimportant whether the gas atoms arrive at the cluster as an interstitials or as part as a gas/vacancy complex. For instance, for a cluster which is a neutral sink for vacancies and SIAs, the interaction

$$X_n V_m + X V_3 \rightarrow X_{n+1} V_{m+3} \tag{5a}$$

will eventually be be balanced (on average) by the acquisition of 3 SIAs:

$$X_{n+1} V_{m+3} + 3I \rightarrow X_{n+1} V_m \tag{5b}$$

Thus point defect production during gas incorporation will have an effect on the bubble evolution only when there is a preferential sink for SIAs preventing their complete recombination with vacancies. The dislocations inevitably observed when the heavier inert gases (and helium at energies above a few hundred eV) are incorporated into metals by ion implantation at low homologous temperatures may well serve as preferential sinks for SIAs to give bias-driven cavity growth. However, as discussed by Evans [17], such cavity growth is suppressed when the cavity/bubble densities are high—as generally found for inert gas implants, see below. Only at high homologous temperatures, when bubble densities drop dramatically, will bias-driven cavity growth take place. A recent example is given in the results of Mitchell *et al* [18] for xenon implanted into tin at room temperature ($T/T_m = 0.58$).

In addition to the loop-punching and simple rate theory approaches, it may be interesting to examine effects due to the mobility of gas atoms, vacancies and gas/vacancy complexes during implantation.

3.1. Bubble growth by agglomeration of gas-vacancy complexes

It is known that the bubble concentration limits itself, at an early stage in the gas incorporation, to $\approx 10^{19}$ bubbles/cm^3—corresponding to nearest neighbour distances of the order of 4 nm. Further implantation simply increases bubble size without increasing the bubble concentration. Defects produced by the irradiation are unlikely, in general, to escape from a volume of $\approx 10^{-19}$ cm^3 because this volume will be surrounded by internal (bubble) surfaces. There is evidence from inert gas desorption experiments (e.g. ref. [19]) that when inert gas atoms are trapped within a few nm of an *external* surface, subsequent inert gas implantation leads to a desorption of the trapped gas atoms. This radiation induced transport of gas atoms is essentially unsurprising as the collision cascades which result from $\approx$keV ion implantation have the dimensions of nm. For damaging events in a small volume surrounded by internal surfaces it seems entirely plausible that a similar radiation induced transport of gas (and vacancies) may take place towards the *internal* surfaces.

As far as thermal (i.e. not radiation enhanced) diffusion processes are concerned, although helium is thought to be interstitially mobile in many metals at low temperatures, this is not necessarily true for the heavier inert gases due to their much larger strain and electronic energies. However, studies by van Veen [20] of the interaction between SIAs and substitutional inert gas atoms in W, Mo and Ni indicate that Kr may be interstitially mobile in W; that Ne, Ar and Kr may be interstitially mobile in Mo and that Ne may be interstitially mobile in Ni but that Kr will not be mobile in Ni.

For the heavier inert gases in the more confined lattices, however, the combination of strain and electronic energies on arriving at an interstitial position may be sufficient for the atom to become substitutional by creating one or more SIAs thus forming a gas/vacancy complex. However, the possibility of radiation induced-migration of the heavier gases to internal surfaces, as discussed above, may result in an insensitivity of the bubble evolution to the gas interstitial migration energy. Provided that preferential sinks exist to absorb the SIAs (e.g. dislocation loops—to be discussed further in the next section) the gas/vacancy complexes which form as the energetic ion comes to rest may be transported by radiation enhanced processes to the bubbles.

In this case, rather than an interstitial inert gas atom, the mobile entity is effectively one (or more) inert gas atoms together with a number of vacancies. Bubble growth will thus occur by the acquisition of vacancies and gas atoms together. If the initial bubble nucleus has a high inert gas atom to vacancy ratio then the acquisition of gas and vacancies will result in a reduction of density with growth—tending to the density of the mobile complexes.

Except for very small bubbles (whose gas density will depend on the details of nucleation), the consequence of this type of growth is bubbles—with a range of diameters—all containing gas at approximately the same density. Depending on the precise details of the mobile complex, bubbles may be overpressurized, underpressurized or at near-equilibrium pressure.

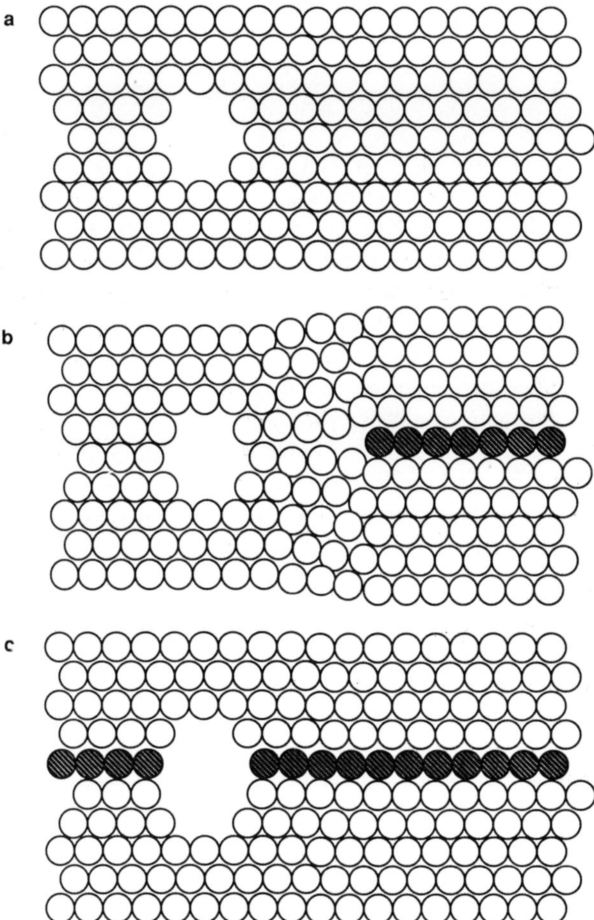

Fig. 7   Proposed athermal growth mechanism. A growing dislocation (loop) intersects a bubble resulting in a volume increase.

3.2. *Bubble growth as a consequence of the growth of interstitial loops*

Theoretical modelling of helium clusters in nickel by Wilson *et al.* [21] indicated that self interstitial atoms (in this case created by overpressure in the clusters) would be expected to be bound to the clusters. A similar calculation by Caspers et al [22] of He clustering in Mo also concluded that an interstitial complex bound to a cluster was always a lower energy configuration than the separate bubble and interstitial cluster. If such a conclusion were also to be applicable to the case of heavy inert gas bubbles then interstitial clusters bound to individual bubbles may be constitute a preferential trap for SIAs. At larger sizes, the interstitial cluster may be expected to collapse to a dislocation loop and, indeed, materials irradiated with inert gases at low temperatures inevitably contain tangles of dislocations amongst the bubbles. It is, at this stage of course impossible to determine whether loops are bound to individual bubbles. However, the existence of growing interstitial loops—whether bound to bubbles or not—during irradiation of a metal with inert gas ions, leads to the possibility of a bubble growth mechanism which, to our knowledge, has not been suggested before. If an interstitial platelet (the edge of which is the dislocation loop) grows so that it passes 'through' or around a bubble then the bubble will increase its volume as illustrated in Fig 7. The resulting bubble growth is not, however, dependent on the degree of overpressure within the bubble. Such growth would obviously occur along specific directions and might be expected to give rise to facetted bubbles. The tensile stress acting at right angles to the plane of the bubble layer may also give rise to stress-induced preferential nucleation of loops parallel, or near parallel, to the bubble layer.

4.   Conclusions

Athermal growth mechanisms considered to date for inert gases in metals generally involve a bubble acquiring gas atoms until some threshold pressure is exceeded whereupon discontinuous growth occurs. We have shown, however, that this type of mechanism does not seem to be consistent with experimental observations of inert gas densities for the heavy inert gases in a variety of metals. Although spectroscopic measurements on helium bubbles indicated extremely high pressures (consistent with a loop punching process), densities corresponding to such pressures have never been measured in electron diffraction from solid heavy inert gas bubbles. As diffraction experiments are not feasible for helium we can not rule out the possibility that loop punching may operate for helium bubbles and that pressures in helium bubbles may be considerably higher than in bubbles of the heavy inert gases.

For the heavy inert gases we have proposed two qualitative mechanisms of bubble growth which do not depend on bubble pressure directly. One of these mechanisms should yield bubbles of different sizes containing gas at the same density—as observed experimentally.

ACKNOWLEDGEMENTS

The authors wish to thank J.H. Evans for invaluable information on bias-driven growth processes, C.A. Faunce for useful discussion and P. J. Grundy for the provision of the microscope facilities. This project was funded by the UK Science and Engineering Research Council.

REFERENCES

1. C. Templier, C. Jaouen, J-P. Delafond and J. Grilhé, Comptes Rendus (Paris) **299**, (1984) 613.
2. A. Vom Felde, J. Fink, Th. Müller-Heinzerling, J. Pflüger, B. Scheerer, and G. Linker, Phys. Rev. Lett. **53**, 922 (1984).
3. J. C. Rife, S.E. Donnelly, A.A. Lucas, J-M. Gilles, J.J Ritsko, Phys. Rev. Letters **46**, 1220 (1981).
4. R. Manzke, W. Jäger, H. Trinkaus, G. Grecelius and R. Zeller, Solid State Comm. **44**, 481 (1982).
5. F. Seitz, Phys. Rev. **79**, 723 (1950).
6. J. S. Lally and P.G. Partridge, Phil. Mag. **13** , 9 (1966).
7. W.R. Wampler, T. Schoeber and B. Lengeler, Phil. Mag. **34**, 129 (1976).

8. J.H. Evans, A. van Veen and L.M. Caspers, Nature **291**, 310 (1981).

9. G.W Greenwood, A.J.E. Foreman and D.E. Rimmer, J. Nucl. Mater. **4**, 305 (1959).

10. H. Trinkaus, Rad. Effects **78**, 189 (1983).

11. D. Vidal, L. Guengant and J. Vermisse, Physica **A116**, 227 (1982).

12. C. Ronchi, J. Nucl. Mater. **96**, 314 (1981).

13. J.H. Evans and D.J. Mazey, J. Nucl. Mater. **138**, 176 (1986).

14. E. Gerritsen, PhD Thesis, Univ. of Groningen, The Netherlands (1990).

15. J.H. Evans, J. Nucl. Mater. **61**, 1 (1976).

16. S.E. Donnelly, Rad. Effects **90**, 1 (1985).

17. J.H. Evans, Nucl. Instr. Methods **B18**, 16 (1986).

18. D.R.G. Mitchell, S. E. Donnelly and J. H. Evans, Phil. Mag. **A61**, 531 (1990).

19. S E Donnelly, R P Webb, D G Armour and G Carter. Proc. of 7th Int. Conf. on Atomic Collisions in Solids, Moscow, USSR, (1977).

20. A van Veen, Mat. Sci. Forum, **15-18**, 3 (1987).

21. W.D. Wilson, C.L. Bisson and M.I. Baskes Phys. Rev. **B24**, 5616 (1981).

22. L.M. Caspers, A, van Veen and T.J. Bullough Rad. Effects **78**, 67 (1983).

POSSIBLE MECHANISMS LIMITING THE PRESSURE IN INERT GAS BUBBLES IN METALS

H. Trinkaus

*Institut für Festkörperforschung des Forschungszentrums Jülich,
Postfach 1913, D 5170 Jülich, BRD and Association KFA-EURATOM*

ABSTRACT

When inert gases are introduced into metals by nuclear transmutation or ion implantation they precipitate as bubbles because of their low solubility. At temperatures below about 0.25 of the melting temperature of the metal, the pressure within these bubbles is limited either by dislocation loop punching or by the displacement damage associated with implantation. Theory predicts that the minimum pressure at which dislocation loop punching can occur is of the order of the theoretical shear strength of the matrix. It is shown that this is confirmed by recent precision dilatometry of tritided metals. The lower pressure values in solid inert gas bubbles forming during ion implantation of heavy inert gases (Ar, Kr, Xe) into metals are attributed to a displacement damage induced creep process which is described by a phenomenological approach. The results of a detailed data analysis are used to calibrate spectroscopic measurements on bubbles forming during helium implantation.

1. Introductory Overview

The most crucial property of inert gases governing their behaviour in metals is their extremely low solubity there [1]. Consequently, when an inert gas is forced into a metal matrix by ion implantation or element transmutation, it strongly tends to precipitate in the form of bubbles and, inversely, its tendency to return into the matrix is extremely low even if it is under very high pressure and high temperature. Therefore, the pressure within bubbles forming during continuous inert gas incorporation into the metal will not be limited by the thermal equilibrium solubility of the inert gas but by material transport within the metal matrix. The latter can occur by the motion of defects of the vacancy type from the matrix to the bubbles or by the motion of defects of the self-interstitial type from the bubbles into the matrix.

The emission of interstitial type defects as in the dislocation loop punching process may be considered to define the mechanical stability limit of the matrix. Thus, inert gases incorporated into metals may be used to study this limit at the low spatial scale of bubbles [2] — provided pressure relaxation by the absorption of vacancy type defects is negligible as expected for low temperatures and low displacement damage. From a more general point of view, such studies are even significant in the context of the question concerning the ultimate maximum pressure attainable in human laboratories.

Dislocation loop punching as a possible mechanism for pressure relieve in bubbles was first discussed in some detail by Greenwood, Foreman, and Rimmer (GFR) [3]. Based on continuum elasticity these authors derived an energetic condition for the minimum pressure, p_{LP}, necessary for this process to occur. They found that at large bubble radii r, p_{LP} should decrease as $1/r$. Unfortunately they did not expressively state that their condition is indeed a necessary but possibly not a sufficient one. For small bubbles and loops the GFR condition becomes doubtful because of the limitation of the elastic continuum theory [4]. For large bubbles and loops on the other hand it becomes questionable to represent a sufficient condition since the elastic interaction between bubbles and loops was ignored in its derivation. Simple estimates of this interaction resulted in the conclusion that p_{LP} should converge with increasing r to a finite asymptotic value [5]. Recently, Wolfer [6] has confirmed this preliminary conclusion by detailed calculations on the basis of the elastic continuum theory and has shown that the asymptotic value should be close to 0.2μ where μ is the shear modulus of the metal. At the same time an analogous result was found in a simulation study of "overpressurized" bubbles in amorphous metals [7].

Concerning experimental tests of these theoretical conclusions there is, apart of the problem how to establish the conditions for loop punching to be the pressure limiting process, the problem how to measure the pressure itself. In fact, the pressure in bubbles can only be measured indirectly either by determining its effect on the matrix, for instance, from the strain contrast in Transmission Electron Microscopy (TEM), or by determining the gas density in the bubbles and by converting this with aid of an appropriate equation of state (EOS) to pressure. The first method is not yet sufficiently accurate to date [8] and even if it were so it would suffer from the lack of knowledge concerning the resistance of the bubble surface against stretching forces. Accordingly, we presently rely on gas density measurements which again can be done more or less directly or indirectly. Unfortunately, quantitative TEM of bubble structures, a relatively direct method frequently used to estimate gas densities in bubbles, is not sufficiently accurate for our purposes.

In fact, the first attempts to quantitatively determine gas densities in bubbles were made on the basis of spectroscopic methods such as Ultra-Violet Absorption Spectroscopy (UVAS) [9–11] and Electron Energy Loss Spectroscopy (EELS) [10–15] applied to helium/metal systems. A crude guideline for estimating He densities in bubbles from the observed large blue shifts of the 1S–2P transition was provided by the known spectral difference between gaseous and liquid He. For a more accurate He density determination, detailed calculations of the 1S and 2P energy levels were required. Unfortunately, the different theoretical methods used in Namur [9–11] and Jülich [12–15] yielded values for the pressure in He bubbles in Al differing by a factor of 2 or even more. This controversy still remains to be resolved.

A substantial advance in determining gas densities in bubbles has been the discovery of electron diffraction from the lattices of heavier inert gases enclosed in bubbles in Al by vom Felde et al. [16] and Templier et al. [17]. Subsequently other inert gas/metal systems have been studied by electron [18–27] as well as X-ray diffraction [28–31]. These studies have demonstrated that inert gases precipitated into bubbles in metals during implantation at ambient temperature are under rather high pressures of the order of several GPa. The pressure values deduced from gas densities by use of appropriate EOSs appeared to be in crude agreement with the GFR condition [32] but were in most cases below the asymptotic value predicted by Wolfer [6]. Should the idea of an elastic bubble-loop interaction resulting in an increased threshold pressure against loop punching be incorrect?

Another important advance in determining gas densities in bubbles is the precision dilatometry of tritided metals [33]. A careful analysis of dilatometric data obtained for several tritided metals has indicated that the pressure converges to constant values around 0.2μ in agreement with Wolfer's conclusion. What is the explanation for the apparent discrepancy between the results obtained from the diffractometry of heavy inert gases implanted into metals and those obtained from the dilatometry of tritided metals?

As documented in the present volume, a number of other methods to determine gas densities and pressures in bubbles have been developed such as Small Angle Neutron Scattering (SANS), Positron Annihilation Techniques (PAT), Nuclear Magnetic Resonance (NMR), Extended X-Ray Absorption Fine Structure (EXAFS), and Mössbauer Spectroscopy. These methods are indirect in the sense that either experimental calibration or detailed modeling is required to deduce gas densities. Moreover, the information provided by these techniques is still too limited to demonstrate systematic trends as can be done by the techniques mentioned above.

The reliability of pressure values deduced from measured densities depends, of course, on the reliability of the equations of state (EOS) employed. At the end of the seventies when interest in quantitatively determining gas pressures in bubbles in metals emerged, the existing experimental data concerning EOSs of inert gases were mainly restricted to pressures below 1 GPa [2, 34]. Whereas pressures in small equilibrium bubble in the nm size range are close to this value, pressures at which loop punching is expected to occur were estimated to range up to 20 GPa. For this range theoretical EOSs were needed at that time such as the ones, for instance, given for He by Young et al. [35] and the author [4] or for Ar, Kr, and Xe by Ronchi [36].

In 1980, Mills et al. [37] published an empirical EOS for He based on detailed measurements in the pressure range from 0.2 to 2 GPa and the temperature range from 75 to 300 K. In 1981, a compilation of room temperature data for all the inert gases was given by Vidal et al. [38]. The range of several GPa relevant for the present problem was first reached for Xe measured at 85 K by Syassen et al. [39] and later for Ne and Ar at ambient temperature by Finger et al. [40]. During recent years, the pressure range in room temperature measurements on He [41], Ne [42], Ar [43], Kr [44], and Xe [45–47] has been extended continuously, for Xe, for instance, up to 200 GPa [47], and are now going to approach the ultimate pressure attainable in laboratories by use of diamond anvil cells.

The most important result of these recent measurements is that above a few GPa inert gases are substantially softer than predicted by previous theoretical EOSs [4, 35, 36]. Thus, for instance, in the 1 to 20 GPa pressure range relevant for the present problem, the pressure values predicted by the Ronchi EOS [36] are almost by a factor of two too high. On the other hand, theoretical EOSs are still needed for interpolating and extrapolating experimental data. In the author's opinion, the presently existing experimental data together with revised theoretical EOSs should suffice to attribute to given density values in the interesting range pressure values with an uncertainty smaller than 10 %.

In the present paper an attempt is made to resolve the (apparent) discrepancy between the theoretical prediction concerning the threshold pressure for dislocation loop punching and its confirmation by the dilatometry of tritided metals on one side, and the observation of lower pressures in bubbles forming in metals during ion implantation of the heavy inert gases Ar, Kr, and Xe on the other hand. For this, possible bubble growth mechanisms and their regimes in the parameter space are discussed first. In order to test the theoretical ideas, the existing dilatrometric data for tritided metals and diffraction data for heavy inert gases implanted into metals are then analysed and correlated with other system parameters. Finally, an attempt is made to calibrate spectroscopic measurements on bubbles forming during helium ion implantation into metals.

2. Bubble Growth Under Gas Supply

2.1. Main mechanisms and their regimes

Under continuous gas supply as occurring during inert gas ion implantation or trititium transmutation into helium, bubbles can grow by two main mechanisms opposite to each other in character: absorption of defects of vacancy type or emission of defects of self-interstitial type. Vacancy absorption may be driven by gas absorption (pressure driven growth) or by an excess of vacancies in the point defect fluxes to bubbles resulting from the preferential absorption of self-interstitial atoms (SIAs) by dislocations (bias driven growth). On the other hand, emission of interstitial-type defects such as dislocation loop punching is always pressure driven. Which case or subcase is dominant in a specific situation depends on the implantation conditions characterized by experimental parameters such as temperature T, gas generation rate P_G, displacement rate P_D, dose D or gas concentration c [48].

In the present paper we are interested in bubble states achieved during the incorporation of inert gases into metals at room temperature which, of course, has different meanings for different melting temperatures T_m of the metals ("homologous temperatures" T/T_m). Typical values of the local gas generation rates during implantation into TEM foils are around 10^{-5} s^{-1} and range from 10^{-11} to $3\ 10^{-9}$ s^{-1} for tritided metals. The number of recoil induced displacements per gas atom increases from about 50 for He to about 3000 for Xe in the case of ion implantation, and is negligible in tritided metals (or about 1 there if the displacements occurring during bubble formation are considered as displacement damage). Atomic gas concentrations suited for gas density measurements in bubbles are between 1 % and 10 % for ion implanted foils and between 0.1 % and 1 % for tritided metals.

For the question concerning the dependence of the bubble growth mechanisms on these parameters, the corresponding dependence of the main contributions from the bubble structure and the concurrently evolving dislocation structure to the total sink strength for the annihilation (and production) of vacancies and SIAs is crucial. The information on these dependences is still limited but there is agreement about the following trends [49,50]. Both the bubble number density C and the dislocation density ρ increase with decreasing homologous temperature and seem to reach upper limits at low homologous temperature. Whereas C increases with increasing gas generation rate P_G, ρ seems to saturate above a relatively low displacement rate. The annihilation of radiation induced defects is recombination controlled at low T and/or high P_D where C and ρ saturate as well as at high T and/or low P_D where C and ρ decrease substantially with increasing T. In the medium range defect annihilation is sink controlled whereby the bubble structure represents the dominant sink at high P_G and/or low P_D/P_G and the dislocation structure in the opposite case.

A detailed discussion of the bubble growth regimes on the basis of these dependence of C and ρ on experimental parameters is beyond the scope of this paper. Here only crude guidelines, similar but somewhat different to the ones presented by Evans [48], will be given. Thus, for instance, gas driven growth at pressures p only slightly above the equilibrium value $p = 2\gamma/r$ (where γ is the surface free energy) occurs at high T and/or low $P_{D,G}$ where the thermally activated production of vacancies by dislocations is larger than the vacancy production by displacements and sufficient to balance the gas production. A crude condition for this is

$$P_G < 10^4\,\text{s}^{-1}\ \exp\left(-13\,T_m/T\right). \tag{1}$$

Bias driven growth occurs in a medium region of the (P_D, P_G, T) space where the production of vacancies by displacements is dominant and the supplying of bubbles with excess vacancies is significantly larger than with gas atoms. This applies under conditions opposite to condition (1) and when

$$P_D/P_G > 100\ \text{ or }\ P_G^2/P_D < 10^5\,\text{s}^{-1}\exp\left(-7.5\,T_m/T\right) \tag{2}$$

For given displacements per gas atom, P_D/P_G, the bubble and dislocation sink strength must not be too different [48] which results in lower and upper bounds of P_G for bias driven growth. In the remaining part of the parameter space at low T and/or high P_G, bubbles are far from equilibrium and grow by some non-equilibrium pressure driven mechanisms.

For implantation into TEM foils eqs. (1) and (2) mean that bubble growth follows closely the equilibrium condition, $p = 2\gamma/r$, above $0.6\,T_m$, may be bias driven between 0.25 and $0.6\,T_m$ if $P_D/P_G \geq 100$ and occurs by some non-equilibrium pressure driven mechanism at lower temperatures, but already below $0.6\,T_m$ for He bubbles since in this case $P_D/P_G < 100$. For tritides the transition between equilibrium and non-equilibrium growth occurs around $0.45\,T_m$.

For the metals for which inert gas implantation at ambient temperature has been considered so far, only Sn is a candidate for bubble growth close to equilibrium. Zn and Al implanted with heavy inert gases represent candidates for bias driven growth. Enhanced recombination and defect clustering within displacement cascades could, however, change them to candidates for non-equilibrium pressure driven growth. A judgement about the growth mechanism in these systems will be made on the basis of the whole data set to be discussed later. In all other metals (Ag, Au, Cu, Ni, Fe, Ti, Lu, Zr, Nb, Mo, Ta) considered so far, bubble growth at the ambient temperature occurs by some non-equilibrium pressure driven mechanism such as dislocation loop punching or displacement damage induced local creep.

In the low T/high $P_{D,G}$ region far from equilibrium on which interest is focused in this paper, atomic transport is probably occurring by non-diffusional processes such as spontaneous defect reactions in a dense defect structure or recoil induced displacements which are particularly efficient under cascade formation conditions. Note that in this region far from equilibrium the term "equilibrium pressure", even if it may be used as a helpful reference, does not have any physical meaning: The system has no chance to probe what "equilibrium" could mean.

2.2. Dislocation loop punching and related processes

Since the emission of self-interstitial type defects is the process defining the ultimate upper limit for pressures in bubbles, it is discussed here in some detail. A necessary energetic condition for such a process to occur is obtained by considering the change in the total free energy, ΔF, associated with a complete separation of the self-interstitial type defect from a bubble. In the continuum approximation, the condition may be written as [5]

$$\Delta F = \Delta F_B + F_D = \tilde{p}\Delta V_B + F_D \leq 0. \tag{3}$$

Here ΔF_B is the change in the bubble free energy consisting of a decrease in the free energy of the compressed gas and an increase in the surface free energy of the bubble, and F_D is the formation free energy of the interstitial type defect; $\tilde{p}$ is the average of the excess pressure $(p- 2\gamma/r)$ over the initial and final states and ΔV_B the change of the bubble volume per dissociation event.

For a single self-interstitial atom (SIA), ΔV_B is equal to the atomic volume of the matrix, Ω, and the formation free energy F_D is between 1/2 and 1 times $\mu\Omega$, where μ is the (average) shear modulus of the metal. With this, eq. (3) yields the pressure $\tilde{p}_{IE}$ for complete dissociation or emission of single SIAs from bubbles

$$\tilde{p}_{IE} \geq (1/2 \text{ to } 1)\mu. \tag{4}$$

Single SIAs, if at all, would only be emitted by atomistically small gas-vacancy clusters. With increasing bubble size, the emission of SIA clusters becomes energetically more favourable than the emission of single SIAs since the binding energy of the SIAs to their clusters is saved in this collective emission process. For clusters consisting of more than about a dozen SIAs, the energetically most favourable structure is that of a dislocation loop. Using the elastic continuum approximation for the formation energy of a dislocation loop and assuming that the loop radius is equal to the bubble radius r, one obtains the GFR condition [3]

$$\tilde{P}_{LP} \geq \mu b \, (\ln r/r_0)/[2\pi(1-v)r], \tag{5}$$

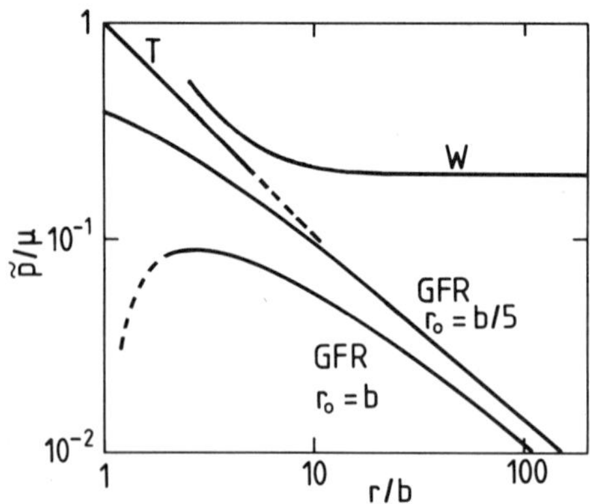

Fig. 1 Gas pressure required for dislocation loop punching by bubbles in metals in units of the shear modulus of the metal, p/μ, versus bubble radius in units of the Burgers vector, r/b, according to GFR [3] (for two values of r_o), Trinkaus (T) [4] and Wolfer (W) [6], respectively.

where b and r_o are the Burgers vector and the effective core radius of the dislocation, respectively, and n is Poisson's ratio. In Fig. 1, p/μ is plotted vs. r/b for $n = 1/3$ and two values of r_o. The increase of p/μ up to a maximum value is an artifact of the elastic continuum approximation which becomes poor for small sizes, $r \leq 10b$. In the size range $2b \leq r \leq 20b$, a reasonable interpolation between computer simulation data for SIA clusters and the elastic continuum approximation for dislocation loops is provided by the simple approximation (straight line labeled by T in Fig. 1) [4].

$$\tilde{p}_{LP} \geq \mu b/r .\tag{6}$$

Equations (4) to (6) represent necessary conditions for the complete separation of a SIA-type defect from a single bubble. An additional condition is that the force acting between the bubble and the emitted defect must be repulsive over the whole reaction path.

The interaction between bubbles and SIA type defects is essentially of elastic nature and consists of first and second order contributions. The interaction between single SIAs and bubbles is probably everywhere dominated by the attractive second order elastic interaction. Therefore, a single SIA produced close to the bubble surface will not be pushed out but stay there as has been shown by computer simulation [51]. The energy change associated with the formation of such a bubble/SIA-complex is equal to the difference between the formation energy of the SIA and its binding energy to the bubble and ranges from about 1/3 to 1/2 of $\mu\Omega$. Hence, the "pressure" $\tilde{p}_{BI}$ in small gas-vacancy clusters will be limited by the formation of such closely bound complexes,

$$\tilde{p}_{BI} \approx (1/3 \text{ to } 1/2) \, \mu,\tag{7}$$

rather than by the complete emission of SIAs as given by eq. (4).

In the interaction of a dislocation loop with a bubble the repulsive first order elastic interaction is dominant at large distances and the attractive second order interaction becomes comparable or larger only for very small distances. Accordingly, the total free energy of bubble/loop complexes is larger for finite distances than for complete separation. From this it follows directly that the GFR criterion given by eq. (5) cannot be a sufficient condition. In

374

fact, the maximum pressure in bubbles seems to be limited by the formation of dislocation loops about one Burgers vector away from the bubble surface. The detailed analysis [6] shows that for small bubble sizes, r ≤ 10b, the pressure required for this process decreases with increasing bubble size but soon converges to a constant value above this size range (curve labeled by W in Fig. 1)

$$\tilde{p}_{LP}\,(r \rightarrow \infty) \rightarrow\, \sim 0.2\,\mu\,. \tag{8}$$

This asymptotic value is close to the theoretical shear strength, commonly estimated by $\mu/2\pi$, a result which is probably not surprising since the loop formation process has been assumed to occur without the assistance of any dislocation source such as pre-existing dislocations. A similar result as given by eq. (8) has been found in a simulation study of "overpressurized" bubbles in amorphous materials [7]. Amazingly the authors of the latter study refer to the GFR condition when concluding (incorrectly) that the maximum pressure in bubbles in amorphous materials were considerably larger than in crystalline materials.

It should be mentioned here that an accumulation of punched out dislocation loops between bubbles results in an increase of the pressure required for further loop punching [52].

For an elastically anisotropic metal it is necessary to specify the value of μ in eqs. (4) to (8). The most convincing choice is the shear modulus controlling shear along the glide cylinder of the expected loop punching process. For the <110> and <111> directions in fcc and bcc metals, respectively, this means using:

$$\mu = C_{44}/2 + (C_{11} - C_{12})/4 \quad \text{for <110> in fcc,} \tag{9a}$$

$$\mu = C_{44}/3 + (C_{11} - C_{12})/3 \quad \text{for <111> in bcc,} \tag{9b}$$

where C_{11}, C_{12}, and C_{44} are the cubic elastic moduli. Values according to eqs. (9a) and (9b) will be used in the subsequent data analysis.

Another question concerns the difference between excess and actual pressure connected with the creation of additional surface during loop punching. The theoretical analysis indicates that the formation of a dislocation loop close to a bubble surface and its detachment from the bubble are continuous rather than abrupt processes. Thus, the necessity of surface creation would not affect the condition for loop punching if the surface creation process were complete before the loop is definitively pushed out. In this case, the actual instead of the excess pressure would have to be used in eq. (8). In the following data correlation the difference between both will be ignored; this is not too important particularly since the specific surface free energy γ is strongly correlated with the shear modulus μ.

An alternative process to the discussed emission of SIA-type defects by bubbles is the small scale transfer of SIAs from a bubble to an attached dislocation. This process implies bubble migration and/or local dislocation climb. For the temperature range of interest here, estimates show that the pressure level required for this process to occur is comparable or even higher than the one required for loop punching.

2.3. A phenomenological approach to pressure driven growth under displacement damage

Except for very small (subthreshold) energies, ion implantation into solids is accompanied by displacement damage. In the low temperature region far from equilibrium, where one may hope to be able to establish the conditions for dislocation loop punching, the gas pressure in bubbles may alternatively be limited by the continuously produced displacement damage. Possible mechanisms for the necessary damage induced local creep have not been discussed so far. The following phenomenological approach is presented here to provide guidelines for the data analysis discussed later.

The simplest assumption is that the local creep rate is proportional to the local stress and the displacement rate as in linear irradiation creep. In this case, overpressurized bubbles subject to displacement damage grow as in a viscous medium. Assuming that the respective creep modulus scales with the shear modulus μ, the relative change of the swelling due to bubbles, S, with time t or displacement dose D may be written as

$$\mathrm{dlnS/dD} = A\ p(n)/\mu, \tag{10}$$

where A is a dimensionless constant characterizing the creep efficiencies of displacement damage and pressure. Apart of the gas EOS, p(n), an additional relationship between swelling, dose, and gas density, n, in the bubbles is required. If most of the implanted gas is contained in bubbles this is simply

$$n = D/(NS\Omega), \tag{11}$$

where N is the average number of displacements per gas atom.

For a realistic gas EOS the solution of eqs. (10) and (11) is complicated. To get an idea of the general trends, we assume a power law behaviour, $p \propto n^\alpha$ with $\alpha = \partial\mathrm{lnp}/\partial\mathrm{lnn}$, which is reasonable within limited gas density ranges. In this case, a special solution of eqs. (10) and (11) starting from (infinitely) high initial pressure is

$$p/\mu = [(\alpha+1)/\alpha]/(AD) \approx 1/(AD). \tag{12}$$

Accordingly, we expect the pressure to decrease inversely proportional with dose D.

3.   Analysis and Correlation of Existing Data

3.1. Dilatometry of tritided metals

An important effect of tritium transmutation into helium in tritided metals is swelling which demonstrates that the ^{3}He atoms precipitating into bubbles require more space than the T atoms from which they originate. Dilatometry of such systems provides direct information on ^{3}He densities and, with aid of an appropriate EOS, on pressures in bubbles. In fact, an increased accuracy of strain gauge and density measurements [33] in combination with an improved EOS [4] (revised according to recent high pressure diffractometry of solid ^{4}He [41] and corrected for the isotope mass difference) has raised the accuracy in determining pressures in bubbles to a level where theoretical models for the maximum possible pressure can be tested. In this section, the main conclusions will be discussed. For the detailed data analysis, refer to reference [33].

The main result is that swelling increases almost linearly with time. This indicates approximately constant ^{3}He densities. For Nb, Ta, and Lu tritides the volume per ^{3}He atom is found to be between 7 and 8 Å^3 at the end of the respective measuring periods, corresponding to pressures around and somewhat above 0.2 μ [33] (Table 1). These values are in agreement with Wolfer's analysis [6] but above the values expected from the GFR condition for dislocation loop punching [3]. The high value of $p_f/\mu \approx 0.264$ for Lu indicates very low bubble sizes, probably in the subnanometer range, corresponding to an extremely high bubble density as expected to occur at the low temperature of 78 K applied in this case.

The conclusions drawn from the absolute values of the pressure are strengthened by its trends following from the slight increase of the observed volume change per decay event with time. The estimated relative decrease, $(-\Delta p_f/p_f)$, of the pressure within a period in which the average bubble radius is expected to have doubled, is obviously correlated with the pressure level: the higher the pressure the stronger is its decrease. This is in agreement with the

376

Table 1 Dilatometric information on the pressure in ^{3}He bubbles forming during T transmutation in tritided metals [33]: Pressure in units of the shear modulus at the end of the measuring periods, p_f/μ; estimated relative decrease $-\Delta p/p_f$ within a period in which the bubble dimension is expected to have doubled; extrapolated asymptotic value, (p_∞/μ).

	Nb $T_{0.0253}$ (299 K)	Ta $T_{0.0744}$ (299 K)	Lu $T_{0.15}$ (78 K)
p_f/μ	0.196	0.203	0.264
$-\Delta p/p_f$	6 %	10 %	30 %
p_∞/μ	0.184	0.183	0.185

behaviour following from Wolfers analysis but opposite to the behaviour following from the GFR condition.

From the absolute values and the trends of p/μ as given in Table 1, the asymptotic value, p_∞/μ, could be estimated if the form of the decrease at small bubble sizes were known. Assuming as the simplest form $p = p_\infty + (p_0 - p_\infty)\, r_0/r$, we find for p_∞ the values given in the last line of Table 1 which are surprisingly close to each other. Considering possible uncertainties in the measurements and their analysis, we conclude that

$$p_\infty/\mu = 0.18 \pm 0.02. \tag{13}$$

3.2. Diffractometry of solid inert gases implanted into metals

For inert gases implanted into thin TEM foils, quantities characterizing the implantation conditions, in particular the displacement dose, must be considered as parameters possibly relevant for pressure relaxation. Diffractometry of solid inert gases is the most direct method to study this.

In Table 2 the available electron diffraction (ED) [16–27] and X-ray diffraction (XD) [28–31] data for a number of inert gas/metal systems are compiled (together with EELS data discussed below). The first group of columns give parameters of the metals relevant for the analysis: the atomic volume Ω, the shear modulus μ (effective for loop punching) and the minimum displacement threshold energy E_D. The second group of columns contains the relevant implantation parameters: energy E_I, fluence ϕ, (local) atomic gas concentration c, and displacement dose D. The third group of columns gives quantities characterizing the implanted inert gas bubbles: the average bubble diameter d, the volume per gas atom v, the related pressure and its ratio to the shear modulus μ. The last columns indicate the method used and the pertinent reference. The metal parameters have been taken from standard tables or deduced from tabulated values; E_I, ϕ, c, d as well as v (except for He) have been taken from the respective reference; D, p, and p/μ have been derived from the preceeding parameters, independently of corresponding derivations in the respective reference.

The displacement dose D, for instance, has been deduced from given values of ϕ or c by employing the TRIM code [53]. For mono-energetic implantation [9–11, 17–31] the value of D given in Table 2 corresponds to the average over that central part of the implantation peak which contains 50 % of the implanted gas. For "homogeneous" implantation [12–16] the given value of D represents the average over the homogeneously implanted region. In a couple of cases [5,6] D could not be determined since neither ϕ nor c were given. For samples prepared by the "Harwell combined implantation and sputtering method" [18] D could not be evaluated since in this method it is unclear how the damage should be characterized. The accuracy of the deduced D values is not better than $\pm$ 50 % mainly because of the poor

Table 2 Compilation and evaluation of spectroscopic data (UVAS, EELS) and diffraction data (ED, XD) on inert gas bubbles forming during ion implantation into metals as given by BJ [25], BL [26], DL [11], EM 1–3 [18–20], F [27], GB [28], GC [23], JM [14,15], KT [30], M [12], RD [10], T [24], TGR [22], TK [29], VFF [16]. The data in parenthesis could not be used in Fig. 2 for the following reasons: [1] ϕ uncertain, EELS peak too broad, [2] no information on ϕ or c, [3] damage unclear.

Gas	Metal	Ω (Å³)	μ(GPa)	E_D (eV)	E_I(keV)	$\phi(10^{20}m^{-2})$	c (%)	D (dpa)	d(Å)	v(Å³)	p(GPa)	p/μ	Method	Ref.
He	Al	16.6	26.2	16.5	5	4.5		3.6	< 20	8.1	8.48	0.323	UVAS	DL
					5	7.0		5.6		8.6	7.04	0.269		DL
					5	10		8.0	~ 20	17.9	0.725	0.0277	EELS	RD,DL
					5	15		12		13.2	1.86	0.0710		
					5	16.5		13.2		11.1	3.19	0.122		
					0.5–8	4.3	3.3	2.3	13	7.53	10.6	0.406	EELS	JM,M
							5.5	3.8		7.74	9.76	0.372		
							6.5	4.5		7.96	8.95	0.341		
							7.7	5.4		8.22	8.10	0.309		
						14	10.4	7.3	18.4	8.77	6.62	0.253		
						17	13	9.1	20	9.12	5.87	0.224		
							16.4	11.5		9.92	4.52	0.173		
						34	26.0	18.2	40	12.9	2.00	0.076		
	(Ni	11.0	94.6	23	0.5–8	10?		4.0?	< 10	5.4?	28?	0.3?	EELS	JM,M
						100?		40?	12.6	6.8?	14?	0.15?		)[1]
Ne	Al				5–40?		3	8.5	26	13.7	4.8	0.19	EELS	vFF
Ar	Al				10–80?		3	16	30	28.6	2.8	0.11	ED	vFF
										25.0	5.6	0.21	EELS	
					(100	< 0.5		< 5		21.6	13.5	0.5	XD	TK)[2]
										26.1	4.4	0.17		
	(Ni				100	> 1		> 16	30?	18	25	0.26	XD	KT)[2]

Gas	Metal	Ω (Å³)	μ(GPa)	E_D (eV)	E_I(keV)	$\phi(10^{20}m^{-2})$	c (%)	D (dpa)	d(Å)	v(Å³)	p(GPa)	p/μ	Method	Ref.
Kr	Al	16.6	26.2	16.5	65	0.9		17	?	36.3	2.2	0.084	ED	BJ
						1.3		25	17	37.5	1.8	0.069		
					10-130?		3	26	35	38.4	1.6	0.061	ED	F
					100	2		44	35	40.5	1.2	0.047	XD	GB
					200	2		46	23	37.6	1.8	0.069		
	Ti	17.6	42.3	19	100	1.6		54	25-50	37.5	1.8	0.043	ED	EM3
	(Ni						5	-	30	31.2	5.3	0.056	ED	EM1[3]
	Ni	11.0	94.6	23	180	1		32		34.1	3.0	0.032	ED	BL
						2		64	22	35.9	2.3	0.024		
						3		96	38	36.8	2.0	0.021		
						4		128		37.9	1.7	0.018		
	(Cu	11.8	54.7	19			3-5	-	30?	35.2	2.6	0.048	ED	EM1[3]
	Mo	15.6	126	36	100		$\geq$ 5	$\geq$ 40	40	33.2	3.7	0.029	ED	EM2
	(Au	17.0	23.3	36	100	2?		100?	< 30	41.6	0.95	0.041	ED	EM1[2]
Xe	Al				20-200		3	17.5	26	43.5	2.9	0.11	ED	vFF
					131	2		60	39	53.5	0.78	0.03	XD	GB
					200	0.2		7	20	41.6	4.16	0.16	ED	TGRD
						0.3		10.5	20	41.6	4.16	0.16		
						0.5		17.5	30	43.9	2.76	0.105		
						0.66		23	40	46.3	2.13	0.081		
						0.8		28	40	48.8	1.60	0.061		
						1		35	50	51.3	1.15	0.044		
	α-Fe	11.8	90	17	390	1		60		43.9	2.76	0.031	ED	T,TGR
	γ-Fe	11.0	94.6	23	230	1		50	23	44.3	2.62	0.028	XD	JG
	Ni	11.0	94.6	23	360	1		50		43.9	2.76	0.029	ED	TGR
	Cu	11.8	54.7	19	360	0.5/1/2		29/58/116		48.8	1.5	0.027	ED	T,TGR,GC
	Zn	15.2	46.5	15	360	1		78	50?	59.1	0.5	0.011	ED	T,TGR
	Zr	23.3	37.7	21	300	1		68		59.1	0.5	0.013	ED	T
	Ag	17.1	33.4	25	360	1		65		51.3	1.1	0.033	ED	T,TGR
	Au	17.0	31.2	36	360	1		64		54.0	0.7	0.022	ED	T,TGR

experimental accuracy of ϕ and c, but also because of uncertainties in E_D and simplifications in the TRIM code. To derive pressure values from gas densities, recent experimental data [37–47] have been interpolated, for He on the basis of the EOS given in ref. 4, for heavier inert gases on the basis of reduced EOS as described in ref. 34.

The inspection of the ED and XD data for Ar, Kr, and Xe compiled in Table 2 reveals the following trends:

1. *All* the p/μ-values deduced are below the ones for He deduced from the dilatometry of tritided metals.

2. As has been shown previously on the basis of a smaller data set [32] the deduced p-values correlate quite well with μ, in particular when the p-μ-values for each inert gas species are considered separately. The corresponding correlation coefficients are between 0.8 and 0.9. It must, however, be mentioned that the correlation between the pressure and other energetic parameters of the matrix such as the vacancy migration energy E^V_M is comparably close which is not surprising in view of the close correlation between μ and such parameters. On the other hand, a correlation between p and μ is consistent with the predictions of the models described in sections 2.2 and 2.3 whereas it is not at all clear why p should scale with a quantity such as E^V_M.

3. Even though the scatter is large, there is a perceptible trend that the p-values decrease with increasing atomic number of the inert gas, in particular when samples containing similar gas concentrations are compared. This indicates that bubble growth is not controlled exclusively by the pressure as expected for dislocation loop punching.

4. There is a noticeable trend that the p-values decrease with increasing bubble diameter d, swelling $S = cv/\Omega$ or displacement dose D which are correlated parameters. This trend becomes even clearer when the p-μ-correlation is accounted for by considering the correlation of the ratio p/μ with d, S or D. The phenomenological approach discussed in section 2.3 suggest that p/μ should be inversely proportional to D. The high correlation coefficient of 0.93 for all ED and XD related p/μ–D^{-1} pairs of values in Table 2 as opposed to values below 0.8 for the corresponding correlations of p/μ with d^{-1} and S^{-1} indicates that the p/μ–D^{-1} correlation is indeed the most direct one among the three.

Moreover, the p/μ–D^{-1} correlation does not only cover the correlation of p with μ but also with the inert gas species: Obviously, the decrease of p with increasing atomic number is due to an increasing damage level. Consequently, all correlations of the inert gases discussed here may be summarized by the linear regression of the p/μ–D^{-1} values which yields

$$p/\mu \approx 1.3/D + 0.01. \tag{14}$$

Within the data scatter, which is mainly determined by the uncertainty in D rather than in p/μ, no difference between different metals implanted with different heavy inert gases is perceptible. This indicates that the same displacement damage induced pressure driven bubble growth mechanism occurs in all of these systems, including the ones where Al $(T/T_m \cong 0.3)$ forms the matrix.

Equations (12) and (14) may be used to define an effective "creep compliance" for the present type of local creep. The resulting value of $A/\mu \approx 0.77$ is significantly lower (almost one order of magitude) than the corresponding ones found in normal irradiation creep [54].

3.3. Spectroscopy of helium implanted into metals

The determination of pressure values in bubbles forming during He implantation into thin TEM foils is more complicated. On one hand, dilatometry of TEM samples is difficult and has not been performed so far. On the other hand, diffractometry of solid He in metals is difficult

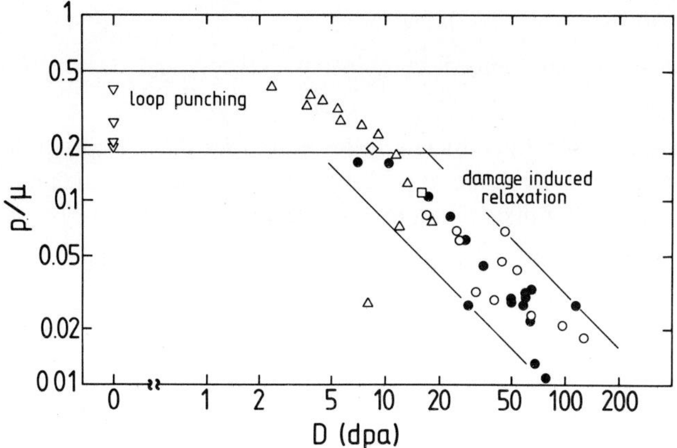

Fig. 2 Pressure in inert gas bubbles in units of the shear modulus of the metal, p/μ, versus displacement dose in dpa according to Table 1 and 2. Values for ^{3}He (∇) from T transmutation derived from dilatometric data of tritided metals; values for ion implanted ^{4}He (Δ) and Ne (◊) derived from spectroscopic data; values for ion implanted Ar (□), Kr (○), And Xe (●) derived from diffraction data.

and in many metals, moreover, loop punching limits the pressure in bubbles to a range where He is fluid at ambient temperature. Thus one depends on more indirect methods for which either experimental calibration or detailed theoretical modelling is required. In the present context, Ultraviolet Absorption Spectroscopy (UVAS) [9–11] and Electron Energy Loss Spectroscopy (EELS) [10–15] are particularly suited since, so far, they represent the only indirect methods for which the dependence of the measured quantity (here the blue-shift) upon implantation fluence or implanted gas concentration has been studied systematically.

In the following, an attempt is made to calibrate these spectroscopic density measurements on ^{4}He implanted into metals with aid of the correlations discussed above. Fig. 2 shows a double logarithmic representation of the correlation between p/μ and D deduced from the dilatometry of tritided metals (∇) and the diffractometry of ion implanted solid Ar(□), Kr(○), and Xe (●). The data deduced for ^{3}He bubbles in tritided metals are considered to define the p/μ-range where loop punching occurs. For ion implantation, the data suggest that this range is only reached for D ≤ 10 dpa whereas for D > 10 dpa the p/μ-values appear to be limited by displacement damage as described by eq. (14).

The data for bubbles forming during He implantation are expected to fit into the frame defined by the two types of relaxation mechanisms. The p/μ-values given in Table 2 and Fig. 2 for He in Al correspond to a "best fit" on the basis of a supposed linear relationship between the energy shift of the 1S-2P transition, ΔE, and the He density, n = 1/v, according to

$$\Delta E = Cn \text{ with } C = (25 \pm 5) \text{ Å}^3 \text{ eV}. \tag{15}$$

The given value of the proportionality constant C agrees with the one derived from the temperature dependence of ΔE for He in Al assuming that during annealing the pressure approaches its equilibrium value, p = 2γ/r. The given uncertainty range of ± 5Å^3 eV covers the somewhat lower theoretical value of 22 Å^3 eV used in ref. 15 but is significantly below the values quoted in ref. 11.

The thus determined p/μ-values given in Table 2 and Fig. 2 for He in Al suggest that in this case a transition from loop punching to damage induced pressure relaxation takes place.

4.   Summary and Outlook

The results of the present data analysis combined with theoretical ideas may be summarized as follows:

(1) The dilatometry of tritided metals yields values for the pressure p in ^{3}He bubbles between 1/2 and 1/6 of the shear modulus μ of the metal, consistent with theoretical predictions for the pressure required for dislocation loop punching.

(2) Electron and X-ray diffraction from solid Ar, Kr, and Xe implanted into thin TEM foils yield pressure values below this range which decrease with increasing displacement dose D indicating that in this case pressure relaxation is due to displacement damage rather than loop punching.

(3) The transition between loop punching and damage induced pressure relaxation seems to occur around a dose of 10 dpa.

(4) A fitting of spectroscopic data for bubbles forming during helium implantation into this data frame confirms a previous calibration based on the assumption that equilibrium bubbles are formed during annealing.

A number of questions remain to be answered. The details of pressure relaxation under displacement damage in the parameter range considered, for instance, is not at all clear. To confirm the stated correlation between p/μ and D, it would be helpful to study the effect of irradiation on the pressures within bubbles having formed during pre-implantation up to a given gas concentration.

The calibration of spectroscopic gas density measurements is still not sufficiently accurate and depends on the assumption of a linear relationship between the energy shift and the gas density. A more reliable calibration, for instance with aid of ^{3}He bubbles forming in tritided metals is desirable. The guidelines provided by the present data analysis appears, however, to be already sufficient to test some other calibration methods. Thus, for instance, the (non-linear) relationship between the energy shift and the He density recently derived from refractive-index measurements of dense He [55] would yield pressure values above 0.5 μ for bubbles forming during He implantation into metals. This can almost certainly be ruled out.

ACKNOWLEDGEMENTS

Assistance in data compilation by Drs. W. Jäger, S. Mantl and T. Schober is gratefully acknowledged.

REFERENCES

1. Proc. Int. Conf. on Helium in Metals, Jülich, Germany (1982), Rad. Effects **78**, (1983).
2. S.E. Donnelly, Rad. Effects **90**, 1 (1985).
3. G.W. Greenwood, A.J.E. Foreman, and D.E. Rimmer, J. Nucl. Mater. **4**, 305 (1959).
4. H. Trinkaus, Ref. 1, p. 189.
5. H. Trinkaus and W.G. Wolfer, J. Nucl. Mater. **122 & 123**, 552 (1984).
6. W.G. Wolfer, Phil. Mag. **A58**, 285 (1988).
7. G. Knuyt, M. D'Olieslaeger, L. De Schepper, and L.M. Stals, Phil. Mag. **A58**, 243 (1988).
8. B. Cochrane and P.J. Goodhew, Phys. Stat. Sol. **A77**, 269 (1983).
9. S.E. Donnelly, J.C. Rife, J.M. Gilles, and A.A. Lucas, IEEE Trans. on Nucl. Sci. **NS28** (2), 1820 (1981).
10. J.C.Rife, S.E. Donnelly, A.A. Lucas, J.M. Gilles, and J.J. Ritsko, Phys. Rev. Lett. **46**, 1220 (1981).
11. S.E. Donnelly, A.A. Lucas, J.P. Vigneron, and J.C. Rife, Ref. 1, p. 337.
12. R. Manzke, PhD-Thesis, University of Cologne, Jül-Report 1814, Jülich (1982).
13. R. Manzke, W. Jäger, H. Trinkaus, G. Grecelius, and R. Zeller, Sol. State Com. **44**, 481 (1982).
14. W. Jäger, R. Manzke, H. Trinkaus, G. Grecelius, R. Zeller, J. Fink, and H.L. Bay, J. Nucl. Mater. **111 & 112**, 674 (1982).
15. W. Jäger, R. Manzke, H. Trinkaus, R. Zeller, J. Fink, and G. Crecelius, Ref. 1, p. 315.

16. A. vom Felde, J. Fink, Th. Müller-Heinzerling, J. Pflüger, B. Scheerer, G. Linker, and D. Kaletta, Phys. Rev. Lett. **53**, 922 (1984); A. vom Felde, Diplomthesis, Univ. Karlsruhe (1984).

17. C. Templier, C. Jaonen, J.P. Rivière, J. Delafond, and J. Grithé, Comptes Rendus (Paris) **99**, 613 (1984).

18. J.H. Evans and D.J. Mazey, J. Phys. F: Met. Phys. **15**, L1 (1985).

19. J.H. Evans and D.J. Mazey, Scripta Met. **19**, 621 (1985).

20. J.H. Evans and D.J. Mazey, J. Nucl. Mat. **138**, 176 (1986).

21. C. Templier, H. Garem, J.P. Rivière, and J. Delafond, Nucl.Instr. & Methods in Phys. Res. **B18**, 24 (1986).

22. C. Templier, H. Garem, and J.P. Rivière, Phil. Mag. **A53**, 667 (1986).

23. J. Guillot, M. Cartraud, H. Garem, C. Templier and J.-C. Desoyer, C.R. Acad. Sci. Paris, **305**, 161 (1987).

24. C. Templier, Thesis, University of Poitiers (1987).

25. R.C. Birtcher and W. Jäger, Ultramicroscopy **22**, 267 (1987).

26. R.C. Birtcher and A.S. Liu, Mat. Res. Soc. Symp. Proc. **74**, 345 (1987).

27. J. Fink, Advances in Electronics and Electron Physics **75**, 121 (1989).

28. L. Gråbaek, J. Bohr, E. Johnson, H.H. Andersen, A. Johansen, and L. Sarholt-Kristensen, Mat. Sci. Eng. **A115**, 97 (1989).

29. A.K. Tyagi, R. Khanna, and G.V.N. Rao, Scripta Met. **20**, 1245 (1986).

30. R. Khanna, A.K. Tyagi, R.V. Nandedkar, and G.V.N. Rao, Scripta Met. **20**, 181 (1986).

31. E. Johnson, E. Gerritsen, N.G. Chechnin, A. Johanson, L. Sarholt-Kristensen, H.A.A. Keetels, L. Grabaek, and J. Bohr, Nucl. Instr.& Meth. **B39**, 573 (1989).

32. J.H. Evans and D.J. Mazey, J. Phys. **F15**, L1 (1985).

33. T. Schober, C. Dieker. R. Lässer, and H. Trinkaus, Phys. Rev. B **40**, 1277 (1989); T. Schober and H. Trinkaus, this volume.

34. M.L. Klein and J.A. Venables, eds., Rare Gases in Solids, Vol. II, Academic Press, London - New York - San Francisco (1977).

35. D.A. Young, A.K. McMahan, and M. Ross, Phys. Rev. **B24**, 5119 (1981).

36. C. Ronchi, J. Nucl. Mater. **96**, 134 (1981).

37. R.L. Mills, D.H. Liebenberg, and J.C. Bronsen, Phys. Rev. **B21**, 5137 (1980).

38. D. Vidal, L. Guengant, and J. Vermesse, Physica **116A**, 227 (1982).

39. K. Syassen and W.B. Holzapfel, Phys. Rev. **B18**, 5826 (1978).

40. L.W. Finger, R.M. Hazen, G. Zou, H.K. Mao, and P.M. Bell, Appl. Phys. Lett. **39**, 892 (1981).

41. H.K. Mao, R.J. Hemley, Y. Wu, A.P. Jephcoat, L.W. Finger, C.S. Zha, and W.A. Bassett, Phys. Rev. Lett. **60**, 2649 (1988).

42. R.J. Hemley, C.S. Zha, A.P. Jephcoat, H.K. Mao, L.W. Finger, and D.E. Cox, Phys. Rev. **B39**, 11 820 (1989).

43. M. Ross, H.K. Mao, P.M. Bell, and J.A. Xu, J. Chem. Phys. **82**, 1028 (1986).

44. A. Polian, J.M. Besson, M. Grimsditch, and W.A. Grosshans, Phys. Rev. **B39**, 1332 (1989).

45. A.N. Zisman, I.V. Aleksandrov, and S.M. Stisho, Phys. Rev.B **32**, 484 (1985).

46. K.A. Goettel, J.H. Eggert, Isaac Silvera, and W.C. Moss, Phys. Rev. Lett. **62**, 665 (1989).

47. R. Reichlin, K.E. Brister, A.K. McMahan, M. Ross, S. Martin, Y.K. Vohra, and A.L. Ruoff, Phys. Rev. Lett. **62**, 669 (1989).

48. J.H. Evans, Nucl. Instr. & Methods B **18**, 16 (1986).

49. H. Trinkaus, J. Nucl. Mater. **174, 178**, (1990).

50. B.N. Singh and H. Trinkaus, this volume.

51. W.D. Wilson, C.L. Bisson, and M.I. Baskes, Phys. Rev. **B24**, 5616 (1981).

52. W.G. Wolfer, Phil. Mag. **A59**, 87 (1989).

53. J.P. Biersack and L.G. Haggmark, Nucl. Inst. & Methods **174**, 257 (1980).

54. P. Jung and M.I. Ansari, J. Nucl. Mater. **138**, 40 (1986).

55. R. Le Toullec, P. Loubeyre, and J.P. Pinceaux, Phys. Rev. **B40**, 2368 (1989).

PARAMETERS AND PROCESSES CONTROLLING HELIUM BUBBLE FORMATION IN METALS AT ELEVATED TEMPERATURES

B.N. Singh[1] and H. Trinkaus[2]

[1] *Materials Department, Risø National Laboratory, DK–4000 Roskilde, Denmark*

[2] *Institut für Festkörperforschung des Forschungszentrums Jülich
D–5170 Jülich, Germany*

ABSTRACT

In an effort to further the understanding of the nucleation mechanisms as well as processes such as gas diffusion and dissociation, the experimental data on temperature dependencies of bubble density obtained under different experimental conditions are compiled and analysed. In particular, the effects of parameters such as helium concentration, helium generation rate, annealing time and displacement damage rate on the measured bubble densities are considered. The results are analysed in terms of nucleation and coarsening models. The analysis shows that the low temperature regime is diffusion controlled whereas the high temperature regime is gas dissociation controlled. In the high temperature regime, the large difference in the observed bubble density between the hot–implantation/irradiation and the cold–implantation followed by annealing experiments seems to be associated with large differences in the concentration (or pressure) of helium within bubbles in the embryonic and developed state. Furthermore, in this temperature regime, the effective generation rate of helium is identified to be the most important parameter for bubble nucleation. The lack of data in the low temperature regime makes it clear that further studies are needed to understand the details of the nucleation and coarsening behaviour in this regime.

1.   Introduction

The problem of transport and accumulation of inert gases in the form of bubbles still remains ellusive. The origin of the problem lies in the lack of knowledge about the nature and the strength of interactions of inert gas atoms in crystalline solids, particularly in the presence of other lattice defects. In modelling bubble formation it is necessary to make certain assumptions regarding fundamental mechanisms. On the basis of appropriate models fundamental quantities such as diffusion and dissociation energies of gas atoms may be obtained from the experimental data.

There are two main types of experiments that are commonly used to study the phenomenon of bubble formation. First, there are annealing experiments where specimens preimplanted with an inert gas (e.g. helium) at relatively low temperatures are subsequently annealed isochronally at elevated temperatures. In these experiments, the initial finely dispersed population of helium–vacancy (He–V) clusters formed during the low temperature implantation coarsens to a bubble population which eventually becomes visible in a transmission electron microscope (TEM). It is worth noting here that during this process the information on the initial state of He–V clusters is quickly lost and no definitive conclusions can be drawn regarding the nucleation process. The details of the coarsening kinetics, on the other hand, can provide valuable information on bubble migration or gas dissociation from bubbles.

Experiments of the second type are those where gas atoms are implanted directly at elevated temperatures. At these high temperatures the continuous gas supply would tend to immobilize and stabilize the bubbles nucleated already during the early stages of the nucleation regime. If bubble nucleation ceases, the evolving bubble population preserves information regarding the nucleation process. An uncertainty arises, however, from the fact that the estimated nucleation dose is invariably found to be considerably lower than the dose at which the bubble density can be reliably determined experimentally. In other words, the experimentally measured bubble density may not represent the density of bubbles at the time of nucleation.

Superposition of additional irradiation (simultaneous or separate) on both types of experiments may be used to examine the specific role of radiation damage on bubble nucleation and coarsening. Radiation damage could affect, for instance, the diffusivity of gas atoms and bubbles and may alter the state of the bubbles.

In the past, a considerable amount of work has been performed to model both types of experiments, annealing after cold–implantation [1–9] and hot–implantation [10–17]. For instance, bubble coarsening during annealing has been modelled in terms of bubble migration and coalescence on the one hand [1, 2] and Ostwald ripening on the other [3–5]. The probable regimes of dominance of these two main coarsening mechanisms [6–8] and the effect of pressure on the corresponding coarsening rates [9] have been discussed. The bubble evolution during hot implantation has been modelled in terms of a gas diffusion controlled diatomic nucleation mechanism [10–17] as well as in terms of a gas dissociation controlled multi–atomic nucleation mechanism [13–15]. It has been argued that migration of bubble nuclei could reduce the bubble densities resulting from the latter nucleation models [16, 17]. Radiation damage may affect bubble nucleation via its effects on gas diffusion [12, 13, 17, 18]. At high doses, irradiation resolution of gas atoms from bubbles may become significant [19]. A concurrent net vacancy supersaturation may result in a transformation of gas–stabilized bubbles to unlimited growing voids [20].

As regards experimental observations, we have collected results on the effects of helium in austenitic stainless steels simply because the largest amount of data are available for these materials both for the annealing after cold–implantation [21–24] and the hot–implantation [25, 26] types of experiments. These results are presented in section 3. In section 3, results are also presented for (a) cold–implantation followed by hot–irradiation [24, 26, 27], (b) hot–implantation followed by hot–irradiation [26, 28, 29] and (c) reactor irradiations [24, 27, 30, 31]. Throughout this section, we concentrate on the temperature dependence of the measured bubble density. For the cold–implantation and annealing type of experiments, we consider the concentration of implanted helium and the annealing time to be the main parameters. For the hot–implantation case, the rate of helium implantation is taken to be the most crucial parameter. The two types of experiments are compared and correlated to test the basic theoretical ideas (section 4).

2.   Theoretical Background

2.1. Cold–implantation and annealing

During cold–implantation, small He–V complexes are formed. Coarsening of these complexes during subsequent annealing at elevated temperatures removes the pre–nucleation structure and leads to the formation of bubble nuclei. Further coarsening of these bubble nuclei will occur either by bubble migration and coalescence [1, 2] or by thermal resolution and reabsorption of gas atoms and vacancies (Ostwald ripening) [3–5]. As to which mechanism is going to be the dominant one, would depend on parameters such as the atomic concentration of implanted He, c_{He}, annealing temperature, T, and annealing time, t. For the present purpose, it is sufficient to consider, as was done in ref. 9, one typical case for each mechanism. In the present work we emphasize the evolution of He bubble number density, C_B, rather than the bubble size.

For bubble coarsening by coalescence we assume that the bubble migration is surface diffusion controlled and that the total volume of the bubble population is conserved. In this case, the finely dispersed structure of He–V complexes coarsens with time, t, according to [2, 9]

$$C_B \approx 0.1 \Omega^{-7/6} (v_{He} c_{He} / D_{sd} t)^{1/2}, \qquad (1)$$

where Ω and v_{He} are the metal and He atomic volumes, respectively, and D_{sd} is the surface self–diffusion coefficient. Note the typical square–root dependence upon $(c_{He}/D_{sd}t)$ which is also found for equilibrium bubbles containing ideal gas [4]. According to eqn. (1), the apparent activation energy for C_B is minus half the surface diffusion energy, E_{sd}. The latter can increase significantly with increasing gas pressure within the bubbles [9].

Ostwald ripening of gas bubbles requires the transport of gas atoms as well as vacancies. For He–dissociation controlled Ostwald ripening (not He diffusion controlled as assumed in ref. 26) of bubbles of constant total volume the corresponding expression for the coarsening of He–V complexes with time is [9]

$$C_B \approx \frac{kT \, c_{He}}{4 \gamma v_{He} \, D_{He} \, \hat{c}_{He} t \, \exp(\mu_{He}/kT)}, \qquad (2)$$

where γ is the surface free energy, D_{He} is the He diffusivity in the matrix, $\hat{c}_{He}$ is the He concentration in solution for the He density at which the He chemical potential μ_{He} vanishes and $\hat{c}_{He} \exp(\mu_{He}/kT)$ is the actual concentration in the matrix. A similar expression is found for vacancy dissociation controlled bubble coarsening [9] whereas Ostwald ripening of equilibrium bubbles containing real gas is more complicated [5].

The most important feature of eq. (2) is the linear dependence of C_B on $c_{He}/D_{He} \hat{c}_{He} t$. Accordingly, the apparent activation energy of C_B is minus that of $D_{He} \hat{c}_{He}$ which is equal to the sum of the He diffusion and solution energies representing the He dissociation energy, E_{He}^{diss}. It is expected that $E_{He}^{diss} >> E_{sd}$.

The coarsening mechanism resulting in the lowest C_B is the dominant one. The difference in the dependencies of C_B upon c_{He}, T and t indicate the regimes of bubble coalescence and Ostwald ripening. Coalescence will be dominant at high c_{He}, low T and short t, and vice versa for Ostwald ripening. The regime boundary in the c_{He}, T, t parameter space is found by equating the C_B values from eqs. (1) and (2), e.g. the "transition temperature", T_{tr},

$$k T_{tr} = \frac{2 (E_{He}^{diss} - \mu_{He}) - E_{sd}}{\ln [(v_{He}^3 / \Omega^{7/3}) (\gamma/kT_m)^2 (D_{Heo}^2 / D_{sdo}) t / c_{He}]}, \qquad (3a)$$

where D_{Heo} and D_{sdo} are the pre–exponential factors corresponding to D_{He} and D_{sd}, respectively. The temperature in the argument of the logarithm has been substituted by half of the melting temperature, T_m. According to eq. (3a), T_{tr} increases with increasing c_{He}/t but only very weakly. The C_B values at the transition temperature which should be directly observable in a C_B vs. $1/T$ plot are obtained by eliminating c_{He}/t from eqs. (1) and (2).

$$C_B^{tr}\,\Omega = \frac{1}{25}\,\frac{v_{He}^2\,\gamma}{\Omega^{4/3}kT}\,D_{sd}^{-1}D_{He}\,\hat{c}_{He}\,\exp\left(\mu_{He}/kT\right) \tag{3b}$$

Accordingly, the apparent activation energy of this bubble density C_B^{tr} is $E_{He}^{diss} - E_{sd}$. In the commonly studied bubble size and temperature ranges, the numerical factor in front of D_{sd}^{-1} is of the order of 1.

2.2. Hot–implantation

Under hot–implantation, a continuous gas supply would immobilize and stabilize bubbles nucleated already during early stages, and hence would tend to prevent further coarsening. If bubble nucleation ceases, the observable bubble densities may be considered to provide information on the embryonic state of the bubbles.

Under the conditions of continuous He generation, the He clustering rate, C_B, first increases with increasing He concentration in the matrix, $\hat{c}_{He}$. Both reach maximum values, say at time $t = t^*$, when the He precipitation rate compensates the He generation rate G_{He}. For negligible mobility of bubble nuclei this instantaneous steady state situation may be described by [13–15]

$$G_{He} = 4\pi r^* D_{He}\,\hat{c}_{He}^*\,C_B^*, \tag{4}$$

where r^* is the He trapping radius of a bubble nucleus (≈ 0.5 nm). To estimate C_B^* information on $\hat{c}_{He}$ is required.

The simplest assumption in the further treatment is that already two He atoms form a stable nucleus (di–atomic nucleation) [10–15]. In this case, the maximum nucleation rate is reached when a newly created He atom is as likely to reach an existing nucleus as to meet another He atom, i.e. when the number densities of He atoms and nuclei are comparable, $\hat{c}_{He}^*/\Omega \approx C_B^*$. This yields the estimate

$$C_B^* \approx G_{He}^{1/2}/(4\pi r^*\Omega D_{He})^{1/2}\quad\text{at}\quad c_{He}^* \approx 3\hat{c}_{He}^* \approx 3C_B^*\Omega. \tag{5}$$

The He concentration at which the nucleation peak occurs, c_{He}^*, may be estimated on the basis of bubble densities found in TEM studies. Thus, bubble densities between 10^{19} and 10^{23} m^{-3} indicate nucleation concentrations between 10^{-10} and 10^{-6} He atoms per host atom.

Equation (5) is analogous to eq. (1). It is characterized by a square root dependence of C_B upon G_{He}/D_{He}. Accordingly, di–atomic nucleation may be called He diffusion controlled. The apparent activation energy of C_B^* is minus half the He diffusion energy.

Beyond the nucleation peak, the nucleation rate decreases monotonically with decreasing atomic He concentration. In the asymptotic regime, a quasi–steady state assumption for the He concentration in the matrix, $\hat{c}_{He}\approx 0$, yields (for negligible bubble mobility), a weak increase in the bubble density according to $t^{1/3}$ and $\ln^{1/2}t$ depending on the assumption about the increase in the bubble sink strength during bubble growth [15]. Even though the nucleation rate decreases after the nucleation peak it can yield bubble densities one to two orders of magnitude above C_B^* until the bubble growth results in observable sizes since this is reached only at He concentrations orders of magnitude above the nucleation concentration. This feature is illustrated in Fig. 1 (for irradiation resolution, see below) on the basis of approximations described in refs. 13 and 14. Accordingly, observable bubble densities would not reflect the early nucleation stage. Fortunately, however, the square root dependence of C_B upon G_{He}/D_{He} is conserved in the post–peak nucleation phase.

Observable bubble densities scale with C_B^* according to eq. (5) and are thus He diffusion controlled only if the He diffusivity is large compared with the diffusivity of bubble nuclei or

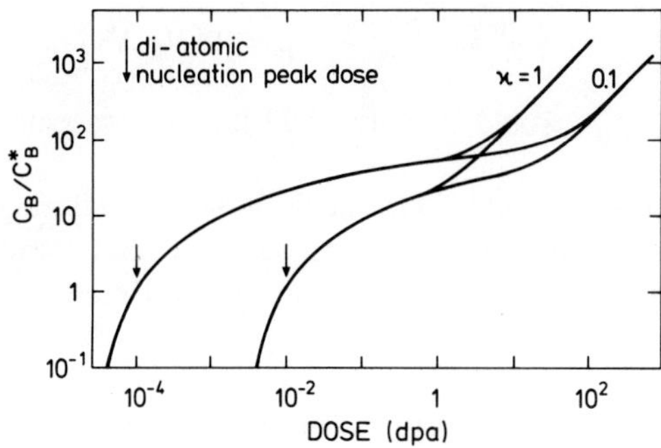

Fig. 1 Dose dependence of bubble density normalized against its value at the peak nucleation dose resulting from di-atomic nucleation for peak nucleation doses of 10^{-4} and 10^{-2} dpa and resolution parameters of $\kappa = 0.1$ and 1. In the intermediate range the bubble density increases only weakly with dose.

bubbles. The latter assumption is certainly justified if He diffuses interstitially and probably correct if He diffuses, under concurrent radiation damage, by the replacement mechanism [12, 13, 18]. Bubble migration by surface diffusion could, however, be faster than He diffusion by the vacancy mechanism. In such a case [17], observable bubble densities would be bubble migration controlled.

At high temperatures and/or low He generation rates small He–V clusters decay before having captured migrating He. Thus, only clusters above a certain critical size (depending on the He concentration in solution) are stabilized by He supply (multi–atomic nucleation). The nucleation rate then depends very sensitively on the He concentration in solution and is significant only around (or above) a critical value of this quantity, $\hat{c}_{He}^*$, which may be identified as the thermal equilibrium He concentration in the presence of critical nuclei. Interpreting eq. (4) in this sense one may write [13–15]

$$C_B^* = G_{He}/(4\pi r^* D_{He} \hat{c}_{He}^*). \tag{6}$$

At the end of the nucleation stage, C_B reaches approximately twice the value of C_B^* [13–15]. In the growth stage, the nucleation rate decreases drastically with decreasing atomic He concentration. The bubble density remains at the level reached shortly after the nucleation peak, i.e. at about $C_B = 2C_B^*$.

Equation (6) is analogous to eq. (2). It predicts a linear dependence of C_B on $G_{He}/(D_{He}\hat{c}_{He}^*)$ and an apparent activation energy for C_B equal to the He dissociation energy, E_{He}^{diss}. Accordingly, multi–atomic nucleation may be called He dissociation limited.

The difference in the dependencies of C_B on G_{He} and T indicate the regimes of di–atomic and multi–atomic bubble nucleation. The di–atomic nucleation will be dominant at high G_{He} and low T, and vice versa for the multi–atomic nucleation. The "transition temperature", T_{tr}, obtained by eliminating C_B^* from eqs. (5) and (6) is given by

$$kT_{tr} = \frac{2(E_{He}^{diss} - \mu_{He}^*) - E_{He}^{diss}}{\ln[(4\pi r^* D_{Heo})/(\Omega G_{He})]}, \tag{7a}$$

where μ_{He}^* is the chemical potential of He in critical bubble nuclei and D_{Heo} is the pre–exponential factor of D_{He}. According to eq. (7a), T_{tr} increases only weakly with G_{He}. The C_B value at the transition temperature obtained by eliminating G_{He} from eqs. (5) and (6) is given by the simple expression

$$C_B^{*tr}\,\Omega = \hat{c}_{He}^{*} = \hat{c}_{He}\exp(\mu_{He}^{*}/kT), \qquad (7b)$$

According to eq. (7b), C_B^{*tr} would directly yield the He concentration in solution at the nucleation peak.

The analogy between eqs. (1) and (2) for bubble coarsening during annealing and eqs. (5) and (6) for bubble nucleation during hot–implantation becomes complete when c_{He}/t appearing in eqs. (1) and (2) is considered an effective average He generation rate. An important difference is that in these two types of experiments very different states of the bubble evolution are probed: during coarsening the late state, during hot implantation the embryonic state. In this respect, the strong decrease of the He chemical potential with decreasing pressure during bubble growth is crucial.

2.3. Radiation damage effects

It is well known that during implantation concurrent displacement damage affects He diffusion and via this bubble nucleation [11–13, 16, 18]. It is also conceivable that radiation enhanced self–diffusion could contribute to bubble migration. To examine this we compare the contributions of surface diffusion and radiation enhanced self–diffusion to bubble migration according to

$$D_B = 3\,\Omega^{4/3}\,D_{sd}/(2\,\pi\,r_B^4) \qquad \text{for surface diffusion} \qquad (8a)$$

$$D_B = 3\,\Omega\,(D_v C_v + D_i C_i)/(4\,\pi\,r_B^3) \qquad \text{for bulk diffusion} \qquad (8b)$$

In Fig. 2, the displacement rate, G_D, required to render both contributions for bubbles with $r_B = 10\,\Omega^{1/3}$ equal is plotted vs. T_m/T. For undisturbed surface diffusion unrealistically high G_D would be needed for the radiation enhanced diffusion to enhance the bubble migration. A high gas pressure can, however, considerably reduce D_{sd} [9]. For pressures at the mechanical stability limit, the radiation enhanced diffusion could contribute significantly to bubble migration for high G_D and/or low T.

At high doses the internal He generation by irradiation resolution from bubbles can become important. In Fig. 1 this effect is accounted for by introducing an internal He generation rate $\kappa\,c_{He}G_D$ where κ is a resolution parameter [19] estimated to be of the order of 1 or smaller. At high temperatures where the thermal resolution of helium from bubbles become significant, the displacement dose for the onset of the radiation resolution effect increases drastically with temperature.

Another effect of irradiation may arise due to the radiation induced effective vacancy supersaturation. This can result in a relaxation of an overpressure in bubbles. But more importantly, it can result in the transformation of bubbles to voids when the bubble size exceeds a certain critical value [20]. Since this transformation process perturbs the information on bubble nucleation, we exclude voids from our considerations.

3. Experimental Data and General Trends

In order to test the validity of theoretical ideas and models, it is relevant to examine the effects of some of the main experimental parameters on the evolution of bubble microstructure during annealing as well as implantation experiments. For this purpose, ideally the results of a systematic study of the effects of various parameters on the evolution of bubble density and size in a given material would have been preferable. Unfortunately, such results are not available in the literature. We have, therefore, chosen to compile the data for austenitic stainless steels for which a reasonable amount of results were available.

390

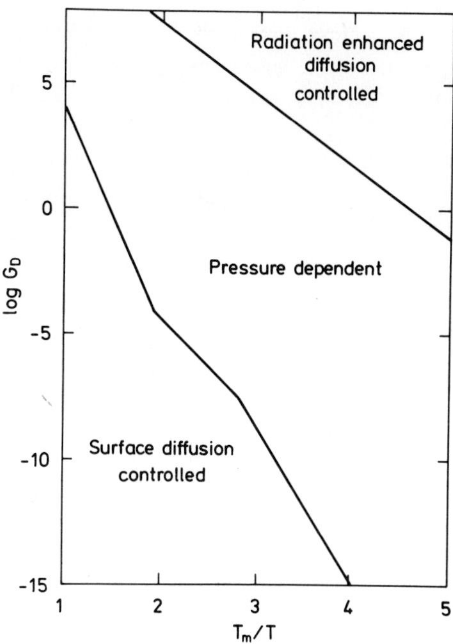

Fig. 2 Mechanism map for bubble migration. Plotted is the displacement rate G_D against T_m/T, for which the irradiation enhanced diffusion, $D_v C_v$, becomes equal to the surface diffusivity, D_s. Upper line: free surface diffusion assuming $D_s = 3 \ 10^{-6} \exp(-7 \ T_m/T) \ m^2 s^{-1}$ and $D_v = 3 \ 10^{-6} \exp(-7.5 \ T_m/T) \ m^2 s^{-1}$; lower line: surface diffusion reduced by a factor of $\exp(-7 \ T_m/T)$ resulting from an extreme gas pressure of $p = 0.2\mu$. For accessible values of G_D, the radiation enhanced diffusion is only important for strongly overpressurized bubbles.

For comparison purposes, results of the following type of experiments are compiled in Figs. 3, 4 and details of the relevant parameters for these experiments are given in Table 1:
- (a) cold–implantation followed by annealing (I_c+A)
- (b) hot–implantation at elevated temperatures (I_h)
- (c) cold–implantation followed by irradiation at elevated temperatures (I_c+R_h)
- (d) hot–implantation followed by irradiation (separately or simultaneously) at elevated temperatures (I_h+R_h)
- (e) irradiation at elevated temperatures where helium is generated via nuclear reactions (R_h)

The main results plotted in Figs. 3, 4 show the temperature dependence of the bubble density (C_B) for commercial steels. In all cases bubble densities have been determined by transmission electron microscopy (TEM). It should be noted that the TEM results may have some uncertainties both at the very low or the very high irradiation or annealing temperatures; at the low temperatures because of the small size and at the high temperatures because of very low densities (and very poor statistics). The following parameters are considered to be the control parameters in different categories of experiments:
- (i) helium concentration (c_{He}) and annealing time (t) in the (I_c+A) type of experiments
- (ii) helium implantation rate (G_{He}) and c_{He} in the I_h type of experiments
- (iii) displacement damage rate (G_D), c_{He} and displacement dose in the (I_c+R_h) type of experiments
- (iv) G_{He}, G_D, c_{He} and dose in the (I_h+R_h) and R_h types of experiments.

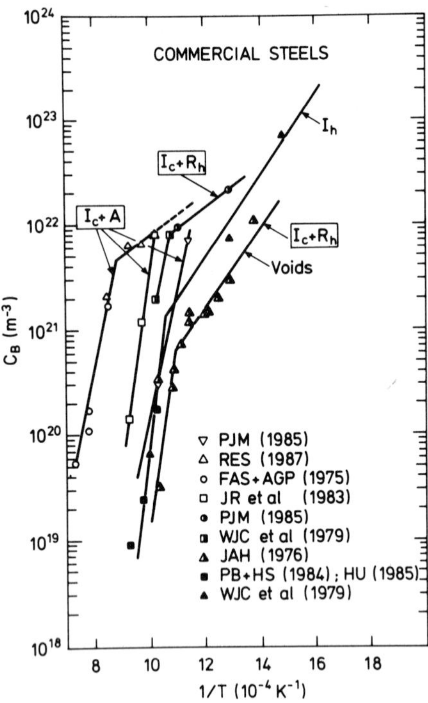

Fig. 3 Temperature dependence of the observed bubble density (C_B) in commercial austenitic stainless steels. Cavities in the (I_c+R_h) experiments are likely to be voids. Note the effect of annealing time on C_B.

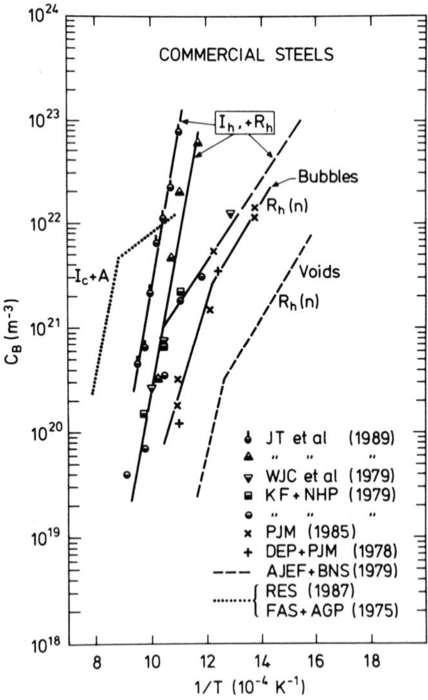

Fig. 4 Same as for Fig. 3. The broken line marked R_h (n) voids represents the average behaviour taken from ref. [31].

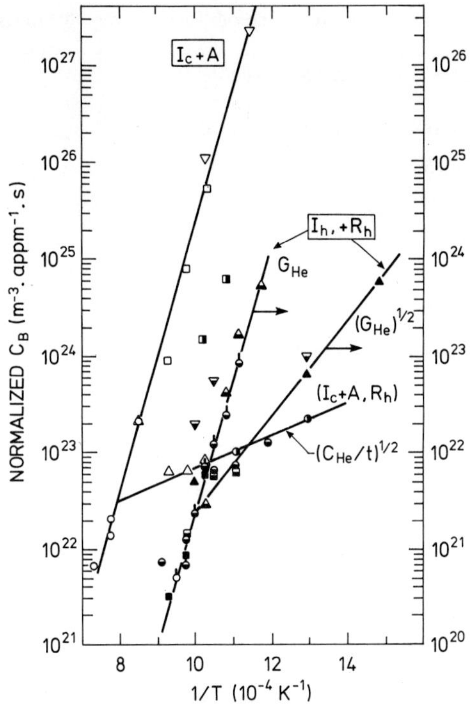

Fig. 5 Temperature dependence of C_B normalized against helium production rate of 1 appm s^{-1}. The C_B values are scaled by c_{He}/t (in the case of $(I_c+A(+R_h))$ and G_{He} (in the case of (I_h+R_h)) equal to 1 appm helium per second, using square root relations for the low temperature regime (eqs. (1) and (5)) and linear relations for the high temperature regime (eqs. (5) and (6)). Note that keys have the same meaning as in Table 1 and Figs. 3,4.

3.1. General trends

The temperature dependence of the measured bubble density (C_B) for austenitic stainless steels and for different experimental conditions has been plotted in Figs. 3,4. In the following, we shall first describe the general trends that emerge from the results plotted in Figs. 3, 4. In addition, when necessary some remarks will be made regarding special effects arising either from one of the experimental parameters or materials variable.

Most of the experimental results shown in Figs. 3, 4 exhibit a general trend that the temperature dependence of C_B has two well–defined regimes: the low temperature regime with a low apparent activation energy (E_B^{app}) and the high temperature regime with a high E_B^{app}.

The temperature at which the transition from the low E_B^{app} to the high E_B^{app} occurs (T_{tr}) is found to be higher at higher values of G_{He} and c_{He}. In the case of (I_c+A) type of experiments, T_{tr} seems to shift towards a lower temperature (Fig. 3) with increasing annealing time. A general tendency appears to be that the higher the C_B values, the higher is the T_{tr}.

For a similar value of c_{He}, T_{tr} is found to be higher in the case of (I_c+A) than (I_c+R_h), (I_h+R_h) or R_h type of experiments (Figs. 3, 4). A comparison of results in Fig. 4 shows that T_{tr} is clearly lower for the (I_h+R_h) than for the (I_c+A) experiments even though G_{He} and c_{He} in the former are considerably higher than in the latter.

Generally (i.e. with and without additional radiation damage, R_h), C_B increases with G_{He} and c_{He}. In the case of I_h experiments, however, the increase in C_B with c_{He} may be mostly due to an implicit dependence of C_B on G_{He}. C_B in the (I_c+A) type of experiments is found to decrease with annealing time. Furthermore, C_B $(I_c+A (+R_h))$ is higher than C_B $(I_h (+R_h))$ particularly at higher temperatures.

Table 1 List of experimental conditions for data presented in Figs. 3-5.

(A). Cold-Implantation (I_c) followed by Annealing (A):

Key	Fig.No.	Material	c_{He} (appm)	t (h)	Ref.
△	3	316 S.S.	32-41	1	21
○	3	"	25-35	1	22
□	3	"	100	185	23
▽	3	"	110	10,000	24

(B) Hot-Implantation (I_h):

Key	Fig.No.	Material	G_{He} (appm/s)	c_{He} (appm)	t_{imp} (h)	Ref.
▪	3	316 S.S.	$2.8\ 10^{-2}$	300	3	25
▲	3	304 S.S.	$1.3\ 10^{-2}$	195	4.17	26

(C) Cold-Implantation (I_c) followed by Hot-Irradiation (I_h):

Key	Fig.No.	Material	G_D (dpa s^{-1})	c_{He} (appm)	Dose (dpa)	Ref.
◑	3	316 S.S.	$5\ 10^{-7}$	115	8.4	24
◪	3	304 S.S.	$2\ 10^{-4}$	195	3	26
▲	3	316 S.S.	$1.3\ 10^{-3}$	10	40	27

(D) Hot-Implantation (I_h) and Hot-Irradiation (R_h):

Key	Fig.No.	Material	G_{He} (appm/s)	G_D (dpa/s)	c_{He} (appm)	Dose (dpa)	Ref.
◓	4	316 S.S.	1.09	$9.1\ 10^{-3}$	2196	18.3	28
▲	4	316 S.S.	$9.2\ 10^{-2}$	$9.1\ 10^{-4}$	1092	9.1	28
▽	4	304 S.S.	$1.3\ 10^{-2}$	$2\ 10^{-4}$	195	3	26
▣	4	316 S.S.	$11.8\ 10^{-2}$	$6\ 10^{-3}$	20	1	29
◒	4	316 S.S.	$11.8\ 10^{-2}$	$6\ 10^{-3}$	1400	70	29

E) Irradiation (R_h) with Neutrons:

Key	Fig.No.	Material	G_{He} (appm/s)	G_D (dpa/s)	c_{He} (appm)	Dose (dpa)	Ref.
x	4	316 S.S.	$4.7\ 10^{-5}$	$\sim 10^{-6}$	380-1020	9-18	24
+	4	316 S.S.	$5.9\ 10^{-7}$	$\sim 10^{-6}$	20-30	31-36	30

As regards the effect of irradiation on the bubble formation behaviour, the lack of data does not allow a proper evaluation. Even in cases where some specific effects may be attributed to radiation damage, no definitive conclusion can be drawn because of the complications arising due to the possibility of bubble–void transformation.

4. Discussion

As indicated in section 2, both annealing after cold–implantation and hot–implantation experiments are characterized by two distinctly different regimes: a gas or bubble diffusion controlled low temperature regime with a relatively low apparent activation energy for the bubble density, and a gas dissociation controlled high temperature regime with a high apparent activation energy. The bubble density is predicted to increase with c_{He}/t and G_{He}, respectively. The same trends, although somewhat weaker in strength, are predicted for the transition temperature. Qualitatively, this theoretical picture is fully confirmed by the data presented in section 3.

394

Theoretically, significant radiation damage effects on bubble formation are expected for low gas to dpa ratios where bubbles are transformed into voids or for extreme conditions such as high overpressure in bubbles. An apparent absence of radiation damage effects in cases where the observed cavities may be identified as bubbles confirms this limitation.

This qualitative agreement may be tested and quantified for cases where sufficient data are available (e.g. in stainless steel). For first estimates of the fundamental energies involved, we consider the transition temperatures as given by eqs. (3a) and (7a). For the investigated experimental parameter ranges, the logarithm in both equations is between 50 and 60. For transition temperatures around 1000 K, this yields estimates of about 5 eV for the energies in the nominators of eqs. (3a) and (7a). For $\mu^{*}_{He} \approx 0$ and diffusion energies around 1 eV this would be consistent with He dissociation energies from bubbles of about 3 eV. The difference in the transition temperatures of the two types of experiments may be attributed to differences in the chemical potential of He in bubble nuclei and well developed annealed bubbles.

In the following we make an attempt to correlate the data for stainless steels on the basis of the functional dependencies predicted in section 2. We exclude all the data for which the cavities are clearly in the void state and reactor data for which the He generation rate is not constant. All the remaining ion implantation data are scaled to standard conditions defined by c_{He}/t and G_{He} equal to 1 appm He/s, using square root relations for the low temperature regime according to eqs. (1) and (5), and linear relations for the high temperature regime according to eqs. (2) and (6), respectively.

The effect shown in Fig. 5 is quite striking. The clouds of data points, in particular for the high temperature regime, condense into a rather narrow bands. Note that the high temperature bands contain data points differing in c_{He}/t or G_{He}, respectively, by more than three orders of magnitude. In view of the limited amount of data, the less convincing effect in the low temperature regime is not surprising.

An important observation is that the band for the hot-implantation data (Fig. 5) covers data for gas to damage ratios ranging from 10 to 10^4 appm He/dpa. This illustrates the fact that for the relative high gas production to displacement rate considered, radiation damage effects are not noticeable which is consistent with our qualitative supposition.

Consequently, the remaining dependencies of C_B are those upon c_{He}/t and G_{He}, respectively, for the (I_c+A) and I_h conditions. Thus, the high temperature data may be summarized as

$$C_B t/c_{He} = 2\ 10^{17}\ m^{-3}\ \exp(2.8\ eV/kT) \text{ for } I_c+A+(R_h)$$

$$C_B/G_{He} = 10^{14}\ m^{-3}\ \exp(2.65\ eV/kT) \text{ for } I_h+(R_h)$$

The data for the low temperature regime, in particular for hot implantation, are not sufficient for such quantitative treatment.

A striking feature revealed by Fig. 5 is the large separation of the high temperature branches for the two types of experiments which amounts to about four orders of magnitude. To examine this, we consider the following theoretical ratio for the two cases according to eqs. (2) and (6)

$$(C_B^a\, t/c_{He})/C_B^h\, /G_{He}) = \frac{\pi r^{*} kT}{2 v_{He}^a \gamma}\ \exp\,[(\mu_{He}^h - \mu_{He}^a)/kT], \qquad (9)$$

where the superfixes a and h refer to I_c+A and I_h experiments, respectively. In the parameter range considered here, the factor in front of the exponential is about 1/4. Hence, the exponential containing the difference in the chemical potentials of He in bubble nuclei and well developed annealed bubbles would have to account for a factor of about $4\ 10^4$ corresponding to $(\mu^h_{He}-\mu^a_{He})/kT \approx 10.6$. This indicates that the pressures in bubble nuclei and annealed bubbles must be substantially different. Using the He equation of state given in ref. [6] and assuming that the pressure in annealed bubbles converges to about 3 GPa as determined by the SANS technique in Ni [32], we estimate pressures in bubble nuclei to be somewhat below 20 GPa in stainless steel which is conceivable. Hence, the difference in the high temperature branches of the two types of experiments may be attributed to differences in the state of He in bubble nuclei and visible bubbles.

To deduce standard He dissociation energies, the apparent activation energies for C_B^a and/or C_B^h must be corrected for the potential energy of He within the corresponding bubbles. We estimate values somewhat above 3 eV. Since this is close to the self-diffusion energy, it is conceivable that coarsening during annealing is controlled by vacancy instead of He dissociation [9]. In view of the relatively simple models used, the deduced absolute values of dissociation energies should, however, not be taken literally.

Even though the low temperature data are not sufficient for a quantitative analysis they provide some information on the underlying mechanisms. For I_c+A the apparent activation energy is about 0.35 eV which would be consistent with surface diffusion controlled bubble migration. The absolute value of the experimentally found bubble densities would, however, require extremely low pre–exponential factor for the surface diffusivity. On the other hand, the relatively high apparent activation energy of about 1 eV for I_h would require an extremely high pre–exponential factor of the underlying diffusivity. It could be that for both cases the activation energies are close to each other.

Some information is provided by the absolute values of the normalized bubble densities in the low temperature regimes. According to eqs. (1) and (5) the I_c+A values should be comparable to the I_h values if both were controlled by the same diffusion mechanism such as surface diffusion. In the investigated temperature range the I_c+A values seem, however, to be significantly above the I_h–values indicating that the latter is controlled by a faster He diffusion process. In fact, the I_h–values are consistent with He diffusion by the He/self–interstitial replacement mechanism for which the He diffusivity becomes equal to the vacancy diffusivity. More information is, however, required to confirm this supposition.

5.   Conclusions

The temperature dependence of the experimentally measured cavity (bubbles and voids) densities exhibit two distinctly different regimes for all experimental conditions considered: (a) a low temperature regime with a low apparent activation energy and (b) a high temperature regime with a high apparent activation energy.

This behaviour is found to be consistent with theoretical considerations in terms of (a) bubble migration and coalescence, and di-atomic type homogeneous nucleation and (b) thermal resolution and multi-atomic type homogeneous nucleation. The low temperature regime is found to be diffusion (gas or bubble) controlled whereas the high temperature regime appears to be gas-dissociation controlled.

A quantitative analysis of the experimental results suggests that the enormous (about four orders of magnitude) difference in C_B between the $(I_c$+A) and $(I_h,+R_h)$ experiments in the high temperature regime may arise from the difference in the state of helium in the bubble nuclei; in $(I_h,+R_h)$ experiments, the gas pressure in the bubble nuclei may be as high as 20 GPa in stainless steel. The lack of data in the low temperature regime makes it very difficult to reach a firm conclusion. However, the difference in C_B $(I_c$+A) and C_B $(I_h,+R_h)$ would suggest a fast diffusion (gas atoms or bubbles) mechanism for $(I_h,+R_h)$ experiments.

The helium generation rate (G_{He} or c_{He}/t) is found to be the most important parameter for bubble formation in the low as well as the high temperature regimes.

Further experimental and theoretical investigations with low G_{He} and c_{He} are needed to understand the nucleation and coarsening behaviour in the low temperature regime which is the most relevant regime from the technological application's point of view.

There is not enough data to allow a proper analysis of the specific effects of irradiation on bubble nucleation. However, the available results indicate that the radiation damage effects would be important for bubble nucleation and coarsening, in particular at lower temperatures, when the gas generation rate is low and the displacement rate is still sufficient for cavity nucleation.

REFERENCES

1. E.E. Gruber, J. Appl. Phys. **38**, 243 (1967); F.A. Nichols, J. Nucl. Mater. **30**, 143, (1969).

2. P.J. Goodhew and S.K. Tyler, Proc. Roy. Soc. London, **A377**, 151 (1981).

3. G.W. Greenwood and A. Boltax, J. Nucl. Mater. **5**, 234 (1962).

4. A.J. Marksworth, Met. Trans. **4**, 2651 (1973).

5. P. Fichtner, H. Schroeder and H. Trinkaus, this volume.

6. H. Trinkaus, Rad. Effects **78**, 189 (1983).

7. P.J. Goodhew, Scripta Met. **18**, 1069 (1984).

8. H. Schroeder, P. Fichtner and H. Trinkaus, this volume.

9. H. Trinkaus, Scripta Met. **23**, 1773 (1989).

10. G.W. Greenwood, A.J.E. Foreman and D.E. Rimmer, J. Nucl. Mater. **4**, 305 (1959).

11. B.N. Singh and A.J.E. Foreman, J. Nucl. Mater. **103 & 104**, 1469 (1981).

12. B.N. Singh and A.J.E. Foreman, J. Nucl. Mater. **122 & 123**, 537 (1984).

13. H. Trinkaus, J. Nucl. Mater. **118**, 39 (1983).

14. H. Trinkaus, J. Nucl. Mater. **133 & 134**, 105 (1985).

15. H. Trinkaus, Rad. Effects **101**, 91 (1986).

16. B.N. Singh and A.J.E. Foreman, in: *Consultant Symposium – The Physics of Irradiation Produced Voids* (ed. R.s. Nelson), A.E.R.E.–R7934, January 1975, p. 205; Scripta Met., **9**, 1135 (1975).

17. A.J.E. Foreman and B.N. Singh, J. Nucl. Mater. **141–143**, 672 (1986).

18. N.M. Ghoniem, S. Sharafat, J.M. Williams and L.K. Mansur, J. Nucl. Mater. **117**, 96 (1983).

19. N.M. Ghoniem, J. Nucl. Mater. **174**, 168 (1990).

20. L.K. Mansur and W.A. Coghlan, J. Nucl. Mater. **119**, 1 (1983).

21. R.E. Stoller, *Microstructural Evolution in Fast–Neutron Irradiated Austenitic Stainless Steel*, ORNL– 6430, December 1987.

22. F.A. Smidt, Jr. and A.G. Pieper, in: *Properties of Reactor Structural Alloys after Neutron or Particle Irradiation*, ASTM STP 570, ASTM, 1975, p. 352.

23. J. Rothaut, H. Schroeder and H. Ullmaier, Phil. Mag. **A47**, 781 (1983).

24. P.J. Maziasz, *Effects of Helium Content on Microstructural Development in Type 316 Stainless Steel under Neutron Irradiation*, ORNL–6121, November 1985.

25. P. Batfalsky and H. Schroeder, J. Nucl. Mater. **122 & 123**, 1475 (1984). H. Ullmaier, J. Nucl. Mater. **133 & 134**, 100 (1985).

26. W.J. Choyke, J.N. McGruer and J.R. Townsend, J. Nucl. Mater. **85 & 86**, 647 (1979).

27. J.A. Hudson, J. Nucl. Mater. **60**, 89 (1976).

28. J. Tenbrink, R.P. Wahi and H. Wollenberger, in: *Effects of Radiation on Materials*, ASTM STP 1046, 1989, p. 543.

29. K. Farrell and N.H. Packan, J. Nucl. Mater., **85 & 86**, 683 (1979).

30. D.F. Pedraza and P.J. Maziasz, in: *Effects of Radiation on Materials.* ASTM STP 955, 1987, p.161.

31. A.J.E. Foreman and B.N. Singh, in: *Irradiation Behaviour of Metallic Materials for Fast Reactor Core Components* (ed. J. Poirier and J.M. Dupouy), CEA–DMECN, Gif–sur–Yvette, 1979, p. 113.

32. H. Ullmaier, 1990, this volume.

INERT GASES IN NUCLEAR FUELS

FUNDAMENTAL ASPECTS OF INERT GAS BEHAVIOUR IN NUCLEAR FUELS: OXIDES, CARBIDES AND NITRIDES

Hj. Matzke

Commission of the European Communities
Joint Research Centre
European Institute for Transuranium Elements
Postfach 2340, D–7500 Karlsruhe
Federal Republic of Germany

ABSTRACT

Fundamental aspects (lattice location, interaction with lattice defects, diffusion, precipitation into bubbles, radiation induced re-solution, thermal solubility, grain boundary bubbles, their venting etc.) of the behaviour of inert gases in ceramic compounds of uranium (oxides, carbides and nitrides) are discussed. Emphasis is placed on heavy rare gases (Kr,Xe) and on UO_2 for which most data and theoretical calculations exist. Both fission and ion implantation are used to produce or incorporate the gases in the ceramics. The experimental techniques for studying inert gas behaviour include release measurements, electron microscopy, damage and lattice location studies by Rutherford-backscattering/channeling, electron microprobe analysis etc. The resultant picture is rather complete for UO_2, but important gaps still remain for carbides and nitrides.

1.   Introduction

The behaviour of the rare gases Kr and Xe formed as fission products in nuclear reactor fuels has been studied from the beginning of the commercial utilization of nuclear power. These inert gases have a large fission yield (≈ 25 %), and therefore large amounts of Kr and, in particular, Xe are created within the fuels. The quantities formed can easily correspond to some cm^3 (NPT)/cm^3 fuel. The early interest in the behaviour of the rare gases was due to the fact that metallic fuels used in the early days swell to a large extent due to gas diffusion, formation and venting of gas-filled bubbles, as well as to concerns that the pressure exerted by released gas might cause failure of the metal clad containing the fuel. In addition, the risk of health hazards due to possible escape of (radioactive) rare gases into the reactor building or into the environment was studied. Finally, one particular rare gas isotope formed in fission, i.e. Xe^{135}, can lead to a temporary "reactor poisoning" because of its large cross section for neutron capture.

As a consequence, both technological and basic aspects of inert gas behavior were studied all over the world, with most emphasis being placed on UO_2. The degree of knowledge achieved was already fairly advanced in the 1970's for this particular nuclear fuel. The present author summarized this knowledge in a critical review in 1980 [1]. Work on the "advanced fuels", i.e. carbides and nitrides continued on a rather large scale until the middle of the 1980s, and the relevant knowledge was described in a monograph [2].

More recently, the question of fission gas behaviour during operational transients is being studied [3]. These transients may be due to temperature or power rises. Also interest is extending to other gaseous and volatile fission products or precursors in radioactive decay chains of fission gases, i.e. to Rb, Cs, I, Te etc. However, basic work on inert gas behavior was, and is, continuously pursued as well. In addition, progress in modelling fission gas behavior with computer codes was achieved [3, 4] and calculations became available concerning aspects of the behavior of inert gases in UO_2 using atomistic simulation studies (e.g. [5], and see also contribution of R.W. Grimes to these proceedings [6]).

In addition to the investigation of reactor irradiated nuclear fuels which are very radioactive and have to be handled in shielded equipment or in hot cells, ion implantation was frequently used to introduce the inert gases into all classes of fuels treated here: oxides, carbides and nitrides. In combination with Rutherford backscattering / channeling methods, this work has led to a deep knowledge of the implantation processes and of defect formation and recovery in these binary compounds consisting of a very heavy (U) and a very light component (O, C or N).

The main points of these more recent studies will be summarized and a review of the previous knowledge will be given, following a short description of the relevant properties of the nuclear fuels treated and of their operating conditions in nuclear reactors. This description is included since the present paper was asked to serve as an introduction to a series of more topical papers dealing with nuclear fuels.

2. Ceramic Nuclear Fuels and their Operating Conditions

This section describes some properties of the ceramics treated here, i.e. oxides, carbides and nitrides of U (and/or Pu), explains typical operating conditions and recalls relevant properties of the fission process.

Today's electricity generating nuclear power stations use almost exclusively uranium dioxide, UO_2, as fuel. In breeder reactors operated with fast neutrons, a mixed oxide $(U_{1-x}Pu_x)O_2$ is used (x $\approx$ 0.2). The carbides and nitrides UC and UN (or (U, Pu)C and (U, Pu)N) were and are being developed as second-generation fast breeder "advanced fuels". The most recent fast reactor, the Indian FBTR, uses a Pu-rich monocarbide as fuel, and the planned multinational European Fast Reactor, EFR, considers (U,Pu)N as a second generation fuel. UC was also used in organic-cooled reactors.

All three classes of materials are ceramics with high melting points (T_m = 2880 °C for UO_2, 2525 °C for UC, 2830 °C for UN). All the materials are nonstoichiometric at high temperatures: UO_2, in fact, is the oxide with the largest degree of nonstoichiometry known. The single phase field $UO_{2\pm x}$ extends, at $\approx$ 2500°C, from $UO_{1.65}$ to $UO_{2.25}$ [7]. The structure is the fluorite structure, the majority defects are oxygen vacancies in UO_{2-x} and (clusters of) oxygen interstitials in UO_{2+x} (see e.g. [7]). For UC, the single-phase field of the NaCl-structured $UC_{1\pm x}$ extends from about $UC_{0.92}$ to about $UC_{1.9}$ at about 2300°C, carbon vacancies being the dominant defect in UC_{1-x} and C-interstitials forming C_2-pairs with normal lattice C-atoms being the majority defect in UC_{1+x}. For the nitride UN, the single phase field (again NaCl-structure) is narrow if expressed in terms of N/U-ratio, but very wide if expressed in terms of nitrogen partial pressures. For instance, at 1400°C, the $UN_{1\pm x}$-phase field extends from $p(N_2)$ $\approx$ 1 bar (decomposition pressure of mononitride in equilibrium with the sesquinitride U_2N_3) to $\approx 10^{-9}$ bar (decomposition pressure of the mononitride in equilibrium with liquid U) (e.g. (2)). The changes in defect structure caused by deviations from stoichiometry can, of course, have consequences on lattice location and diffusion mechanism of the inert gases, as is, e.g., known for $UO_{2\pm x}$ (see below).

The bonding conditions vary widely between the three materials treated here: from largely ionic (UO_2) through predominantly covalent (UC) to largely metallic (UN).

For reactor use, the materials are fabricated as sintered pellets of densities of typically 95 % of theoretical for UO_2 and, for fast breeder application, even lower densities ($\approx$ 85 % of theor.)

for the advanced fuels. The pellets vary in diameter from about 6 to 13 mm, and are stacked upon one-another in thin metal sheaths, also called clad. The clad is a Zr-alloy (Zircaloy) for water-cooled reactors and stainless steel for liquid metal-cooled fast reactors. The fuel rods formed in this way are filled with He (pressure $\approx$ 30 bar), welded and assembled into fuel elements which may be up to 4.5 m long.

The physical process on which nuclear electricity production is based is the fission of U or Pu, the slowing down of the two fission products formed and transfer of the heat thus produced into conventional generators. Fission leads to the formation of a light fission product, lfp, and a heavy fission product, hfp. The median lfp has an atomic mass A of 96, an atomic number Z of 42 (Mo), an original energy of 95 MeV, an original velocity of $1.4\ 10^9$ cms^{-1} and a most probable charge state of 21. The corresponding numbers for the median hfp are A = 137, Z = 56 (Ba), E_0 = 67 MeV, v_0 = $1\ 10^9$ cms^{-1} and a most probable charge state of 24. The group of lfp's comprises some Se, Br, Kr, Rb, Sr, Nb, Y, Ag, Cd with concentrations between 0.2 to 3.9 atoms per 100 fissions, and mainly Tc, Rh, Ru, X, Pd, Zr and Mo with between 5 and 22 atoms per 100 fissions. The corresponding groups for hfp's are Sb, Te, I, Pm, Sm, Eu, Gd with small and Pr, Ce, Nd, La, Ba, Cs and Xe with large fission yields. For instance, for fission of Pu-239, the most abundant f.p. is Xe with 22.8 Xe-atoms per 100 fission events being formed (though not directly but rather by β-decays via so-called precursors).

In contrast to ion implanted specimens as treated in most of the other contributions to these proceedings, we therefore deal with materials constantly changing their impurity concentrations by growing in some 30 different elements at rates of about 50 to 500 ppm per day (normalized to U-atoms, and depending on the type of the reactor) or about 25 to 250 ppm (weight). Their concentration will therefore surpass that of the pre-existing impurities (typically about 100 ppm in single crystals, some 10^2 ppm in laboratory-grade samples and 1000 – 2000 ppm in technological grade samples, including non-metallic impurities such as N and O in a carbide) already after a few days of irradiation.

At the same time, the high fission density creates high fuel temperatures. This leads to pronounced temperature gradients (see Fig. 1a), in particular for oxide fuels which have low thermal conductivities. These gradients can amount to 10^4 K/cm and provide therefore a significant driving force for matter transport, both for thermal diffusion of atoms, for pore migration by an evaporation-condensation mechanism and also for mobility of inert gas bubbles. It is well known and mechanistically understood that (and why) U and/or Pu atoms as well as oxygen interstitials and vacancies migrate up the thermal gradient, and that (and why) pores migrate as well up the thermal gradient, actually forming a cylindrical "central void" in fast breeder fuels, as indicated in Fig. 1a. Migrating pores can sweep inert gas atoms, clusters or bubbles, as can moving grain boundaries, e.g. during grain growth which can occur in the central part of the fuel because of the high fuel temperatures. Since both oxygen interstitials (in UO_{2+x}) and oxygen vacancies (in UO_{2-x}) migrate up the thermal gradient [8], a radial gradient in composition exists affecting inert gas atom location, diffusion, trapping and bubble behavior not only because of different temperatures along the pellet radius, but frequently also because of the chemical gradient in oxygen content. The starting composition of UO_2 for water reactors is stoichiometric, for fast breeders it is substoichiometric, typically $(U,Pu)O_{1.97}$. Since the fission process is oxidizing, the O/M-ratio increases during operation. Though two fission products are formed per fissioned U or Pu atom, the f.p.'s use up slightly less than two oxygen atoms since most of them are rare gases or noble metals.

Carbides are chosen to be hyperstoichiometric at the beginning. Since both carbides and nitrides have a high thermal conductivity, central temperatures are generally lower than in oxides and temperature gradients are much smaller as well, thus the above processes are less significant in these advanced fuels.

The high local temperatures in the centre of the oxide fuels cause insoluble impurities to precipitate. The precipitates most frequently found are fission gas bubbles and spherical

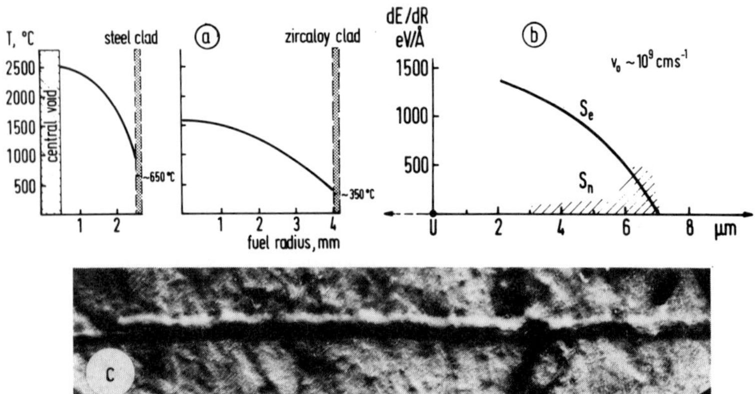

Fig. 1 a) Typical radial temperature profiles in oxide fuel elements;
b) Electronic stopping power, S_e, and nuclear stopping power, S_n, for the median heavy fission fragment, hfp, in UO_2;
c) A fission track at the surface of UO_2 seen in a replica [12]. (Total length: 1.5 μm).

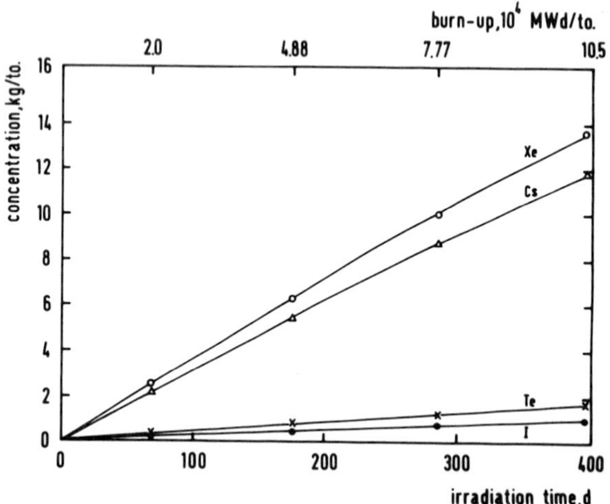

Fig. 2 Ingrowth of volatile fission products in the fuel of a fast reactor as a function of time, showing that predominantly Xe (and Cs) are formed as gaseous or volatile fission products.

precipitates of insoluble metals, rich in noble metals. These are often called "five metal particles" since they contain Ru, Rh, Pd, Tc and Mo. Ceramic precipitates are also found, containing Ba, Zr and Sr and being of the perovskite-type. Details concerning these precipitates are given in two subsequent contributions to these proceedings [9, 10]. Since the "real" fuels are very radioactive, sophisticated techniques with shielded equipment and the use of hot cells are necessary for their study. To avoid these difficulties, chemically "simulated high burnup" fuel, so-called SIMFUEL is increasingly used [11] to replicate the chemical state and microstructure of irradiated fuel.

In addition to precipitation of insoluble f.p.'s, there is constant re-solution by interaction of the fuel lattice with the slowing-down process of f.p.'s causing fission spikes or tracks to be formed. Fig. 1b shows the large energy density deposited along the track of the median hfp (hence a Ba-ion of 67 MeV energy) in UO_2. The spike has a length of ≈ 7 μm, it is characterized by a very high electronic stopping at its beginning, and a very high nuclear stopping peaking at its end. About 10^5 defects are produced in each fission event. The spike can be seen at surfaces [12], e.g. in replica electron microscopy (Fig. 1c). Stereo microscopy shows that the width of the permanently disturbed zone is $\approx$ 7 nm and the height of the spike (at surfaces as in Fig. 1c) is $\approx$ 20 nm. Significant temperature increases along the axis occur causing large hydrostatic pressure gradients [13], leading e.g. to a separation of vacancies from interstitials and hence a largely temperature-independent, athermal radiation-enhanced diffusion as well as e.g. re-solution of inert gas atoms from bubbles or even complete destruction of bubble nuclei or small bubbles. Each atom is affected by a fission spike at a rate of once in a few hours to once in a day, depending on the reactor type. Fission spike effects are less important in the carbides and nitrides than in the oxides because of their larger electrical and thermal conductivities [1] .

At low fuel temperatures, hence at the edge of the fuel pellets, thermal diffusion and thermally activated precipitation are largely absent. The state of the material resembles to some extent that of high-dose ion-implanted specimens.

An obvious difficulty in understanding inert gas bubble behavior in ceramic nuclear fuels, in addition to the constantly acting re-solution, is the fact that other volatile f.p. elements are formed which tend to precipitate as well. Fig. 2 shows the ingrowth of volatile elements with time in a fast breeder fuel, and Fig. 3 demonstrates that Rb and I tend to precipitate into bubbles as well. The TEM pictures in Fig. 3 a,b show UO_2 ion implanted with only rubidium and iodine; in Fig. 3c, five metal particles are shown in a reactor irradiated fuel to be located at inert gas bubbles.

A final difference between producing inert gases in nuclear fuels and labelling specimens by ion implantation consists in the fact that the inert gases Kr and Xe are practically not formed directly by fission. Rather, precursors are formed, as shown in Fig. 4 for mass 133 for fission of U and Pu. Xe-133 is a frequently used tracer, but it originates by β-decays from the precursors Sb, Te and I-133. Many of such precursors have a half-life long enough to allow diffusional mobility of the precursor; thus the gas to be studied may be located at a site some distance away from where it was formed. Also, the lattice location may be affected.

The above rather long general description of the particular conditions of fission inert gases is meant to improve understanding of the following sections, as well as of the following papers dealing with topical problems of inert gases in nuclear fuels.

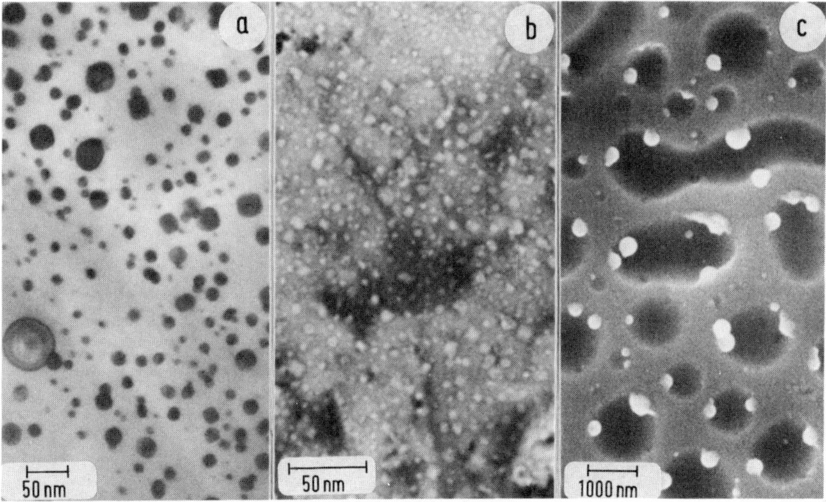

Fig. 3 Transmission electron micrographs of a) Rb-ion implanted UO_2 annealed to 1500°C (5 10^{15} Rb-ions/cm^2, 45 keV); b) Iodine-implanted UO_2 annealed to 900°C (10^{16} I-ions/cm^2, 40 keV); c) Scanning electron micrograph of reactor-irradiated UO_2 showing 5-metal particles at (grain boundary) inert gas bubbles.

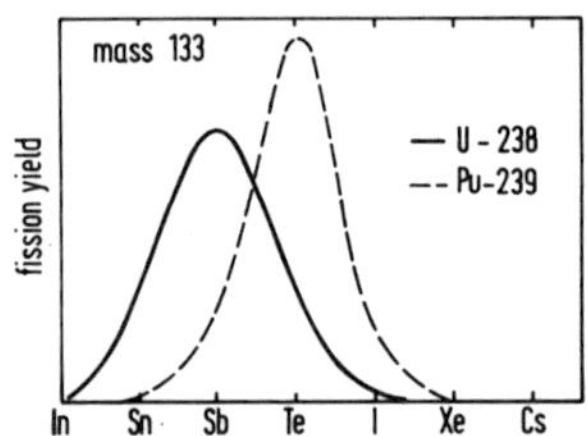

Fig. 4   Yields for fast fission of U-238 and Pu-239 for the mass chain 133 leading to the only stable Cs-isotope, Cs-133.

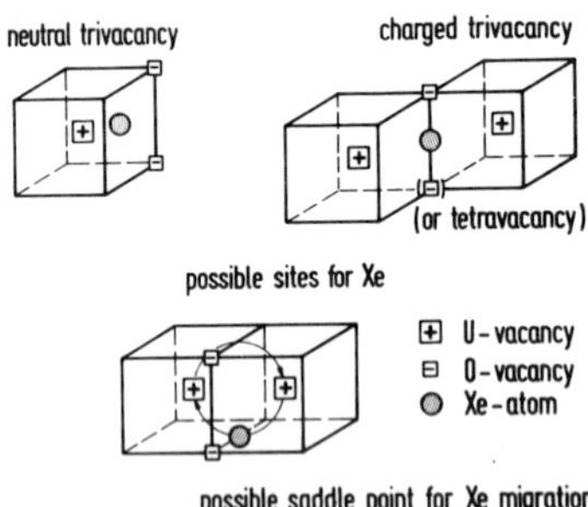

Fig. 5   Suggested sites of Xe in UO_2 and a possible saddle point for Xe diffusion.

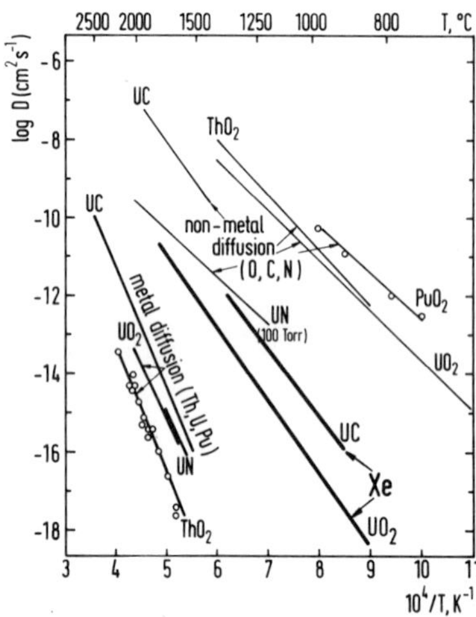

Fig. 6   Schematic Arrhenius diagram showing non-metal atom and metal atom diffusion in nuclear oxides, carbides and nitrides. Experimental points are excluded for clarity except for the lowest (metal atom, U and Th, diffusion in ThO_2) and the fastest diffusion (O in PuO_2). The thick lines show the recommended temperature dependences for diffusion of single Xe atoms in UO_2 and in UC.

3.   Diffusion of Single Gas Atoms

At low damage and gas concentration, unperturbed diffusion of single fission inert gas atoms (Kr, Xe) can be observed. For UO_2, largely agreeing data exist for 3 different ways of introducing the gas (Xe): controlled ion implantation [14], fission [15] and labelling with fission recoils [16]. The recommended Arrhenius relation [1] is

406

$$D_{UO_2}^{Xe} = 0.5 \exp\left(-375\,000/RT\right) \text{ cm}^2\,\text{s}^{-1} \tag{1}$$

hence an activation enthalpy of 375 kJ/mol, or $\approx$ 90 kcal/mol, or 3.9 eV/atom. Doping the largely ionic UO_2 with heterovalent impurities leads to large changes in the concentration of either oxygen or uranium vacancies. Early experiments [14] showed that such doping does not affect diffusion of single Xe-atoms thus excluding single O- or U-vacancies as lattice sites. This conclusion was confirmed by the absence of a blocking effect of <100> atomic rows on the α-particle emission of the heaviest rare gas Rn or Em which also excludes the centre of the large interstitial $(\frac{1}{2},\frac{1}{2},\frac{1}{2})$ sites of the fluorite structure as equilibrium site for lattice location [17]. The most probable alternative for lattice location of Xe-atoms are trivacancies, or Schottky trios, hence molecular vacancies in the UO_2 structure [1, 14, 17]. This suggestion was later confirmed by calculational work by Catlow, Grimes *et al.* [5, 6]. This site, i.e. the neutral cluster of a U vacancy and two O vacancies, the alternative of a tetravacancy and the saddle point of a U-vacancy-assisted diffusion mechanism are shown in Fig. 5.

Fig. 6 shows the Arrhenius line of the data for Xe-release from low dose ion-implanted UO_2 [14] which were confirmed on many specimens, including those containing heterovalent impurities and other ones which were ion-implanted at 400 or 1000°C [18]. These latter temperatures were chosen as being above the recovery stages of O-defects or U-defects, respectively. The same sets of data were also obtained for Xe and the isostructural ThO_2 [19]. Fig. 6 shows also that in all three classes of nuclear ceramics, the nonmetal atoms (O, C, N) are fast diffusors whereas the metal atoms (U, Pu and also Th in ThO_2) are diffusing at very low rates. Metal atom diffusion is thus rate-controlling for bulk matter transport in all 3 classes of materials, hence for e.g. grain growth and creep, but also sintering and inert gas bubble mobility.

Fig. 6 includes also diffusion of Xe at low concentration in UC, considered to correspond to unperturbed diffusion of single gas atoms, and described by the relation

$$D_{UC}^{Xe} = 0.3 \exp\left(-350\,000/RT\right) \text{ cm}^2\,\text{s}^{-1} \tag{2}$$

(ΔH = 84.5 kcal/mol, or 3.65 eV/atom). As with UO_2, the scatter is $\pm$ 5 kcal/mol or $\pm$ 0.2 eV. Fig. 7 shows the reliable available data with a band indicating the scatter and labelled with D (see below). Both reactor-irradiated UC [20–23] and ion-implanted UC were used [24]. Similar arguments, however based on less experimental evidence than for UO_2, lead to the suggestion of a vacancy cluster as site for Xe, e.g. of the type XeV_4 where V are (U and C) vacancies. The situation for UN is unsatisfactory. Data from 4 different sources (see ref. [2]) show significant scatter. By analogy with the isostructural UC, a similar mechanism with slightly lower D-values (in agreement with the higher melting point of UN) may be suggested. Atomic mobilities are generally slower in the nitride than in the carbide, as shown in Fig. 8 for carbon (nitrogen) and metal atom (U and Pu) diffusion, for grain growth and for fission gas release from operating fuel elements in the sequence MC-M(C,N)-MN where M = U and/or Pu.

A question of both basic and technological interest is that of the diffusion rate and the diffusion mechanism of Xe in hyperstoichiometric UO_{2+x}. UO_{2+x} has a much larger concentration of U-vacancies than UO_2 and all mass-transport processes are greatly enhanced [8]. The calculations of Catlow, Grimes *et al.* [5, 6] show the probability of both diffusion rates and mechanisms being different from those in UO_2 and earlier measurements showed largely enhanced release of Xe from UO_{2+x} as compared with UO_2. These results are critically reviewed in ref. [1] pointing out a number of weak points in the early experiments. We have

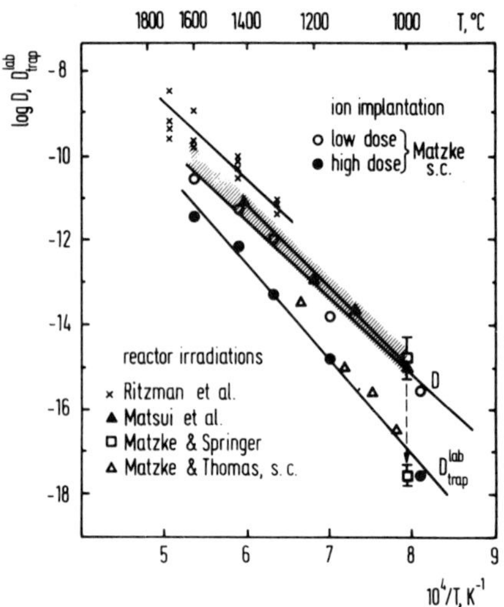

Fig. 7 Arrhenius diagram for the diffusion of Xe in reactor irradiated [20-23] or ion bombarded UC [24]. The shaded band shows the recommended values [2] for single gas atom diffusion. At high gas concentrations and high damage levels, trapping reduces D to D_{trap}^{lab}.

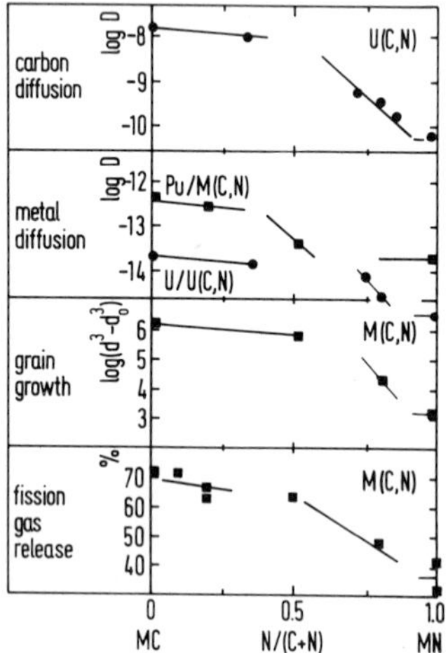

Fig. 8 Typical trend of decreasing atomic mobilities in the sequence carbide-carbonitrides-nitride, shown for C (and N) diffusion and U and Pu diffusion (D at 1700°C in cm^{2}s^{-1}), for grain growth at 1700°C and for fission gas release from a fast irradiation at 1.2 at% burn-up (for refs. see [2]).

planned experiments at controlled x-values but these have yet to be performed. Recent ion-implantation work of the author has clearly shown that trapping (see below) is very much reduced in UO_{2+x}, but this work was so far not extended to very low gas concentrations. Such crucial experiments are planned for the near future.

4. Trapping of Inert Gas Atoms

As in most other structurally stable substances, trapping of gas atoms occurs in the ceramic nuclear fuels, trapping sites being radiation damage in the form of larger vacancy clusters, stacking fault tetrahedra, the strain field of dislocations etc., but also pre-existing sintering pores and, at high gas concentrations, other gas atoms. Such gas-gas encounters lead to the formation of nuclei for gas bubbles, and, as final step, to gas-filled intra- and intergranular bubbles. Other volatile fission products such as Rb, Cs, Te and I can also form bubbles or also liquid-filled precipitates (see Fig. 3). It is not known at present whether or not bubble nuclei become stabilized by the presence of these volatiles. The trapping retards release (in time in isothermal experiment and in temperature in isochronal experiments) and thus reduces the effective diffusion coefficients. This is shown by an arrow at $T = 1000$ °C for UC in Fig. 7. It was also observed in UN [2], and extensively studied in UO_2 and ThO_2 [e.g. 14, 19, 25, 26]. Fig. 9 shows the diffusion coefficent D for Xe to be constant up to a gas concentration of $\approx 10^{-6}$ at%, or a fission dose of about 10^{16} fissions/cm^3. A dramatic decrease in D caused by trapping sets in at higher gas and/or damage concentrations. The early data (circles and squares) indicated a saturation in this decrease, corresponding to a low D measured in the laboratory, hence in the absence of radiation induced re-solution, called D_{trap}^{lab} (see also Table 1). Later ion implantation work at intermediate dose levels [26] indicated that this picture may have been too simple. The much more complicated behavior shown by the dashed curve can be interpreted as being due to a complex interplay of i) formation of gas-filled bubbles of different pressure and ii) gas brought into re-solution or being thermally dissolved in the lattice whenever the pressure approaches high values.

For the oxides, trapping is related to metal atom diffusion (D^M). It is more pronounced in substoichiometric UO_{2-x}, e.g. produced by doping with Y_2O_3 (reducing D^M) and it is less pronounced (or largely absent) in UO_{2+x} or UO_2 doped with Nb_2O_5 (with largely increased D^M) [14]. D_{trap}^{lab} is thus affected by small amounts of doping additives, in contrast to D. Since trapping sets in at rather low irradiation levels (gas concentrations), misinterpretations of data have frequently been made when the radiation doses and the related diffusion phenomena were not correctly defined [e.g. 27]. Trapping increases the activation enthalpy ΔH of inert gas diffusion by about 0.5 to 0.8 eV, hence a value of about 4.4 eV is reached for UO_2 and UC.

Due to the interaction with fission spikes and due to radiation-induced re-solution of gas from traps, the trapping is less effective during irradiation (where, for UO_2, the O/U-ratio may increase as well, adding to the reduced trapping). The effective diffusion coefficient under irradiation (see Table 1) is thus larger than that observed in the laboratory. The pressure effects caused by the thermal spike cause metal interstitials to move away from the spike axis thus explaining the observed rather large athermal radiation (fission) induced U and Pu diffusion in all three fuels [2, 28]. The effect is largest for the oxides because of their low thermal conductivity, but it is also observed in carbides and nitrides. Fig. 10 shows the suggested temperature dependence of D_{eff} for Xe in UC, including also the radiation-enhanced diffusion of U and Pu, the thermally activated Xe diffusion at low dose (D) and at high

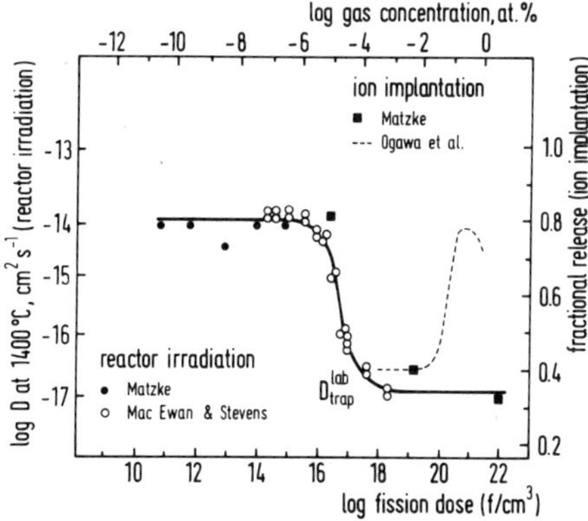

Fig. 9   Trapping of Xe in reactor irradiated (full and open circles) and ion-implanted UO_2 (squares) and ion implanted ThO_2 (dashed curve) [14, 25, 26].

damage levels without re-solution, hence D_{trap}^{lab}. A lower limit of low temperature diffusion for the inert gases is given by the atomic mixing by fission spikes. Whether or not low temperature mobilities in the pressure gradient of the spikes contribute to radiation-enhanced gas diffusion is unknown so far. Since iodine and, in particular, Te diffuse faster than Xe, precursor diffusion is certainly of importance for a number of the inert gas nuclides formed by fission.

As mentioned in Section 3 and above, trapping of Xe in UO_{2+x} is much less pronounced than in UO_2, related to the greatly enhanced U-diffusion rates [8]; it may even be absent. Electron microscopy to investigate in detail the differences of damage behavior and of gas-damage interactions in UO_{2+x} have yet to be done.

Table 1   Definition of diffusion coefficients

D	for intrinsic diffusion of single rare gas atoms in the undamaged lattice; possibly affected by deviations from stoichiometry and by accumulation of dissolved fission products.
D_{trap}^{lab}	for diffusion in the presence of traps, but in the absence of fission (irradiation). D_{trap}^{lab} = f (nvt) between about 10^{16} and 10^{20} nvt. At higher dose (nvt), saturation is indicated with a very low diffusion coefficient.
$D_{trap}^{in-pile}$	for diffusion in the presence of both traps and irradiation, describing gas mobility between fission gas bubbles.
D_{eff}	$D_{trap}^{in-pile}$ x b/(g+b), describing effective mobility including precipitation in bubbles; b is the re-solution probability, g the capture probability.
D_{bubble}	for diffusion of the bubble itself, describing gas transport in the absence of re-solution. Proceeds via surface or volume diffusion of matrix atoms (the metal atoms (U,Pu) being rate controlling) or via evaporation-condensation.

5. Bubble Formation and Mobility

Bubble formation has been observed in all three fuels, both following reactor irradiation (see the two following papers by Thomas [10] and Ray, Thiele and Matzke [9] on UO_2). For UO_2, it has also extensively been studied following ion implantation [see e.g. Fig. 3]. *Fission induced re-solution* of inert gases from bubbles is a well documented phenomenon which has been part of physically based fuel modelling codes for long. Re-solution of rare gas bubbles in UO_2 by ion implantation was first shown during this meeting [29]; previously, Fe precipitates in UO_2 had been shown to be destroyed by Xe-implantation. Whether or not *thermal re-solution* contributes to inert gas release is still an open question. The thermal solubility of Xe in UO_2, UC and UN is certainly very low. A certain solubility of Ar in UO_2 was recently proven to exist in unpublished work of the author. Grain boundary bubbles tend to grow larger than intragranular bubbles and are less subject to fission-induced re-solution. Their venting to the gap between fuel and clad is often rate-determining for final release. This release is thus due to gas diffusion perturbed by a multiple precipitation re-solution process, formation of grain-boundary bubbles, formation of interconnected tunnels, e.g. at triple points of grains of the fuel, and venting of these bubbles. Venting can also occur by grain boundary fracture due to thermal stresses during temperature changes, e.g. during start-up or shut-down of the reactor. A good example of such "bursts" is shown in Fig. 11 for UO_2 containing grain boundary cavities filled with high pressure Ar. At high resolution (T-rise of 3 Ks^{-1}), the "peak" in this type of thermal desorption spectrum seen at fast anneal rates (30 Ks^{-1}) is seen to be resolved in a larger number of individual instantaneous "burst" events.

410

The bubbles in irradiated fuel probably contain not only inert gases but also other volatiles of low solubility (Rb, Cs, Te, I). As shown in Fig. 3, these volatiles can also form bubble-like features in the absence of inert gases in ion-implanted specimens. A direct quantitative analysis of bubbles in reactor-irradiated fuels has only been done for certain UO_2-fuels, without any evidence of the presence of volatiles [10].

Most observed bubbles in UO_2 are located at dislocation lines, and, in irradiated UO_2, invariably "5-metal particles" are attached to both grain boundary and in-grain bubbles (see e.g. Fig. 3c). This interaction reduces bubble mobilities. Bubble diffusion has therefore been seen for individual bubbles only in irradiated UO_2 [1]. In contrast, bubble diffusion showing a clear indication of surface diffusion-controlled bubble mobility has been reported for He-implanted UC and UN (see [2]).

6. Recent Results on Inert Gas Implantated UO_2 and UN. Is there any Indication of Solid Inert Gas Precipitates in UO_2?

Ion implantation results on ceramic nuclear fuels have been mentioned frequently in the previous sections. Recent work applying Rutherford backscattering and the channeling effect technique have provided deeper insight in the physics of ion implantation in these ceramics consisting of a very heavy (U) and a very light component, (O, C, or N). Fig. 12 shows the (calculated) profiles of 40 keV Xe-ions (range profile) and of the damage produced (damage profile) in UO_2. Most of the implanted ions leave their damage behind. Incongruent sputtering, i.e. preferential loss of the light component leads to a substoichiometric surface layer, also known as "altered surface layer" at high doses. Rutherford-backscattering and channeling measurements of inert gas implanted UN [32] and UO_2 [33] single crystals have recently yielded the number of displaced U-atoms per incident ion: most U-defects annihilate in the collision cascade. Even at implantation at 5 K in UO_2, only about 20 U-defects per 40 keV Xe-ion (total dose 1 10^{15} ions/cm^2) survive instantaneous recombination, whereas about 300 defects/ion are formed. In UN (largely metallic bonding) but not in UO_2 (largely ionic bonding), long range migration of U-interstitials is seen below room temperature. The annealing stages of the defects have also been measured. Fig. 13 shows the two main stages for Xe-implanted UO_2: at 77 to 110 K, U interstitials are thought to move, whereas U vacancies become mobile at and above 600 K. The degree of understanding the physical processes of ion implantation in the nuclear ceramics is thus rather advanced.

It is thus obvious, for the present workshop, to ask for the possibility, or even evidence, of the formation of solid inert gas precipitates in these ceramics. Solid rare gas precipitates can

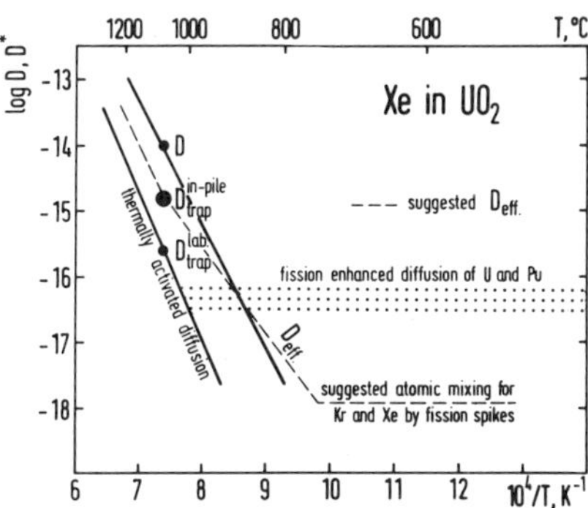

Fig. 10 Schematic presentation of the possible temperature dependence of the effective in-pile gas diffusion coefficient, D_{eff} for Xe in UC, as a function of temperature; the figure shows also radiation-enhanced metal atom diffusion, the upper and lower bound (D and D_{trap}^{lab} for Xe diffusion) and the athermal contribution of atomic mixing for Kr and Xe by fission spikes.

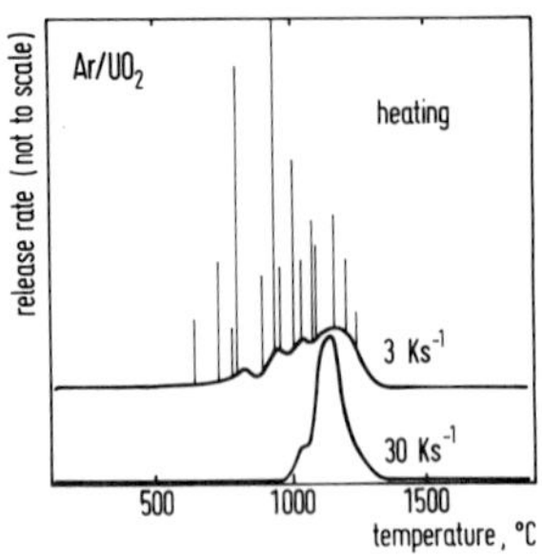

Fig. 11   Release spectra of Ar-containing UO_2 for two heating rates [30].

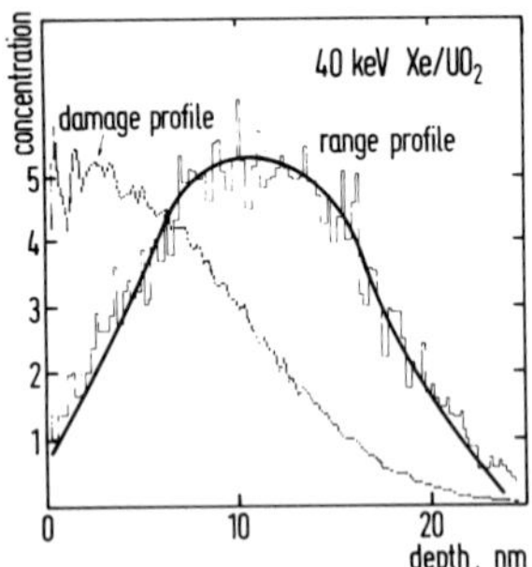

Fig. 12   Damage and range profile of 40 keV Xe-ions in UO_2, calculated with the TRIM-code [31].

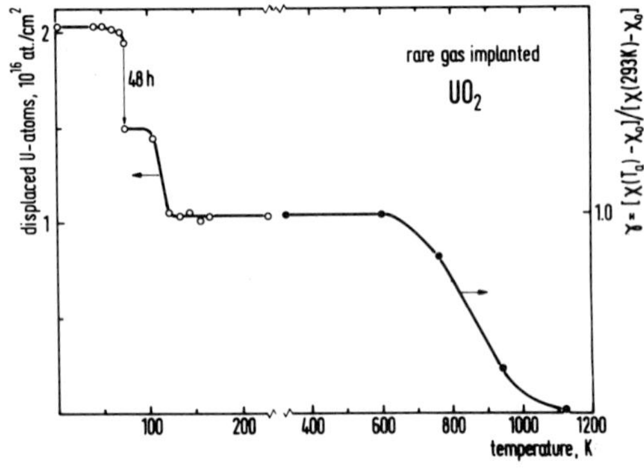

Fig. 13   Damage recovery of inert gas implanted UO_2 (10^{15} Xe-ions/cm^2) as a function of temperature, expressed as number of displaced U-atoms for low temperatures (T ≤ 293 K) and as dechanneling parameter γ for higher temperature [33].

be formed at or above room temperatures in a number of metals, as reported in many contributions to these proceedings. They have never been reported for any ceramic so far. Since there is significant defect recombination during implantation in UO_2, the implanted inert gas might be thought to athermally form precipitates. Indications of "inert gas bubbles" have indeed been observed in as-implanted UO_2 in high resolution transmission electron microscopy [34] (see Fig. 14). Electron diffraction gave no evidence of solid Xe-precipitates, however. The Rutherford backscattering (RBS)-channeling work mentioned above has provided further interesting evidence [35]: Fig. 15 shows RBS spectra (aligned and random) of a UO_2 single crystal implanted with Xe-ions at 77 K. A damage peak (displaced surviving U-atoms) and a Xe-impurity peak are seen. In all evaluated spectra so far, the Xe-peaks were

412

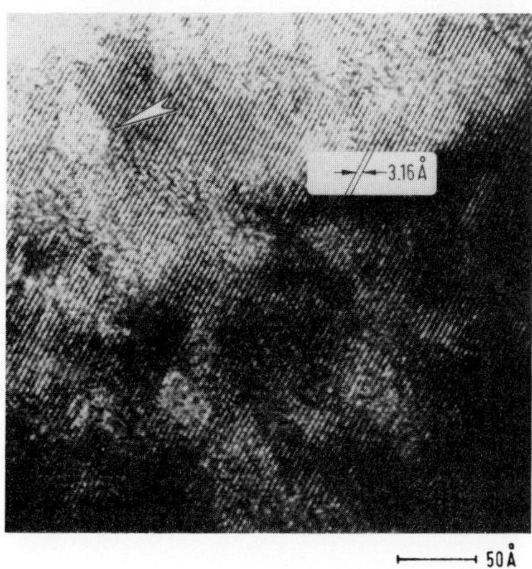

Fig. 14 High resolution transmission electron microscopy of Xe-implanted UO_2 showing images of <111> lattice planes and indicating inert gas bubble formation during implantation (see arrow) [34].

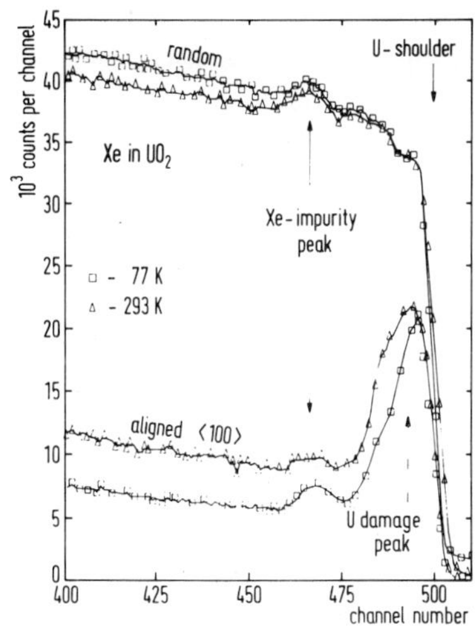

Fig. 15 Channeling-Rutherford backscattering spectra of UO_2 single crystals implanted at 77 K with Xe-ions, and subsequently warmed-up to room temperature [35]

identical in aligned and random crystals, in agreement with the statement of section 3, that Xe-atoms do not occupy substitutional sites in UO_2. All previous implants were done at room temperature, and the same result was obtained for the 77 K implant of Fig. 15. For the first time, however, did one RBS spectrum show an alignment of Xe-atoms with <100> atomic rows in UO_2 when the crystal was warmed up to room temperature. Also, the "square shape" of the impurity peak at 293 K was never seen before. Obviously, the results indicate that when the low temperature recovery stages of U-defects are passed (see Fig. 13), the Xe-atoms can occupy sites of lower energy. These may be sites of individual Xe-atoms along <100> rows; the results are, however, also compatible with epitaxial precipitates of solid Xe in UO_2. Electron microscopy and diffraction studies on such specimens have not been performed yet, but are planned for the future. The question of solid rare gas precipitates in ceramics thus remains open.

413

7. Summary

Extensive experimental, and more recently also calculational work on heavy inert gases in UO_2 has yielded a rather complete picture on Xe (and to a smaller extent Kr) diffusion, trapping, bubble formation, re-solution and release. Much less information is available on UC and UN.

A good understanding exists also on the physical processes during ion implantation of UO_2 and UN. Attempts to search for solid rare gas precipitates were made, but no conclusive results have been obtained so far.

REFERENCES

1. Hj. Matzke, Radiation Effects **53**, 219 (1980).

2. Hj. Matzke, *Science and Technology of Advanced LMFBR Fuels, A Monograph on Solid State Physics. Chemistry and Technology of Carbides, Nitrides and Carbonitrides of Uranium and Plutonium*, North Holland, Amsterdam, 740 pages.

3. Hj. Matzke, H. Blank, M. Coquerelle, I.L.F. Ray, C. Ronchi and C.T. Walker, J. Nucl. Mater. **166**, 165 (1989).

4. K. Lassmann, C. Ronchi and G.J. Small, J. Nucl. Mater. **166**, 112 (1989).

5. R.A. Jackson and C.R.A. Catlow, J. Nucl. Mater. **127**, 161 (1985).

6. R.W. Grimes, this volume.

7. Hj. Matzke, *Diffusion in Nonstoichiometric Oxides*, Chapter 4 in *Nonstoichiometric Oxides*, Ed. O.T. Sorensen, Academic Press p. 155 (1981).

8. Hj. Matzke, J. Chem. Soc., Faraday Transactions **86**, 1243 (1990).

9. I.L.F. Ray, H. Thiele and Hj. Matzke, this volume.

10. L.E. Thomas, this volume.

11. P. G. Lucuta, R.A. Verrall, Hj. Matzke and B.J. Palmer, J. Nucl. Mater., in print.

12. C. Ronchi, J. Appl. Phys. **44**, 3575 (1973).

13. H. Blank and Hj. Matzke, Radiation Effects **17**, 57 (1973).

14. Hj. Matzke, Nucl. Applications **2** 131 (1966).

15. W. Miekeley and F. Felix, J. Nucl. Mater. **42**, 297 (1972).

16. J. C. Carter, E.J. Driscoll and T.S. Elleman, Phys. Stat. Sol. **A14**, 673 (1972).

17. Hj. Matzke and J. A. Davies, J. Appl. Phys. **38**, 805 (1967).

18. Hj. Matzke and J.R. MacEwan, J. Nucl. Mater. **28**, 316 (1968).

19. Hj. Matzke, J. Nucl. Mater. **21**, 190 (1967).

20. R.L. Ritzmann, A.J. Markworth, W. Oldsfield and W. Chubb, Nucl. Applic. Technol. **12**, 436 (1975).

21. H. Matsui, K. Sakanishi and T. Kirihara, J. Nucl. Sci. Technol. **9**, 167 (1970).

22. Hj. Matzke and R. Thomas, 1962, cited in Hj. Matzke and R. Lindner, Atomkernenergie **9**, 2 (1964) (in German) and Canadian Report NRC-TT 1329 (in English).

23. Hj. Matzke and F. Springer, Radiation Effects **2**, 11 (1962).

24. Hj. Matzke, J. Nucl. Mater. **30**, 110 (1969).

25. J.R. MacEwan and W.H. Stevens, J. Nucl. Mater. **11**, 77 (1964).

26. T. Ogawa, R.A. Verrall, D.M. Schreiter and O.W. Westcott, Canada Report AECL-9475 (1987) p. 543.

27. K. Une, I. Tanabe and M. Oguma, J. Nucl. Mater. **150**, 93 (1987).

28. Hj. Matzke, Radiation Effects **75**, 317 (1983).

29. J. H. Evans, this volume.

30. P. Bailey, S.E. Donnelly, D.G. Armour and Hj. Matzke, J. Nucl. Mater. **158**, 19 (1988).

31. J.P. Biersack, in Ion Beam Modification of Insulators, eds. G. Arnold and P. Mazzoldi, Elsevier, Amsterdam (1987) p. 1.

32. A. Turos, S. Fritz and Hj. Matzke, Phys. Rev. **B41**, 3968 (1990).

33. Hj. Matzke, O. Meyer and A. Turos, Radiation Effects and Defects in Solids, in print.

34. C. Rossouw, S.E. Donnelly and Hj. Matzke, unpublished results.

35. Hj. Matzke and A. Turos, unpublished results.

SIMULATING THE BEHAVIOUR OF INERT GASES IN UO_2

R.W. Grimes

Davy Faraday Research Laboratory
The Royal Institution of Great Britain
21 Albemarle Street, London, W1X 4BS, U.K.

ABSTRACT

The behaviour of He, Ne, Ar, Kr and Xe in UO_2 has been investigated using the semi-classical Mott-Littleton simulation technique. The interactions of the gas atoms with the lattice show two extremes of behaviour characterised by the properties of Xe and He. Xenon is very insoluble—its retention in UO_2 is a consequence of the large activation energy necessary for diffusion (>3eV). The most stable solution site for Xe is a function of stoichiometry: in UO_{2-x}, the solution energy is lowest at a tri-vacancy; in UO_2, at a di-vacancy and in UO_{2+x}, at a uranium vacancy. In comparison with xenon, helium is small and readily accommodated at either a defect or interstitial site (solution energy ~0.1eV). Despite this, in the perfect lattice, He exhibits a barrier of ~4eV to diffusion between interstitial sites. However, in the presence of defects, diffusion barriers are reduced to ~0.2eV. These results are discussed with respect to the range of vacancy and vacancy clusters (solution sites) of the type expected in a highly defective or radiation damaged material.

1.   Introduction

The inert gas atoms Xe and Kr account for approximately 15% of the total fission yield [1]. Thus, with burn-up proceeding to ~4%, it is possible that the concentration of fission gas will be large enough to exert a significant influence on fuel properties. Unfortunately, to investigate the behaviour of fission gas experimentally is both difficult and expensive. Therefore, computer simulation techniques offer a particularly valuable alternative or supplement to experimental studies.

In this work, we shall investigate the stability of gas atoms at a variety of solution sites within a defective lattice. Possible mechanisms by which the inert gas atoms migrate through the fuel will be studied by considering low energy migration pathways between solution sites. First, we shall discuss the results of previous computational and experimental studies.

2.   Review of Previous Work

The inert gas atoms Xe and Kr are insoluble in UO_2, and are only found in nuclear fuels as a consequence of fission. Thus, at sufficiently high reactor operating temperatures, gas atoms migrate (possibly via radiation-induced resolution [2]) to grain boundaries, dislocation loops or pre-existing pores where they aggregate into bubbles [3] or can vent to the fuel-clad gap. Bubble formation is important since it degrades mechanical properties, leads to fuel swelling and as such must be considered as a performance limiting factor. In addition, although Xe and Kr are chemically inert, recent work by MacInnes & Winter [4] suggests that these inert gases will reduce the oxygen chemical potential of UO_2 since oxygen ions will bind to defect clusters in which fission gas is trapped.

Fundamental Aspects of Inert Gases in Solids
Edited by S.E. Donnelly and J.H. Evans, Plenum Press, New York, 1991

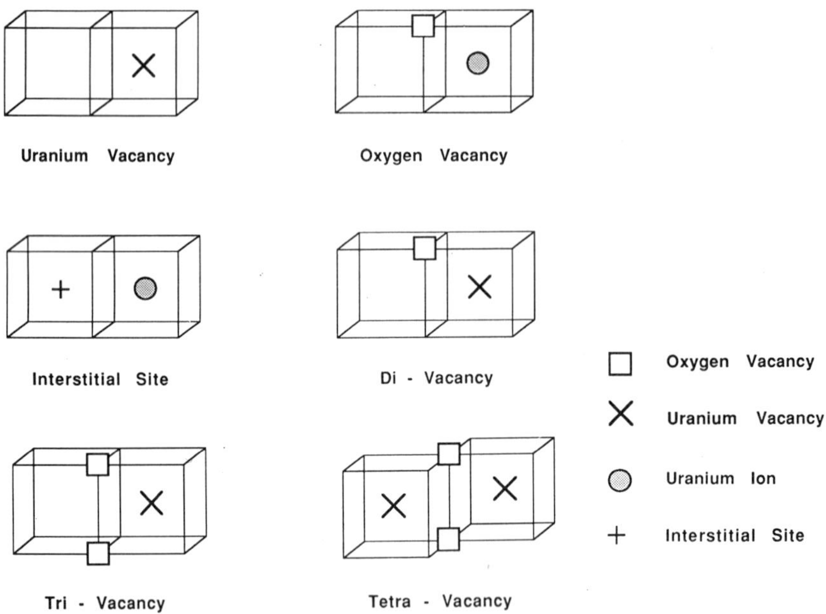

Fig. 1   Solution sites for gas atoms

At the atomic level, channelling experiments carried out on UO_2 single crystals containing Rn-222 show that radon does not occupy normal lattice sites [5]. By inference, it was suggested that xenon would also not occupy a lattice site. This assumption is supported by the observation that xenon diffusion in UO_2 is unaffected by doping with either penta- or tri-valent ions. Such a result is consistent with xenon diffusion via neutral tri-vacancy clusters [3] (see Fig. 1).

The solution and migration behaviour of Xe in UO_2 has also been investigated using simulation techniques. Previous computer modelling studies of UO_2 have concentrated on the derivation of adequate interatomic potentials and on the description of defect structures and non-stoichiometry [6–8]. These studies have clearly established the viability of computer modelling techniques for calculating defect formation, clustering and migration energies in UO_2. Detailed studies of the behaviour of Xe by Jackson and Catlow [9,10], following the earlier work of Catlow [7], determined the sites occupied by the inert gas atoms, established the dependence of site occupancy on stoichiometry and showed that the solution energy of Xe in UO_2 is large and positive. In more recent work, Grimes, Catlow and Stoneham [11] reconsidered the solubility of Xe and calculated solution energies for Ne, Ar, Kr; Ball and Grimes [12] also considered possible migration pathways for the diffusion of Xe.

The properties of He have been discussed by Grimes, Miller and Catlow [13]. In oxide fuels, intra-granular helium is formed as a decay product of fission. In addition, due to its high thermal conductivity, helium is used to fill the space between the uranium dioxide fuel rods and the cladding materials. Consequently, helium is the most abundant gas phase species in the fuel rod assembly.

Under normal operating conditions, intra-granular helium will diffuse from within the fuel to the fuel/clad gap. Results of a recent theoretical model [14] predict that the fraction of helium released is proportional to the fractional release of other fission gases. However, due presumably to its small size, the total amount of helium lost from the fuel is much greater.

Although the amount of Ne and Ar created by fission is negligible, it is still useful to investigate the behaviour of these species since they are often used (via ion implantation) to

416

probe the structure of materials. In addition, they provide a link between the properties of large atoms such as Xe and the smaller He atom.

In discussing the behaviour of inert gases in nuclear fuels we must consider the effect of both accident [15,16] and normal operating conditions. In the case of normal operating conditions, we expect fuel to become more hyper-stoichiometric as actinide burn-up continues [17]. During accidents such as a breach of the cladding material, fuel may be subjected to very oxidising conditions as is the case in a water cooled reactor or very reducing atmosphere if the reactor is cooled by sodium. Since in one case a breach of the cladding causes the fuel to become hypo-stoichiometric and in the other to become hyper-stoichiometric the effect on fission product stability will be quite different. Any useful model of fuel behaviour must therefore include the effect of non-stoichiometry.

3.   Crystallography

UO_2 adopts the fluorite structure and exhibits a lattice parameter of 5.47 Å. The uranium 4+ ions are in the centre of a cube of oxygen ions while the oxygen 2- ions are tetrahedrally coordinated by uranium. Interstitial ions may be accomodated at vacant cube centre sites as shown in Fig. 1.

The intrinsic defect structure of UO_2 is dominated by anion Frenkel disorder with Frenkel pair formation energies in the range of 4–5 eV [8,20]. Schottky trio energies are higher (8–10 eV) while cation Frenkel energies are sufficiently high for these defects to be of negligible importance. However, minority defects can play a dominant rôle in fission product/lattice interactions if fission product solubility is much greater at the minority defect site. Such processes are encouraged by the continuous formation of thermodynamically less favourable defects through radiation damage and high temperatures effects. The defects and defect clusters we shall consider in this study are shown in Figure 1; these solution sites provide the range in size and effective charge necessary to accommodate the variety of fission products we wish to consider. They were also calculated to be the most stable defects involving less than five lattice components.

It is important to recognise that UO_2 can exhibit extensive deviations from the stoichiometric composition. In hyper-stoichiometric UO_{2+x}, at elevated temperatures, x may attain a value of 0.25 with a maximum possible deviation of 0.22 at 1400K [18] (a typical operating temperature for a conventional reactor). For hypo-stoichiometric UO_{2-x}, lower deviations from stoichiometry are observed with x reaching a maximum value of only 0.02 at 1800K, although this may increase to 0.30 by 2800 K [19].

4.   Computational Technique

4.1   Mott-Littleton methodology

This procedure is based upon a description of the lattice in terms of effective potentials. The crystal lattice is partitioned into two regions: an inner region I that includes a defect at its centre and an outer region II which extend to infinity. In region I, interactions are calculated explicitly and all ions are relaxed to zero force. We consider interactions due to long-range coulombic effects (assuming formal charges on all ions) and also short-range forces that are modelled using parameterised pair potentials. The response of region II is treated using the Mott-Littleton approximation [21]. In this work, the CASCADE code [12] was employed for all calculations.

The relaxed positions of the ions in region I are determined using a Newton-Raphson minimisation technique as outlined by Norgett and Fletcher [22]. To ensure a smooth transition between regions I and II, we incorporate an interfacial region IIa in which the ion displacements are determined via the Mott-Littleton approximation but in which the interactions with the ions in region I are calculated by explicit summation. In the present calculations

Table 1   Short-range pair potential parameters

Interacting ions	Potential parameters		
	A (eV)	ρ (Å)	C (eV/Å^6)
$U^{4+} - U^{4+}$	18600.00	0.27468	32.64
$U^{4+} - O^{2-}$	2494.20	0.34123	40.16
$O^{2-} - O^{2-}$	108.00	0.38000	56.06
$Xe - U^{4+}$	6139.16	0.33950	71.84
$Xe - O^{2-}$	598.00	0.42570	108.38
$Kr - U^{4+}$	5912.78	0.31910	50.34
$Kr - O^{2-}$	800.38	0.38880	55.13
$Ar - U^{4+}$	5020.60	0.3091	35.81
$Ar - O^{2-}$	736.86	0.3762	47.32
$Ne - U^{4+}$	6722.66	0.2587	11.59
$Ne - O^{2-}$	952.08	0.3249	13.91
	A (eV)	B (eV)	
$He - U^{4+}$	500.43	7.366	
$He - O^{2-}$	2247.84	11.762	

region I has a radius of 4.2 lattice units (445 ions) and region IIa extends out to 7.2 lattice units and incorporates an additional 1800 ions. Region sizes were chosen to be large enough to ensure that no appreciable change in defect formation energy occurs if the region sizes are increased further. The effect of region size on defect energy has been discussed in previous studies [6,8]. Long-range coulombic interactions are summed using Ewald's method [23]. Short-range repulsive interactions between the perfect lattice and Ne, Ar, Kr and Xe were calculated using electron-gas methodology [24], and fitted to the A and ρ parameters of Buckingham potentials [12] (see Table 1). The attractive portion of the short-range interactions that represent the correlation energy were calculated using the Slater-Kirkwood formulae [25] and incorporated into the appropriate Buckingham potentials as the C term (see Table 1).

Repulsive interactions between He and the lattice were determined from embedded quantum cluster calculations [26]. Attractive correlation terms were again determined using the Slater-Kirkwood formulae [25]. Both components of the short-range energy were fitted to a single Lennard-Jones potential by varying the repulsive (A) and attractive (B) parameters (see Table1). The short range potentials model the effect of electron cloud overlap and dispersion interactions, both of which are negligible beyond a few lattice spacings. This has been discussed in detail in previous work [27,28].

4.2   The shell model

Ion polarization effects are described using the shell model of Dick and Overhauser [29], which has been widely used in both lattice dynamical [30] and defect calculations [6,27]. The model describes an ion in terms of a massless shell of charge Y surrounding a massive core of charge X. The formal charge state of an ion is therefore equal to (X+Y). The core and shell charges are coupled by means of an isotropic harmonic spring of force constant k so that Y^2/k is the atom polarisability (see Table 2). Polarization of an ion can then occur through the displacement of the shell relative to the core. This allows experimental lattice properties such as dielectric constants and phonon dispersion curves to be reproduced [8,31]. In this study the shell model parameters for O^{2-} and U^{4+} were determined by fitting to the experimental dielectric constants for UO_2. Parameters for the inert gases were chosen so that Y^2/k reproduced the experimental polarisabilities.

Ultimately, justification for this simulation technique comes from its success in investigating a diverse range of problems in oxide materials. Of relevance to the present study is the

418

Table 2   Shell model parameters

Ion	Y(e)	k (eV/Å^2)
U^{4+}	−6.54	98.24
O^{2-}	−4.4	296.8
Xe	−11.3	460.8
Kr	−9.9	573.7
Ar	−8.9	674.2
Ne	−7.9	2269.5
He	−1.0	72.7

modelling of extensive non-stoichiometry in $Fe_{1-x}O$ [32] and TiO_{2-x} [33]. Also, there have been successful predictions of the structure and stability of zeolites [34] and other alumino-silicates such as clays and micas [35] all of which are expected to show significant covalency.

5. The Stability of Fission Gas in the Lattice

5.1   *The energy to incorporate gas atoms at pre-existing trap sites*

Definition of incorporation energy

We wish to determine the stability of gas atoms substituted at pre-existing trap sites. This is achieved by calculating incorporation energies which are defined to be the energy to trap a gas atom at a pre-existing trap site.

The Mott-Littleton methodology requires that we calculate both the substitution energy of the fission product at the trap site and the energy to form the trap site so that:

$$\begin{array}{ccccc} \text{Incorporation} & = & \text{The fission product} & - & \text{Trap formation} \\ \text{energy} & & \text{substitution energy} & & \text{energy} \end{array}$$

Thus, a positive result means that energy is required to incorporate the atom in UO_2 whereas a negative energy implies that incorporation is energetically favourable. In this definition, the thermodynamic zero is a gas-phase atom. This is appropriate since the gas atoms assume a gas phase state outside the fuel in the fuel-clad gap.

A comparison of incorporation energies is the most simple way by which fission product stability may be assessed. However, this measure of fission product stability is limited since incorporation energies do not take into account any equilibrium between trap sites and are insensitive to changes in fuel stoichiometry. Nevertheless, the incorporation energy can be used to predict the most stable solution site for a gas atom provided that trap sites are available for occupation. This will be the situation when the concentration of gas atoms is low enough that incorporation proceeds through occupation of intrinsic defect sites.

Results for Xe, Kr and Ar

Incorporation energies for the gas atoms Xe, Kr and Ar are reported in Table 3. For these three atoms, the incorporation energy at any trap site is positive indicating that they are less stable in UO_2 than in the free gas phase. Thus, the calculations predict that Xe, Kr and Ar remain within the fuel only when diffusion barriers are high enough to prevent migration. The insolubility is a consequence of large size of the atoms. We should note that unlike many charged fission products, since the inert gases are neutral they are not stabilised by the Madelung potential of the defect site [11].

The most stable trap site for Xe, Kr and Ar is the neutral tri-vacancy as it is the vacancy cluster with the largest effective radius (the tetra-vacancy cluster has more components but its

419

Table 3 Incorporation energies of fission products (eV)

	He	Ne	Ar	Kr	Xe
Oxygen Vacancy	–0.12	3.24	7.35	9.93	13.34
Uranium Vacancy	–0.05	1.18	2.65	3.79	4.99
Di-vacancy	–0.09	0.47	1.69	2.39	2.84
Neutral Tri-vacancy	–0.08	–0.13	0.67	1.09	1.16
Charged Tetra-vacancy	–0.03	0.22	0.80	1.33	2.00
Interstitial	–0.13	4.62	9.83	13.31	17.23

shape is less accommodating to the spherical gas atoms). At all trap sites, the incorporation energies are ordered as Xe > Kr > Ar. This is a consequence of the greater electron-electron repulsions offered by larger atoms.

Results for He

The incorporation energies for helium are presented in Table 3. At all trap sites, the incorporation energies of helium are negative. Thus, helium enters the lattice exothermally. This is in contrast to the results determined for Xe, Kr and Ar (see Table 3). The low incorporation energies for He are a consequence of a favourable balance between repulsive electron-electron interactions and attractive dispersive terms. The small attractive terms dominate electron-electron repulsion because helium is small enough to be easily accommodated at the large trap sites. For larger atoms, such as Xe, the extensive electron density overlap between the atoms and the lattice causes electron-electron repulsions to be significantly larger than dispersive terms. We note that in most metals helium solution energies are positive [36], due to the smaller size of interstitial sites.

Although there are some differences between incorporation energies of He at different trap sites, these are, by and large, insignificant. Therefore we expect only a slight preference for helium to be located at any particular site. However, at all stoichiometries, the number of interstitial sites is much greater than all other potential trap sites. Consequently, we should expect to find that the majority of He atoms in UO_2 occupy interstitial sites.

Results for Ne

Although Ne is a larger atom than He, it is still small enough to be readily accommodated in the large tri-vacancy trap site. Thus, the dispersive interactions are larger than the electron-electron repulsive terms and the incorporation energy of Ne at this site is negative (see Table 3). In fact, this negative incorporation energy for Ne is not significant since the tri-vacancy site is a minority defect in the lattice and the population of trapped Ne will therefore be small.

The other trap sites investigated in the study are smaller than the Ne atom and the corresponding incorporation energies are positive. In particular, we predict large unfavourable energies at sites such as the uranium vacancy and the interstitial site, both of which can have large intrinsic populations in the lattice. Therefore, to accommodate significant amounts of Ne in UO_2, it will be necessary to create trap sites as will be discussed in the section 5(b).

<u>Quantum mechanical calculations on incorporation energies</u>

Recently, a comparison of classical Mott-Littleton and quantum mechanical Hartree-Fock methods has been carried out by calculating the formation energy of point defects in MgO [37] and the incorporation energies of inert gas atoms in UO_2 [11]. An acceptable level of agreement was found between these different methods thereby lending support to the classical approach.

Of particular significance to the present study are the results for the incorporation of Ne, Ar, Kr and Xe at uranium vacancies in UO_2 [11]. It was found that incorporation energies calculated by both methods predicted the same trend, that is, of a decreasingly favourable energy for larger atom sizes. Since the differences between energies calculated by the two methods are small, we believe the short-range potentials used in our studies are of good quality [11].

5.2   Accounting for the equilibrium between trap sites: Solution Energies

The results of section 5.1 are only useful for describing site stability when all the gas atoms can be accommodated at trap sites usually present in the lattice (ie. intrinsic trap sites). This will only occur if 1) gas atom concentrations are very low (for Xe and Kr, this implies a low fuel burn-up) or, 2) the most stable trap site is the interstitial site (effectively the case for He). Thus it is usually necessary for trap sites to be formed to accommodate the fission products (ie. extrinsic trap sites). This implies that the equilibrium between trap sites will be an important feature of the calculations.

By combining the energy to form the trap site with the fission product incorporation energy, we may define a solution energy,

Solution energy	=	Incorporation	+	Equilibrium solution
(in equilibrium with trap sites)		energy		site formation energy

The type of trap site present in the highest concentrations will be that with the lowest formation energy. The trap formation energy is a function of fuel stoichiometry and can be derived by considering the various defect equilibria appropriate to the stoichiometry in question. Since the intrinsic disorder in UO_2 is of the oxygen Frenkel type, with a minority concentration of Schottky defects, the defect equilibria controlling the formation of the trap sites are (in Kröger-Vink notation):

i) Oxygen Frenkel formation

$$O_o^x \leftrightarrow O_i'' + V_o^{\bullet\bullet}$$

for which the equilibrium constant, K_f, is given by:

$$K_f = [O_i''] [V_o^{\bullet\bullet}] = \exp(S_f/k_B) . \exp(-E_f/k_BT)$$

and,

ii) Schottky trio formation

$$2O_o^x + U_u^x \leftrightarrow 2V_o^{\bullet\bullet} + V_u'''' + 2O_o^x + U_u^x$$

with the equilibrium constant, K_s, given by:

$$K_s = [V_u''''] [V_o^{\bullet\bullet}]^2 = \exp(S_s/k_B) . \exp(-E_s/k_BT).$$

As discussed in previous work [7,9], the formation energy of the trap site, E_T, can be related to the Frenkel (E_f) and Schottky (E_s) energies and to the defect cluster binding energies. Variation of the formation energy with stoichiometry is essentially a consequence of the different rôles that Frenkel defects play in maintaining a constant stoichiometry. For example, formation of a uranium vacancy in UO_{2-x} is accompanied by the formation of two oxygen vacancies. This is simply the Schottky equilibrium so that, $E_T = E_s$. However, in UO_2, the thermal equilibrium between interstitial oxygen and oxygen vacancies means that uranium vacancy formation proceeds via,

$$O_i'' + O_o^x + U_u^x \leftrightarrow V_u'''' + V_o^{\bullet\bullet} + 2O_o^x + U_u^x$$

and the equilibrium constant K_u is given by:

$$K_u = \frac{[V_u''''][V_o^{\bullet\bullet}]}{[O_i'']} = \frac{[V_u''''][V_o^{\bullet\bullet}]^2}{[O_i''][V_o^{\bullet\bullet}]} = \frac{K_s}{K_f}$$

so that $E_T = E_s - E_f$. Lastly, in UO_{2+x}, we can write:

$$2O_i'' + U_u^x \leftrightarrow V_u'''' + 2O_o^x + U_u^x$$

which leads to the relationship $E_T = E_s - 2E_f$.

The formation energies and expressions have been reported in previous work [7,9]. They suggest that in UO_{2-x} the minority vacancy or vacancy cluster defect concentration will be dominated by the neutral tri-vacancy (the majority defect is of course the oxygen vacancy; the uranium interstitial may also be an important minority defect). In UO_2, both the uranium vacancy and the di-vacancy cluster are the important minority defects with the uranium vacancy becoming the most populous minority defect in UO_{2+x}.

The solution energies for the inert gases Ne, Ar, Kr and Xe are reported in Table 4. For He, as described in section 5(a), the results in Table 3 suggest that solution will be dominated by interstitial occupancy since a very large number of interstitial sites are available for He solution. Therefore, the solution energy of He is ~−0.1 eV at all stoichiometries.

<u>Solution energies for Xe and Kr</u>

The results presented in Table 4, suggest that in UO_{2-x} the tri-vacancy trap provides the lowest solution energies for Xe and Kr. This is a consequence of the low solution energy at a tri-vacancy site (see Table 2) and the relatively low formation energy of the tri-vacancy in UO_{2-x} (see reference 9).

In UO_2 the tri-vacancy site is much more difficult to form relative to the di-vacancy. As a result, although the incorporation energy of Xe at a tri-vacancy is lower than at a di-vacancy, we predict that solution of Xe in UO_2 can occur at both di- and tri-vacancy sites. In this context we note that solution at either the di- or tri-vacancy sites will result in Xe being displaced from a regular cation site. This is in agreement with the conclusion of the experimental work [3,4] which suggests that Xe will not occupy a lattice site.

The situation for Kr in UO_2 is different from that for Xe in so far as solution at the di-vacancy site is favoured over that for the tri-vacancy site. However, this still constitutes solution at a non-lattice site.

Further oxidation of the fuel to UO_{2+x} is accompanied by an increase in di- and tri-vacancy formation energies relative to the uranium vacancy site [9]. As a consequence Xe and Kr will occupy single vacancy sites (see Table 4).

422

In previous work [11] the possibility for oxidation of Xe trapped at the uranium vacancy site was discussed. More recently, Grimes and Catlow [38] concluded that only in UO_{2+x}, could Xe possibly exist as Xe^+ and that further oxidation to Xe^{2+} would not occur. This work has also shown that even with the formation of Xe^+, the solution site characteristics of Xe would not be altered. Since the ionization energy of Kr is larger than for Xe, a positive charge state for Kr is never predicted [38].

In conclusion, we note that for both Xe and Kr, solution energies decrease as the fuel is oxidised. In other words, the lowest solution energy in UO_{2-x} is greater than (or the same as) it is in UO_2 which is greater than the lowest solution energy in UO_{2+x}.

Solution energies for Ne and Ar

The solution characteristics of Ne and Ar are significantly different from those of Kr and Xe. For example, the lowest energy solution site for Ne and Ar in UO_{2-x} is the oxygen vacancy whereas for Kr and Xe solution most readily occured at the tri-vacancy site. Conversely, solution for Ne, Ar, Kr and Xe is most stable in the uranium vacancy in UO_{2+x}. For stoichiometric UO_2, we predict a difference between Ne and Ar: Ne, the smaller atom, is most readily accommodated at the interstitial site whereas Ar is most stable at the di-vacancy, (which is characteristic of larger atoms such as Kr). These predictions could be tested using the combination of ion implantation and channeling experiments which has been employed in

Table 4 Solution energies of fission products (eV); the lowest energies are underlined.

(a) Xenon

	Stoichiometry		
	UO_{2-x}	UO_2	UO_{2+x}
Oxygen Vacancy	13.34	16.75	20.16
Uranium Vacancy	18.32	11.50	_4.68_
Di-vacancy	12.93	_9.52_	6.11
Neutral Tri-vacancy	_9.57_	_9.57_	9.57
Charged Tetra-vacancy	19.78	12.96	6.13
Interstitial	17.23	17.23	

(b) Krypton

	Stoichiometry		
	UO_{2-x}	UO_2	UO_{2+x}
Oxygen Vacancy	9.93	13.34	16.75
Uranium Vacancy	17.23	10.31	_3.48_
Di-vacancy	12.49	_9.08_	5.67
Neutral Tri-vacancy	_9.49_	_9.49_	9.49
Charged Tetra-vacancy	19.11	12.29	5.47
Interstitial	13.31	13.31	13.31

(c) Argon

	Stoichiometry		
	UO_{2-x}	UO_2	UO_{2+x}
Oxygen Vacancy	_7.35_	10.76	14.17
Uranium Vacancy	15.99	9.16	_2.34_
Di-vacancy	11.78	_8.37_	4.96
Neutral Tri-vacancy	9.08	9.08	9.08
Charged Tetra-vacancy	18.58	11.76	4.93
Interstitial	9.83	9.83	9.83

(d) Neon

	Stoichiometry		
	UO_{2-x}	UO_2	UO_{2+x}
Oxygen Vacancy	_3.25_	6.65	10.06
Uranium Vacancy	14.52	7.69	_0.87_
Di-vacancy	10.56	7.15	3.74
Neutral Tri-vacancy	8.55	8.55	8.55
Charged Tetra-vacancy	18.00	11.18	4.35
Interstitial	4.62	_4.62_	4.62

previous studies of the solution sites for Rb, Te and Cs in UO_2 [39].

Not only do Ne and Ar exhibit different solution site preferences from Kr and Xe, they also show different trends in their variation of solution energy with stoichiometry. For both Ne and Kr, we observe an increase in solution energy upon oxidation from UO_{2-x} to UO_2. This is in contrast to the results for Kr and Xe (see Table 4). Further oxidation to UO_{2+x} brings about a lowering of the solution energy as was apparent for Kr and Xe. Thus, we predict that the solution energy of Ne and Ar is highest in stoichiometric UO_2 and lower in either of the defective lattices. Consequently, the equilibrium concentration of Ne and Ar will be lowest in the stoichiometric material.

The solution energies calculated for Ne, Ar, Kr and Xe are large enough for the equilibrium concentration of gas atoms to be small in all cases. Larger concentrations of gas atoms can of course be created by ion implantation. In such cases, the concentration of gas atoms remaining in the lattice will be decided not only by the solution energies but also by the ease with which the gas atoms can diffuse out of the lattice. We shall consider possible migration mechanisms for gas atoms in the next section.

6.   Migration Energies

6.1   Migration mechanisms for He

In a non-defective lattice, if He is trapped at an interstitial site, transport will be controlled by the activation energy for migration between interstitial sites. The barrier for this process has been calculated to be 3.8 eV [13] (see Table 3 and Figure 2). The significance of such a high value is that migration of helium in this manner will be slow even at elevated temperatures.

Alternatively, He diffusion may be vacancy assisted. Energy barriers to helium migration between interstitial sites via oxygen and uranium vacancies are calculated to be 0.38 and 0.24 eV respectively (see Table 5 and Fig. 2). However, as part of the vacancy assisted mechanism, the activation energy for vacancy migration must also be considered.

In UO_{2-x}, by virtue of the non-stoichiometry, oxygen vacancies are available to assist He migration. The activation energy for oxygen vacancy migration is 0.51 eV [40]. For UO_2, oxygen vacancy assisted migration can only occur once the oxygen vacancies have been formed. The energy to form an oxygen vacancy is half the Frenkel energy [9]. By comparing the component energies we predict that in UO_{2-x} and UO_2, He migration is controlled by the activation of oxygen vacancy migration.

In UO_{2+x}, uranium vacancies assist in the transport of He. The activation for uranium vacancy migration is 2.1 eV. This includes the uranium vacancy formation energy in UO_{2+x} ($E_S - 2E_F$: see section 5.2). Therefore, in UO_{2+x} the limiting factor for He diffusion is the cation vacancy migration activation energy.

In active fuel, defects are continuously created by radiation damage. This provides helium atoms with defect assisted pathways by which they can migrate. In addition to simple vacancies, radiation damage also creates more complex defects such as di- or tri-vacancy clusters which are also available for helium migration. Preliminary calculations suggested that these larger defects present even lower barriers to migration. Therefore, during normal reactor operating conditions we expect UO_2 to maintain an appreciable and relatively mobile concentration of helium. This could have important implications for processes in fuel such as thermal conductivity.

6.2   Migration mechanisms for Xe

<u>In UO_2 and UO_{2-x}</u>

Solution energy calculations predict that Xe will be trapped at tri-vacancies in UO_{2-x} and UO_2. The mechanism investigated for the movement of the Xe from the tri-vacancy trap involves the association of a second uranium vacancy to form a larger tetra-vacancy cluster.

Table 5 Energy barriers to He migration

	Activation energy for He migration	*Activation/formation energy for vacancy migration*
Between Interstitial Sites in a Perfect Lattice	3.80	—
Between Interstitial Sites: Oxygen Vacancy Assisted	0.38	0.51 (for UO_{2-x})[§] 2.41 (for UO_2)[§]
Between Interstitial Sites: Uranium Vacancy Assisted	0.24	2.1 (for UO_{2+x})[†]

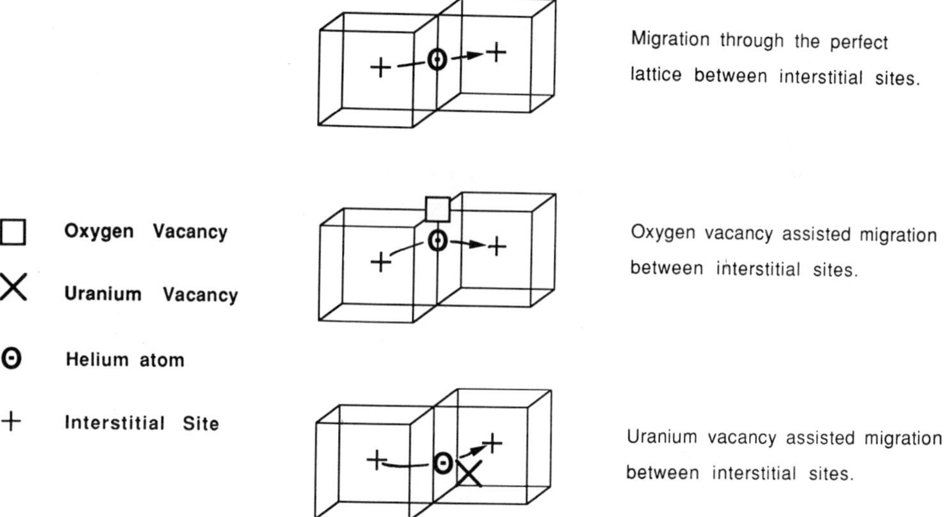

Fig. 2   Helium migration pathways in UO_2. In each case He is shown in the saddle-point configuration.

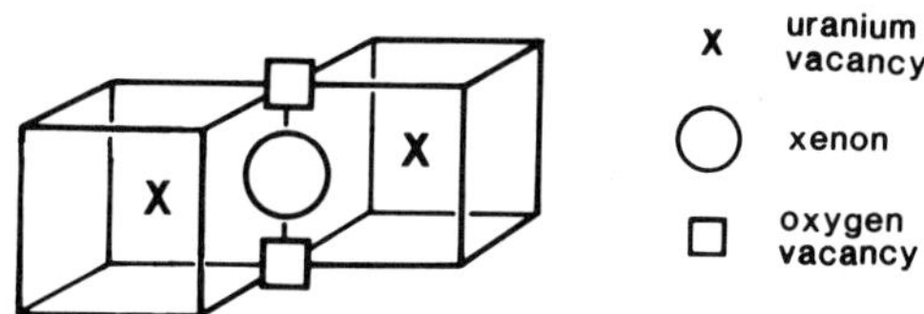

Fig. 3   Xenon migration pathway in UO_{2-x} and UO_2. Xenon is at the saddle-point

§ see reference [40]   † see reference [41]

Table 6   Migration energies for Xe Diffusion

	Stoichiometry		
	UO_{2-x}	UO_2	UO_{2+x}
Trapping energy of second cation vacancy to solution site	–3.12	–3.12	4.48
Activation energy for Xe migration between uranium vacancies	0.11	0.11	0.11
Activation/formation energy for uranium cation migration §	6.1	5.6	2.1
Experimental activation energies for Xe migration †	6.0±0.1	3.9±0.4	1.7±0.4

The gain in energy for the system is 3.12 eV (see Table 6). The Xe atom may then move across the new trap between equilibrium positions (calculated to be approximately at uranium lattice sites). The barrier to this process is 0.11 eV (see Table 6). The saddle point for Xe migration occurs in the centre of the cluster (see Figure 3). It is important to note that at the saddle-point, we fix the position of the Xe atom but allow all of the surrounding ions to relax. After the Xe has migrated, the now vacant uranium site moves away from the defect complex. This is the reverse of the first step and therefore requires 3.12 eV. Rearrangment of the tri-vacancy: Xe complex can occur through interactions with oxygen vacancies. We note that in this respect, oxygen vacancy migration energies are 0.51 eV [40]. This is significantly lower than the dissociation energy for the tetra-vacancy complex (3.12 eV).

The energy components reported in Table 6 suggest that energy is gained in creating the tetra-vacancy from the tri-vacancy and uranium vacancy components. This is a consequence of the electrostatic attraction between the positively charged oxygen vacancies of the tri-vacancy cluster and the negatively charged uranium vacancy. However, tetra-vacancies will still be a minority defect in UO_{2-x} and UO_2 since the energy gain when a tri–vacancy traps a uranium vacancy is less than that for the formation energy of a uranium vacancy [6,9].

To compare migration processes for Xe diffusion with measured Arrhenius energies it is necessary to account for the cation vacancy formation and activation energies (ie. the activation energy for cation migration). This has been determined from experimental data [41,42] by fitting to a model that considers the equilibrium between cation and anion vacancies, di-vacancies and interstitials [43]. The activation energies for cation migration are included in Table 6. They show that in both UO_{2-x} and UO_2 the vacancy activation energy is larger than the barrier to Xe migration. Therefore, in this case, and for any even more favourable vacancy assisted pathway, we can conclude that Xe migration is limited by cation diffusion. This model is supported by experimentally determined activation energies for Xe diffusion [44] (see Table 6).

In UO_{2+x}

The lowest energy migration mechanism was found to occur with the association of another cation vacancy to the trapped Xe. However, in UO_{2+x}, the Xe is trapped at a cation vacancy and so the new trap contains two adjacent uranium vacancies between which the Xe moves

---

§ See references 41, 42 and 43   † See reference 44

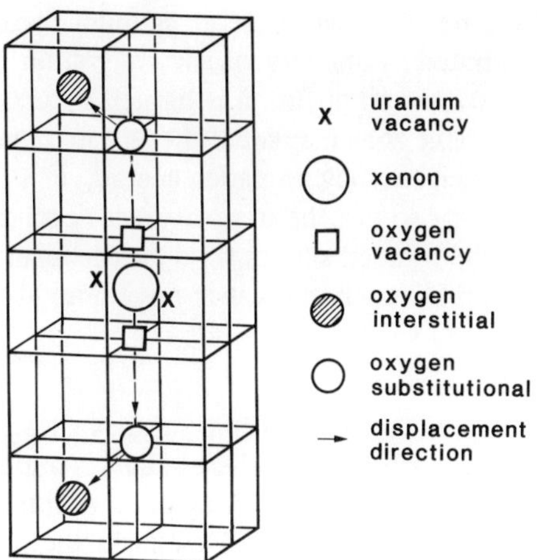

Fig. 4   Xenon migration in UO_{2+x}. Xe is shown in the saddle-point configuration.

(see Figure 4). The association of the second uranium vacancy to the trap site requires 4.48 eV.

During migration of the Xe atom between uranium vacancy sites, the saddle-point configuration (see Figure 4) is stabilized by the relaxation of oxygen ions from lattice to interstitial sites. Therefore, we effectively form a charged tetravacancy cluster with two associated interstitial oxygen ions. As such, the migration of the Xe has occured via the same defect complex as was suggested in the cases of UO_2 and UO_{2-x}.

The activation energy for uranium vacancy migration has been determined to be 2.1 eV [41–43]. Since the vacancy assisted activation energy for Xe migration in UO_{2+x} is larger than the uranium vacancy migration activation energy (ie. 4.48 eV, see Table 6) we conclude that the diffusion of Xe by this pathway would be controlled by the migration of the Xe within the trap.

The experimental activation energy for Xe diffusion [44] is much lower than the calculated value for Xe migration in the defect complex. This suggests that either we have not modelled the same pathway for Xe migration that was apparent in the experimental study or other factors such as radiation damage helped to achieving the low Xe activation energy. Obviously additional work will be nesessary before this inconsistency is properly resolved. To this end, it might be important to consider how U^{5+} (trapped holes) might interact with trapped Xe in UO_{2+x}. However, it has been possible to discount a number of factors such as migration via an ionised intermediary [12] and non-vacancy assisted mechanisms [7]. In addition, we should note that the stoichiometry excludes the possibility of oxygen vacancy assisted mechanisms since the oxygen vacancy formation energy (=E_F in UO_{2+x}) is too high.

Lastly, we should note that the activation energy might be affected by the trapping of Xe atoms in bubbles [4] so that only a fraction of the Xe will be mobile at any one time. A full treatment of the entire diffusion problem would require an understanding of the equilibrium between trapped and mobile gas atoms.

6.3   Activation energies for Ne, Ar and Kr

The barriers to migration of Xe within vacancy clusters are an order of magnitude larger than the equivalent energies calculated for He (see Tables 5, 6 and 7). Nevertheless, an important factor in both cases is the activation energy for the vacancy migration. If we assume that diffusion of Ne, Ar and Kr occurs through vacancy assisted pathways, the diffusion of these atoms will also be limited by the activation of vacancy migration.

427

In UO_{2+x}, Ne, Ar and Kr are all most stable at uranium vacancy sites (see Table 4). Therefore, diffusion will be limited by uranium vacancy migration. In UO_2, the solution site for Ar and Kr is the di-vacancy and so diffusion is limited by uranium vacancy migration. However, in UO_2, Ne is most stable at an interstitial site so that migration between interstitial sites can be limited by the oxygen vacancy migration energy. In UO_{2-x}, the solution site for Ne and Ar is the oxygen vacancy so that the oxygen vacancy migration energy will be the limiting energy. Conversely, the solution site for Kr is the tri-vacancy so that as with Xe, the migration energy will be limited by uranium vacancy migration.

7. Summary

The solution energy of helium at both interstitial and vacancy sites in UO_2 is small and negative (~–0.1 eV). However, in the perfect lattice where helium is trapped at interstitial sites, the barrier to migration is large. Conversely, when migration occurs via lattice ion vacancies diffusion barriers are significantly reduced (~0.3 eV). In these cases, diffusion will be controlled by oxygen vacancies in UO_{2-x} and UO_2, and by uranium vacancies in UO_{2+x}. Vacancy controlled migration pathways are also postulated for Xe atoms. However, since Xe is trapped at a uranium vacancy in UO_{2+x}, at a tri-vacancy cluster in UO_{2-x} and at a di- or tri-vacancy in UO_2 our mechanism for diffusion only considers the association of a cation vacancy to the trap sites. For all stoichiometries the local environment of the migrating Xe becomes the charged tetra-vacancy but the activation energy for migration does depend on the initial trap because of the differences in the formation of the tetra-vacancy. The uranium vacancy activation energy will be a deciding factor in Xe diffusion.

Solution energies were also calculated for Ne, Ar and Kr. These atoms exhibited properties intermediate between He and Xe. The solution energies for Ne and Ar are highest in UO_2 while those for Kr and Xe are highest for UO_{2-x}. These differences are a consequence of solution site preferences.

ACKNOWLEDGEMENTS

This work has been funded by the corporate research program of AEA Technology at Harwell Laboratory. I wish to express my thanks to those with whom I have collaborated at various stages of this work; Prof. C. R. A. Catlow, Prof. A. M. Stoneham, Dr. A. H. Harker Dr. J. H. Harding, R. G. J. Ball and R. H. Miller.

REFERENCES

1. H. Kleykamp, J. Nucl. Mater. **131**, 221 (1985) and S. Imato, J. Nucl. Mater. **140**, 19 (1986).
2. J.A. Turnbull and R.M. Cornell, J. Nucl. Mater. **41**, 156 (1971).
3. Hj. Matzke, Rad. Effects, **53**, 219 (1980).
4. D.A. MacInnes and P.W. Winter, J. Phys. Chem. Solids, **49**, 143 (1988).
5. Hj. Matzke and J.A. Davies, J. Appl. Phys. **38**, 805 (1967).
6. C.R.A. Catlow, Proc. Roy. Soc. Lond. A. **353**, 533 (1977).
7. C.R.A. Catlow, Proc. Roy. Soc. Lond. A. **364**, 473 (1978).
8. R.A. Jackson, A.D. Murray, J.H. Harding and C.R.A.Catlow, Phil. Mag. A. **53**, 27 (1986).
9. R.A. Jackson and C.R.A. Catlow, J. Nucl. Mater. **127**, 161 (1985).
10. R.A. Jackson and C.R.A. Catlow, J. Nucl. Mater.**127**, 167 (1985).
11. R. W. Grimes, C. R. A. Catlow and A. M. Stoneham, J. Am. Cer. Soc. **72**, 1856 (1989).
12. R. G. J. Ball and R. W. Grimes, J. Chem. Soc. Faraday Trans. **86**, 1257 (1990).
13. R.W. Grimes, R.E. Miller and C.R.A. Catlow, J. Nucl. Mater. **172**, 123 (1990).
14. M. Billaux, M. Lippens, D. Boulanger and H. Nidifi, IWGFPT/32, p.182, International Atomic Energy Agency, Vienna, 1989.
15. D. Cubicciotti, J. Nucl. Mater. **154**, 53 (1988).
16. R.G.J. Ball, W.G. Burns, J. Henshaw, M.A. Mignanelli and P.E. Potter, J. Nucl. Mater. **167**, 191 (1989).
17. F.T. Ewart, R.G. Taylor, J.M. Horspool and G. James, J. Nucl. Mater. **61**, 254 (1976).

18. L.M. Kovba, Dokl. Chem. (Engl. Trans.), **194**, 632 (1970).

19. R.E. Latta and R.E. Fryxell, J. Nucl. Mater. **35**, 195 (1970).

20. Hj. Matzke, J. Chem. Soc. Faraday Trans. **83**, 1121 (1987).

21. N.F. Mott and M.J. Littleton, Trans. Faraday Soc. **34**, 485 (1938).

22. M.J. Norgett and R. Fletcher, J. Phys. C **3**, L190 (1970).

23. M.J. Norgett, United Kingdom Atomic Energy Authority Report AERE-R. 7650 (1974).

24. J.H. Harding and A.H. Harker, United Kingdom Atomic Energy Authority Report AERE-R. 10425 (1982).

25. J. C. Slater and J. G. Kirkwood, Phys. Rev. **37**, 682 (1931).

26. M. F. Guest, J. Kendrick and S. A. Pope, *Program GAMESS Documentation*, SERC Daresbury Laboratory, (1983).

27. C.R.A. Catlow and W.C. Mackrodt (eds), *Computer Modelling of Solids*, Springer Lecture Notes in Physics, vol 166 Springer, Berlin Ch. 1, Ch.10 (1982).

28. R.W. Grimes, Mol. Sim. **5**, 9 (1990).

29. B.G. Dick and A.W. Overhauser, Phys. Rev. **112**, 90 (1958).

30. W. Cochran, *The Dynamics of Atoms in Crystals*, p.55. Arnold, London (1973).

31. J.H. Harding, P. Masri, and A.M. Stoneham, J. Nucl. Mater. **92**, 73 (1980).

32. C.R.A. Catlow and B.E.F. Fender, J. Phys. C, **8**, 3267 (1975).

33. C.R.A. Catlow and R. James, Proc. Roy. Soc. Lond. A. **384**, 157 (1982).

34. R.A. Jackson and C.R.A. Catlow, Mol. Sim. **1**, 207 (1988).

35. D.R. Collins and C.R.A. Catlow, submitted to Am. Min.

36. See for example, D. E. Rimmer and A. H. Cottrell, Phil. Mag. **2**, 1345 (1957) or C. F. Melius, W. D. Wilson and C. L. Bisson, Rad. Effects, **53**, 111 (1980).

37. R.W. Grimes, C.R.A. Catlow and A.M. Stoneham, J. Phys.: Condens. Matter. **1**, 7367 (1989).

38. R.W. Grimes and C.R.A. Catlow, Philos. Trans. R. Soc. Lond. (1991), in press.

39. Hj. Matzke and H. Blank, J. Nucl. Mater. **166**, 120 (1989).

40. K.C. Kim and D.R. Olander, J. Nucl. Mater. **102**, 192 (1982).

41. M.S. Seltzer, J.S. Perrin, A.H. Clauer and B.A. Wilcox, Reactor Tech. **14**, 99 (1971).

42. J.L. Routbort, N.A. Javed and J.C. Voglewede, J. Nucl. Mater. **42**, 297 (1972).

43. J.H. Harding, private communication.

CONDENSED-PHASE XENON AND KRYPTON IN UO$_2$ SPENT FUEL

L.E. Thomas

Pacific Northwest Laboratory
Richland, Washington, USA

ABSTRACT

Solid or near-solid xenon-krypton particles in light-water-reactor spent fuels having a wide range of burnup and gas release characteristics were studied by analytical transmission electron microscopy. The fission-gas particles formed at high temperatures near the fuel pellet centers and were associated with metallic ε-phase particles and high local strains inside the UO$_2$ grains. Distinctive angular pits, formed by gas escape from surface-intersected particles in thin-foil samples, also indicated the presence of the particles. Analyses of individual 20-to 100-nm particles by energy-dispersive X-ray spectrometry indicated compositions near Xe-6 wt% Kr and densities appropriate to solid xenon, although the particles appeared noncrystalline. The existence of highly pressurized Xe-Kr particles distinct from gas bubbles in reactor spent fuels suggests the possibility of producing inert-gas solids in other ceramics.

1.   Introduction

Inert gases such as neon, argon, krypton and xenon in ion implanted metals tend to condense at room temperature as liquid and solid crystalline precipitates having high internal pressures [1–4]. The only nonmetal in which inert-gas solids have been found after ion implantation is silicon [5,6], but there is no known reason why such solids cannot form in UO$_2$ or other ceramics. Recent microstructural characterization studies of spent fuels from commercial light-water reactors (LWRs) by transmission electron microscopy (TEM) have revealed that the fission gases sometimes precipitate within the UO$_2$ grains as dense, internally pressurized particles rather than as gas bubbles [7–9]. Outside the reactor fuels speciality, the formation of condensed xenon-krypton (Xe-Kr) phase in spent fuel is not widely known. This paper describes the occurrence and characteristics of the Xe-Kr particles in UO$_2$ spent fuels as well as the X-ray spectrometric method used to determine the gas densities in individual particles.

Precipitation of inert fission gases in reactor spent fuels involves conditions greatly different from those of ion bombarded metals. Fuels for commercial pressurized-water and boiling-water reactors (PWRs and BWRs) consist of UO$_2$ pellets in which the uranium is enriched a few percent in ^{235}U. Fuel rods about 4 m in length contain the 10-mm-diameter, 15-mm-long cylindrical fuel pellets in zirconium alloy cladding. During burnup, i.e. power generation by fissioning the ^{235}U and some of its transmutation products, numerous fission products are produced. Although xenon is the most abundant fission product in LWR fuels, it constitutes only about 0.5 wt% of the fuel at a typical burnup of 30 MWd/kgU. Krypton is only about 0.03 wt% of the fuel. A calculated composition for typical LWR spent fuel listing the most abundant fission products is given in Table 1. In total, the fuel contains about 96 wt% UO$_2$, 1% transmutation products (mostly Pu), and 3% fission products.

Fundamental Aspects of Inert Gases in Solids
Edited by S.E. Donnelly and J.H. Evans, Plenum Press, New York, 1991

Table 1. Composition (wt%) of LWR Spent Fuel; 30 MWd/kgU burnup after 10 year decay, calculated by ORIGEN-2 code [11].

U	Pu	Xe	Nd	Mo	Cs	Ce	Ru	Ba	Pd	La
95.9	0.88	0.49	0.36	0.30	0.22	0.21	0.20	0.16	0.14	0.11

Pr	Sm	Tc	Sr	Am	Te	Rh	Np	Y	Zr	Kr
0.10	0.07	0.07	0.07	0.06	0.05	0.04	0.04	0.04	0.03	0.03

Depending on local fuel operating temperatures, the fission products in LWR fuels may remain in solution, precipitate in the UO_2 grains or along grain boundaries, or be released into the gas-filled gap between the fuel pellets and cladding. The final state of the fission products involves a complex interaction of fission product generation, migration, radioactive decay, precipitation, and re-solution under conditions of irradiation-enhanced and thermal mobility. Local concentrations of a fission product element in the fuel may, for example, reflect the mobility of its radioactive precursors during burnup. These dynamic conditions in reactor fuels also affect the final state of the fission gases.

The most important factor affecting fission-product precipitation and release in LWR fuels is the local fuel temperature or, better stated, the fuel temperature history. Fuel temperatures vary with the heat generation during burnup and, as the heat flows to the water-cooled cladding, are highest at the fuel pellet centers. Peak temperatures in LWR fuels fluctuate widely with power demands and other factors, and usually decrease toward the end of fuel life as the ^{235}U is consumed. In most PWR and BWR fuels, operating temperatures average near 1000°C near the pellet centers and decrease to about 450°C near the pellet outer edges. Fuel rods operated at these temperatures show little release of the fission gases (<0.3%) to the fuel-cladding gap. Higher peak temperatures to about 1500°C at the pellet centerlines may also occur in LWR fuels during normal operation, and these temperature transients are associated with increased gas release, UO_2 grain growth, and gas bubble formation along grain boundaries near the pellet centers. As a rule, fission gas release of more than 5% measured by sampling gases from the fuel-cladding gap indicates higher than normal temperatures. Fuel restructuring, recrystallization, macroscopic precipitation, and temperature gradient phenomena found in fuels operated at higher temperatures [10] do not normally occur in LWR fuels.

Although the fission-product precipitates in LWR fuels are mostly too small to observe by optical microscopy, TEM examinations [7] have shown that high number densities of 1- to 100-nm gas bubbles and particles exist throughout the spent fuels. Insoluble fission gases and metals tend to coprecipitate in the UO_2 grains and form bubble-particle pairs near the pellet edges or particle-particle aggregates toward the pellet centers. Depending on local fuel temperatures, the nature of the fission-gas aggregates changes at some distance from the fuel pellet edge, from 1- to 10-nm bubbles to larger, internally pressurized particles. Except for the Xe-Kr particles, the fission-product particles in LWR spent fuels are nearly all ε-ruthenium phase, a solid-solution alloy of molybdenum, ruthenium, technetium, palladium and rhodium. The other fission products in the low-gas-release fuels have not been found in the particles or segregated along the grain boundaries, and thus apparently remain in the UO_2 matrix [9].

Xe-Kr particles were observed in numerous PWR and BWR fuels with a wide range of burnup and gas release. The fuels, fuel-rod-average burnups, percentage of produced gas released into the fuel-cladding gap, and the pellet radial locations in which the particles were found are given in Table 2. Xe-Kr particle formation shifted radially outward in the fuels with increasing gas release, i.e., with increasing temperature during fuel operation.

Table 2. LWR spent fuels containing Xe-Kr particles

Reactor	Type MWd/kgU	Burnup, %	Gas Release	Location in Fuel
H.B. Robinson (ATM-101)	PWR	33	0.2	center
Turkey Point	PWR	27	<0.3	center
Calvert Cliffs (ATM-103)	PWR	33	0.3	center to mid-radius
Calvert Cliffs (ATM-104)	PWR	44	1.1	center to mid-radius
Cooper ATM-105	BWR	34	0.6	center
Calvert Cliffs ATM-106	PWR	48	11.2 17.7	mid-radius to edge

2.   Experimental Methods

To examine highly radioactive fuels by TEM, samples were prepared from 200- to 400-μm fragments taken from selected radial locations of transversely sectioned fuel rods. The transverse sections were from peak-power regions of the fuel rods, as determined by gamma scanning. The spent fuels were also extensively characterized by other methods [11]. Sample preparation for TEM [12] involved embedding a single fragment in epoxy, grinding it to a thickness of 15 μm, and thinning it to perforation by argon ion micromilling. Sample heating and other preparation artifacts during ion milling were avoided by cooling the sample holder with liquid nitrogen. Finished samples contained less than 10 μg of fuel in an epoxy/metal composite and were suitable for examination in a self-shielded TEM. Besides reducing hazards associated with handling, the limited sample activity avoided γ-interference effects during energy-dispersive X-ray spectrometry (EDS).

Examinations were conducted in a 200 KV analytical TEM equipped with a "high-takeoff" Si(Li) X-ray detector. EDS analysis with a 20-nm convergent electron probe allowed detection of the major fission-product constituents of embedded particles as small as 10 nm. Fission products in particle-free regions of the UO_2 matrix were usually undetectable by EDS because the bulk concentrations (Table 1) were at or below the detection limits of 0.3 to 0.5 wt%.

3.   Xenon-Krypton Particle Observations

Particles containing high xenon and krypton concentrations were found in all the fuels listed in Table 2. The particles formed inside UO_2 grains near the fuel pellet centers but were not found on or within about 0.5 μm from the grain boundaries. Microstructural characteristics of the particles observed by TEM included 1) sizes of 20 to 100 nm, 2) association with dislocation tangles and other evidence of high local strains in the surrounding UO_2, 3) occurrence in contact with similar-size particles of metallic ε-phase, 4) lack of strongly orientation-dependent image contrast, and 5) presence of distinctive, angular surface pits left by gas escape from surface-intersected particles. These characteristics made it possible to quickly identify the Xe-Kr particles by examination. The Xe-Kr particles could be found by identifying those particles associated with local dislocation tangles but not containing pits or bubbles, and particle analyzing regions that lacked apparent diffraction contrast. Particles that were fully contained within the UO_2 foils and retained the Xe-Kr phase showed dark image contrast.

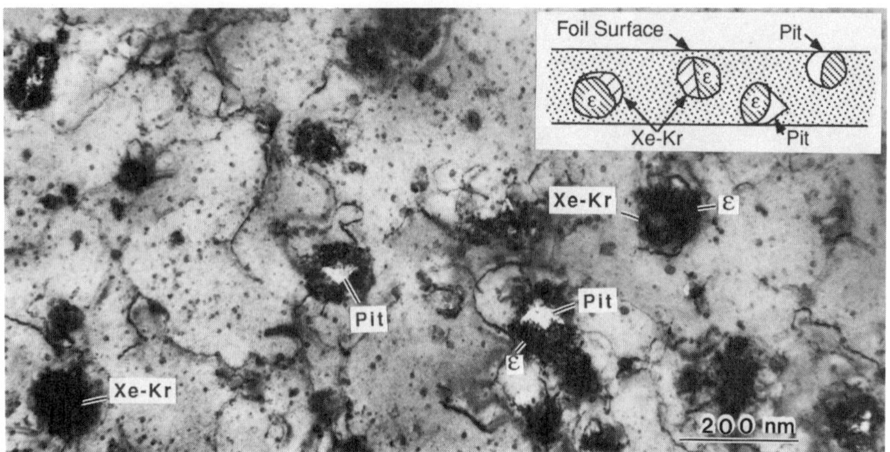

Fig. 1. Xenon-containing particles and surrounding dislocations in UO_2 spent fuel (ATM-103). Pits result from gas release at surface-intersected particles. Inset shows schematic side view of thin foil region.

Fig. 1 shows the particles in low-gas-release ATM-103 fuel with the dislocations imaged in strong diffraction contrast. Although high local strains and inherently strong electron scattering from UO_2 tended to obscure individual dislocations, the strained regions surrounding the particles appeared to contain high densities of small loops as well as dislocation lines. The dislocation tangles were always associated with Xe-Kr particles or with relict pits formed where the particles were intersected by foil surfaces during sample preparation. Stereoscopic TEM observations confirmed the association of the pits with foil surfaces and were used to sketch the schematic side view of a thin foil region shown in the inset. Judging from the sizes and shapes of the pits formed where the gas phase escaped, the original Xe-Kr particles were 20 to 100 nm in size and had irregular rounded and faceted shapes. Fig. 1 also illustrates the scarcity of obvious gas bubbles and the presence of many smaller, unstrained particles in regions containing the Xe-Kr particles.

EDS and electron diffraction analyses of the particles associated with dislocation tangles detected only two phases: metallic ε-phase and the Xe-Kr phase. The ε-phase, also known as the "five-metal" or "4-d" metal phase, is a hexagonal-close-packed solid-solution alloy found throughout UO_2 spent fuels. Although EDS spectra from the Xe-Kr particles always contained contributions from adjacent ε-phase particles, the existence of the separate Xe-Kr phase was confirmed by high-resolution spot analyses and X-ray mapping of individual particles. Fig. 2 shows EDS spectra obtained by fine-spot analyses of adjacent particle regions. Xenon from the Xe-Kr particles was identified from its characteristic L and K X-ray peaks near 4 KeV and 30 KeV and, judging from the X-ray intensities, was present at high concentrations; krypton was detected at much lower concentrations. The original spectra from the particles embedded in the UO_2 matrix were processed to remove the matrix contributions by stripping (channel-by-channel subtraction) a spectrum collected from an immediately adjacent particle-free UO_2 region. The data processing also removed minor contributions from the molybdenum metal washers used in the composite TEM samples. This processing method was based on the absence of detectable fission products in the particle-free matrix regions and the apparent absence of uranium in the particles.

Semiquantitative EDS analyses of the ε-phase particles gave an approximate composition of 40 wt% Mo, 25% Ru, 15% Pd, 10% Tc, and 10% Rh, roughly in proportion to the fission yields. Only ε-phase was found among the smaller (<20-nm-diameter) particles not associated with dislocation tangles, although other phases could have been present.

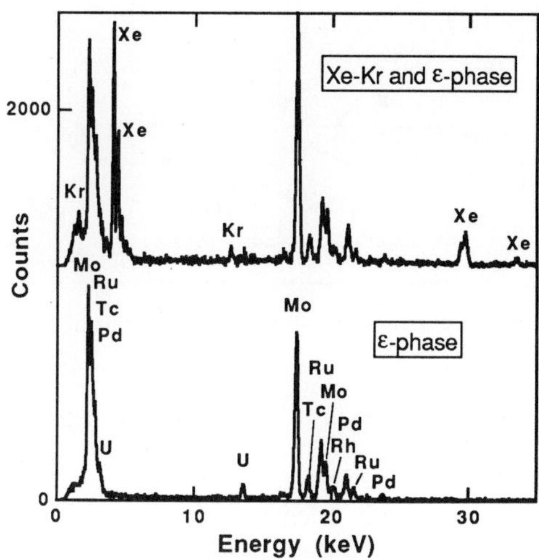

Fig. 2. EDS spectra from particles. Top: Overlapped Xe-Kr and ε-phase; matrix contribution removed. Bottom: ε-phase particle.

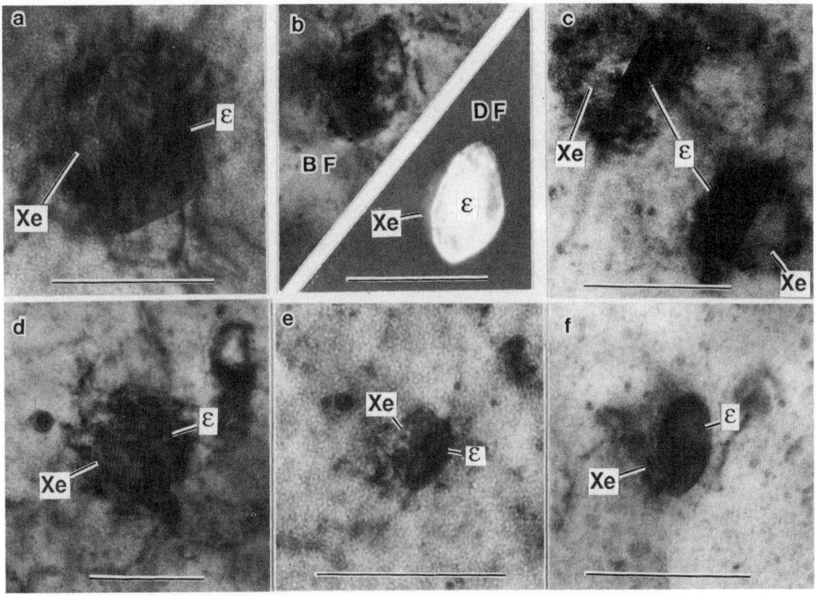

Fig. 3. Xenon-containing particles in LWR spent fuels. Marker = 100 nm.
(a) ATM-103 pellet mid-radius, (b) ATM-103 center (brightfield/darkfield pair),
(c) Turkey Point—location not known, (d) ATM-101 center, (e) ATM-105 center.

A gallery of fission-product particles from several different spent fuels (Fig. 3) shows the irregular shapes and weak image contrast from the Xe-Kr particles in contact with ε-phase. The particles were imaged with minimal strain contrast from the UO_2 matrix but still show interference effects. Shapes of the Xe-Kr particles varied from angular to rounded, and the particles sometimes appeared extended along dislocations (e.g., in Fig. 3a).

Experiments with electron diffraction and diffraction image contrast failed to detect crystallinity from the Xe-Kr particles. Although observations of image fringe patterns at some

435

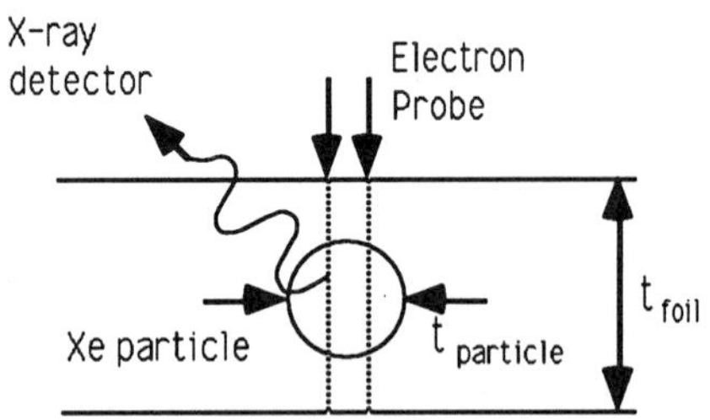

Fig. 4. Analysis geometry for density determination of Xe-Kr particles.

particles, as in Figs. 3d and 3f, left some doubt that all the particles were noncrystalline, the fringes could be interpreted as an interference effect from superimposed particle/matrix interfaces or possibly indicating an interfacial layer structure. Diffraction patterns taken by selected-area and microbeam diffraction techniques showed only reflections from the hcp ε-phase and the UO_2 matrix. Patterns taken with samples tilted to low-index zones of the matrix and ε-phase particles as well as to random orientations of these phases showed no identifiable diffraction spots from face-centered-cubic or hcp xenon with lattice parameters similar to those known from ion bombarded metals. Local matrix strains and strong diffraction contrast from the matrix and ε-phase particles tended to obscure details of the Xe-Kr particles, but the image contrast of the Xe-Kr particles appeared essentially unchanged by sample tilting. Beam-centered darkfield imaging at random incident beam tilts also yielded no diffraction contrast from the Xe-Kr particles. These experiments indicated that the Xe-Kr particles were noncrystalline.

4. Density Determination of Xe-Kr Particles

Because the Xe-Kr particles in LWR fuels gave no detectable electron diffraction spots to allow a density determination by measuring lattice parameters, the densities were determined by EDS microanalysis of individual particles. The method is a simple modification of the Cliff-Lorimer ratio analysis of thin foils [13].

The simplified geometry used for this analysis, shown in Fig. 4, is based on the following assumptions:

1) The particles are sufficiently large relative to the electron beam, and the beam spreading sufficiently small, that excitation volumes in the particle and the UO_2 matrix can be approximated by the particle and matrix thicknesses along the beam direction. Allowing for beam spreading in the sample, the beam size containing 90% of the electrons was at most about half the typical particle diameter. In accord with the assumptions, a Monte Carlo calculation of the broadening for a 20-nm-diameter incident beam of 200 kV electrons in a 100 nm thick UO_2 foil predicted an exit beam diameter of about 23 nm.

2) All detected xenon is from the particle, and all uranium is from the UO_2 matrix. At an average bulk concentration of 0.5 wt% given by the fission yield (Table 1), xenon in the matrix is at the minimum detectability limit for EDS and therefore can be neglected. The ε-phase particles, which inevitably contributed to the analysis by their adjacency to the Xe-Kr particles, contained neither xenon nor uranium.

3) The thickness ratio of the particle and foil can be determined.

For a sufficiently thin sample that X-ray absorption and fluorescence can be neglected, the concentration ratio of of two elements, A and B, is given by the relation

$$\frac{C_A}{C_B} = k_{AB}\frac{I_A}{I_B} \tag{1}$$

where C_A is the weight fraction of element A, I_A is its characteristic X-ray intensity, and k_{AB} is a constant that can be measured if standards exist or calculated as discussed in [13]. Analyses of particles by using an appropriately calculated k-factor for krypton and xenon X-rays ($k_{Kr-Xe} = 0.95$ for $Kr_{K\alpha}$ X-rays and $Xe_{L\alpha}$ X-rays) gave krypton/xenon ratios about equal to the fission yield ratio of two elements, i.e. about 6% Kr and 94% Xe. Therefore the krypton was neglected in density determinations.

For the analysis geometry as shown in Fig. 4, the uranium and xenon X-rays come from separate volumes of UO_2 and Xe-Kr phase. Rewriting equation 1 in terms of densities p and thicknesses t along the beam direction for the two phases gives

$$\frac{\rho_{Xe}}{\rho_{UO_2}} = k_{Xe-U}\frac{I_{Xe}}{I_U}\frac{t_{UO_2}}{t_{Xe}} \tag{2}$$

X-ray absorption for the analysis geometry and UO_2 thickness of the analyzed regions was <10% and therefore negligible. The calculated k-factor used for $Xe_{L\alpha}$ and $U_{L\alpha}$ X-rays was $k_{Xe-U} = 0.524$.

Particle and foil thicknesses for the analysis were determined by several methods including stereoscopy, measurement of X-ray intensities from the UO_2, and particle diameter measurements. Stereoscopic image measurements gave accurate local foil thicknesses, but were inaccurate for the particles because the particle-matrix interfaces generally lacked suitable reference features. For a uniformly thick foil in the vicinity of an enclosed particle, measurements of the uranium X-ray intensities with the electron beam on and off the particle indicated the change in UO_2 thickness caused by the particle. Particle thicknesses obtained by this method were consistent with the results of particle diameter measurements. Density results from several particles, determined with an estimated accuracy of ±50% are given in Table 3. The densities ranged from approximately 2 to 4 g/cm^3. For comparison, the density of solid face-centered cubic xenon with a lattice parameter of 0.6 nm is about 4 g/cm^3.

An alternative determination of the xenon density in the particles was attempted by cooling a sample with liquid nitrogen to crystallize the Xe-Kr phase so that the lattice parameter could be measured by electron diffraction. The Xe-Kr particles did not appear to crystallize on cooling to -170°C, but the result was considered inconclusive because of experimental limitations.

Table 3. Density analysis of Xe-Kr particles

Particle Diameter, nm	UO_2/particle thickness ratio I_{Xe}/I_U	X-ray Intensity,	Density g/cm^3
55	1.18	0.375	2.5
60	1.0	0.660	3.8
80	0.75	0.364	1.6
50	1.0	0.474	2.7
55	1.0	0.296	1.7

5. Discussion

Greenwood *et al.* observed in their analysis of swelling by bubble growth in metallic α-uranium fuel that pressurized fission-gas bubbles could form in reactor fuels [14]. The authors also described the dislocation punching mechanism later used to explain the formation of inert-gas solids in ion bombarded metals. Early TEM observations of irradiated UO_2 fuels [15, 16] concentrated on bubble formation and fission-driven resolution of the gases in bubbles, but lacked the analytical tools needed to recognize solid fission-gas particles. Only the recent development of methods and facilities for preparing and examining suitably small fuel samples in analytical TEMs has made it possible to identify the fission-gas particles in highly radioactive spent fuel. Other investigators have not yet confirmed the observations. However, the particles are found in fuels having a wide range of burnup and gas release characteristics, and should be easily detected once their characteristics are recognized.

The Xe-Kr particles in spent fuel appear in TEM as irregular, 20- to 100-nm, noncrystalline particles surrounded by dislocation tangles and are associated with similar-size particles of metallic ε-phase away from grain boundaries near the fuel pellet centers. The gas-containing particles appear dark unless intersected by a foil surface, and are often partly obscured by diffraction contrast from the UO_2 matrix. Bubble-like features or pits associated with the dislocation-enveloped particles mark locations where the gases have escaped at foil surfaces. Although the gases are missing from intersected particles, observation of the distinctive pits usually indicates the presence of the Xe-Kr particles in the region. Because the fission gas concentrations in the particles are so large, the predominant xenon in individual particles is readily identified by EDS electron microbeam analysis or by EDS X-ray mapping. The less abundant krypton is harder to detect. In principle, electron energy-loss spectrometry (EELS) could be used to identify the gas-containing particles, confirm the crystalline or noncrystalline state of the Xe-Kr phase, and measure the gas density in the particles. Initial attempts to observe the core-loss fine structure of the xenon M-edge from individual particles by parallel-detection EELS in a 120-KV TEM were unsuccessful due to sample thickness limitations of the EELS method.

Despite apparent similarities between the fission-gas particles in spent fuel and the inert-gas solids in ion-bombarded metals, there are major differences between the particle characteristics and formation conditions. In both cases, the highly pressurized particles form by precipitation of insoluble inert gases in the presence of irradiation-induced point-defect fluxes. Inert-gas solids in metals have been known for several years and their formation is comprehensively reviewed in this conference [17]. The xenon and krypton particles in metals are only about one-tenth the size of the fission-gas particles in UO_2 fuels (i.e, about 5 nm), have fcc or hcp crystal structures with usually parallel orientations in the host lattice, near-equilibrium pressures, and apparently form independently of other precipitate phases. The fission gases in fuels coprecipitate with insoluble fission-product metals and, whether precipitated as bubbles or particles, are always associated with metallic ε-phase particles.

The lack of crystallinity in the FG particles is difficult to prove with certainty. Some of the particles might be crystalline or contain thin crystalline layers at the particle matrix interfaces, but diffraction experiments based on the likely orientation relationships with the UO_2 matrix and the associated ε-phase particles gave no evidence of crystallinity in the Xe-Kr particles. Compared to the inert-gas particles in metals, the lack of crystallinity in the much larger fission-gas particles could be interpreted as a consequence of insufficient pressurization. As a result of the atomic mixing and re-solution caused by fission events, the Xe-Kr phase in LWR spent fuel could be an amorphous solid rather than a dense fluid. The gas density in the particles determined by EDS analysis is near that expected for solid xenon, and there is no known requirement that even under equilibrium conditions an inert-gas phase in fuel should crystallize with the physical epitaxy found in metals. The fission-gas phase could also be amorphized under irradiation. In-reactor amorphization of crystalline phases is a well known phenomenon. For example, in zirconium alloy cladding used in LWR fuels, the intermetallic Laves phase becomes amorphous under neutron irradiation [18].

One of the requirements for solid xenon at room temperature is the presence of high internal pressures in the confining solid matrix. If the dislocation tangles surrounding the xenon-containing particles arise by loop punching, limiting pressures in "large" particles can be calculated from [14]:

$$p = \frac{2\gamma}{r} + \frac{Gb}{r} .$$

(3)

The two terms are the equilibrium pressure for a bubble of radius r and the additional pressure needed to deform the confining material by dislocation punching. In UO_2, the lowest temperature for measurable plastic strain is about 1000°C [10]. Taking the surface energy γ as 1 N/m, the shear modulus G as 69 GPa at 1000°C [10], and b as 0.39 nm, equation 3 gives a punching limited pressure of about 1 GPa for a 60-nm-diameter particle. For xenon at this pressure, the Ronchi equation of state [19] gives a density of about 3.3 g/cm^3 at this temperature. This is about the density expected for solid xenon.

Dislocation punching implies particle pressures well above the equilibrium values limited by surface tension forces, and temperatures high enough to allow plastic deformation of the matrix. A creep mechanism operating at temperatures below that needed for punching to occur might also account for the dislocation tangles associated with Xe-Kr particles in spent fuel. For example, the preferential absorption of irradiation-induced vacancies could allow the particle growth and generate the dislocations by precipitating the excess interstitials.

At least two conditions seem needed to produce inert gas solids in UO_2. These conditions are the presence of the insoluble gases and appropriate temperatures during gas precipitation. As indicated by the occurrence of the particles near the pellet centers in low-gas-release fuels, and toward the cooler pellet edges in higher-release fuels, the fission-gas particles apparently form over a limited range of fuel operating temperatures. Temperature calculations for the fuels in which the power history was reasonably well known indicated that the Xe-Kr particles formed between 700 and 1100°C. A modified version of the GAPCON-THERMAL-2 (GT-2) fuel performance code [20] with input parameters adjusted to predict the observed gas releases was used to calculate the fuel temperature histories. The exact temperatures of particle formation are poorly known because of uncertainties in the calculations (estimated to be ± 100°C), and because the temperatures varied widely during fuel life. Solid xenon is unlikely to form at such high temperatures, but the final form of the particles could be attained at lower temperatures near the end of fuel life or after the fuel was removed from reactor. High fuel temperatures may be required to provide sufficient gas atom or gas-vacancy mobility for particle growth, to allow particle ripening to occur in the presence of fission-induced resolution, or to allow particle growth by dislocation punching. The observation that only the larger, 20- to 100-nm particles in spent fuel show evidence of pressurization and Xe-Kr concentrations also suggests that resolution and ripening play a role in Xe-Kr particle formation.

Another possible requirement for forming Xe-Kr particles in LWR fuels may be the presence of ε-phase or other second-phase particles. The ε-phase particles may simply provide nucleating sites for fission-gas aggregates, or fulfill an additional requirement for coprecipitation of other insolubles. Although second-phase particles are not needed to form inert-gas solids in ion bombarded metals, their effects on IG precipitation could be worth investigating. Dispersed particles of ε-phase closely resembling those in actual LWR spent fuels have been produced in simulated UO_2 fuel [21], and thus are not unique to actual spent fuel.

The above observations suggest that fission xenon and krypton form solids under high pressure in irradiated UO_2 reactor fuel. The implications of this finding relate to the fission gas behavior under various conditions in operating and spent fuels, as well as to possibilities for producing inert-gas solids in other ceramics.

6.   Conclusions

Fission xenon and krypton precipitate as dense, highly pressurized particles within the UO_2 grains in LWR spent fuels. Although the existence of fission-gas solids in spent fuel is little known, having been overlooked earlier due to a lack of suitable analytical methods applicable to highly radioactive materials, the Xe-Kr particles are probably present in most LWR spent fuels. Unlike the inert-gas solids in ion bombarded metals, the Xe-Kr particles in LWR fuels are about 10 times larger, noncrystalline, always occur in contact with another precipitate phase (the well-known "five-metal" phase in spent fuels, ε-ruthenium), and are overpressurized. The amorphous nature of the Xe-Kr particles precludes measuring their densities by electron diffraction. However, the highly concentrated fission-gas elements in the particles were detected by X-ray spectrometry in the TEM, and were nearly dense enough for solid xenon. Precipitation of the fission gases in reactor fuel is likely to be a fruitful area for further research since the phenomena and their consequences for fuel operation and processing are not well understood. Based on the observations, it should be possible to produce inert-gas solids in other, nonradioactive ceramic materials by gas ion bombardment at appropriate temperatures, particularly if insoluble metals are also present in the materials.

ACKNOWLEDGEMENTS

Research was supported by the U. S. Department of Energy under contract DE-AC06-76 RLO-1830 with Pacific Northwest Laboratory which is operated with Battelle Memorial Institute. The author would like to acknowledge the efforts of L.A. Charlot in the difficult task of TEM sample preparation and C.E. Beyer for fuel temperature calculations and helpful discussions.

REFERENCES

1. A. Vom Felde, J. Fink, Th-Müller-Heinzerling, J. Pflüger, B. Scherrer, G. Linker and D. Kaletta, Phys. Rev. Lett. **53**, 922 (1984).

2. G. Templier, H. Garem, and J.P. Rivière, Phil. Mag. **53**, 667 (1986).

3. J.H. Evans and D.J. Mazey, J. Phys. F **15**, L1 (1985).

4. G. Deconninck and A. Lefebvre, Mater. Sci and Engr. **90**, 167 (1987).

5. C. Templier, B. Boubecker, H. Garem, E.L. Mathé, and J.C. Desoyer, Phys Stat. Sol. (a) **92**, 511 (1985).

6. G. Faraci, A.R. Pennisi, A. Terraci and S. Mobilio, Phys. Rev. B **38**, 13468 (1988).

7. L.E. Thomas and R.J. Guenther in *Scientific Basis for Nuclear Waste Management XII*, Symposium Proc Vol 127, W. Lutze, ed., Materials Research Soc., Pittsburgh, PA (1989) 293.

8. L.E. Thomas and R.J. Guenther, in *EMSA Proceedings 1988*, G.W. Bailey, ed., San Francisco Press, San Francisco, CA (1988) 512.

9. L.E. Thomas and L.A. Charlot, Chem. Trans. **9**, 397 (1990).

10. D.R. Olander, *Fundamental Aspects of Nuclear Reactor Fuel Elements*, Technical Information Center, Springfield, VA, TID 26711-P1 (1974), Ch. 14 and 16.

11. J.O. Barner, Characterization of LWR Spent Fuel MCC-Approved Testing Material—ATM-101, PNL-5109 Rev. 1, Pacific Northwest Laboratory, 1985.

12. J.M. McCarthy and L.E. Thomas, in *EMSA Proceedings 1985*, G.W. Bailey, ed., San Francisco Press, San Francisco, CA (1985) 184.

13. J.I. Goldstein, D.B. Williams, and G. Cliff, in *Principles of Analytical Electron Microscopy*, Eds. D.C. Joy, A.D. Romig, and J.I. Goldstein, Plenum Press, New York, 1986.

14. G.W. Greenwood, A.J.E. Foreman, and D.E. Rimmer, J. Nucl. Mater. **4**, (1959) 305.

15. A.D. Whapham and B.E. Sheldon, Phil Mag. **12**, (1965) 1179.

16. A.J. Manley, J. Nucl. Mater. **27**, (1968) 216.

17. C. Templier, this volume.

18. W.J.S. Yang, R.P. Tucker, B. Cheng, R.B. Adamson, J. Nucl. Mater. **138**, 185 (1986).

19. C. Ronchi, J. Nucl. Mater. **96**, 314 (1981).

20. M.E. Cunningham and C.E. Beyer, *GT2R2: an Updated Version of GAPCON-THERMAL 2*, NUREG/CR-3907 PNL-5178. Pacific Northwest Laboratory, 1984.

21. P.B. Lucuta, B.J. Palmer, Hj Matzke, and D.S. Hartwig, Proc. Second International Conference on CANDU Fuel, Ed. I.J. Hastings, 1989, 132.

KINETICS OF RECRYSTALLIZATION AND FISSION-GAS-INDUCED SWELLING IN HIGH BURNUP UO_2 AND U_3Si_2 NUCLEAR FUELS*

J. Rest and G. L. Hofman

*Materials and Components Technology Division,
Argonne National Laboratory, Argonne, IL 60439 USA*

ABSTRACT

Recent observations on experimental low-temperature swelling of irradiated uranium silicide dispersion fuels have indicated that the growth of fission-gas bubbles appears to be affected by fission rate. The swelling curve of the material exhibits a distinct knee that shifts to higher fission density with increased fission rate due to higher enrichments. Current state-of-the-art models of fission-gas behavior do not predict such a dependence. Grain "subdivision" has been observed in high-burnup uranium oxide. This observation and other indirect evidence from various experiments led the authors to speculate that a dense network of grain boundaries forms in uranium silicide at a dose corresponding to the knee in the swelling curve; fission-gas-bubbles nucleate at the boundaries and then grow at an accelerated rate relative to that of fission-gas bubbles in the bulk material. A theoretical formulation is presented wherein the stored energy in the material is concentrated on a network of recrystallization sites that diminish with dose due to interaction with radiation-produced defects (vacancy-solute pairs). Recrystallization is induced by statistical fluctuations when the energy per site is high enough that the creation of grain boundary surfaces is offset by the creation of strain-free volumes with, a resultant net decrease in the free energy of the material. This formulation, applied within the context of a mechanistic treatment of gas-bubble behavior, is shown to provide a plausible interpretation of the observed phenomena.

1.   Introduction

Recent experimental observations on low-temperature swelling of irradiated uranium silicide dispersion fuels (U_3Si, U_3Si_2) have indicated that the growth of fission-gas bubbles appears to be affected by fission rate [1]. The swelling behavior of U_3Si_2 is illustrated in detail in Fig. 1. The swelling curve of this material exhibits a distinct knee that shifts to a higher fission density with increased fission rate due to higher enrichments. Below the knee, no gas-bubbles can be detected by scanning electron microscopy (SEM), and if present they must be below the resolution limit of the instrument, i.e., smaller than ≈ 0.04 μm in diameter. Just at the knee, gas-bubbles are first seen to form in a heterogeneous fashion, as shown in Fig. 2. Above the knee, the bubble population rapidly multiplies and the bubble size increases linearly with fission density for the three fission rates shown in Fig. 3.

As illustrated in Fig. 4, the bubble morphology retains its stable character up to the highest fission densities, i.e., bubble size and spacing were rather uniform with no sign of bubble interlinking. The bubbles also form short, linear, intersecting patterns suggestive of association with an underlying microstructural feature.

---

* Work supported by the U.S. Department of Energy, Office of International Affairs and Energy Emergencies, under Contract W-31-109-Eng-38, and by the U.S. Arms Control and Disarmament Agency.

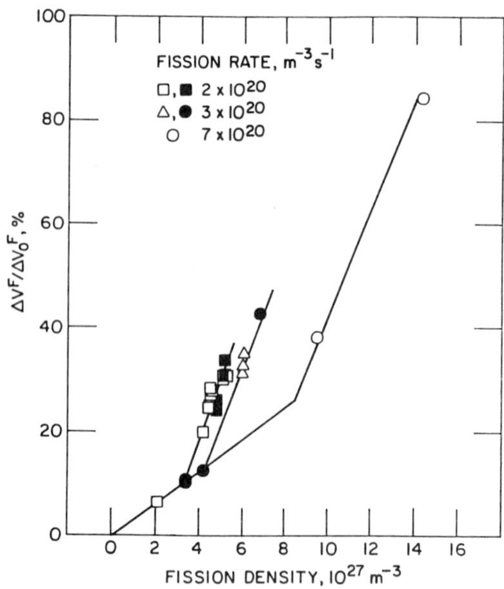

Fig. 1 Swelling of U_3Si_2 for three fission rates as a function of fission density.

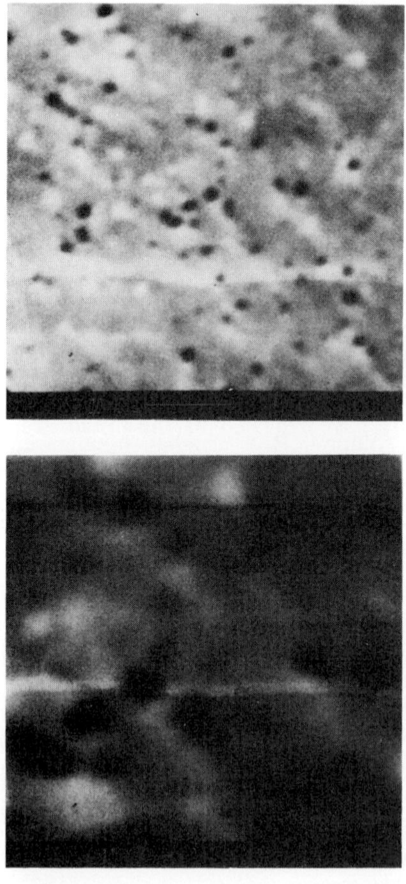

Fig. 2 First evidence of fission-gas bubbles in U_3Si_2, irradiated at 3×10^{27} fissions $m^{-3}s^{-1}$ just at the knee in the swelling curve.

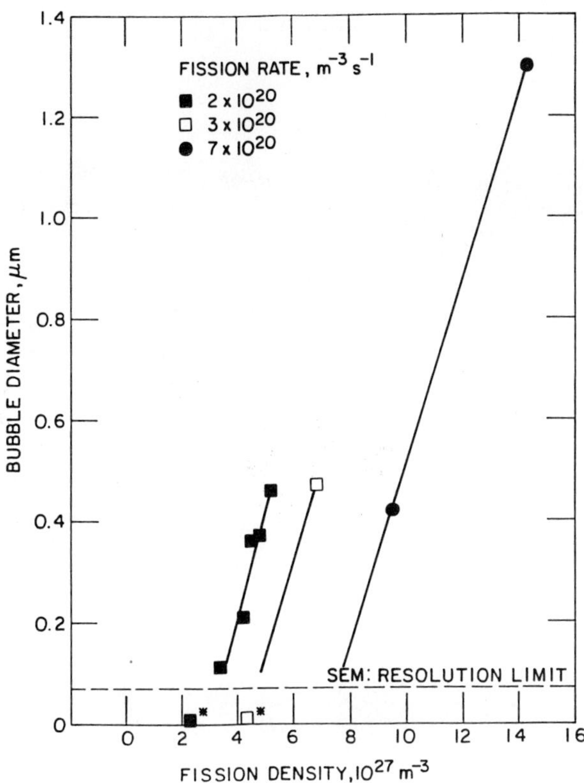

Fig. 3 Average fission-gas bubble diameter in U_3Si_2 as a function of fission density for three values of the fission rate.

Current state-of-the-art models for fission-gas behavior do not predict a dependence of bubble growth on fission rate. Calculations [2] with GRASS-SST [3] have interpreted the swelling as due to the combined effects of a population of bubbles below the limits of experimental resolution and a distribution of larger, visible bubbles attached to dislocations. The position of a peak in the bubble size distribution is determined by the offset between the growth of the bubbles due to diffusion of gas atoms and shrinkage due to fission-fragment-induced re-solution. Both irradiation-enhanced diffusion and gas-atom re-solution have an approximately linear dependence on the fission rate. Therefore, an increase in the rate alone would not significantly affect the position of a bubble peak and thus gas-bubble swelling. Sensitivity studies have also indicated that a large change (hypothetically many orders of magnitude instead of approximately linear) in bubble nucleation rate at higher dose rates would not affect fuel swelling appreciably.

A bubble population with an observed bubble diameter can only be calculated if microstructural features such as grain boundaries or dislocation networks are introduced (the original grain size of the fuel is large, i.e., on the order of the fuel particle size). Indeed, the short, linear intersecting pattern of bubbles shown in Fig. 2 is suggestive of bubbles associated with grain boundaries. In addition, to provide for an interpretation of the observed rate dependence of swelling, the formation of these defect structures should be a function of dose rate as well as dose. Grain "subdivision" was observed by Bleiberg [4] in uranium oxide. He showed that original 10–20-μm grains subdivided into unit sizes of less than one micrometer at $\approx$2 10^{21} fissions cm^{-3} while retaining their crystalline structure. UO_2 swelling also increased in rate at this point and the bubble morphology was strikingly similar to that observed in U_3Si_2.

445

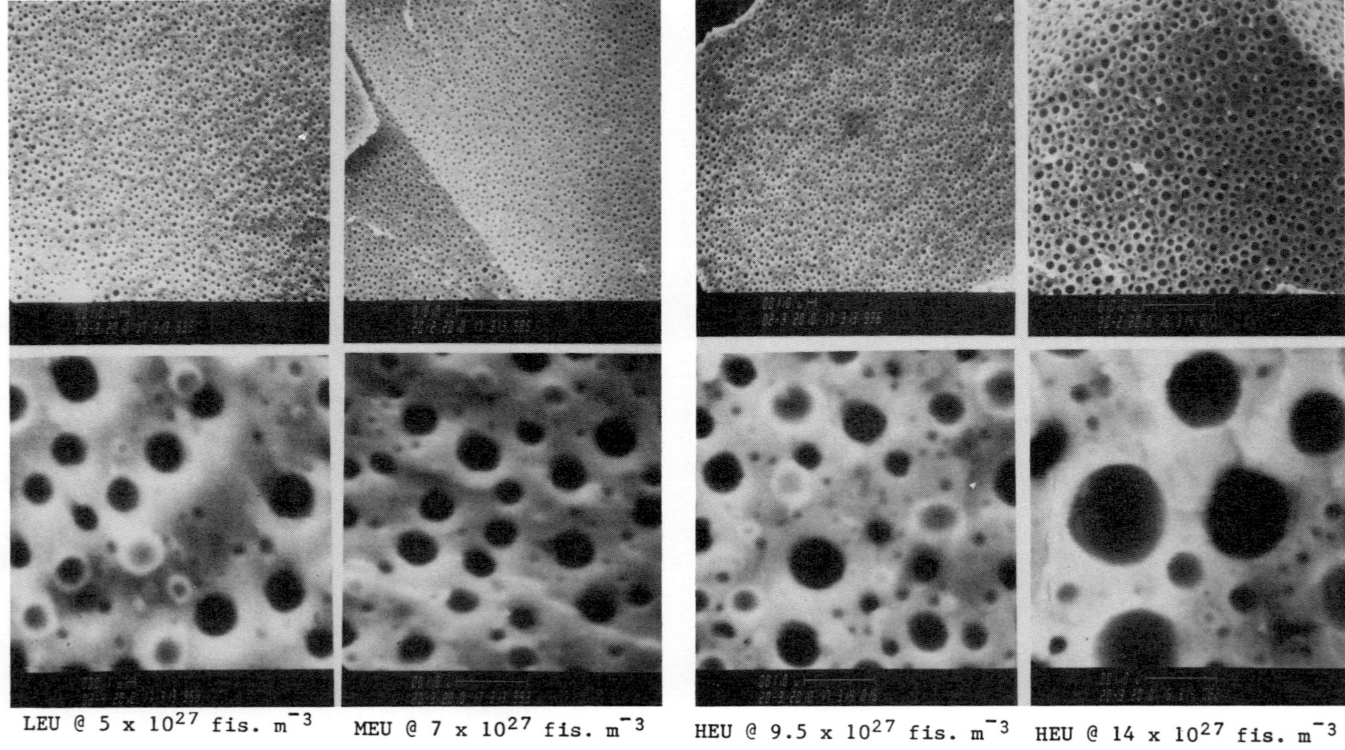

Fig. 4 Fission-gas bubble morphology in U_3Si_2 irradiated to various fission densities at three different fission rates.

One of the present authors (G.L.H.) has also observed transformation of irradiated UO_2 to very small grains, as shown in the SEM fractograph in Fig. 5. Further evidence of this phenomenon has been found in ion bombardment studies on U_3Si above the glass transition temperature [5]. In these studies, recrystallized grain sizes on the order of 100 Å were observed directly in a high-voltage electron microscope. It seems plausible that formation of small grains also occurs in U_3Si_2.

At this time, there is no definitive evidence for such a restructuring of crystalline U_3Si_2 undergoing low-temperature irradiation to high doses. However, preliminary observations indicate that a subgrain-like structure exists in U_3Si_2 above the knee. Detailed examinations of the irradiated material are in progress. The preliminary observations, as well as indirect evidence from the experiments mentioned above lead the authors to speculate that a dense network of grain boundaries forms at a dose corresponding to the knee in the swelling curve, upon which gas-bubbles nucleate and then grow at an accelerated rate relative to that in the bulk material.

In Section 2, a theoretical formulation is presented wherein the energy stored in the material is concentrated in a network of recrystallization sites that diminish with dose due to interaction with radiation-produced defects (vacancy-solute pairs). Recrystallization is in-

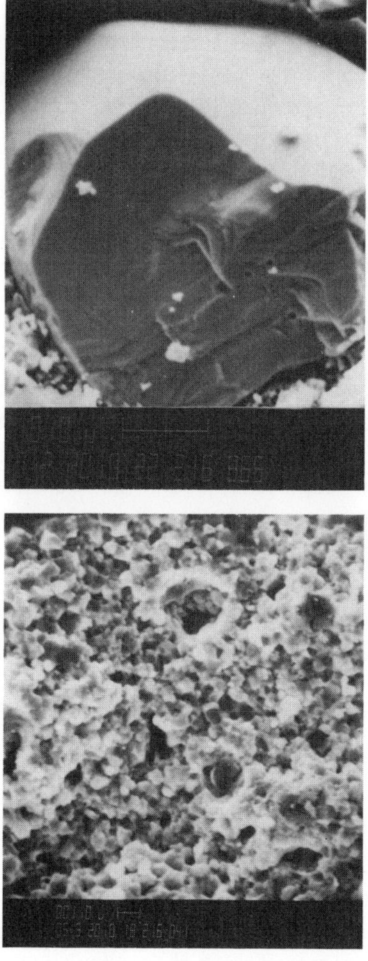

Fig. 5   Fracture surface of UO_2 particles before and after irradiation, showing grain refinement.

duced by statistical fluctuations when the energy per site is high enough that the creation of grain boundary surfaces is offset by the creation of strain-free volumes, with a resultant net decrease in the free energy of the material. In Sections 3 and 4, the theory is applied, within the context of a mechanistic treatment of gas-bubble behavior, to the interpretation of swelling in U_3Si_2 and UO_2, respectively. Finally, the conclusions are summarized in Section 5.

2.   Theory

The dispersion analysis research tool (DART) is a mechanistic model for the prediction of fission-product-induced swelling in dispersion fuels [1]. DART calculates the size distributions of irradiation-induced fission-gas-bubbles as a function of fuel morphology, as well as solid fission product swelling. The gas-driven swelling models in DART are based on those in GRASS-SST [3] and FASTGRASS [6]. The DART mechanical/stress model consists of spherical fuel particles surrounded by matrix aluminum. The model treats the inner fuel sphere as an elastically deformable body and the surrounding shell of Al-matrix material as perfectly plastic. During irradiation, the generation of fission products induces elastic deformation of the fuel particles. First, the as-fabricated pores in the matrix material close in response to expansion of fuel particle volume; then plastic flow/swelling of the aluminum occurs.

Recrystallization is a common phenomenon in cold-worked metals [7]. The initiation of recrystallization usually (but not always) may be traced to the "abnormal" growth of some particular cell formed by deformation and/or recovery. Thus, potential sites where recrystallization can occur are not formed by statistical fluctuations, but are, in fact, already present in the structure of the material. In general, the recrystallization process has as its driving force the increased energy of a matrix that has been plastically deformed. The newly evolving grains are strain-free and the migrating boundaries can absorb dislocations as they move through the strained material.

The concept that recrystallization nuclei [8] could originate from small pre-formed blocks of "strain-free" crystal within the microstructure of a deformed state was proposed by Cahn [9] and Beck [10]. These "strain-free" blocks within a crystal are formed by the spatial rearrangement of dislocations into lower energy arrays by polygonization. This model for recrystallization nuclei was further modified by Cottrell [11] who suggested than an additional necessity for the formation of a successful recrystallization nucleus was likely to be the development of a highly mobile and thus, in general, high angle bounding interface. These ideas have led to the current understanding that a viable recrystallization nucleus will have a size advantage compared to the neighboring subgrain structure and will be surrounded, at least in part, by a highly mobile interface. Doherty and Cahn [12] analyzed a two-dimensional situation where a "subgrain" becomes bounded by a high angle (mobile) interface but lacks a necessary size advantage. By assuming that the surface energy of the high angle boundary was three times that of the surrounding subgrain boundaries, these authors noted that, for a situation involving equilibriated triple junction angles, the linear size of the potential nucleus would have to be approximately three times that of the rest of the population of subgrains.

Based on the above discussion, assume the existence of c_s recrystalization sites per unit volume of material. These sites may exist in the as-fabricated material and/or form relatively early in the irradiation period. The recrystallization sites act as sinks for irradiation-produced defects. Radiation-induced compositional changes in the vicinity of voids, dislocation loops, and external surfaces have been observed in several irradiated metals and alloys [13]. Frequently, segregation leads to the precipitation of a second phase in voids and dislocation loops. It is assumed here that the solute atoms require considerable energy to become part of a dumbbell-shaped interstitial and therefore do not migrate via an interstitial mechanism [14]. It is further assumed that long-range diffusion of vacancy-solute pairs to the immobile sites eliminates the sites at a rate given by

$$\frac{dc_s}{dt} = -K_{sm}\, c_s c_m, \tag{1}$$

where K_{sm} is the reaction rate for the immobilization of recrystallization sites by vacancy-solute pairs, and c_m is the pair concentration. If one assumes that the concentration of vacancy-solute pairs is in steady state with the concentrations of vacancies, interstitials, and solute atoms, then, [14]

$$c_m = R_v c_v c_I, \tag{2}$$

where c_v is the vacancy concentration, c_I the solute concentration, and R_v the reaction rate for the formation of vacancy-solute pairs in thermal equilibrium.

An equilibrium concentration of mobile defects is reached relatively early in the irradiation [15]. Studies made by the present authors utilizing a chemical rate theory of solute segregation [16] for an impurity concentration which is dependent on the fission (or dpa) rate indicate that the above conclusion for a fixed impurity concentration also holds for the irradiation conditions discussed in this paper. The equilibrium concentration of mobile defects within the bulk material (the vacancy concentration, c_v) can be determined from the rate equations describing point-defect behavior. At the low irradiation temperature of the U_3Si_2 fuels, mutual recombination will be dominant over annihilation at internal sinks, and the appropriate solution of the rate equations is [17]

$$c_v = \left(\frac{D_i}{D_v K_{iv}}\right)^{1/2} K^{1/2}, \tag{3}$$

where D_i and D_v are the random-walk diffusion coefficients of vacancies and interstitials, respectively, K is the production rate of Frenkel pairs, and K_{iv} is the reaction rate coefficient for mutual recombination. The rate coefficients are given by

$$K_{iv} = 4\pi r_{iv}(D_i + D_v)/\Omega \cong 4\pi r_{iv} D_i / \Omega, \tag{4}$$

$$K_{sm} = 4\pi r_{sm} D_v / \Omega, \tag{4b}$$

where Ω is the atomic volume, r_{iv} is the radius of the recombination volume, and r_{sm} is the annihilation radius of a recrystallization site/vacancy-solute pair.

Using Eqs. 2 and 3 in eqn. 1 gives

$$\frac{dc_s}{dt} = -K_{sm}\, fK^{1/2} c_s, \tag{5}$$

where

$$f = \left(\frac{D_i}{D_v K_{iv}}\right)^{1/2} R_v c_I. \tag{6}$$

Upon integrating eqn. 5,

$$c_s \,/\, c_s^o = e^{-K\,sm\,f\,K^{1/2}t},$$ (7)

where c_s^o is the concentration of recrystallization sites at $t = 0$.

For a given value of stored energy E_s, the formation of a new crystal of material results in a net change in free energy given by

$$\Delta G = -E_s V + 8\,\pi\gamma\left(\frac{3}{4\,\pi}\,V^{2/3}\right),$$ (8)

where V is the volume of the newly formed crystal (a spherical crystal shape has been assumed for simplicity). The first term on the right side of Eq. 8 is the decrease in free energy due to the creation of a strain-free volume, V, and the second term is the work required to create the boundary surface with surface energy density γ. ΔG has a maximum at a value of V given by

$$V_{max} = \frac{4\,\pi}{3}\left(\frac{4\gamma}{E_s}\right)^3,$$ (9)

where the value of ΔG at V_{max} is

$$(\Delta G)_{max} = \frac{128\,\pi}{3}\frac{\gamma^3}{E_s^2}.$$ (10)

From Eqs. 9 and 10, it is evident that as E_s increases, V_{max} and $(\Delta G)_{max}$ shift to smaller values. When $(\Delta G)_{max}$ decreases to a value on the order of the thermal energy, kT, a relatively small energy fluctuation can allow the system to jump over the "barrier" $(\Delta G)_{max}$, and a new crystal of material will be created.

As $(\Delta G)_{max}$ approaches kT, the stored energy, which is taken to be concentrated in the network of recrystallization sites with density c_s, is assumed to have a rate of change with respect to a change in c_s given by Boltzman's law, i.e.,

$$\frac{dE_s}{dc_s} = -\frac{kT}{c_s}.$$ (11)

Integrating Eq. 11 results in

$$c_s = e^{-E_F/kT},$$ (12)

where E_s has been formally identified with the formation energy, E_F, of the new crystal. In principle, irradiation creep processes can provide additional contributions to the stored energy term in Eq. 8. However, materials such as U_3Si_2 and UO_2 have been observed to have a high degree of plasticity under irradiation [18]. Plastic flow processes during irradiation inhibit the buildup of significant strain energy densities.

450

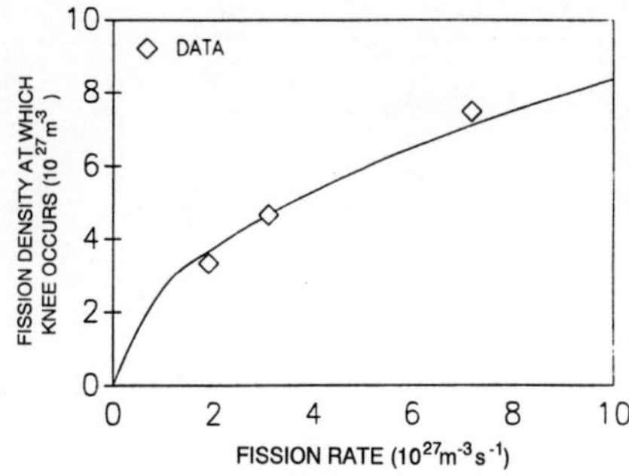

Fig. 6 Comparison of the curve of (Eq. 13) with data obtained from Fig. 1.

Equating the expression obtained for c_s by a kinetic analysis, Eq. 7, to the expression obtained by a thermodynamic analysis, Eq. 12, results in an expression for the fission density at which the knee occurs (onset of grain boundary formation) in terms of the fission rate:

$$(Nt)_{knee} = \frac{E_F' N^{1/2} \beta^{1/2}}{K_{sm} \, fkT} , \tag{13}$$

where b is a conversion factor (1 dpa = 6 10^{17} fissions cm^{-3}), and $E_F' = E_F + E_F^s$; E_F^s is the formation energy of a recrystallization site. Eq. 13 is the basic result of the above analysis. In principle, the material constants E_F', K_{sm}, and f can be determined by experiment.

3. Recrystallization and Swelling in U_3Si_2

Fig. 6 shows a comparison of the curve of Eq. 13 obtained with the values of the material constants listed in Table 1 and data obtained from Fig. 1. Due to the unavailability of many of the materials constants for U_3Si_2, estimates of various parameters were taken from the references listed in Table 1. As is evident from Fig. 6, the above analysis provides a plausible interpretation of the data.

Fig. 7 shows DART-calculated results for fuel particle swelling of low-enriched (LEU) U_3Si_2-Al fuel plates as a function of fission density. The solid-line calculations were made in the spirit of the theory presented in Section 2 by assuming that fuel recrystallization occurred at $\approx 3 \ 10^{21}$ fissions cm^{-3} (Fig. 6) and that grain boundary formation was characterized by an average grain size of 1.0 μm. Subsequently, gas-atom diffusion to the grain boundaries, bub-

Table 1 Values of various materials constants used in the calculation for U_3Si_2.

Parameter	Value	Reference
E_F'	2.0 eV	1
D_v^0	$5 \ 10^{-3}$ cm^{-2} s^{-1}	1*
ε_v	1.28 eV	8,9*
r_{iv}	$2 \ 10^{-8}$ cm	11
r_{sm}	$3 \ 10^{-8}$ cm	11
c_I	$1 \ 10^{-3}$	1
R_v	10	9

$^*D_v = D_v^0 \exp(-\varepsilon_v / kT)$

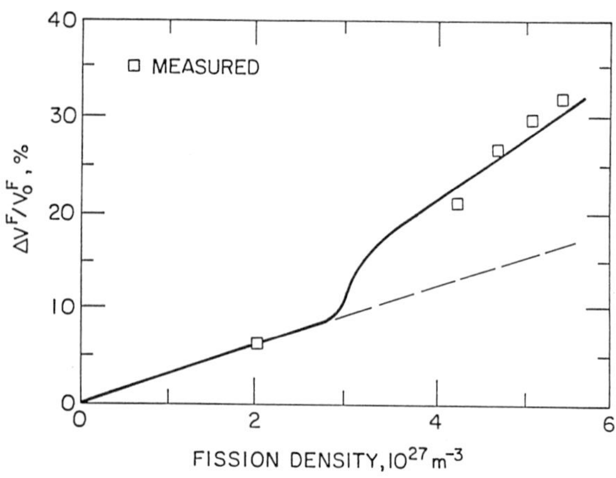

Fig. 7   DART-calculated swelling of U_3Si_2 (for four values of the recrystallized grain size compared with data).

ble nucleation, and accelerated growth (relative to that of bubbles in the bulk material) result in an increased swelling rate, as shown in Fig. 7. In general, bubbles on grain boundaries are thought to grow to larger sizes than those in the bulk material, due to a reduced effect of gas-atom re-solution [3, 19, 21]. (At the relatively low temperatures characteristic of the U_3Si_2 irradiations, the effects of re-solution dominate bubble growth as compared to enhance gas-atom diffusion along the grain boundaries.) Physically, this result is interpreted as due to a small value of gas-atom penetration depth. The dotted-line fuel swelling calculations in Fig. 7 were assuming that no grain refinement occurred. The calculated sizes of intergranular bubbles in U_3Si_2 are consistent with those that have been observed [2]. On average, ejected gas atoms remain within the influence of the boundary and are quickly recaptured. Also shown in Fig. 7 are the measured swelling values, which show that the assumption of grain boundary formation provides calculated swelling values consistent with those observed.

Fig. 8 shows DART-calculated total fuel swelling per 10^{21} fissions/cm^3 subsequent to grain recrystallization as a function of the recrystallized grain size for LEU fuel. The maximum in the curve occurs when grain size is about 0.5 μm. For smaller grain sizes, the increased grain boundary area per unit volume leads to reduced bubble interaction and, thus, smaller values of fuel swelling. For larger grain sizes, increased diffusion distances for the bulk gas lead to reduced amounts of gas reaching the grain boundaries and, thus, to relatively small bubble sizes and fuel swelling values. The calculations represented in Figs. 7 and 8 indicate that if recrystallization occurs during in-reactor, low-temperature irradiation of U_3Si_2, grain size should be on the order of 0.25-1 μm in diameter.

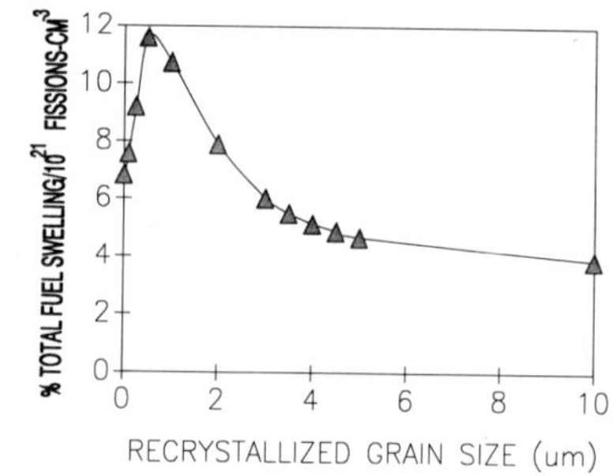

Fig. 8   DART-calculated total fuel swelling rate for U_3Si_2 as a function of recrystallized grain size.

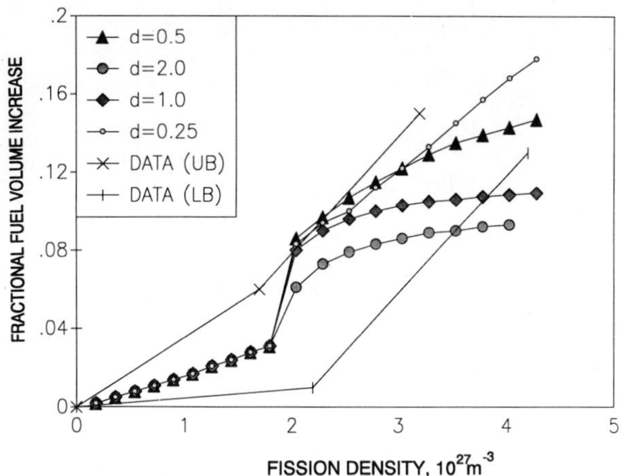

Fig. 9   DART-calculated fractional fuel volume increase in bulk UO$_2$ as a function of fission density for four values of recrystallized grain size.

4.   Recrystallization, Swelling, and Gas Release in UO$_2$

In this section, the assessment of the theory for grain recrystallization is directed to the case of high-burnup UO$_2$ where grain subdivision has been observed directly.

Postirradiation examinations [4] of bulk UO$_2$ wafers irradiated in pressurized hot-water loops to high burnup have revealed that grain subdivision from $\approx$15 to <1 μm diameter occurred in the burnup range of 24-31 10^{20} fissions/cm^3. Changes in fuel volume determined by wet-density measurements also showed that the swelling rate of the material changed from an initial value of about 0.16% to 0.7% ΔV per 10^{20} fissions/cm^3 in the burnup range of 17-36 10^{20} fissions/cm^3. These observations of a knee in the swelling curve and its association with grain recrystallization are qualitatively similar to that shown for U$_3$Si$_2$ in Fig. 1.

Fig. 9 shows DART-calculated fractional fuel volume increase in bulk UO$_2$ as a function of fission density for four values of recrystallized grain size. By theory, with the materials properties shown in Table 2, grain recrystallization is predicted to occur at about 18 10^{20} fissions cm^{-3}. As shown in Fig. 10, the predicted onset of recrystallization is very sensitive to vacancy migration energy. The vacancy migration energy used in the UO$_2$ calculation shown in Fig. 9 is within the range of values reported in Ref. 22. Both ε_v and E_f can be roughly obtained by scaling the values used for U$_3$Si$_2$ by the ratio of the melting temperature of UO$_2$ and U$_3$Si$_2$. Fig. 9 also shows the bounds on the data from Ref. 4 and reveals that a grain size of

Table 2   Values of various materials constants used in the calculation for UO$_2$.

Parameter	Value	Reference
E'_F	5 eV	Present Work
D^0_v	5 10^{-3} cm^{-2} s^{-1}	Present Work*
ε_v	3 eV	16*
r_{iv}	2 10^{-8} cm	11
r_{sm}	3 10^{-8} cm	11
c_I	1 10^{-3}	Present Work
R_v	10	9

$^*D_v = D^0_v \exp(-\varepsilon_v / kT)$

453

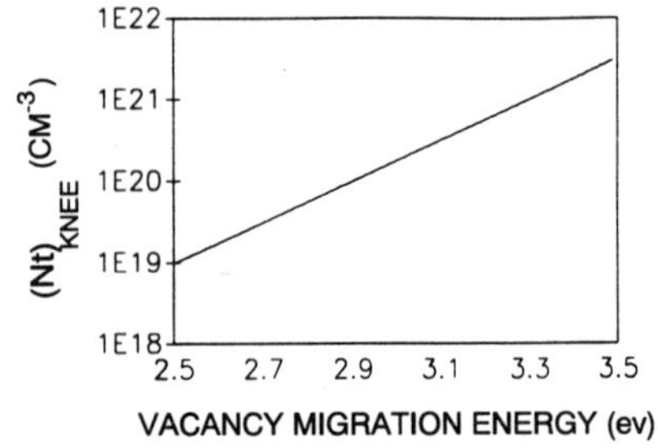

Fig. 10 Calculated fission density at which grain recrystallization is initiated in bulk UO$_2$ as a function of vacancy migration energy.

0.25-0.5 µm best reflects the trends of the data. For recrystallized grain sizes of less than ≈0.5 µm in diameter, the retained gas exists, for the most part, in relatively isolated grain face bubbles (Fig.11). These bubbles are relatively widely spaced because of the large value of the grain boundary area per unit volume. Thus, the bubbles are noninteracting and grow by accumulation of gas atoms. It is this particular growth mechanism that leads to the linear swelling behavior shown in Fig. 9. This predicted behavior is consistent with the observation of regularly spaced, noninteracting bubbles in the U$_3$Si$_2$ shown in Figs. 2 and 4.

Fig. 12 shows DART-calculated percent fission-gas release for the bulk UO$_2$ irradiations as a function of fission density for four values of recrystallized grain size. The experimental bounds from Ref. 4 are also shown in Fig. 12, which reveals that larger values of grain size (in the range of 0.25-0.5 µm) result in higher percentages of fission-gas release. Larger grain size results in smaller values for the grain boundary area per unit volume, and hence higher densities of gas on the boundaries. Higher densities of gas-bubbles result in increased bubble interaction and an increased level of interlinkage to the fuel surface. A recrystallized grain size of ≈0.5 µm best reflects the trend of the data shown in Fig. 12. This grain size value is consistent with the range of grain sizes that best reflects the trend of the swelling data shown in Fig. 9. The DART-calculated releases shown in Fig. 12 tend to underpredict the data for gas releases less than ≈3%. This discrepancy could be due to the lack of direct-recoil and knockout gas release models in DART. In addition, small levels of gas release can occur through cracks that develop upon heating.

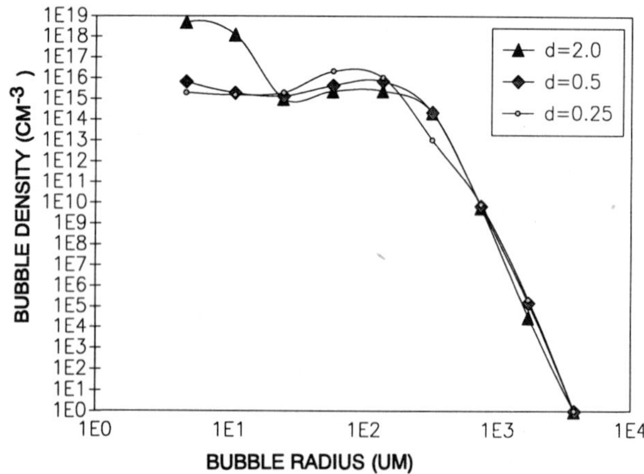

Fig. 11 DART calculated distribution of fission–gas bubble size in bulk UO$_2$ for three values of recrystallized grain size.

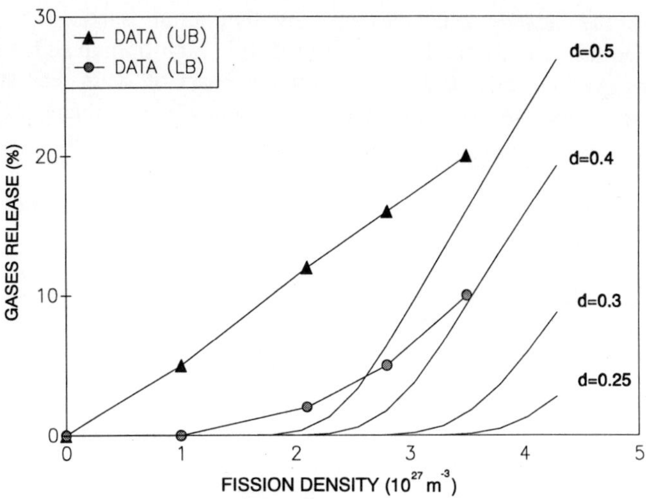

Fig. 12   DART calculated percent gas release from bulk UO_2 as a function of fission density for four values of recrystallized grain size.

5.   Conclusions

The observed fission-rate dependence of low-temperature swelling of irradiated uranium silicide dispersion fuels is ascribed to the formation, at high burnup, of grain boundaries in the material. Subsequently, fission-gas atoms diffuse to the boundaries and nucleate fission-gas bubbles that grow at an accelerated rate relative to that in the bulk material. We propose the existence of recrystallization sites upon which stored energy density is concentrated. Grain boundary formation occurs when the density of recrystallization sites decreases to a level where the net change in free energy due to the creation of strain-free volumes (decrease in free energy) and the creation of boundary surfaces (increase in free energy) is on the order of the thermal energy, kT. It is proposed that the decrease in recrystallization site density is due to interaction with mobile, irradiation-produced defects (vacancy-solute pairs).

The proposed kinetics of grain recrystallization described in this paper are similar to a description of radiation-induced amorphization [23] that is thought to occur in a two step process: introduction of chemical disorder via point-defect recombination, followed by stabilization and accumulation of point defects. Although chemical disordering increases the system energy and volume, the crystalline-to-amorphous transition is only initiated during the second step, i.e., the buildup of point defects is necessary for amorphization. In the case of grain recrystallization, the second step is the elimination of potential recrystallization sites by vacancy-solute pairs.

The theory of grain recrystallization and gas-driven fuel swelling presented in this paper has been applied to the interpretation of the observed low-temperature swelling of irradiated uranium silicide dispersion fuels. The swelling rate of the material exhibits a distinct knee that shifts to higher fission density with increased fission rate. We propose that the basis for this kinetic phenomenom is the formation of vacancy-solute pairs that, upon migrating to potential recrystallization sites, immobilize the sites and thereby increase the average value of the stored energy per site. Although definitive evidence for the occurrence of grain recrystallization in low-temperature in-reactor irradiation of uranium silicide is currently unavailable, the theory provides a plausible interpretation of the phenomenon. (Preliminary observations indicate that a subgrainlike structure exists in U_3Si_2 above the knee. Grain recrystallization in U_3Si_2 under ion bombardment at temperatures above the amorphization temperature is described in Ref. 5). The nature of the recrystallization sites (e.g., triple points) and the identity of the vacancy-solute pair are currently unestablished.

We also applied the theory to grain recrystallization in high-burnup UO_2 where grain subdivision was observed directly. Calculations for the onset of grain recrystallization, fuel swelling rate, and gas release are in reasonable agreement with the trends of the data.

REFERENCES

1. J. Rest and G. L. Hofman, presented at 15th ASTM Symposium on Effects of Radiation on Materials, Nashville, TN, June 17-21, 1990.

2. J. Rest et al., Effects of Radiation on Materials, 14th International Symp. (Vol. II), ASTM STP 1046, N. H. Packan, R. E. Stoller, and A. S. Kumar, eds., American Society for Testing and Materials, Philadelphia, 789 (1990).

3. J. Rest, GRASS-SST: *A Comprehensive Mechanistic Model for the Prediction of Fission-Gas Behavior in UO_2-Base Fuels During Steady-State and Transient Conditions*, NUREG/CR-0202, ANL-78-53, Argonne National Laboratory Report, Argonne, IL (1978).

4. M. L. Bleiberg, R. M. Berman, and B. Lustman, *Symposium on Radiation Damage in Solid and Reactor Materials*, Proc. Series, IAEA, Venice, 319 (1963).

5. R. C. Birtcher, and L. M. Wang, presented at 7th International Conf. on Ion Beam Modification of Materials, IBMM90, Sept. 9-14, 1990, Knoxville, TN.

6. J. Rest and A. W. Cronenberg, J. Nucl. Mater. **150**, 203 (1987).

7. W. D. Kingery, H. R. Bowen, and D. R. Uhlman, Introduction to Ceramics, 2nd Ed., John Wiley and Sons, 449 (1976).

8. A. R. Jones, *Grain Boundary Phenomena During the Nucleation of Recrystallization*, Grain Boundary Structure and Kinetics, American Society of Metals, 379 (1980).

9. R. W. Cahn, Proc. Roy. Soc. **60A**, 323 (1950).

10. P. A. Beck, J. Appl. Phys. **20**, 633 (1949).

11. A. H. Cottrell, Prog. Met. Phys. **4**, 255 (1953).

12. R. D. Doherty and R. W. Cahn, J. Less Comm. Met. **28**, 279 (1972).

13. P. R. Okamoto, N. Q. Lam, and H. Wiedersich, Proc. Workshop on Correlation of Neutron and Charge Particle Damage, CONF-760673, 111 (1976).

14. S. M. Murphy, J. Nucl. Mater. **169**, 31 (1989).

15. S. J. Rothman, N. Q. Lam, R. Sizmann, and H. Bisswanger, Rad. Effects, **20**, 223 (1973).

16. R. A. Johnson and N. Q. Lam, Phys. Rev. B. **13**, 4364 (1976).

17. N. Q. Lam, S. J. Rothman, and R. Sizmann, Rad. Effects, **23**, 53 (1974).

18. W. K. Barney, Geneva Conference Paper, **P/615**, 269 (1958).

19. P. T. Elton, and D. A. MacInnes, *A View of Fission-Induced Resolution in Oxide Fuel*, UKAEA Report, **SRD R 321** (Jan., 1985).

20. D. M. Cowling, R. J. White and M. O. Tucker, J. Nucl. Mater. **110**, 37(1982).

21. A. D. Brailsford, Harwell Report, AERE-R 7466 (June 1973).

22. H. Matzke, Adv. Ceram. **17**, 1 (1986).

23. M. J. Sabochick and N. Q. Lam, submitted to Phys. Rev. B (July 1990).

FISSION GAS BEHAVIOUR DURING POWER TRANSIENTS IN HIGH BURN-UP LWR NUCLEAR FUELS STUDIED BY ELECTRON MICROSCOPY

I.L.F. Ray, H.Thiele and Hj. Matzke

Commission of the European Communities, Joint Research Centre
Karlsruhe Establishment, European Institute for Transuranium Elements
Postfach 2340, D-7500 Karlsruhe, Federal Republic of Germany

ABSTRACT

A Transmission Electron Microscope study has been made of UO_2 nuclear fuel samples which have been subjected to short term in-reactor power transients, involving an increase of the fuel temperature of about 300°C. Under steady state operating conditions most of the fission gas (Xe, Kr) is retained in solution in the oxide fuel matrix, or is precipitated into a population of very small fission gas bubbles (<3nm diameter). This bubble population is continuously subjected to re-solution and re-nucleation, and large fission gas bubbles are not able to grow. In contrast, in fuels with about 4.5% FIMA burn up, containing about 0.5 at% Xe and 0.06 at% Kr as fission gases, subjected to an increase in temperature through a power transient, a population of large fission gas bubbles grows, which is almost invariably associated with the dislocation networks. The implications of the growth of this fission gas bubble population is discussed.

1.   Introduction

A study has been made of specimens taken from light water reactor (LWR) fuel pins which have been subjected to in-reactor power transients, using high resolution Transmission Electron Microscopy (TEM) and high resolution Scanning Electron Microscopy (SEM). Specimens from two equivalent base-irradiated fuel pins which had not been transient tested were also examined to characterise the pre-transient microstructure. These base-irradiated pins had been in reactor for about three years and accumulated a total burn up of about 4.5% FIMA. In the transient tests the fuel pins were subjected to a power increase of about 25% over a period of two days. Details of the transient tests, burn up and pin designations are given in Table 1.

A second technique, Replica Electron Microscopy (REM) in which a replica is made of a prepared fuel pin surface allowing a full fuel cross section to be examined indirectly, was also used.

The purpose of this analysis by electron microscopy was to examine the ways in which the transient affects the fuel microstructure, with particular reference to the behaviour of the fission gases. Some observations on the solid fission products are also reported in so far as they are relevant to the fission gas behaviour.

Under steady state operating conditions most of the fission gas (principally 90% xenon with about 10% krypton) is retained in solution in the fuel matrix or precipitated into a high density homogeneous population of very small bubbles (<3 nm diameter). This bubble population is continuously subjected to re-solution and re-nucleation, and larger fission gas

Table 1 Details of the transient tests, burn up and pin designations

Specimen	Burn up (%FIMA)	Test Details
B1	4.5	Base irradiation. No transient.
B2	4.5	Base irradiation. No transient.
T1	4.5	Transient Test TTL[*]= 400 W/cm.
T2	4.5	Transient Test TTL = 420 W/cm.
T3	4.5	Transient Test TTL = 420 W/cm.
T4	2.5	Transient Test TTL = 450 W/cm

[*] TTL = Transient Terminal Level of Linear Power.

bubbles are not expected to grow. The radial temperature profile in the fuel under these conditions shows a relatively steep gradient at the periphery from the fuel surface temperature of about 500°C ($T/T_m \approx 0.25$), followed by a much flatter rise to the centre where the temperature, depending upon the linear power, is around 1200°C ($T/T_m \approx 0.47$).

However, when a fuel undergoes a power transient the resulting increase in local temperature, which is small at the periphery but about 300°C at the centre, leads to changes in the equilibrium conditions and large (10 to 500 nm diameter) fission gas bubbles can grow. The growth of this population of large bubbles leads to local swelling of the fuel, and it is thus of considerable importance technologically to understand the nucleation and growth of fission gas bubbles under various transient conditions in-reactor. Somewhat analogous behaviour can also be found with the solid fission products, which also tend to form large precipitates under transient conditions.

A more general discussion of the behaviour of inert gases in irradiated nuclear fuels is given by Matzke [1] elsewhere in this volume.

2. Sample Preparation and Examination

2.1. Transmission electron microscopy

Two techniques have been used to prepare fuel samples suitable for TEM. Both of these start from a polished cross sectional slice of a fuel pin which has first been examined by optical microscopy, and in chosen cases also by Electron Probe Microanalysis (EPMA). Such sections are cut in the hot cell facility of the Transuranium Institute and polished down to a thickness of about 0.3 mm.

These thin fuel sections are almost invariably heavily cracked, and when the cladding is removed the fuel breaks into small pieces defined by the pattern of cracks. At this stage the optical micrographs are required so that each piece can be related from its shape back to its original position on the fuel cross section. Two different methods are now used to complete the preparation depending on the primary aim of the examination.

In the first method a chosen piece of fuel is electropolished by remote manipulation in a lead shielded glovebox until a small hole is formed. The areas around the periphery of the hole are usually thin enough (< 150 nm) for TEM. The electropolishing is performed in a Unithin double jet apparatus using Lenoir's solution at a polishing potential of 40 V. After perforation the samples are usually treated for one hour in an ion-beam thinning apparatus to clean the surfaces, which are frequently covered with a thin contamination layer from the polishing.

The advantage of this method of preparation is that very high positional accuracy can be obtained for the areas examined by TEM, enabling reliable estimates of the fuel temperature to be made. The disadvantage is that the bulk of the sample, and thus its activity, is relatively high which complicates specimen handling and prevents the local chemical composition from being determined in the microscope using Energy Dispersive X-ray Analysis (EDAX). In addition a proportion of any metallic precipitates formed in the fuel is lost during the electropolishing.

The second method of preparation involves taking a small piece of the fuel from a chosen radial position, and crushing this to a fine powder. The crushing is performed under methanol, and drops of the resulting suspension are allowed to dry on carbon-film coated copper support grids. The small fuel fragments remaining supported on the film are usually thin enough for TEM. This method has the advantages of low specimen activity, enabling easy handling and the application of EDAX analysis, but does not have the high positional accuracy of the electropolishing technique. Any precipitates of the metallic fission products are, however, retained and can be analysed by EDAX.

The specimens were examined in a 200 kV Hitachi H700 HST Transmission Electron Microscope specially modified for handling radioactive materials by direct connection to a glovebox system [2,3], and equipped with secondary electron imaging facilities and a Tracor Northern TN5500 energy dispersive X-ray analysis system. With this microscope a resolution of better than 0.5 nm by TEM and 5 nm by SEM could be attained on samples with activities up to 3 Rem/h at 10 cm measuring distance.

2.2. *Replica electron microscopy*

This technique also starts from a fuel cross section which has been cut, polished and then lightly etched to reveal the grain boundary structure and any large fission gas bubbles which may be present. A plastic replica is made of the full cross section using a plastic film (Triafol foil) lightly wetted with acetone. The resulting negative plastic replicas are decontaminated, then a germanium shadowed carbon/formvar positive replica is prepared from each by vacuum deposition. The carbon replicas are photographed to enable positional information to be obtained, and small pieces are then carefully cut and mounted on support grids for examination in the TEM.

Being an indirect replication technique the resolution of detail which can be obtained is limited to about 30 nm, but the method complements the TEM technique in enabling bubbles in the size range 30 nm up to 500 nm to be imaged and measured, with the big advantage that the full fuel diameter of approximately 9 mm can be examined instead of just a limited area. In addition the samples have a very low activity and can be handled easily outside a glovebox, and the statistical significance of the results is good.

The replicas were examined in a Siemens Elmiskop IA TEM, at an accelerating voltage of 60 kV to optimise the contrast from the replicas.

3. Experimental Results—Characterisation of the Base-Irradiated Non-Transiented Fuel Samples

For each of the base-irradiated non-transiented samples two specimens were examined by TEM, one from the periphery of the fuel ($T/T_m \approx 0.25$) and one from the centre ($T/T_m \approx 0.47$). Both fuels showed effectively the same microstructural characteristics.

3.1. *The peripheral specimens*

The microstructure at the periphery was characterised by the following features:

    i)   A high density dislocation network.

    ii)   A high density of very small fission gas bubbles and solid fission product precipitates.

These were the only significant features of the microstructure at the periphery and there was no development of a population of large fission gas bubbles.

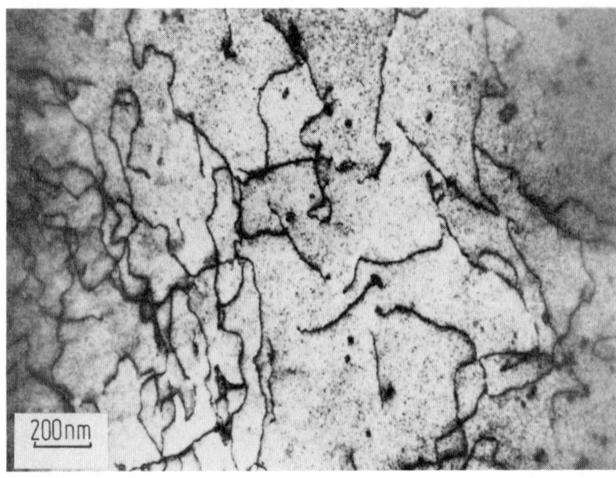

Fig. 1   Typical dislocation structure as observed by TEM at the periphery of the base-irradiated fuel samples.

A mean dislocation density of 2.2 10^{10}cm.cm^{-3} was determined within the grains. This value was relatively constant. The dislocation lines were frequently jogged and were characteristic of networks formed by climb-induced growth in the high point defect flux. There was no evidence for plastic deformation having occurred. An example of the typical dislocation structure is shown in Fig. 1. The very high density of small dislocation loops as seen in this micrograph was also a characteristic feature of the microstructure at the periphery.

Extensive dislocation networks forming low angle grain boundaries were also frequently found. The dislocation density in these boundaries reached local high values of up to 1.4 10^{11} cm.cm^{-3}.

At the periphery a very high density of small fission gas bubbles was found, with an average diameter of 8 nm, and a narrow size distribution. The bubbles were homogeneously distributed with an average density of 1.2 10^{16} cm^{-3}. The small fission gas bubbles were almost always linked to small precipitates of approximately the same size distribution.

An example of this fission gas bubble population is shown in Fig. 2 from fuel sample B1. The local microswelling value due to this population of small fission gas bubbles is negligibly small at about 0.1%.

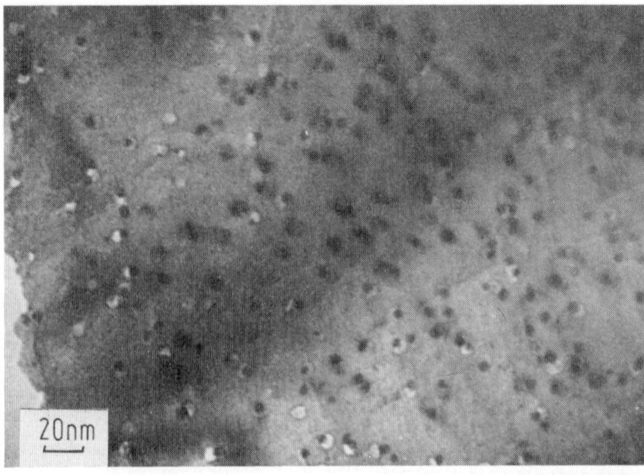

Fig. 2   An example of the population of very small fission gas bubbles observed in the base-irradiated fuel sample.

460

3.2. *The central specimens*

In contrast to specimens from fuel pins which had been transient tested, the central and peripheral microstructures of the base-irradiated fuel were very similar, the main difference being the occasional presence of a low density of rather larger fission gas bubbles at the centre. Both the dislocation density and character were very similar to those observed at the periphery, the measured density being equal at $2.2\ 10^{10}$ cm.cm^{-3}.

A population of very small fission gas bubbles was found at the centre of the fuel, again almost invariably associated with small precipitates of approximately the same size. The size distribution was narrow with a mean diameter of 7.5 nm, and the density was uniform with an average value of $1.9\ 10^{16}$ cm^{-3}.

Rather unexpectedly, areas showing larger fission gas bubbles in the size range 5 nm to 30 nm were found in the central regions of the base-irradiated fuels. The distribution of these areas was always very inhomogeneous, and the bubbles were invariably linked to dislocation networks. The local density of this bubble population reached a value of $2.4\ 10^{15}$ cm^{-3}, but the overall density was very low because the bubble population was completely absent over 90% of the area examined.

The growth of fission gas bubbles of this size on a highly localised basis under steady state irradiation conditions is not expected and the mechanism for their formation is not understood. One possible explanation for their occurrence is that even during steady state irradiation sufficiently large power increases sometimes occur which, coupled with local temperature fluctuations due to cracks or local porosity, may have effects similar to those produced by a small power transient.

The microswelling contribution from this bubble population is however negligible, reaching local high values up to 1%, but a total of less than 0.1% when averaged over the full fuel cross section.

4.   Experimental Results —Transiented Samples

The primary microstructural change found to have occurred in the post transiented fuel was the growth of a population of large fission gas bubbles accompanied by a similar growth of large precipitates. Evidence was also found that plastic deformation had occurred.

4.1. *The peripheral samples*

The microstructures of the transiented fuel samples at the pin periphery were characterised by the following features:

i)   A high density dislocation network.
ii)   A high density of very small fission gas bubbles and solid fission product precipitates.
iii)   A low density population of large fission gas bubbles.

A mean dislocation density of $2.2\ 10^{10}$ cm.cm^{-3} was determined within the fuel grains at the periphery, and this was constant over all the four transiented samples, including the lower burn up specimen T4.

Within the individual grains the dislocation distribution was uniform and extensive network formation had occurred. An example of a typical dislocation structure from sample T2 is shown in Fig. 3. The dislocation lines were not found to be heavily jogged, and the networks are typical of those formed by plastic deformation.

Although within the matrix the dislocation density was quite uniform, local very high values were reached in the immediate neighbourhood of the large fission gas bubbles and solid fission product precipitates. Values up to $8\ 10^{11}$ cm.cm^{-3} were measured. This will be illustrated later.

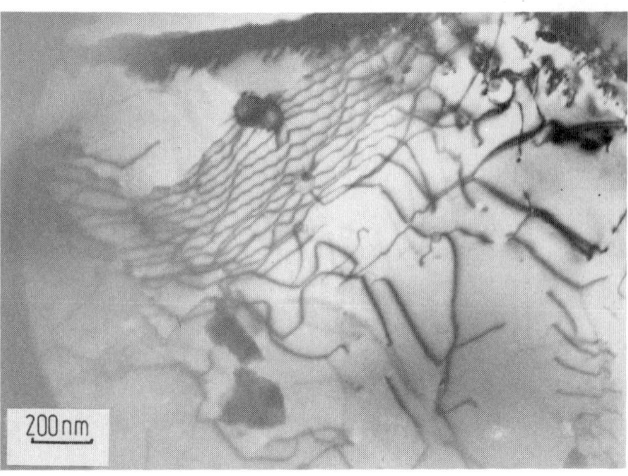

Fig. 3 Dislocation structures observed at the periphery of the transient-tested fuel samples, showing evidence for plastic deformation.

The relevance of the observations on the dislocation networks is that all the large fission gas bubbles and solid fission product precipitates appear to have a strong association with the networks.

At the periphery there was a very high density of small fission gas bubbles, with an average diameter of 8 nm and a narrow size distribution. The bubbles were homogeneously distributed, with an average density of $1.2 \ 10^{16} \ cm^{-3}$. This bubble population was common to all the samples independently of burn up and transient conditions.

The small fission gas bubbles are almost invariably associated with precipitates of the same size range. Fig. 4 shows an example of this small fission gas bubble population, imaged on sample T3.

A low density of large fission gas bubbles was found at the periphery in the transiented samples, associated invariably with the dislocation networks. The bubbles ranged in size from 50 nm to 300 nm and their distribution was relatively inhomogeneous, with an average density of $6.0 \ 10^{13} \ cm^{-3}$. An example of this bubble population is shown in Fig. 5. The bubbles are frequently found in association with precipitates of about the same size.

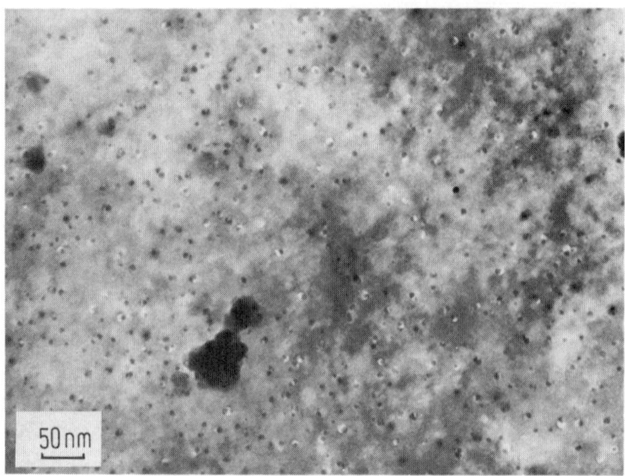

Fig. 4 Population of small fission gas bubbles linked to small precipitates as seen at the periphery of the transient tested fuel samples.

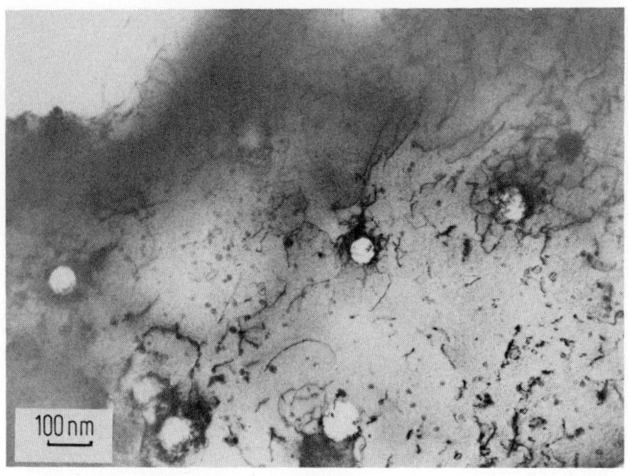

Fig. 5 Large fission gas bubbles formed on dislocation networks at the periphery of the transient-tested samples. These have a low density and inhomogeneous distribution.

The highest value of local swelling measured in any of the samples due to this bubble population at the periphery was 0.4%, which is a negligible contribution to the overall fuel swelling.

4.2. The central specimens

The microstructure in the central regions of the transiented fuels differed significantly from that observed at the periphery. The effects of the transient are much more pronounced at the fuel centre than at the periphery, with a temperature increase of about 300°C, to give central values of about $T/T_m = 0.56$. The central microstructure was characterised by the following features:

i)   A much lower average dislocation density within the grains, but extensive dislocation networks.

ii) A total absence of the population of small fission gas bubbles (diameter < 10 nm) as observed at the periphery, but a similar population of very small bubbles localised in the immediate neighbourhood of the large fission gas bubbles and pores.

iii) A population of large fission gas bubbles (diameters in the range 50 to 500 nm) invariably linked to the dislocation networks and often attached to metallic precipitates of about the same size distribution.

The dislocation density within the grains was significantly lower in the central regions of the fuels, with a mean value of $6.1\ 10^9$ cm.cm^{-3}. This was relatively constant from sample to sample. A specimen prepared from the mid-radial position of sample T4 gave a mean dislocation density of $7.0\ 10^9$ cm.cm^{-3}. There was extensive network and sub-grain boundary formation, with local network dislocation densities up to $5\ 10^{10}$ cm.cm^{-3}.

The homogeneously distributed population of small fission gas bubbles found at the periphery was entirely absent in the fuel matrix at the pin centre. However large bubbles (diameter > 100nm) were often found to be surrounded byan atmosphere of very small fission gas bubbles. These bubbles appear to form a distinct population with a mean diameter of 4.5 nm and a local density above 10^{17} cm^{-3}, and are confined to a shell with a thickness of around 100 nm from the bubble surface.

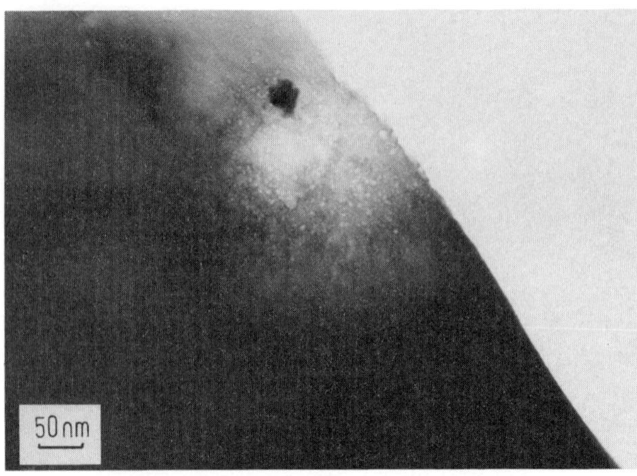

Fig. 6   An example of an "atmosphere" of very small fission gas bubbles as frequently found surrounding large fission gas bubbles at the centre of the transient-tested fuel samples.

An example of this effect from sample T2 is shown in Fig. 6. A high density of precipitates was also found in conjunction with these large bubbles.

The existence of this bubble population combined with a high density of precipitates in the immediate neighbourhood of the large fission gas bubbles formed during the transient is not understood, though it is clearly a consequence of the mechanism responsible for the rapid bubble growth.

The primary microstructural difference between the base-irradiated and the transiented samples was the growth of a population of large fission gas bubbles (> 50nm) during the transient. The distribution of this bubble population is most conveniently analysed as a function of radius using the REM technique.

Fig. 7 shows a series of REM micrographs taken at various radial positions showing the development of the population of large fission gas bubbles at various radial positions for the sample T2. The following features can be seen from these micrographs:

Periphery	$R/R_0 = 1.0$	Grain boundaries visible from the etching. No grain boundary bubbles or intragranular bubbles visible.
	$R/R_0 = 0.74$	Start of grain boundary decoration by small fission gas bubbles. No intra-granular bubbles.
	$R/R_0 = 0.72$	Strong grain boundary decoration. Start of intragranular bubble formation.
	$R/R_0 = 0.55$	Peak position for intragranular swelling. Denuded zones visible adjacent to grain boundaries.
Centre	$R/R_0 = 0$	Very low density of large grain boundary and intragranular bubbles.

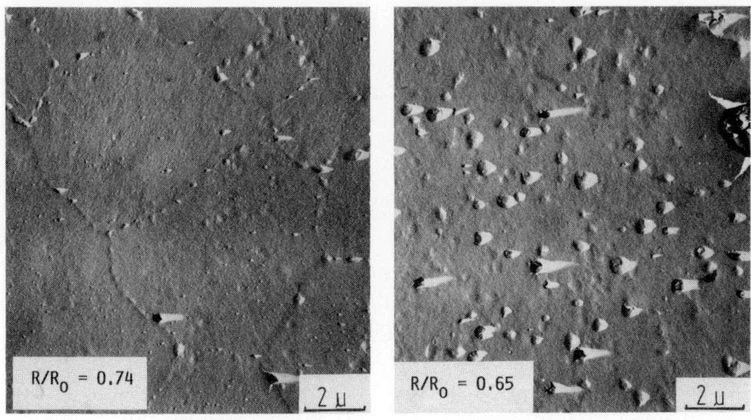

Fig. 7 Replica Electron Micrographs showing the growth of grain boundary and intragranular fission gas bubbles as a function of radial position in a transient-tested fuel sample.

The microswelling resulting from the growth of this intragranular fission gas population averaged over the full fuel radius amounted to about 1.0%. All the transient tested fuels showed a peak in the radial microswelling profile at a position of about $R/R_o = 0.5$.

Transmission electron microscopy showed that these large fission gas bubbles were invariably associated with dislocation networks, and were also usually surrounded by a shell containing a very high density of dislocations and dislocation loops. an example is shown in Fig. 8 for fuel sample T1. Very large spherical precipitates of the metallic fission products were also found within the grains, often in close association with large bubbles.

At the centre of the fuels the large fission gas bubbles ranged in size from 50 nm to 300 nm, with an average density of $7.0 \ 10^{12} \ cm^{-3}$, and a microswelling contribution of approximately 0.9%.

The formation of a high dislocation density shell round these large bubbles is thought to be due to dislocation loop punching to relieve the bubble overpressure during growth. The local dislocation density in a shell of thickness about 450 nm round the large bubbles reaches values

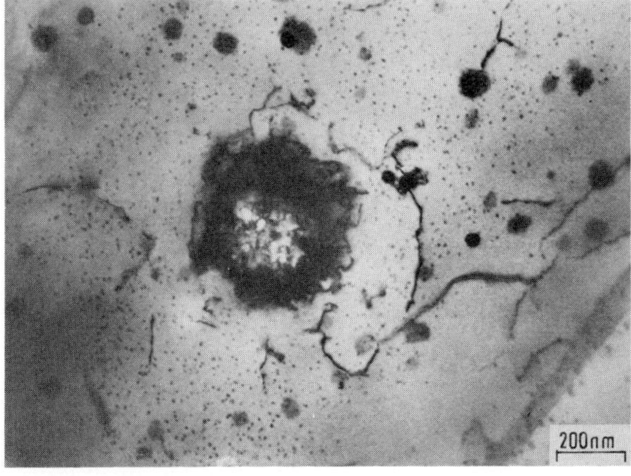

Fig. 8 An example of the high density local dislocation network formed in the vicinity of large fission gas bubbles at the centre of transient-tested fuel samples. (Sample T1).

of around 10^{11} cm.cm^{-3}. A simple model (4) shows that such dislocation loop punching could occur at an overpressure of 0.5 kbar for bubbles with diameters around 300 nm, a value which is easily attainable under the conditions of the transient.

5.　Conclusions

The following conclusions can be drawn about the behaviour of the inert gas fission products in LWR UO$_2$ nuclear fuels submitted to power transients:

1.　The microstructure of the base irradiated fuels is characterised by a uniform dislocation density, on average 2.2 10^{10} cm.cm^{-3}, largely climb induced, and a high density of very small fission gas bubbles. There is very little variation in microstructure between the periphery and the centre of the fuel.

2.　The nature of the dislocation structures observed in the transiented samples indicates that plastic deformation has occurred during the transient, particularly towards the centre of the fuel. The dislocation density at the fuel centre is lower than at the periphery, and extensive network and low angle grain boundary formation is observed.

3.　The population of small fission gas bubbles observed in the base-irradiated sample was also found with the same density and size distribution at the periphery of the transiented samples, but was completely absent at the centre of the fuel.

4.　A population of large fission gas bubbles grows during the transient associated with the dislocation networks, and leads to local microswelling. The large bubbles are often surrounded by dense dislocation tangles, probably resulting from dislocation loop punching to relieve bubble overpressure. The density of these bubbles is low at the periphery and centre of the fuels, and peaks at a radial position around R/Ro = 0.5. The microswelling contribution from this bubble population reaches values of about 1.5% averaged over the total fuel cross section.

5.　The present study shows that an increase in temperature from about 0.47 T_m to about 0.56 T_m in the centre of the fuel, over a period of two days, has very important and rapid consequences for the distribution of the fission gases which have accumulated in the fuel lattice over an irradiation period of three years.

REFERENCES

1. Hj. Matzke, this volume.
2. I.L.F. Ray, H. Thiele, and H. Blank, J. de Physique **45**, C2 - 849 (1984).
3. I.L.F. Ray and H. Thiele, 20th Meeting of the Hot Cells Working Group, Karlsruhe (1981).
4. D.A. Jones, and J.W. Mitchell, Phil. Mag. **3**, 334 (1958).

INDEXES

AUTHOR INDEX

SUBJECT INDEX[*]

---

[*]Page numbers identify the first page of the paper in which the topic is to be found.